APPROXIMATION BY COMPLEX BERNSTEIN AND CONVOLUTION TYPE OPERATORS

SERIES ON CONCRETE AND APPLICABLE MATHEMATICS

ISSN: 1793-1142

Series Editor: Professor George A. Anastassiou
Department of Mathematical Sciences
The University of Memphis
Memphis, TN 38152, USA

Series on Concrete and Applicable Mathematics – Vol. 8

APPROXIMATION BY COMPLEX BERNSTEIN AND CONVOLUTION TYPE OPERATORS

Sorin G Gal

University of Oradea, Romania

 **World Scientifi**

NEW JERSEY · LONDON · SINGAPORE · BEIJING · SHANGHAI · HONG KONG · TAIPEI · CHENNAI

Published by

World Scientific Publishing Co. Pte. Ltd.

5 Toh Tuck Link, Singapore 596224

USA office: 27 Warren Street, Suite 401-402, Hackensack, NJ 07601

UK office: 57 Shelton Street, Covent Garden, London WC2H 9HE

British Library Cataloguing-in-Publication Data
A catalogue record for this book is available from the British Library.

APPROXIMATION BY COMPLEX BERNSTEIN AND CONVOLUTION TYPE OPERATORS
Series on Concrete and Applicable Mathematics — Vol. 8

Copyright © 2009 by World Scientific Publishing Co. Pte. Ltd.

ISBN-13 978-981-4282-42-0
ISBN-10 981-4282-42-1

Printed in Singapore.

To my wife Rodica and children Ciprian and Gratiela

Preface

The *Bernstein* polynomials attached to $f : [0,1] \to \mathbb{R}$ and given by

$$B_n(f)(x) = \sum_{k=0}^{n} p_{n,k}(x) f(\frac{k}{n}), \quad p_{n,k}(x) = \binom{n}{k} x^k (1-x)^{n-k}, x \in [0,1],$$

probably are the most famous algebraic polynomials in Approximation Theory and were introduced in 1912 by S.N. Bernstein [38] in order to give the first constructive (and simple) proof to the Weierstrass' approximation theorem. Many books and papers were dedicated to their study, probably the most known being the book of Lorentz [125]. The importance of Bernstein polynomials also consists in the fact that their form has suggested and still continues to suggest to mathematicians the construction of a great variety of other approximation operators, like the *Schurer* polynomials, *Kantorovich* polynomials, *Stancu* polynomials, *q-Bernstein* polynomials, *Durrmeyer* polynomials, *Favard-Szász-Mirakjan* operators, *Baskakov* operators, and the list can continue with very many others.

A natural question concerning the Bernstein polynomials of a real variable (and by analogy, concerning any Bernstein-type operator of a real variable x) is the following : if in the expression of $B_n(f)(x)$ one replaces $x \in [0,1]$ by z in some regions in $\mathbb{C}$ (containing $[0,1]$) where f is supposed to be analytic, (a process we call *complexification*), then what convergence properties have the complex Bernstein polynomials

$$B_n(f)(z) = \sum_{k=0}^{n} p_{n,k}(z) f(\frac{k}{n}), \quad p_{n,k}(z) = \binom{n}{k} z^k (1-z)^{n-k}, z \in \mathbb{C} ?$$

In other words, the problem is to study the *overconvergence phenomenon* (not in the sense known as Wash's overconvergence in the interpolation of functions !) for the Bernstein polynomials, that is to extend their convergence properties and orders of these convergencies to larger sets in the complex plane than the real interval $[0,1]$.

The first goal of the present book is to give some answers in Chapter 1 to the above question of *overconvergence*, for several classes of Bernstein-type operators. In essence, it will be shown that for all the Bernstein-type operators, the orders of approximation from the real axis are preserved in complex domains too.

We recall that concerning the complex Bernstein polynomials, Wright [199], Kantorovich [113], Bernstein [39; 40; 41], Lorentz [125] and Tonne [190] have given interesting answers to this question. It is worth noting that an entire Chapter 4 of 38 pages is dedicated to it in the book of Lorentz [125]. In that book interesting convergence properties of $B_n(f)(z)$ and of its so-called degenerate form, in various domains in $\mathbb{C}$, like compact disks, ellipses, loops, autonomous sets are presented. Notice that in the above mentioned papers no quantitative estimates of these convergence results were obtained. Also, convergence results without any quantitative estimate were obtained for the complex Favard-Szász-Mirakjan operators by Dressel-Gergen and Purcell [65] and for the complex Jakimovski-Leviatan operators by Wood [200]. The above qualitative results are theoretically based on the "bridge" made by the classical result of Vitali (see Theorem 1.0.1), between the (well-established) approximation results for the Bernstein-type operators of real variable and those for the Bernstein-type operators of complex variable.

It is worth noting that in the other books or long surveys dealing with complex approximation, like those of Sewell [162], Dzjadyk [69], Gaier [76], Suetin [186], Andrievskii-Belyi-Dzjadyk [26], [27], or the surveys of Dyn'kin [62] and Andrievskii [25], the topic of the present book is not considered. Also, in other books like Lorentz [125] (Chapter 4) and Gal [77] (Chapters 3 and 4), the topic of the present book is attended in a tangential (and somehow different) way only.

In the Preface of their important book [60] in 1993 concerning the Approximation Theory of functions of real variable, DeVore and Lorentz note that, I cite "the Approximation Theory of functions of complex variables would require new books". The present book seeks to be one among the answers to this requirement and can briefly be described as follows.

In Chapter 1 one deepens the study of the approximation properties for the complex Bernstein polynomials $B_n(f)(z)$ in compact disks and in some special compact subsets of $\mathbb{C}$. In addition, similar results for other Bernstein-type polynomials/operators including those mentioned above are presented.

In detail, Chapter 1 can be described as follows :

– Section 1.0 contains the main results and concepts in complex analysis required for the proofs of the results in this book. For example, we mention here the Vitali's theorem, Cauchy's formula, Bernstein's inequality, Faber polynomials associated to a domain in $\mathbb{C}$, Faber series, Faber coefficients, Faber mapping.

– in Section 1.1 the exact orders in simultaneous approximation by $B_n(f)(z)$ and its derivatives, Voronovskaja's result with quantitative upper estimate and shape preserving properties of $B_n(f)(z)$ are obtained ; Subsection 1.1.1 contains the results on compact disks centered at origin while Subsection 1.1.2 contains some approximation results on compact sets in $\mathbb{C}$ for the so-called Bernstein-Faber polynomials ;

– Section 1.2 contains convergence results with quantitative estimates of the iterates of $B_n(f)(z)$, connected with the theory of the semigroups of operators and the shape preserving properties of these iterates, in the sense that beginning with an index they preserve some properties of f in Geometric Function Theory, like the starlikeness, convexity and spirallikeness ;

– in Section 1.3 the exact order in the generalized Voronovskaja's theorem for $B_n(f)(z)$ is obtained ;

– Section 1.4 presents the exact orders of approximation by Butzer's linear combinations of complex Bernstein polynomials and of Bernstein-Faber polynomials in compact disks and in compact Faber sets, respectively ;

– in Sections 1.5, 1.6, 1.7, 1.8, 1.9 and 1.10 we prove some similar properties for the complex q-Bernstein polynomials, Bernstein-Stancu polynomials, Bernstein-Kantorovich polynomials, Favard-Szász-Mirakjan operators, Baskakov operators and Balázs-Szabados operators, respectively. Besides the approximation results in compact disks for all these complex Bernstein-type operators, it is worth mentioning here the approximation results for the Bernstein-Stancu-Faber polynomials in compact sets in $\mathbb{C}$, a weighted-kind approximation result for the Favard-Szász-Mirakjan operator in strips and the study of two kinds of complex Baskakov operators generated by the real one ;

– Section 1.11 contains bibliographical notes and some open problems. The open problems mainly consist in proposals of similar studies for other types of complex Bernstein-type operators too.

In Chapter 2 we extend some of the results in Chapter 1 to the case of several complex variables. In Section 2.1 the concepts and results in the complex analysis of functions of several complex variables that we need for this chapter are presented. Section 2.2 deals with the approximation by two kinds of bivariate complex Bernstein polynomials, while Sections 2.3 and 2.4 treat the bivariate case of complex Favard-Szász-Mirakjan and Baskakov operators, respectively. All the approximation results are obtained in compact polydisks. Section 2.5 contains some bibliographical notes and open problems.

Chapter 3 deals with the approximation and geometric properties of several types of complex convolutions. Section 3.1 contains the approximation properties of some complex linear convolution polynomials : of de la Vallée Poussin, Fejér, Riesz-Zygmund, Jackson and Rogosinski kinds. More exactly, for these complex linear convolutions Voronovskaja-type results and the exact orders of approximation in compact disks are proved. Section 3.2 studies several kinds of linear non-polynomial convolutions. Thus, in Subsection 3.2.1 one studies the approximation properties (including the exact orders of approximation) in compact disks and compact sets in $\mathbb{C}$ of the non-polynomial complex convolutions of Picard, Poisson-Cauchy, and Gauss-Weierstrass. Also, their geometric properties are studied and applications to PDE in complex setting (i.e to heat and Laplace equations of complex spatial variable) are presented. In the Subsections 3.2.2, 3.2.3, 3.2.4 and 3.2.5 the approximation and geometric properties in compact disks of the complex q-Picard and q-Gauss-Weierstrass convolutions, Post-Widder complex convolution, rotation-invariant complex convolution and Sikkema complex convolution, respectively are presented. Section 3.3 contains the approximation and geometric properties of a nonlinear-type complex convolution in compact disks.

Finally, in Chapter 4 one presents several related topics : approximation by Bernstein polynomials of quaternion variable in Section 4.1, approximation of vector-valued functions of real and complex variables by operators of the type introduced in the previous chapters in Section 4.2 and strong approximation by Taylor series in the unit disk in Section 4.3.

Let us mention that most of the results in this book have been obtained by the author of this monograph : in a series of papers, single or jointly written with other researchers (as can be seen in the bibliography) and as new results that appear for the first time here.

It is important to note that the present book suggests for further research similar studies for other complex linear and nonlinear convolutions and for the complex forms of other Bernstein-type operators in approximation theory, like those of Durrmeyer-type, Meyer-König-Zeller-type, Jakimovski-Leviatan-type, Bleimann-Butzer-Hahn-type, Gamma-type, beta-type, to mention only a few.

The book mainly is addressed to researchers in the fields of complex approximation of functions and its applications, mathematical analysis and numerical analysis.

Also, since most of the proofs use elementary complex analysis, it is accessible to graduate students and suitable for graduate courses in the above domains.

Sorin G. Gal
Department of Mathematics and Computer Science
University of Oradea
Romania

Contents

Chapter 1

Bernstein-Type Operators of One Complex Variable

In the Sections 1.1-1.4 first we obtain the exact degrees in approximation of analytic functions in compact.disks by complex Bernstein polynomials and their Butzer's linear combination and in generalized Voronovskaja's results. These sections include approximation results on compact sets in $\mathbb{C}$ for the so-called Bernstein-Faber polynomials and their Butzer's linear combination in compact Faber sets. Convergence results on compact disks for the iterates of $B_n(f)(z)$ connected with the theory of the semigroups of operators and shape preserving properties of these iterates (in the sense that beginning with an index they preserve some properties of f in Geometric Function Theory, like the starlikeness, convexity and spirallikeness) also are proved.

Then in the next Sections 1.5-1.10 some similar properties for the complex q-Bernstein polynomials, Bernstein-Stancu polynomials, Bernstein-Kantorovich polynomials, Favard-Szász-Mirakjan operators, Baskakov operators and Balázs-Szabados operators are obtained.

For all kinds of Bernstein operators, the exact degrees of approximation mainly are obtained by three steps : 1) upper estimates ; 2) quantitative Voronovskaja-type formula ; 3) lower estimates by using step 2.

1.0   Auxiliary Results in Complex Analysis

In order to make the book more self-contained, in this section we briefly present the main known results and methods in Complex Analysis we use in our study.

The first one is called Vitali's theorem and can be stated as follows.

Theorem 1.0.1. (Vitali, see e.g. Kohr-Mocanu [118], p. 112, Theorem 3.2.10) *Let Ω be a domain in $\mathbb{C}$ and $F \subset \Omega$ a set having at least one accumulation point in Ω. If the sequence $(f_n)_{n \in \mathbb{N}}$ of analytic functions in Ω is bounded in each compact in Ω and $(f_n(z))_n$ is convergent for any $z \in F$, then $(f_n)_{n \in \mathbb{N}}$ is uniformly convergent in any compact of Ω.*

In our applications, in general $\Omega = \mathbb{D}_R = \{z \in \mathbb{C}; |z| < R\}$ with $R > 1$, F is a segment included in $\mathbb{D}_R$ and the compact subsets considered will be the closed disks $\mathbb{D}_r = \{z \in \mathbb{C}; |z| \leq r\}$ with $1 \leq r < R$.

1

The second important result in Complex Analysis we use is the Cauchy's formula for disks.

Theorem 1.0.2. (Cauchy, see e.g. Kohr-Mocanu [118], p. 28, Theorem 1.2.20) *Let $r > 0$ and $f : \overline{\mathbb{D}}_r \to \mathbb{C}$ be analytic in $\mathbb{D}_r$ and continuous in $\overline{\mathbb{D}}_r$. Then, for any $p \in \{0, 1, 2, ..., \}$ and all $|z| < r$ we have*

$$f^{(p)}(z) = \frac{p!}{2\pi i} \int_\Gamma \frac{f(u)}{(u-z)^{p+1}} du,$$

where $\Gamma = \{z \in \mathbb{C}; |z| = r\}$ and $i^2 = -1$.

An immediate consequence of the Cauchy's formula is the so-called Weierstrass's theorem used in the proofs of shape preserving properties.

Theorem 1.0.3. (Weierstrass, see e.g. Kohr-Mocanu [118], p. 18, Theorem 1.1.6) *Let $G \subset \mathbb{C}$ be an open set. If the sequence $(f_n)_{n\in\mathbb{N}}$ of analytic functions on G converges to the analytic function f, uniformly in each compact in G, then for any $p \in \mathbb{N}$, the sequence of pth derivatives $(f_n^{(p)})_{n\in\mathbb{N}}$ converges to $f^{(p)}$ uniformly on compacts in G.*

Indeed, note that by the above Cauchy's formula we can write

$$f_n^{(p)}(z) - f^{(p)}(z) = \frac{p!}{2\pi i} \int_\Gamma \frac{f_n(u) - f(u)}{(u-z)^{p+1}} du,$$

from which by passing to modulus the theorem easily follows. In our applications, $G = \mathbb{D}_R$ with $R > 1$ and the compact subsets in G are $\mathbb{D}_r$ with $1 \le r < R$.

Another well-known result used in the proof of shape preserving properties is the following.

Theorem 1.0.4. (see e.g. Graham-Kohr [105], Theorem 6.1.18) *If $f_n, f : \Omega \to \mathbb{C}$, $n \in \mathbb{N}$ are analytic in the domain Ω, f is univalent in Ω and $f_n \to f$ uniformly in the compact $K \subset \Omega$, then there exists $n_0(K)$ such that for all $n \ge n_0$, f_n is univalent in K.*

The classical so called Maximum Principle (or Maximum Modulus Theorem) will be frequently used in the proofs of error estimates.

Theorem 1.0.5. (see e.g. Kohr-Mocanu [118], p. 2, Corollary 1.1.20) *If $\Omega \subset \mathbb{C}$ is a bounded domain and $f : \overline{\Omega} \to \mathbb{C}$ is analytic in Ω and continuous in $\overline{\Omega}$, then denoting by Γ the boundary of Ω we have*

$$\max\{|f(z)|; z \in \overline{\Omega}\} = \max\{|f(z)|; z \in \Gamma\}.$$

For our applications again Ω will be an open disk centered at origin.

Useful in some of our proofs will be the well-known so called theorem on the zeroes of analytic functions, which in essence says that the zeroes of an analytic function (non-identical null) necessarily are isolated points. More exactly we can state the following.

Theorem 1.0.6. (see e.g. Kohr-Mocanu [118], p. 20, Theorem 1.1.12) *Suppose that f is analytic in the domain Ω and that f is not identical null in Ω. If a is a zero for f then there exists $r = r(a) > 0$ such that $\mathbb{D}(a, r) = \{z \in \mathbb{C}; |z - a| < r\} \subset \Omega$ and $f(z) \neq 0$, for all $z \in \mathbb{D}(a, r) \setminus \{a\}$.*

Also, we will use the classical so called theorem on the identity of analytic functions.

Theorem 1.0.7. (see e.g. Kohr-Mocanu [118], p. 21, Theorem 1.1.14) *Let $\Omega \subset \mathbb{C}$ be a domain. If $f, g : \Omega \to \mathbb{C}$ are analytic in Ω then $f \equiv g$ on Ω is equivalent with the fact that the set $\{z \in \Omega; f(z) = g(z)\}$ has at least one accumulation point in Ω.*

Finally, we state a basic result very useful in the proofs of the approximation results and called Bernstein's inequality for complex polynomials in compact disks.

Theorem 1.0.8. (Bernstein [43], p. 45, relation (80) for general $r > 0$, see also e.g. Lorentz [126], p. 40, Theorem 4, for $r = 1$) *Let $P(z) = \sum_{k=0}^{n} a_k z^k$ be with $a_k \in \mathbb{C}$, for all $k \in \{0, 1, 2, ...,\}$ and for $r > 0$ denote $\|P_n\|_r = \max\{|P_n(z)|; |z| \leq r\}$.*

(i) For all $|z| \leq 1$ we have $|P_n'(z)| \leq n\|P_n\|_1$;

(ii) If $r > 0$ then for all $|z| \leq r$ we have $|P_n'(z)| \leq \frac{n}{r}\|P_n\|_r$.

One observes that (ii) immediately follows from (i). Indeed, denoting $Q_n(z) = P_n(rz)$, $|z| \leq 1$, by (i) applied to $Q_n(z)$ it easily follows $r|P_n'(rz)| \leq n\|P_n\|_r$, for all $|z| \leq 1$, which proves (ii).

Concerning the approximation of analytic functions by sequences of complex polynomials, as it will be seen in the next sections of this chapter and in the next chapters, the main results one refer to approximation in compact disks centered at origin (in particular in the compact unit disk). The advantage consists in the fact that in these kinds of disks constructive methods can be indicated. But of course that it is very important to obtain approximation results in more general domains in the complex plane. In what follows we briefly present the standard method based on the so-called Faber polynomials introduced by Faber [70], which allows to extend all the constructive methods from the closed unit disk to more general domains. The method is less constructive because a generally unknown mapping function (generated from the Riemann's mapping theorem) enters into considerations. For all the details below on this method see e.g. the book of Gaier [76], pp. 42-54. Also, for other important contributions to the topic of constructive complex approximation see the book of Dzjadyk [69].

Definition 1.0.9. (i) $\gamma : [a, b] \to \mathbb{C}$ is called Jordan curve if it is closed (i.e. $\gamma(a) = \gamma(b)$) and simple (i.e. injective). The length of the curve γ is defined by

$$L(\gamma) = \sup\{\sum_{i=1}^{n} |\gamma(t_i) - \gamma(t_{i-1})|; n \in \mathbb{N}, a = t_0 < ... < t_n = b\}.$$

γ is called rectifiable if $L < +\infty$.

The interior of a Jordan curve is called Jordan domain and the curve is called boundary curve of that domain.

(ii) (Radon [158]) Suppose that $\gamma : [a, b] \to \mathbb{C}$ is a rectifiable Jordan curve. Because $L < +\infty$, it is known that γ has a tangent γ' almost everywhere. Then γ is called of bounded rotation if γ' can be extended to a function of bounded variation on the whole curve.

Remark. Simple examples of Jordan curve of bounded rotation can be made up of finitely many convex arcs (where corners are permitted).

Now, if G is a Jordan domain, then (by the Riemann's mapping theorem) let us denote by Ψ the conformal mapping of $\mathbb{C} \setminus \overline{\mathbb{D}}_1$ onto $\mathbb{C} \setminus \overline{G}$, normalized at ∞, that is $0 < \lim_{w \to \infty} \frac{\Psi(w)}{w} < \infty$. Also, denote by Φ the inverse function of Ψ. Obviously that Ψ and Φ depend on $\overline{G}$, but for the simplicity of notation we will not write them as $\Psi_{\overline{G}}$ and $\Phi_{\overline{G}}$, considering in our presentation that $\overline{G}$ is arbitrary but fixed.

For a Jordan domain G, denote by $A(\overline{G})$ the class of all functions continuous in $\overline{G}$ and analytic in G. In what follows we sketch a method by which any $f \in A(\overline{G})$ can be approximated by polynomials. For our considerations, it is sufficient to suppose that the boundary curve of G is rectifiable and of bounded rotation.

First, one considers the Laurent expansion of $[\Phi(z)]^n$, $n \in \mathbb{N} \cup \{0\}$, valid for large z

$$[\Phi(z)]^n = a_0^{(n)} + ... + a_n^{(n)} z^n + \sum_{k=1}^{\infty} a_{-k}^{(n)} / z^k.$$

Definition 1.0.10. (Faber [70]) (i) The polynomial $F_n(z) = a_0^{(n)} + ... + a_n^{(n)} z^n$, $n \in \mathbb{N} \cup \{0\}$ is called the Faber polynomial of degree n attached to the domain G. (Note that for $z \in \mathbb{D}_R$, $R > 1$ we can write

$$F_n(z) = \frac{1}{2\pi i} \int_{|u|=R} \frac{[\Phi(u)]^n}{u - z} du.)$$

(ii) If $f \in A(\overline{G})$ then

$$a_n(f) = \frac{1}{2\pi i} \int_{|u|=1} \frac{f[\Psi(u)]}{u^{n+1}} du = \frac{1}{2\pi i} \int_{-\pi}^{\pi} f[\Psi(e^{it})] e^{-int} dt, n \in \mathbb{N} \cup \{0\}$$

are called the Faber coefficients of f and $\sum_{n=0}^{\infty} a_n(f) F_n(z)$ is called the Faber series attached to f on G. (Here $i^2 = 1$.) The Faber series represent a natural generalization of Taylor series when the unit disk is replaced by an arbitrary simply connected domain bounded by a "nice" curve.

(iii) The mapping T defined by $T[P_n](z) = \sum_{k=0}^{n} c_k F_k(z)$, where $P_n(w) = \sum_{k=0}^{n} c_k w^k$ is called the Faber mapping.

Remark. By Definition 1.0.10, (iii), the Faber mapping T is linear and defined on the set of all polynomials $\mathcal{P}$ defined on $\overline{\mathbb{D}}_1$ and with values in the set of polynomials $\mathcal{P}$ defined on $\overline{G}$. In some cases it can be extended as a linear and bounded mapping between the Banach spaces $A(\overline{\mathbb{D}}_1)$ and $A(\overline{G})$ (both endowed with the corresponding uniform norms). Below we briefly point out this extension (for full details see e.g.

the book of Gaier [76], pp. 48-49) under the hypothesis that the boundary of G is a rectifiable Jordan curve of bounded rotation. In this case for G, first it follows that $\|T(P)\| \leq C\|P\|$ for each $P \in \mathcal{P}$, where $C > 0$ depends only on G. Then T can be extended to the closure of $\mathcal{P}$ and since $\overline{\mathcal{P}} = A(\overline{\mathbb{D}}_1)$, T can be extended as a linear and bounded operator from $A(\overline{\mathbb{D}}_1)$ into $A(\overline{G})$, with the property that $\|T(f)\| \leq C\|f\|$ for each $f \in A(\overline{\mathbb{D}}_1)$.

Now, since the Faber mapping has the integral representation

$$T[P_n](z) = \frac{1}{2\pi i} \int_C \frac{P_n[\Phi(u)]}{u - z} du,$$

valid for each polynomial P_n, by passing to limits we obtain the formula

$$T[F](z) = \frac{1}{2\pi i} \int_C \frac{F[\Phi(u)]}{u - z} du, \ z \in G, \ F \in A(\overline{\mathbb{D}}_1).$$

Also, the converse formula

$$F(w) = \frac{1}{2\pi i} \int_{|u|=1} \frac{T[F](\Psi(u))}{u - w} du, \ w \in \mathbb{D}_1$$

holds.

The following two known results are of great importance for approximation.

Theorem 1.0.11. *(more precisely see e.g. Gaier [76], p. 50, Theorem 3) If $F \in A(\overline{\mathbb{D}}_1)$, $F(w) = \sum_{n=0}^{\infty} c_n w^n$ then the Faber coefficients of $T[F]$ are c_n.*

Theorem 1.0.12. *(more precisely see e.g. Theorem 4 in Gaier [76], p. 51) Suppose that the boundary of G is a rectifiable Jordan curve of bounded rotation and let $f \in A(\overline{G})$. There exists $F \in A(\overline{\mathbb{D}}_1)$ with $f = T[F]$ if and only if as function of $w \in \overline{\mathbb{D}}_1$, the Cauchy integral $\int_{|u|=1} \frac{f[\Psi(u)]}{u-w} du$ belongs to $A(\overline{\mathbb{D}}_1)$ and in this case we have*

$$F(w) = \frac{1}{2\pi i} \int_{|u|=1} \frac{f[\Psi(u)]}{u - w} du, \ w \in \overline{\mathbb{D}}_1,$$

(F is extended by continuity on $\partial \mathbb{D}_1$).

Remark. Theorem 1.0.12 allows to reduce the approximation of $f \in A(\overline{G})$ to the approximation of $F \in A(\overline{\mathbb{D}}_1)$. Indeed, let $(S_n(F)(w))_{n \in \mathbb{N}}$, $S_n(F)(w) = \sum_{k=0}^{m_n} a_k(F)w^k$, be an approximation sequence for F. Then $T[S_n(F)](z) = \sum_{k=0}^{m_n} a_k(F)F_k(z)$, $n \in \mathbb{N}$ will represent an approximation sequence for f in the set $\overline{G}$ (here $F_k(z), k \in \mathbb{N}$ denote the Faber polynomials attached to G). Indeed, denoting the uniform norms by $\| \cdot \|$, this follows from the relation (6.19), p. 51 in Gaier [76],

$$\left\| f - \sum_{k=0}^{m_n} a_k(F)F_k \right\|_{\overline{G}} = \left\| T(F - \sum_{k=0}^{m_n} a_k(F)e_k) \right\|_{\overline{G}} \leq \|\|T\|\| \cdot \left\| F - \sum_{k=0}^{m_n} a_k(F)e_k \right\|_{\overline{\mathbb{D}}_1},$$

where $e_k(w) = w^k$ and $\|\|T\|\| \leq M < \infty$, because of the hypothesis on the boundary of G.

1.1   Bernstein Polynomials

In this section, we find the exact orders in simultaneous uniform approximation of analytic functions by complex Bernstein polynomials in closed disks, an upper estimate in Voronovskaja's result and we prove that the complex Bernstein polynomials attached to an analytic function, preserve the univalence, starlikeness, convexity and spirallikeness. Also, to Jordan domains Bernstein-type polynomials are attached and approximation results on connected compact sets with estimates are obtained.

1.1.1   *Bernstein Polynomials on Compact Disks*

Concerning the approximation properties (uniform convergence), the results in Wright [199], Kantorovich [113], Bernstein [39; 40; 41], Lorentz [125] and Tonne [190] are well-known. It is worth nothing that an entire Chapter 4 of 38 pages is dedicated to this topic in the book of Lorentz [125]. In that book interesting convergence properties of $B_n(f)(z)$ and of its so-called degenerate form, in various domains in $\mathbb{C}$, like compact disks, ellipses, loops, autonomous sets are presented.

For example, the following three approximation results due to Bernstein, Tonne and Kantorovich concerning the uniform approximation of Bernstein polynomials in the unit disk and in an ellipse hold.

Theorem 1.1.1. *(i) (Bernstein, see e.g. Lorentz [125], p. 88) For the open $G \subset \mathbb{C}$, such that $\overline{\mathbb{D}}_1 \subset G$ and $f : G \to \mathbb{C}$ is analytic in G, the complex Bernstein polynomials $B_n(f)(z) = \sum_{k=0}^{n} \binom{n}{k} z^k (1-z)^{n-k} f(k/n)$, uniformly converge to f in $\overline{\mathbb{D}}_1$. Here $\mathbb{D}_1$ denotes the open unit disk.*

(ii) (Tonne [190]) If $f(z) = \sum_{k=0}^{\infty} c_k z^k$ is analytic in the open unit disk $\mathbb{D}_1$, $f(1)$ is a complex number and there exist $M > 0$ and $m \in \mathbb{N}$ such that $|c_k| \le M(k+1)^m$, for all $k = 0, 1, 2, ...$, then $B_n(f)(z)$ converges uniformly (as $n \to \infty$) to f on each closed subset of $\mathbb{D}_1$.

(iii) (Kantorovich, see e.g. Lorentz [125], p. 90) If f is analytic in the interior of an ellipse of foci 0 and 1, then $B_n(f)(z)$ converges uniformly to $f(z)$ in any closed set contained in the interior of ellipse.

But in all the previous mentioned work no quantitative estimates of these convergence results were obtained. In what follows, first we obtain upper quantitative estimates on compact disks. For this purpose, denote $\mathbb{D}_R = \{z \in \mathbb{C}; |z| < R\}$.

Theorem 1.1.2. *(Gal [77], p. 264, Theorem 3.4.1, (iii)-(v)) Suppose that $R > 1$ and $f : \mathbb{D}_R \to \mathbb{C}$ is analytic in $\mathbb{D}_R$, that is $f(z) = \sum_{k=0}^{\infty} c_k z^k$, for all $z \in \mathbb{D}_R$.*

(i) Let $1 \le r < R$ be arbitrary fixed. For all $|z| \le r$ and $n \in \mathbb{N}$, we have

$$|B_n(f)(z) - f(z)| \le \frac{C_r(f)}{n},$$

where $0 < C_r(f) = \frac{3r(1+r)}{2} \sum_{j=2}^{\infty} j(j-1)|c_j| r^{j-2} < \infty$.

(ii) For the simultaneous approximation by complex Bernstein polynomials, we have : if $1 \leq r < r_1 < R$ are arbitrary fixed, then for all $|z| \leq r$ and $n, p \in \mathbb{N}$,

$$|B_n^{(p)}(f)(z) - f^{(p)}(z)| \leq \frac{C_{r_1}(f)p!r_1}{n(r_1 - r)^{p+1}},$$

where $C_{r_1}(f)$ is given as at the above point (i).

Proof. (i) Denoting $e_k(z) = z^k$ and $\pi_{k,n}(z) = B_n(e_k)(z)$, we evidently have $B_n(f)(z) = \sum_{k=0}^{\infty} c_k \pi_{k,n}(z)$ and we get

$$|B_n(f)(z) - f(z)| \leq \sum_{k=0}^{\infty} |c_k| \cdot |\pi_{k,n}(z) - e_k(z)|,$$

so that we need an estimate for $|\pi_{k,n}(z) - e_k(z)|$.

For this purpose we use the recurrence proved for the real variable case in Andrica [24]

$$\pi_{k+1,n}(z) = \frac{z(1-z)}{n}\pi'_{k,n}(z) + z\pi_{k,n}(z),$$

for all $n \in \mathbb{N}$, $z \in \mathbb{C}$ and $k = 0, 1, \dots$. Since the relationship in Andrica [24] proved for the real case is a simple algebraic manipulation, it is valid for complex variable as well. Taking into account that the paper Andrica [24] is less accessible, let us reproduce here the idea of proof. It consists of the simple algebraic relationship

$$S'_{k,n}(z) = \frac{S_{k+1,n}(z)}{z(1-z)} - n\frac{S_{k,n}(z)}{1-z},$$

which is divided by n^k, where

$$S_{k,n}(z) = \sum_{j=0}^{n} j^k \binom{n}{j} z^j (1-z)^{n-j}.$$

(Note that the cases $z = 0$ and $z = 1$ are trivial in the recurrence for $\pi_{k,n}(z)$.)

From this recurrence, we easily obtain that degree $(\pi_{k,n}(z)) = \min\{n, k\} \leq k$. Also, it easily implies the next recurrence

$$\pi_{k,n}(z) - z^k$$

$$= \frac{z(1-z)}{n}[\pi_{k-1,n}(z) - z^{k-1}]' + \frac{(k-1)z^{k-1}(1-z)}{n} + z[\pi_{k-1,n}(z) - z^{k-1}].$$

Denoting with $\| \cdot \|_r$ the norm in $C(\overline{\mathbb{D}}_r)$, where $\overline{\mathbb{D}}_r = \{z \in \mathbb{C}; |z| \leq r\}$, one observes that by a linear transformation the Bernstein's inequality in the closed unit disk becomes $|P'_k(z)| \leq \frac{k}{r}\|P_k\|_r$, for all $|z| \leq r$, $r \geq 1$, where P_k represents an algebraic polynomial of degree $\leq k$. Therefore, from the above recurrence we get

$$|\pi_{k,n}(z) - e_k(z)| \leq (k-1)\frac{r(1+r)}{rn}\|\pi_{k-1,n} - e_{k-1}(z)\|_r$$
$$+\frac{r^{k-1}(1+r)(k-1)}{n} + r|\pi_{k-1,n}(z) - e_{k-1}(z)|$$
$$\leq (k-1)\frac{(1+r)}{n}\cdot[\|\pi_{k-1,n}\|_r + \|e_{k-1}\|_r]$$
$$+\frac{r^{k-1}(1+r)(k-1)}{n} + r|\pi_{k-1,n}(z) - e_{k-1}(z)|$$
$$\leq r|\pi_{k-1,n}(z) - e_{k-1}(z)|$$
$$+[2(1+r)r^{k-1} + (1+r)r^{k-1}]\frac{k-1}{n}.$$

Above we used that for all $k, n \in \mathbb{N}$ and $|z| \leq r$, $r \geq 1$, we have $|\pi_{k,n}(z)| \leq r^k$ (see relation (4) in the proof of Theorem 4.1.1 in Lorentz[125], p. 88) and $|e_k(z)| \leq r^k$.

Now, by taking $k = 1, 2, ...$, in the inequality

$$|\pi_{k,n}(z) - e_k(z)| \leq r|\pi_{k-1,n}(z) - e_{k-1}(z)| + 3(1+r)r^{k-1}\frac{k-1}{n},$$

by recurrence we easily obtain the required inequality

$$|\pi_{k,n}(z) - e_k(z)| \leq \frac{3(1+r)}{n}[r^{k-1} + 2r^{k-1} + ... + (k-1)r^{k-1}]$$
$$= \frac{3(1+r)}{n}\cdot\frac{k(k-1)}{2}r^{k-1} \leq \frac{3r(1+r)}{2n}\cdot k(k-1)r^{k-2}.$$

This immediately implies the estimate in (i).

Note that since by hypothesis, $f(z) = \sum_k^\infty c_k z^k$ is absolutely and uniformly convergent in $|z| \leq r$, for any $1 \leq r < R$, it follows that the power series obtained by differentiating twice, i.e. $f''(z) = \sum_{k=2}^\infty k(k-1)c_k z^{k-2}$, also is absolutely convergent for $|z| \leq r$, which implies $\sum_{k=2}^\infty k(k-1)|c_k|r^{k-2} < +\infty$.

(ii) Denoting by γ the circle of radius $r_1 > 1$ and center 0, since for any $|z| \leq r$ and $v \in \gamma$, we have $|v - z| \geq r_1 - r$, by the Cauchy's formulas it follows that for all $|z| \leq r$ and $n \in \mathbb{N}$, we have

$$|B_n^{(p)}(f)(z) - f^{(p)}(z)| = \frac{p!}{2\pi}\left|\int_\gamma \frac{B_n(f)(v) - f(v)}{(v-z)^{p+1}}dv\right|$$
$$\leq \frac{C_{r_1}(f)}{n}\frac{p!}{2\pi}\frac{2\pi r_1}{(r_1-r)^{p+1}} = \frac{C_{r_1}(f)}{n}\frac{p!r_1}{(r_1-r)^{p+1}},$$

which proves the theorem. $\square$

Remarks. 1) An analogue to Theorem 1.1.2, (i), case $r = 1$, has been obtained by a different method in Ostrovska [146].

2) Let us give a proof of the relationship $B_n(f)(z) = \sum_{k=0}^\infty c_k B_n(e_k)(z)$ used at the beginning of the proof of Theorem 1.1.2, (i). Denoting $f_m(z) = \sum_{j=0}^m c_j z^j$, $|z| \leq r$, $m \in \mathbb{N}$, since from the linearity of B_n we obviously have $B_n(f_m)(z) =$

$\sum_{k=0}^{m} c_k B_n(e_k)(z)$, it suffices to prove that for any fixed $n \in \mathbb{N}$ and $|z| \le r$ with $r \ge 1$, we have $\lim_{m \to \infty} B_n(f_m)(z) = B_n(f)(z)$. But this is immediate from $\lim_{m \to \infty} \|f_m - f\|_r = 0$ and from the inequality

$$|B_n(f_m)(z) - B_n(f)(z)| \le \sum_{k=0}^{n} \binom{n}{k} |z^k (1-z)^{n-k}| \cdot \|f_m - f\|_r \le M_{r,n} \|f_m - f\|_r,$$

valid for all $|z| \le r$.

In what follows we present the Voronovskaja-type formula with a quantitative upper estimate.

Theorem 1.1.3. (Gal [78]) *Let $R > 1$ and suppose that $f : \mathbb{D}_R \to \mathbb{C}$ is analytic in $\mathbb{D}_R$, that is we can write $f(z) = \sum_{k=0}^{\infty} c_k z^k$, for all $z \in \mathbb{D}_R$.*

(i) The following Voronovskaja-type result in the closed unit disk holds

$$\left| B_n(f)(z) - f(z) - \frac{z(1-z)}{2n} f''(z) \right| \le \frac{|z(1-z)|}{2n} \cdot \frac{10M(f)}{n},$$

for all $n \in \mathbb{N}, z \in \overline{\mathbb{D}}_1$, where $0 < M(f) = \sum_{k=3}^{\infty} k(k-1)(k-2)^2 |c_k| < \infty$.

(ii) Let $r \in [1, R)$. Then for all $n \in \mathbb{N}, |z| \le r$, we have

$$\left| B_n(f)(z) - f(z) - \frac{z(1-z)}{2n} f''(z) \right| \le \frac{5(1+r)^2}{2n} \cdot \frac{M_r(f)}{n},$$

where $M_r(f) = \sum_{k=3}^{\infty} |c_k| k(k-1)(k-2)^2 r^{k-2} < \infty$.

Proof. (i) Denoting $e_k(z) = z^k$, $k = 0, 1, ...$, and $\pi_{k,n}(z) = B_n(e_k)(z)$, we can write $B_n(f)(z) = \sum_{k=0}^{\infty} c_k \pi_{k,n}(z)$, which immediately implies

$$\left| B_n(f)(z) - f(z) - \frac{z(1-z)}{2n} f''(z) \right|$$

$$\le \sum_{k=3}^{\infty} |c_k| \cdot \left| \pi_{k,n}(z) - e_k(z) - \frac{z^{k-1}(1-z)k(k-1)}{2n} \right|,$$

for all $z \in \overline{\mathbb{D}}_1, n \in \mathbb{N}$.

In what follows, we will use the recurrence obtained in the proof of Theorem 1.1.2, (i)

$$\pi_{k+1,n}(z) = \frac{z(1-z)}{n} \pi'_{k,n}(z) + z\pi_{k,n}(z),$$

for all $n \in \mathbb{N}, z \in \mathbb{C}$ and $k = 0, 1, ...$

If we denote

$$E_{k,n}(z) = \pi_{k,n}(z) - e_k(z) - \frac{z^{k-1}(1-z)k(k-1)}{2n},$$

then it is clear that $E_{k,n}(z)$ is a polynomial of degree $\le k$ and by a simple calculation and the use of the above recurrence we obtain the following relationship

$$E_{k,n}(z) = \frac{z(1-z)}{n} E'_{k-1,n}(z) + zE_{k-1,n}(z)$$
$$+ \frac{z^{k-2}(1-z)(k-1)(k-2)}{2n^2} [(k-2) - z(k-1)],$$

for all $k \geq 2$, $n \in \mathbb{N}$ and $z \in \overline{\mathbb{D}}_1$.

According to the Bernstein's inequality $\|E'_{k-1,n}\| \leq (k-1)\|E_{k-1,n}\|$, the above relationship implies for all $|z| \leq 1$, $k \geq 2$, $n \in \mathbb{N}$ that

$$
|E_{k,n}(z)| \leq \frac{|z| \cdot |1-z|}{2n} [2\|E'_{k-1,n}\|]
$$

$$
+|E_{k-1,n}(z)| + \frac{|z| \cdot |1-z|}{2n} \cdot \frac{|z|^{k-3}(k-1)(k-2)}{n}(2k-3)
$$

$$
\leq |E_{k-1,n}(z)| + \frac{|z| \cdot |1-z|}{2n} \left[2\|E'_{k-1,n}\| + \frac{2k(k-1)(k-2)}{n} \right]
$$

$$
\leq |E_{k-1,n}(z)| + \frac{|z| \cdot |1-z|}{2n} \left[2(k-1)\|E_{k-1,n}\| + \frac{2k(k-1)(k-2)}{n} \right]
$$

$$
\leq |E_{k-1,n}(z)| + \frac{|z| \cdot |1-z|}{2n} \left[2(k-1)\|\pi_{k-1,n} - e_{k-1}\| \right.
$$

$$
\left. + 2(k-1) \left\| \frac{(k-1)(k-2)[e_{k-2} - e_{k-1}]}{2n} \right\| + \frac{2k(k-1)(k-2)}{n} \right],
$$

where $\| \cdot \|$ denotes the uniform norm in $C(\overline{\mathbb{D}}_1)$.

Also, taking $r = 1$ in the inequality obtained in the proof of Theorem 1.1.2, (i), it follows

$$
\|\pi_{k,n} - e_k\| \leq \frac{3}{n}(k-1)k.
$$

As a consequence, we get

$$
|E_{k,n}(z)| \leq |E_{k-1,n}(z)| + \frac{|z| \cdot |1-z|}{2n} \left[2(k-1)\frac{3(k-1)(k-2)}{n} \right.
$$

$$
\left. + 2(k-1) \left\| \frac{(k-1)(k-2)[e_{k-2} - e_{k-1}]}{2n} \right\| + \frac{2k(k-1)(k-2)}{n} \right],
$$

which by simple calculation implies

$$
|E_{k,n}(z)| \leq |E_{k-1,n}(z)| + \frac{|z| \cdot |1-z|}{2n} \cdot \frac{10}{n} \cdot k(k-1)(k-2).
$$

Since $E_{0,n}(z) = E_{1,n}(z) = E_{2,n}(z) = 0$, for any $z \in \mathbb{C}$, it follows that the last inequality is trivial for $k = 0, 1, 2$.

By writing the last inequality for $k = 3, 4, ...$, we easily obtain, step by step the following

$$
|E_{k,n}(z)| \leq \frac{|z| \cdot |1-z|}{2n} \cdot \frac{10}{n} \cdot \sum_{j=3}^{k} j(j-1)(j-2) \leq \frac{|z| \cdot |1-z|}{2n} \cdot \frac{10}{n} \cdot k(k-1)(k-2)^2.
$$

In conclusion,

$$
\left| B_n(f)(z) - f(z) - \frac{z(1-z)}{2n} f''(z) \right|
$$

$$
\leq \sum_{k=3}^{\infty} |c_k| \cdot |E_{k,n}(z)| \leq \frac{|z| \cdot |1-z|}{2n} \cdot \frac{10}{n} \cdot \sum_{k=3}^{\infty} |c_k| k(k-1)(k-2)^2.
$$

Note that since $f^{(4)}(z) = \sum_{k=4}^{\infty} c_k k(k-1)(k-2)(k-3)z^{k-4}$, and the series is absolutely convergent in $\overline{\mathbb{D}}_1$, it easily follows that $\sum_{k=3}^{\infty} |c_k| k(k-1)(k-2)^2 < \infty$.

(ii) We will use the relationship obtained in the proof of Theorem 1.1.2, (i)

$$|\pi_{k,n}(z) - e_k(z)| \leq \frac{3r(1+r)}{2n} \cdot k(k-1)r^{k-2},$$

for all $k, n \in \mathbb{N}$, $|z| \leq r$, with $1 \leq r$.

Let us consider the relationship proved at the above point (i) given by

$$E_{k,n}(z) = \frac{z(1-z)}{n}E'_{k-1,n}(z) + zE_{k-1,n}(z)$$
$$+ \frac{z^{k-2}(1-z)(k-1)(k-2)}{2n^2}[(k-2) - z(k-1)],$$

for all $k \geq 2$, $n \in \mathbb{N}$ and $z \in \mathbb{C}$, and let us restrict it only for $|z| \leq r$. For all $k, n \in \mathbb{N}$, $k \geq 2$ and $|z| \leq r$, it implies

$$|E_{k,n}(z)| \leq \frac{r(1+r)}{n}|E'_{k-1,n}(z)| + r|E_{k-1,n}(z)|$$
$$+ \frac{(1+r)r^{k-2}(k-1)(k-2)}{2n^2}[(k-2) + r(k-1)].$$

Now we will estimate $|E'_{k-1,n}(z)|$, for $k \geq 3$. Taking into account that $E_{k-1,n}(z)$ is a polynomial of degree $\leq (k-1)$, we obtain

$$|E'_{k-1,n}(z)| \leq \frac{k-1}{r}\|E_{k-1,n}(z)\|_r$$
$$\leq \frac{k-1}{r}\left[\|\pi_{k-1,n} - e_{k-1}\|_r + \left\|\frac{(k-1)(k-2)(e_{k-1} - e_{k-2})}{2n}\right\|_r\right]$$
$$\leq \frac{k-1}{r}\left[\frac{3r(1+r)(k-1)(k-2)r^{k-3}}{2n} + \frac{r^{k-2}(r+1)(k-1)(k-2)}{2n}\right]$$
$$\leq \frac{k(k-1)(k-2)}{2n}\left[3(1+r)r^{k-3} + r^{k-3}(r+1)\right]$$
$$\leq \frac{2k(k-1)(k-2)(1+r)r^{k-3}}{n}.$$

This implies

$$\frac{r(1+r)}{n}|E'_{k-1,n}(z)| \leq \frac{2r(1+r)^2k(k-1)(k-2)r^{k-3}}{n^2},$$

and

$$|E_{k,n}(z)| \leq r|E_{k-1,n}(z)| + \frac{2r(1+r)^2k(k-1)(k-2)r^{k-3}}{n^2}$$
$$+ \frac{(1+r)(k-1)(k-2)r^{k-2}}{2n^2}[(k-2) + r(k-1)] = r|E_{k-1,n}(z)|$$
$$+ \frac{(1+r)(k-1)(k-2)r^{k-2}}{2n^2}[4k(1+r) + (k-2) + r(k-1)]$$

$$= r|E_{k-1,n}(z)| + \frac{(1+r)(k-1)(k-2)r^{k-2}}{2n^2}[(5k-2) + r(5k-1)]$$

$$\leq r|E_{k-1,n}(z)| + \frac{(1+r)(k-1)(k-2)r^{k-2}5k(1+r)}{2n^2}$$

$$= r|E_{k-1,n}(z)| + \frac{5(1+r)^2k(k-1)(k-2)r^{k-2}}{2n^2}.$$

But $E_{0,n}(z) = E_{1,n}(z) = E_{2,n}(z) = 0$, for any $z \in \mathbb{C}$.

By writing the last inequality for $k = 3, 4, ...$, we easily obtain, step by step the following

$$|E_{k,n}(z)| \leq \frac{5(1+r)^2r^{k-2}}{2n^2}\left[\sum_{j=3}^{k} j(j-1)(j-2)\right]$$

$$\leq \frac{5(1+r)^2k(k-1)(k-2)^2r^{k-2}}{2n^2}.$$

As a conclusion, we obtain

$$\left|B_n(f)(z) - f(z) - \frac{z(1-z)}{2n}f''(z)\right| \leq \sum_{k=3}^{\infty} |c_k| \cdot |E_{k,n}(z)|$$

$$\leq \frac{5(1+r)^2}{2n^2} \sum_{k=3}^{\infty} |c_k|k(k-1)(k-2)^2r^{k-2}.$$

Note that since $f^{(4)}(z) = \sum_{k=4}^{\infty} c_k k(k-1)(k-2)(k-3)z^{k-4}$, and the series is absolutely convergent in $|z| \leq r$, it easily follows that $\sum_{k=3}^{\infty} |c_k|k(k-1)(k-2)^2r^{k-2} < \infty$. Therefore the theorem has been proved. $\square$

Remark. By Gonska-Piţul-Raşa [102], p. 68, Proposition 7.2, for the real Bernstein polynomials

$$B_n(f)(x) = \sum_{k=0}^{n} \binom{n}{k} x^k(1-x)^{n-k} f(k/n), x \in [0,1]$$

attached to a function $f \in C^2[0,1]$, for all $x \in [0,1]$ and $n \in \mathbb{N}$ it holds

$$\left|B_n(f)(x) - f(x) - \frac{x(1-x)}{2n}f''(x)\right| \leq \frac{x(1-x)}{2n}\tilde{\omega}_1(f''; \frac{1}{3\sqrt{n}}),$$

where $\tilde{\omega}_1$ denotes the least concave majorant of the modulus of continuity ω_1 and

$$C^2[0,1] = \{f : [0,1] \to \mathbb{R}; f \text{ is twice continuously differentiable on } [0,1]\}.$$

Now, if $f \in C^3[0,1]$ then we immediately get that the best quantitative uniform estimate in the real Voronovskaja's result is of order $O(1/n^{3/2})$, which is essentially worst that the order $O(1/n^2)$ in Theorem 1.1.3. This suggest that in the real case, the order of approximation could be improved, for example that maybe $\tilde{\omega}_1$ could be replaced by ω_2.

In what follows we will prove that the orders of approximation in Theorem 1.1.2, (i) and (ii) are exactly $\frac{1}{n}$.

We present

Theorem 1.1.4. (Gal [79]) *Let $R > 1$, $\mathbb{D}_R = \{z \in \mathbb{C}; |z| < R\}$ and let us suppose that $f : \mathbb{D}_R \to \mathbb{C}$ is analytic in $\mathbb{D}_R$, that is we can write $f(z) = \sum_{k=0}^{\infty} c_k z^k$, for all $z \in \mathbb{D}_R$. If f is not a polynomial of degree ≤ 1, then for any $r \in [1, R)$ we have*

$$\|B_n(f) - f\|_r \geq \frac{C_r(f)}{n}, n \in \mathbb{N},$$

where $\|f\|_r = \max\{|f(z)|; |z| \leq r\}$ and the constant $C_r(f)$ depends only on f and r.

Proof. For all $z \in \mathbb{D}_R$ and $n \in \mathbb{N}$ we have

$$B_n(f)(z) - f(z)$$
$$= \frac{1}{n} \left\{ \frac{z(1-z)}{2} f''(z) + \frac{1}{n} \left[n^2 \left(B_n(f)(z) - f(z) - \frac{z(1-z)}{2n} f''(z) \right) \right] \right\}.$$

Since by hypothesis $f''(z)$ is not identical zero in $\mathbb{D}_R$, there exists $0 < r_0 < 1$ such that $M_0 = \inf_{|z|=r_0} |f''(z)| > 0$. Indeed, let us suppose the contrary. Then, choosing a sequence $0 < r_n < 1$, $n \in \mathbb{N}$ such that $r_n \searrow 0$, the continuity of f'' on the compact set $\{z \in \mathbb{C}; |z| = r_n\}$, implies that there exists z_n with $|z_n| = r_n$ and $f''(z_n) = 0$. It follows $z_n \to 0$ and by the continuity of f'' in $\mathbb{D}_R$ we get $f''(0) = 0$. The analyticity of f'' implies that 0 is an isolated zero, therefore there exists $r' > 0$ such that $f''(z) \neq 0$ for all $z \in \mathbb{D}_{r'}, z \neq 0$. But this is a contradiction because for sufficiently large n we have $z_n \in \mathbb{D}_{r'}$.

Now let $r \geq 1$ be arbitrary. We obviously have $\|B_n(f) - f\|_r \geq \|B_n(f) - f\|_{r_0}$ and by the Maximum Principle, there exists a point z_0 (depending on n, f and r_0) with $|z_0| = r_0$, such that $\|B_n(f) - f\|_{r_0} = |B_n(f)(z_0) - f(z_0)|$. We get

$$\|B_n(f) - f\|_r \geq |B_n(f)(z_0) - f(z_0)| = \left| \frac{1}{n} \left\{ \frac{z_0(1-z_0)}{2} f''(z_0) \right. \right.$$
$$\left. \left. + \frac{1}{n} \left[n^2 \left(B_n(f)(z_0) - f(z_0) - \frac{z_0(1-z_0)}{2n} f''(z_0) \right) \right] \right\} \right|$$
$$\geq \frac{1}{n} \left| \left| \frac{z_0(1-z_0)}{2} f''(z_0) \right| - \frac{1}{n} \left[n^2 \left| B_n(f)(z_0) - f(z_0) - \frac{z_0(1-z_0)}{2n} f''(z_0) \right| \right] \right|.$$

But $\left| \frac{z_0(1-z_0)}{2} f''(z_0) \right| \geq \frac{r_0(1-r_0)}{2} M_0 > 0$ and by Theorem 1.1.3 we have

$$n^2 \left| B_n(f)(z_0) - f(z_0) - \frac{z_0(1-z_0)}{2n} f''(z_0) \right|$$
$$\leq n^2 \left\| B_n(f) - f - \frac{e_1(1-e_1)}{2n} f'' \right\|_r$$
$$\leq n^2 \frac{5K_r(f)(1+r)^2}{2n^2} = \frac{5K_r(f)(1+r)^2}{2}.$$

Therefore, there exists an index n_0 depending only on f and r, such that for all $n \geq n_0$ we have

$$\left| \frac{z_0(1 - z_0)}{2} f''(z_0) \right| - \frac{1}{n} \left[n^2 \left| B_n(f)(z_0) - f(z_0) - \frac{z_0(1 - z_0)}{2n} f''(z_0) \right| \right]$$
$$\geq \frac{r_0(1 - r_0)}{4} M_0 > 0,$$

which immediately implies

$$\| B_n(f) - f \|_r \geq \frac{1}{n} \cdot \frac{r_0(1 - r_0)}{4} M_0, \forall n \geq n_0.$$

For $n \in \{1, 2, ..., n_0 - 1\}$ we obviously have $\| B_n(f) - f \|_r \geq \frac{M_{r,n}(f)}{n}$ with $M_{r,n}(f) = n \cdot \| B_n(f) - f \|_r > 0$, which finally implies $\| B_n(f) - f \|_r \geq \frac{C_r(f)}{n}$ for all $n \in \mathbb{N}$, where $C_r(f) = \min\{M_{r,1}, M_{r,2}(f), ..., M_{r,n_0-1}(f), \frac{r_0(1-r_0)}{4} M_0\}$. $\square$

Combining now Theorem 1.1.4 with Theorem 1.1.2, (i), we immediately get the following.

Corollary 1.1.5. (Gal [79]) *Let $R > 1$, $\mathbb{D}_R = \{z \in \mathbb{C}; |z| < R\}$ and let us suppose that $f : \mathbb{D}_R \to \mathbb{C}$ is analytic in $\mathbb{D}_R$. If f is not a polynomial of degree ≤ 1, then for any $r \in [1, R)$ we have*

$$\| B_n(f) - f \|_r \sim \frac{1}{n}, n \in \mathbb{N},$$

where the constants in the equivalence depend on f and r.

In the case of simultaneous approximation we present the following.

Theorem 1.1.6. (Gal [79]) *Let $\mathbb{D}_R = \{z \in \mathbb{C}; |z| < R\}$ be with $R > 1$ and let us suppose that $f : \mathbb{D}_R \to \mathbb{C}$ is analytic in $\mathbb{D}_R$, i.e. $f(z) = \sum_{k=0}^{\infty} c_k z^k$, for all $z \in \mathbb{D}_R$. Also, let $1 \leq r < r_1 < R$ and $p \in \mathbb{N}$ be fixed. If f is not a polynomial of degree $\leq \max\{1, p - 1\}$, then we have*

$$\| B_n^{(p)}(f) - f^{(p)} \|_r \sim \frac{1}{n},$$

where the constants in the equivalence depend on f, r, r_1 and p.

Proof.   Taking into account the upper estimate in Theorem 1.1.2, (ii), it remains to prove the lower estimate for $\| B_n^{(p)}(f) - f^{(p)} \|_r$. Firstly, denoting by Γ the circle of radius $r_1 >$ and center 0 (where $r_1 > r \geq 1$), we have the inequality $|v - z| \geq r_1 - r$ valid for all $|z| \leq r$ and $v \in \Gamma$.

As in the proof of Theorem 1.1.4, for all $v \in \Gamma$ and $n \in \mathbb{N}$ we have

$$B_n(f)(v) - f(v)$$
$$= \frac{1}{n} \left\{ \frac{v(1 - v)}{2} f''(v) + \frac{1}{n} \left[n^2 \left(B_n(f)(v) - f(v) - \frac{v(1 - v)}{2n} f''(v) \right) \right] \right\},$$

which replaced in the Cauchy's formula for derivatives implies

$$B_n^{(p)}(f)(z) - f^{(p)}(z) = \frac{1}{n}\left\{\frac{p!}{2\pi i}\int_\Gamma \frac{v(1-v)f''(v)}{2(v-z)^{p+1}}dv\right.$$

$$+ \frac{1}{n}\cdot\frac{p!}{2\pi i}\int_\Gamma \frac{n^2\left(B_n(f)(v) - f(v) - \frac{v(1-v)}{2n}f''(v)\right)}{(v-z)^{p+1}}dv\right\}$$

$$= \frac{1}{n}\left\{\left[\frac{z(1-z)}{2}f''(z)\right]^{(p)}\right.$$

$$+ \frac{1}{n}\cdot\frac{p!}{2\pi i}\int_\Gamma \frac{n^2\left(B_n(f)(v) - f(v) - \frac{v(1-v)}{2n}f''(v)\right)}{(v-z)^{p+1}}dv\right\}.$$

Passing now to absolute value, for all $|z| \leq r$ and $n \in \mathbb{N}$ it follows

$$|B_n^{(p)}(f)(z) - f^{(p)}(z)| \geq \frac{1}{n}\left\{\left|\left[\frac{z(1-z)}{2}f''(z)\right]^{(p)}\right|\right.$$

$$- \frac{1}{n}\left|\frac{p!}{2\pi}\int_\Gamma \frac{n^2\left(B_n(f)(v) - f(v) - \frac{v(1-v)}{2n}f''(v)\right)}{(v-z)^{p+1}}dv\right|\right\},$$

where by using Theorem 1.1.3, (ii), for all $|z| \leq r$ and $n \in \mathbb{N}$ we get

$$\left|\frac{p!}{2\pi}\int_\Gamma \frac{n^2\left(B_n(f)(v) - f(v) - \frac{v(1-v)}{2n}f''(v)\right)}{(v-z)^{p+1}}dv\right|$$

$$\leq \frac{p!}{2\pi}\cdot\frac{2\pi r_1 n^2}{(r_1-r)^{p+1}}\left\|B_n(f) - f - \frac{e_1(1-e_1)}{2n}f''\right\|_{r_1}$$

$$\leq \frac{5K_{r_1}(f)(1+r_1)^2}{2}\cdot\frac{p!r_1}{(r_1-r)^{p+1}}.$$

Denoting now $F_p(z) = \left[\frac{z(1-z)}{2}f''(z)\right]^{(p)}$, by the hypothesis on f it follows that F_p is analytic and is not identically zero in $\mathbb{D}_R$. Reasoning exactly as in the proof of Theorem 1.1.4, there exists $0 < r_0 < 1$ such that $C_0 = \inf_{|z|=r_0}|F_p(z)| > 0$. Continuing exactly as in the proof of Theorem 1.1.4 (with $\|B_n(f) - f\|_r$ replaced by $\|B_n^{(p)}(f) - f^{(p)}\|_r$), finally there exists an index $n_0 \in \mathbb{N}$ depending on f, r, r_1 and p, such that for all $n \geq n_0$ we have

$$\|B_n^{(p)}(f) - f^{(p)}\|_r \geq \frac{1}{n}\cdot\frac{C_0}{2}.$$

The cases when $n \in \{1, 2, ..., n_0 - 1\}$ are similar with those in the proof of Theorem 1.1.4. □

Remark. Let us suppose that $f^{(p)} \in C[0,1]$, $p \in \mathbb{N}$. By taking $r = 1$ and $\lambda = 1$ in Xie [203], Theorem 2, we immediately obtain the following upper estimate for the derivatives of the real Bernstein polynomials attached to f, valid for all $n \geq n_p$

$$\|B_n^{(p)}(f) - f^{(p)}\| \leq A_p[\omega_1(f^{(p)}; 1/n) + \omega_2^\varphi(f^{(p)}; 1/\sqrt{n}) + \|f^{(p)}\|/n],$$

where $\| \cdot \|$ denotes the uniform norm on $C[0,1]$, $n_p \in \mathbb{N}$ depends only on p, ω_1 denotes the uniform modulus of continuity, $\varphi(x) = \sqrt{x(1-x)}$ and ω_2^φ denotes the Ditzian-Totik second order modulus of smoothness defined in Ditzian-Totik [64].

Then, the above Theorem 1.1.6 suggests the following open question : for any $p \in \mathbb{N}$, there exist the positive constants C_p and n_p depending only on p, such that for all $n \geq n_p$

$$C_p[\omega_1(f^{(p)}; 1/n) + \omega_2^\varphi(f^{(p)}; 1/\sqrt{n}) + \|f^{(p)}\|/n] \leq \|B_n^{(p)}(f) - f^{(p)}\|.$$

The geometric properties of Bernstein polynomials are consequences of Theorem 1.1.2 and can be expressed by the following.

Theorem 1.1.7. (Gal [77], pp. 268-269, Theorem 3.4.2) *Let us suppose that $G \subset \mathbb{C}$ is open, such that $\overline{\mathbb{D}}_1 \subset G$ and $f : G \to \mathbb{C}$ is analytic in G.*

(i) If f is univalent in $\overline{\mathbb{D}}_1$, then there exists an index n_0 depending on f, such that for all $n \geq n_0$, the complex Bernstein polynomials $B_n(f)(z) = \sum_{k=0}^n \binom{n}{k} z^k(1-z)^{n-k} f(k/n)$ are univalent in $\overline{\mathbb{D}}_1$.

(ii) If $f(0) = f'(0) - 1 = 0$ and f is starlike in $\overline{\mathbb{D}}_1$, that is

$$Re\left(\frac{zf'(z)}{f(z)}\right) > 0, \text{ for all } z \in \overline{\mathbb{D}}_1,$$

then there exists an index n_0 depending on f, such that for all $n \geq n_0$, the complex Bernstein polynomials are starlike in $\overline{\mathbb{D}}_1$.

If $f(0) = f'(0) - 1 = 0$ and f is starlike only in $\mathbb{D}_1$, then for any disk of radius $0 < r < 1$ and center 0 denoted by $\mathbb{D}_r$, there exists an index $n_0 = n_0(f, \mathbb{D}_r)$, such that for all $n \geq n_0$, the complex Bernstein polynomials $B_n(f)(z)$ are starlike in $\overline{\mathbb{D}}_r$, that is,

$$Re\left(\frac{zB_n'(f)(z)}{B_n(f)(z)}\right) > 0, \text{ for all } z \in \overline{\mathbb{D}}_r.$$

(iii) If $f(0) = f'(0) - 1 = 0$ and f is convex in $\overline{\mathbb{D}}_1$, that is

$$Re\left(\frac{zf''(z)}{f'(z)}\right) + 1 > 0, \text{ for all } z \in \overline{\mathbb{D}}_1,$$

then there exists an index n_0 depending on f, such that for all $n \geq n_0$, the complex Bernstein polynomials are convex in $\overline{\mathbb{D}}_1$.

If $f(0) = f'(0) - 1 = 0$ and f is convex only in $\mathbb{D}_1$, then for any disk of radius $0 < r < 1$ and center 0 denoted by $\mathbb{D}_r$, there exists an index $n_0 = n_0(f, \mathbb{D}_r)$, such that for all $n \geq n_0$, the complex Bernstein polynomials $B_n(f)(z)$ are convex in $\overline{\mathbb{D}}_r$, that is,

$$Re\left(\frac{zB_n''(f)(z)}{B_n'(f)(z)}\right) + 1 > 0, \text{ for all } z \in \overline{\mathbb{D}}_r.$$

(iv) If $f(0) = f'(0) - 1 = 0$, $f(z) \neq 0$, for all $z \in \overline{\mathbb{D}}_1 \setminus \{0\}$ and f is spirallike of type $\gamma \in (-\pi/2, \pi/2)$ in $\overline{\mathbb{D}}_1$, that is

$$Re\left(e^{i\gamma}\frac{zf'(z)}{f(z)}\right) > 0, \text{ for all } z \in \overline{\mathbb{D}}_1,$$

then there exists an index n_0 depending on f and γ, such that for all $n \geq n_0$ we have $B_n(f)(z) \neq 0$, for all $z \in \overline{\mathbb{D}}_1 \setminus \{0\}$, and $B_n(f)(z)$ are spirallike of type γ in $\overline{\mathbb{D}}_1$.

If $f(0) = f'(0) - 1 = 0$, $f(z) \neq 0$ for all $z \in \mathbb{D}_1 \setminus \{0\}$ and f is spirallike of type γ only in $\mathbb{D}_1$, then for any disk of radius $0 < r < 1$ and center 0 denoted by $\mathbb{D}_r$, there exists an index $n_0 = n_0(f, \mathbb{D}_r, \gamma)$, such that for all $n \geq n_0$, the Bernstein polynomials $B_n(f)(z) \neq 0$ for all $z \in \overline{\mathbb{D}}_r \setminus \{0\}$ and they are spirallike of type γ in $\overline{\mathbb{D}}_r$, that is,

$$Re\left(e^{i\gamma}\frac{zB_n'(f)(z)}{B_n(f)(z)}\right) > 0, \text{ for all } z \in \overline{\mathbb{D}}_r.$$

Proof. (i) It is immediate from the uniform convergence in Theorem 1.1.2 and a well-known result concerning sequences of analytic functions converging locally uniformly to an univalent function (see e.g. Kohr-Mocanu [118], p. 130, Theorem 4.1.17 or Graham-Kohr [105], Theorem 6.1.18).

For the proof of next points (ii), (iii) and (iv), let us observe that by Theorem 1.1.2, (i) and (ii) we get that for $n \to \infty$, we have $B_n(f)(z) \to f(z)$, $B_n'(f)(z) \to f'(z)$ and $B_n''(f)(z) \to f''(z)$, uniformly in $\overline{\mathbb{D}}_1$. In all what follows, denote $P_n(f)(z) = \frac{B_n(f)(z)}{nf(1/n)}$.

By $f(0) = f'(0) - 1 = 0$ and the univalence of f, we get $nf(1/n) \neq 0$, $P_n(f)(0) = \frac{f(0)}{nf(1/n)} = 0$, $P'(f)(0) = \frac{B_n'(f)(0)}{nf(1/n)} = 1$, $n \geq 2$, $nf(1/n) = \frac{f(1/n) - f(0)}{1/n}$ converges to $f'(0) = 1$ as $n \to \infty$, which means that for $n \to \infty$, we have $P_n(f)(z) \to f(z)$, $P_n'(f)(z) \to f'(z)$ and $P_n''(f)(z) \to f''(z)$, uniformly in $\overline{\mathbb{D}}_1$.

(ii) By hypothesis we get $|f(z)| > 0$ for all $z \in \overline{\mathbb{D}}_1$ with $z \neq 0$, which from the univalence of f in $\mathbb{D}_1$, implies that we can write $f(z) = zg(z)$, with $g(z) \neq 0$, for all $z \in \overline{\mathbb{D}}_1$, where g is analytic in $\mathbb{D}_1$ and continuous in $\overline{\mathbb{D}}_1$.

Writing $P_n(f)(z)$ in the form $P_n(f)(z) = zQ_n(f)(z)$, obviously $Q_n(f)(z)$ is a polynomial of degree $\leq n - 1$.

Let $|z| = 1$. We have

$$|f(z) - P_n(f)(z)| = |z| \cdot |g(z) - Q_n(f)(z)| = |g(z) - Q_n(f)(z)|,$$

which by the uniform convergence in $\overline{\mathbb{D}}_1$ of $P_n(f)$ to f and by the maximum modulus principle, implies the uniform convergence in $\overline{\mathbb{D}}_1$ of $Q_n(f)(z)$ to $g(z)$.

Since g is continuous in $\overline{\mathbb{D}}_1$ and $|g(z)| > 0$ for all $z \in \overline{\mathbb{D}}_1$, there exist an index $n_1 \in \mathbb{N}$ and $a > 0$ depending on g, such that $|Q_n(f)(z)| > a > 0$, for all $z \in \overline{\mathbb{D}}_1$ and all $n \geq n_0$.

Also, for all $|z| = 1$, we have

$$|f'(z) - P_n'(f)(z)| = |z[g'(z) - Q_n'(f)(z)] + [g(z) - Q_n(f)(z)]|$$
$$\geq |\ |z| \cdot |g'(z) - Q_n'(f)(z)| - |g(z) - Q_n(f)(z)|\ |$$
$$= |\ |g'(z) - Q_n'(f)(z)| - |g(z) - Q_n(f)(z)|\ |,$$

which from the maximum modulus principle, the uniform convergence of $P_n'(f)$ to f' and of $Q_n(f)$ to g, evidently implies the uniform convergence of $Q_n'(f)$ to g'.

Then, for $|z| = 1$, we get

$$
\begin{aligned}
\frac{zP_n'(f)(z)}{P_n(f)} &= \frac{z[zQ_n'(f)(z) + Q_n(f)(z)]}{zQ_n(f)(z)} \\
&= \frac{zQ_n'(f)(z) + Q_n(f)(z)}{Q_n(f)(z)} \to \frac{zg'(z) + g(z)}{g(z)} \\
&= \frac{f'(z)}{g(z)} = \frac{zf'(z)}{f(z)},
\end{aligned}
$$

which again from the maximum modulus principle, implies

$$
\frac{zP_n'(f)(z)}{P_n(f)} \to \frac{zf'(z)}{f(z)}, \text{ uniformly in } \overline{\mathbb{D}}_1.
$$

Since $Re\left(\frac{zf'(z)}{f(z)}\right)$ is continuous in $\overline{\mathbb{D}}_1$, there exists $\alpha \in (0, 1)$, such that

$$
Re\left(\frac{zf'(z)}{f(z)}\right) \geq \alpha, \text{ for all } z \in \overline{\mathbb{D}}_1.
$$

Therefore

$$
Re\left[\frac{zP_n'(f)(z)}{P_n(f)(z)}\right] \to Re\left[\frac{zf'(z)}{f(z)}\right] \geq \alpha > 0
$$

uniformly on $\overline{\mathbb{D}}_1$, i.e. for any $0 < \beta < \alpha$, there is n_0 such that for all $n \geq n_0$ we have

$$
Re\left[\frac{zP_n'(f)(z)}{P_n(f)(z)}\right] > \beta > 0, \text{ for all } z \in \overline{\mathbb{D}}_1.
$$

Since $P_n(f)(z)$ differs from $B_n(f)(z)$ only by a constant, this proves the first part in (ii).

For the second part, the proof is identical with the first part, with the only difference that instead of $\overline{\mathbb{D}}_1$, we reason for $\overline{\mathbb{D}}_r$.

(iv) Obviously we have

$$
Re\left[e^{i\gamma} \frac{zP_n'(f)(z)}{P_n(f)(z)}\right] \to Re\left[e^{i\gamma} \frac{zf'(z)}{f(z)}\right],
$$

uniformly in $\overline{\mathbb{D}}_1$. We also note that since f is univalent in $\overline{\mathbb{D}}_1$, by the above point (i), there exists n_1 such that $B_n(f)(z)$ is univalent in $\overline{\mathbb{D}}_1$ for all $n \geq n_1$, which by $B_n(f)(0) = 0$ implies $B_n(f)(z) \neq 0$, for all $z \in \overline{\mathbb{D}}_1 \setminus \{0\}$, $n \geq n_1$. For the rest, the proof is identical with that from the above point (ii).

(iii) For the first part, by hypothesis there is $\alpha \in (0, 1)$, such that

$$
Re\left[\frac{zf''(z)}{f'(z)}\right] + 1 \geq \alpha > 0,
$$

uniformly in $\overline{\mathbb{D}}_1$. It is not difficult to show that this is equivalent with the fact that for any $\beta \in (0, \alpha)$, the function $zf'(z)$ is starlike of order β in $\overline{\mathbb{D}}_1$ (see e.g. Mocanu-Bulboacă-Sălăgean [138], p. 77), which implies $f'(z) \neq 0$, for all $z \in \overline{\mathbb{D}}_1$, i.e. $|f'(z)| > 0$, for all $z \in \overline{\mathbb{D}}_1$. Also, by the same type of reasonings as those from the above point (ii), we get

$$Re\left[\frac{zP_n''(f)(z)}{P_n'(f)(z)}\right] + 1 \to Re\left[\frac{zf''(z)}{f'(z)}\right] + 1 \geq \alpha > 0,$$

uniformly in $\overline{\mathbb{D}}_1$. As a conclusion, for any $0 < \beta < \alpha$, there is n_0 depending on f, such that for all $n \geq n_0$ we have

$$Re\left[\frac{zP_n''(f)(z)}{P_n'(f)(z)}\right] + 1 > \beta > 0, \text{ for all } z \in \overline{\mathbb{D}}_1.$$

The proof of second part in (iii) is similar, which proves the theorem. $\qquad\square$

1.1.2 *Bernstein-Faber Polynomials on Compact Sets*

In this subsection $G \subset \mathbb{C}$ will be considered a compact set such that $\tilde{\mathbb{C}} \setminus G$ is connected. In this case, according to the Riemann Mapping Theorem, a unique conformal mapping Ψ of $\tilde{\mathbb{C}} \setminus \overline{\mathbb{D}}_1$ onto $\tilde{\mathbb{C}} \setminus G$ exists so that $\Psi(\infty) = \infty$ and $\Psi'(\infty) > 0$.

By using the Faber polynomials $F_p(z)$ attached to G (see Definition 1.0.10), for $f \in A(\overline{G})$ we can introduce the Bernstein-Faber polynomials given by the formula

$$\mathcal{B}_n(f; \overline{G})(z) = \sum_{p=0}^{n} \binom{n}{p} \Delta_{1/n}^p F(0) \cdot F_p(z), z \in G, \ n \in \mathbb{N},$$

where

$$\Delta_h^p F(0) = \sum_{k=0}^{p} (-1)^{p-k} \binom{p}{k} F(kh), \ F(w) = \frac{1}{2\pi i} \int_{|u|=1} \frac{f(\Psi(u))}{u-w} du, \ w \in \mathbb{D}_1.$$

Here, since $F(1)$ is involved in $\Delta_{1/n}^n F(0)$ and therefore in the definition of $\mathcal{B}_n(f; G)(z)$ too, in addition we will suppose that F can be extended by continuity on the boundary $\partial \mathbb{D}_1$.

Remarks. 1) For $G = \overline{\mathbb{D}}_1$ it is easy to see that the above Bernstein-Faber polynomials one reduce to the classical complex Bernstein polynomials given by

$$B_n(f)(z) = \sum_{p=0}^{n} \binom{n}{p} \Delta_{1/n}^p f(0) z^p = \sum_{p=0}^{n} \binom{n}{p} z^p (1-z)^{n-p} f(p/n).$$

2) It is known that, for example, $\int_0^1 \frac{\omega_p(f \circ \Psi; u)_{\partial \mathbb{D}_1}}{u} du < \infty$ is a sufficient condition for the continuity on $\partial \mathbb{D}_1$ of F in the above definition of the Bernstein-Faber polynomials (see e.g. Gaier [76], p. 52, Theorem 6). Here $p \in \mathbb{N}$ is arbitrary fixed.

The first main result one refers to approximation on compact sets without any restriction on their boundaries and can be stated as follows.

Theorem 1.1.8. *Let G be a continuum (that is a connected compact subset of $\mathbb{C}$) and suppose that f is analytic in G, that is there exists $R > 1$ such that f is analytic in G_R. Here recall that G_R denotes the interior of the closed level curve Γ_R given*

by $\Gamma_R = \{z; |\Phi(z)| = R\} = \{\Psi(w); |w| = R\}$ *(and that $G \subset \overline{G}_r$ for all $1 < r < R$).*
Also, we suppose that F given in the definition of Bernstein-Faber polynomials can be extended by continuity on $\partial \mathbb{D}_1$.

For any $1 < r < R$ the following estimate

$$|\mathcal{B}_n(f; \overline{G})(z) - f(z)| \le \frac{C}{n}, \ z \in \overline{G}_r, \ n \in \mathbb{N},$$

holds, where $C > 0$ depends on f, r and G_r but it is independent of n.

Proof.   First we note that since G is a continuum then it follows that $\tilde{\mathbb{C}} \setminus G$ is simply connected. By the proof of Theorem 2, p. 52 in Suetin [186], for any fixed $1 < \beta < R$ we have $f(z) = \sum_{k=0}^{\infty} a_k(f)F_k(z)$ uniformly in $\overline{G}_\beta$, where $a_k(f)$ are the Faber coefficients and are given by $a_k(f) = \frac{1}{2\pi i}\int_{|u|=\beta}\frac{f[\Psi(u)]}{u^{k+1}}du$. Note here that $G \subset \overline{G}_\beta$.

First we will prove that

$$\mathcal{B}_n(f; \overline{G})(z) = \sum_{k=0}^{\infty} a_k(f)\mathcal{B}_n(F_k; \overline{G})(z),$$

for all $z \in G$. (Note here that by hypothesis we have $\overline{G} = G$). For this purpose, denote $f_m(z) = \sum_{k=0}^{m} a_k(f)F_k(z)$, $m \in \mathbb{N}$.

Since by the linearity of $\mathcal{B}_n$ we easily get

$$\mathcal{B}_n(f_m; \overline{G})(z) = \sum_{k=0}^{m} a_k(f)\mathcal{B}_n(F_k; \overline{G})(z), \text{ for all } z \in G,$$

it suffices to prove that $\lim_{m\to\infty} \mathcal{B}_n(f_m; \overline{G})(z) = \mathcal{B}_n(f; \overline{G})(z)$, for all $z \in G$ and $n \in \mathbb{N}$.

First we have

$$\mathcal{B}_n(f_m; \overline{G})(z) = \sum_{p=0}^{n}\binom{n}{p}\Delta_{1/n}^p G_m(0)F_k(z),$$

where $G_m(w) = \frac{1}{2\pi i}\int_{|u|=1}\frac{f_m(\Psi(u))}{u-w}du$ and $F(w) = \frac{1}{2\pi i}\int_{|u|=1}\frac{f(\Psi(u))}{u-w}du$.

Note here that since by Gaier [76], p. 48, first relation before (6.17), we have

$$\mathcal{F}_k(w) = \frac{1}{2\pi i}\int_{|u|=1}\frac{F_k(\Psi(u))}{u-w}du = w^k, \text{ for all } |w| < 1,$$

evidently that $\mathcal{F}_k(w)$ can be extended by continuity on $\partial\mathbb{D}_1$. This also immediately implies that $G_m(w) = \frac{1}{2\pi i}\int_{|u|=1}\frac{f_m(\Psi(u))}{u-w}du$ can be extended by continuity on $\partial\mathbb{D}_1$, which means that $\mathcal{B}_n(F_k; G)(z)$ and $\mathcal{B}_n(f_m; G)(z)$ are well defined.

Now, taking into account the Cauchy's theorem we also can write

$$G_m(w) = \frac{1}{2\pi i}\int_{|u|=\beta}\frac{f_m(\Psi(u))}{u-w}du \text{ and } F(w) = \frac{1}{2\pi i}\int_{|u|=\beta}\frac{f(\Psi(u))}{u-w}du.$$

For all $n, m \in \mathbb{N}$ and $z \in G$ it follows

$$|\mathcal{B}_n(f_m; \overline{G})(z) - \mathcal{B}_n(f; \overline{G})(z)|$$

$$\leq \sum_{p=0}^{n} \binom{n}{p} |\Delta_{1/n}^p (G_m - F)(0)| \cdot |F_k(z)|$$

$$\leq \sum_{p=0}^{n} \binom{n}{p} \sum_{j=0}^{p} \binom{p}{j} |(G_m - F)((p-j)/n)| \cdot |F_k(z)|$$

$$\leq \sum_{p=0}^{n} \binom{n}{p} \sum_{j=0}^{p} \binom{p}{j} C_{j,p,\beta} \|f_m - f\|_{\overline{G}_\beta} \cdot |F_k(z)|$$

$$\leq M_{n,p,\beta,G_\beta} \|f_m - f\|_{\overline{G}_\beta},$$

which by $\lim_{m \to \infty} \|f_m - f\|_{\overline{G}_\beta} = 0$ (see e.g. the proof of Theorem 2, p. 52 in Suetin [186]) implies the desired conclusion. Here $\|f_m - f\|_{\overline{G}_\beta}$ denotes the uniform norm of $f_m - f$ on $\overline{G}_\beta$.

Consequently we obtain

$$|\mathcal{B}_n(f; \overline{G})(z) - f(z)| \leq \sum_{k=0}^{\infty} |a_k(f)| \cdot |\mathcal{B}_n(F_k; \overline{G})(z) - F_k(z)|$$

$$= \sum_{k=0}^{n} |a_k(f)| \cdot |\mathcal{B}_n(F_k; \overline{G})(z) - F_k(z)|$$

$$+ \sum_{k=n+1}^{\infty} |a_k(f)| \cdot |\mathcal{B}_n(F_k; \overline{G})(z) - F_k(z)|.$$

Therefore it remains to estimate $|a_k(f)| \cdot |\mathcal{B}_n(F_k; \overline{G})(z) - F_k(z)|$, firstly for all $0 \leq k \leq n$ and secondly for $k \geq n + 1$, where

$$\mathcal{B}_n(F_k; \overline{G})(z) = \sum_{p=0}^{n} \binom{n}{p} [\Delta_{1/n}^p \mathcal{F}_k(0)] \cdot F_p(z).$$

First it is useful to observe that by Gaier [76], p. 48, combined with the Cauchy's theorem, for any fixed $1 < \beta < R$ we have

$$\mathcal{F}_k(w) := \frac{1}{2\pi i} \int_{|u|=\beta} \frac{F_k[\Psi(u)]}{u - w} du = w^k = e_k(w), \text{ for all } |w| < \beta.$$

Denote

$$D_{n,p,k} = \binom{n}{p} \Delta_{1/n}^p e_k(0) = \binom{n}{p} [0, 1/n, ..., p/n; e_k] \cdot (p!)/n^p.$$

It follows

$$\mathcal{B}_n(F_k; \overline{G})(z) = \sum_{p=0}^{n} D_{n,p,k} \cdot F_p(z).$$

Since for the classical complex Bernstein polynomials attached to a disk of center in origin we can write $B_n(e_k)(z) = \sum_{p=0}^{n} D_{n,p,k} z^p$, since each e_k is convex of any order and $B_n(e_k)(1) = e_k(1) = 1$ for all k, it follows that all $D_{n,p,k} \geq 0$ and $\sum_{p=0}^{n} D_{n,p,k} = 1$, for all k and n. Also, note that $D_{n,k,k} = \frac{n(n-1)...(n-k+1)}{n^k}$.

In the estimation of $|a_k(f)| \cdot |\mathcal{B}_n(F_k; \overline{G})(z) - F_k(z)|$ we distinguish two cases :
1) $0 \leq k \leq n$; 2) $k > n$.

Case 1. We have

$$|\mathcal{B}_n(F_k; \overline{G})(z) - F_k(z)| \leq |F_k(z)| \cdot |1 - D_{n,k,k}| + \sum_{p=0}^{k-1} D_{n,p,k} \cdot |F_p(z)|.$$

Fix now $1 < r < \beta$. By the inequality (13), p. 44 in Suetin [186] we have

$$|F_p(z)| \leq C(r)r^p, \text{ for all } z \in \overline{G}_r, p \geq 0,$$

which immediately implies

$$|\mathcal{B}_n(F_k; \overline{G})(z) - F_k(z)| \leq 2C(r)[1 - D_{n,k,k}]r^k \leq C(r)\frac{k(k-1)}{n}r^k,$$

for all $z \in \overline{G}_r$. Here we used the inequality $1 - \Pi_{i=1}^{k} x_i \leq \sum_{i=1}^{k}(1 - x_i)$ (valid if all $x_i \in [0,1]$) which implies the inequality

$$1 - D_{n,k,k} = 1 - \frac{n(n-1)...(n-k+1)}{n^k} = 1 - \Pi_{i=1}^{k-1}\frac{n-i}{n}$$

$$\leq \sum_{i=1}^{k-1}(1 - (n-i)/n) = \frac{1}{n}\sum_{i=0}^{k-1} i = \frac{k(k-1)}{2n}.$$

Also by the above formula for $a_k(f)$ we easily obtain $|a_k(f)| \leq \frac{C(\beta,f)}{\beta^k}$, for all $k \geq 0$. Note that $C(r), C(\beta, f) > 0$ are constants independent of k.

For all $z \in \overline{G}_r$ and $k = 0, 1, 2, ...n$ it follows

$$|a_k(f)| \cdot |\mathcal{B}_n(F_k; \overline{G})(z) - F_k(z)| \leq \frac{C(r,\beta,f)}{n}k(k-1)\left[\frac{r}{\beta}\right]^k,$$

that is

$$\sum_{k=0}^{n} |a_k(f)| \cdot |\mathcal{B}_n(F_k; \overline{G})(z) - F_k(z)| \leq \frac{C(r,\beta,f)}{n}\sum_{k=2}^{n} k(k-1)d^k, \text{ for all } z \in \overline{G}_r,$$

where $0 < d = r/\beta < 1$. Also, clearly we have $\sum_{k=2}^{n} k(k-1)d^k \leq \sum_{k=2}^{\infty} k(k-1)d^k < \infty$ which finally implies that

$$\sum_{k=0}^{n} |a_k(f)| \cdot |\mathcal{B}_n(F_k; \overline{G})(z) - F_k(z)| \leq \frac{C^*(r,\beta,f)}{n}.$$

Case 2. We have

$$\sum_{k=n+1}^{\infty} |a_k(f)| \cdot |\mathcal{B}_n(F_k; \overline{G})(z) - F_k(z)| \leq \sum_{k=n+1}^{\infty} |a_k(f)| \cdot |\mathcal{B}_n(F_k; \overline{G})(z)|$$

$$+ \sum_{k=n+1}^{\infty} |a_k(f)| \cdot |F_k(z)|.$$

By the estimates mentioned in the case 1), we immediately get

$$\sum_{k=n+1}^{\infty} |a_k(f)| \cdot |F_k(z)| \leq C(r, \beta, f) \sum_{k=n+1}^{\infty} d^k, \text{ for all } z \in \overline{G}_r,$$

with $d = r/\beta$.

Also,

$$\sum_{k=n+1}^{\infty} |a_k(f)| \cdot |\mathcal{B}_n(F_k; \overline{G})(z)| = \sum_{k=n+1}^{\infty} |a_k(f)| \cdot \left| \sum_{p=0}^{n} D_{n,p,k} \cdot F_p(z) \right|$$

$$\leq \sum_{k=n+1}^{\infty} |a_k(f)| \cdot \sum_{p=0}^{n} D_{n,p,k} \cdot |F_p(z)|.$$

But for $p \leq n < k$ and taking into account the estimates obtained in the Case 1) we get

$$|a_k(f)| \cdot |F_p(z)| \leq C(r, \beta, f) \frac{r^p}{\beta^k} \leq C(r, \beta, f) \frac{r^k}{\beta^k}, \text{ for all } z \in \overline{G}_r,$$

which implies

$$\sum_{k=n+1}^{\infty} |a_k(f)| \cdot |\mathcal{B}_n(F_k; \overline{G})(z) - F_k(z)| \leq C(r, \beta, f) \sum_{k=n+1}^{\infty} \sum_{p=0}^{n} D_{n,p,k} \left[\frac{r}{\beta} \right]^k$$

$$= C(r, \beta, f) \sum_{k=n+1}^{\infty} \left[\frac{r}{\beta} \right]^k$$

$$= C(r, \beta, f) \frac{d^{n+1}}{1-d},$$

with $d = r/\beta$.

In conclusion, collecting the estimates in the Cases 1) and 2) we obtain

$$|\mathcal{B}_n(f; \overline{G})(z) - f(z)| \leq \frac{C_1}{n} + C_2 d^{n+1} \leq \frac{C}{n}, \ z \in \overline{G}_r, \ n \in \mathbb{N}.$$

This proves the theorem. $\qquad\qquad\qquad\qquad\qquad\qquad\qquad\qquad\qquad\qquad\square$

Remarks. 1) Simultaneous upper and lower estimates in Theorem 1.1.8 (that is a similar result to Corollary 1.1.5) can easily be obtained under some restrictions on the boundaries. For that purpose firstly we recall some useful concepts and results. The first important concept is that of *Faber set*. Thus, suppose that $G \subset \mathbb{C}$ is compact. If the Faber mapping T (given by Definition 1.0.10, (ii)) defined on the set of all polynomials $\mathcal{P}(\overline{\mathbb{D}}_1)$ and with values in the set of polynomials $\mathcal{P}(G)$ is continuous, then G is called a Faber set. In this case, T admits a unique extension to a linear and bounded mapping between the Banach spaces $A(\overline{\mathbb{D}}_1)$ and $A(\overline{G})$ (see the Remark after Definition 1.0.10, or for full details see e.g. the book of Gaier [76], pp. 48-49). For example, if G is a compact set which is a Jordan domain whose boundary Γ is a rectifiable curve of bounded rotation, then G is a Faber set (see e.g. Gaier [75], p. 51, Theorem 2). Also, Theorem 1 in Frerick-Müller [74] gives other sufficient conditions on the boundary of G which assures that the compact set G is a Faber set. As a consequence, a compact set G whose boundary consists of piecewise convex curves also is a Faber set (see Frerick-Müller [74], p. 429). By Lemma 1 in Frerick-Müller [74], if G is a compact Faber set then the Faber mapping $T : A(\overline{\mathbb{D}}_1) \to A(\overline{G})$ is injective.

2) Another important concept is that of *inverse Faber set*. Thus, according to Anderson-Clunie [22], p. 546, a Faber set G is called inverse Faber set if the Faber operator T is bijective, which implies that $T^{-1} : A(\overline{G}) \to A(\overline{\mathbb{D}}_1)$ given by (see Theorem 1.0.12 or e.g. Anderson-Clunie [22], relation (1.2))

$$T^{-1}(f)(\xi) = \frac{1}{2\pi i} \int_{|w|=1} \frac{f[\Psi(w)]}{w - \xi} dw,$$

also is linear and bounded. An important result is Theorem 2 in Anderson-Clunie [22], p. 548, which says that if G is the closure of a Jordan domain whose boundary Γ is rectifiable and of boundary rotation, and in addition Γ is free of cups, then G is an inverse Faber set. Let us also recall that if the compact set G is a Faber set, then for any $1 < r$, $\overline{G}_r$ is an inverse Faber set, where $\overline{G}_r$ denotes the closure of the Jordan domain bounded by the analytic simple curve $\Gamma_r = \{\Psi(w); |w| = r\}$ (see the Remark on page 434 in Frerick-Müller [74] or Anderson-Clunie [22]). Also, in this case by Theorem 3 in Frerick-Müller [74], for $f \in A(\overline{G}_r)$ we have $T^{-1}(f) \in A(\overline{\mathbb{D}}_r)$.

As a consequence of the considerations in the above two remarks, we can state the following result.

Theorem 1.1.9. *Let G be a compact Faber set such that $\tilde{\mathbb{C}} \setminus G$ is simply connected. If f is analytic on G, that is there exists $R > 1$ such that f is analytic in G_R and if f is not a polynomial of degree ≤ 1, then for any $1 < r < R$ we have*

$$\|\mathcal{B}_n(f; \overline{G}) - f\|_{\overline{G}_r} \sim \frac{1}{n}, \ n \in \mathbb{N},$$

where the constants in the equivalence depend on f, r and G_r but are independent of n. Here $\|f\|_{\overline{G}_r} = \sup_{z \in \overline{G}_r} |f(z)|$.

Proof.   According to the above considerations, there exists g analytic in $\overline{\mathbb{D}}_r$ such that $f = T(g)$, that is $g = T^{-1}(f)$ (therefore F can be extended by continuity on $\partial \mathbb{D}_1$. By hypothesis on f it follows that f cannot be of the form $f(z) = c_0 F_0(z) + c_1 F_1(z)$ where F_0 and F_1 are the Faber polynomials of degree 0 and 1 respectively and $c_0, c_1 \in \mathbb{C}$. This immediately implies that g is not a polynomial of degree ≤ 1.

First we have $B_n(T^{-1}(f)) = T^{-1}[\mathcal{B}_n(f; \overline{G})]$. Indeed,

$$B_n(T^{-1}(f))(z) = \sum_{p=0}^{n} \binom{n}{p} \Delta_{1/n}^p T^{-1}(f)(0) z^p = \sum_{p=0}^{n} \binom{n}{p} \Delta_{1/n}^p F(0) z^p,$$

since $T^{-1}(f)(\xi) = \frac{1}{2\pi i} \int_{|w|=1} \frac{f[\Psi(w)]}{w-\xi} dw = F(\xi)$, and

$$T^{-1}[\mathcal{B}_n(f;\overline{G})](z) = \frac{1}{2\pi i} \int_{|w|=1} \frac{\mathcal{B}_n(f;\overline{G})[\Psi(w)]}{w-z} dw$$

$$= \sum_{p=0}^{n} \binom{n}{p} \Delta_{1/n}^p F(0) \frac{1}{2\pi i} \int_{|w|=1} \frac{F_p[\Psi(w)]}{w-z} dw$$

$$= \sum_{p=0}^{n} \binom{n}{p} \Delta_{1/n}^p F(0) z^p,$$

since according to Gaier [76], p. 48, first relation before (6.17), we have

$$\frac{1}{2\pi i} \int_{|w|=1} \frac{F_p[\Psi(w)]}{w-z} dw = z^p.$$

Then by Corollary 1.1.5 and by the linearity and continuity of T^{-1} we get

$$\frac{C}{n} \le \|B_n(g) - g\|_r = \|B_n(g) - T^{-1}(f)\|_r = \|T^{-1}[\mathcal{B}_n(f;\overline{G})] - T^{-1}(f)\|_r$$

$$\le \||T^{-1}\|| \cdot \|\mathcal{B}_n(f;\overline{G}) - f\|_{\overline{G}_r} \le M\|\mathcal{B}_n(f;\overline{G}) - f\|_{\overline{G}_r},$$

which proves the lower estimate.

On the other hand we have $T[B_n(g)] = \mathcal{B}_n(T(g);\overline{G})$. Indeed,

$$T[B_n(g)](z) = \sum_{p=0}^{n} \Delta_{1/n}^p g(0) F_p(z),$$

and

$$\mathcal{B}_n(T(g);\overline{G})(z) = \sum_{p=0}^{n} \binom{n}{p} \Delta_{1/n}^p H(0) F_p(z),$$

where according to Gaier [76], p. 49. relation (6.17') we have

$$H(w) = \frac{1}{2\pi i} \int_{|u|=1} \frac{T(g)[\Psi(u)]}{u-w} du = g(w).$$

Therefore by the same Corollary 1.1.5 and by the linearity and continuity of T we obtain

$$\|\mathcal{B}_n(f;\overline{G}) - f\|_{\overline{G}_r} = \|\mathcal{B}_n(T(g);\overline{G}) - T(g)\|_{\overline{G}_r} = \|T[B_n(g)] - T(g)\|_{\overline{G}_r}$$

$$\le \||T\|| \cdot \|B_n(g) - g\|_r \le \frac{C}{n},$$

which proves the upper estimate and the theorem. $\qquad\square$

1.2   Iterates of Bernstein Polynomials

First we deal with the approximation properties of the iterates of complex Bernstein polynomials and their relationship with the theory of the semigroups of operators.

For $R > 1$, let us define by $\mathbb{A}_R$ the space of all functions defined and analytic in the open disk of center 0 and radius R denoted by $\mathbb{D}_R$. Denoting $r_j = R - \frac{R-1}{j}$, $j \in \mathbb{N}$ and for $f \in \mathbb{A}_R$, $\|f\|_j = max\{|f(z)|; |z| \le r_j\}$, since $r_1 = 1$ and $r_j \nearrow R$, it is well-known that $\{\|\cdot\|_j, j \in \mathbb{N}\}$ is a countable family of increasing semi-norms on $\mathbb{A}_R$ and that $\mathbb{A}_R$ becomes a metrizable complete locally convex space (Fréchet space), with respect to the metric

$$d(f,g) = \sum_{j=1}^{\infty} \frac{1}{2^j} \cdot \frac{\|f-g\|_j}{1 + \|f-g\|_j}, f,g \in \mathbb{A}_R.$$

It is well-known that $\lim_{n\to\infty} d(f_n, f) = 0$ is equivalent to the fact that the sequence $(f_n)_{n\in\mathbb{N}}$ converges to f uniformly on compacts in $\mathbb{D}_R$. Details about the space $\mathbb{A}_R$ and the metric d can be found in e.g. Kohr-Mocanu [118], pp. 104-107.

Now, for $f \in \mathbb{A}_R$, that is of the form $f(z) = \sum_{k=0}^{\infty} c_k z^k$, for all $z \in \mathbb{D}_R$, let us define the iterates of complex Bernstein polynomial $B_n(f)(z)$, by $B_n^{(1)}(f)(z) = B_n(f)(z)$ and $B_n^{(m)}(f)(z) = B_n[B_n^{(m-1)}(f)](z)$, for any $m \in \mathbb{N}$, $m \ge 2$. Since we have (see e.g. Lorentz [125], p. 88, proof of Theorem 4.1.1), $B_n(f)(z) = \sum_{k=0}^{\infty} c_k B_n(e_k)(z)$, by recurrence for all $m \ge 1$, we easily get that $B_n^{(m)}(f)(z) = \sum_{k=0}^{\infty} c_k B_n^{(m)}(e_k)(z)$, with $e_k(z) = z^k$.

The first main result of this section is the following.

Theorem 1.2.1. (Gal [78]) *Let $f \in \mathbb{A}_R$ with $R > 1$, that is $f(z) = \sum_{k=0}^{\infty} c_k z^k$, for all $z \in \mathbb{D}_R$.*

(i) For any $n \in \mathbb{N}$, we have

$$\lim_{m\to\infty} d[B_n^{(m)}(f), B_1(f)] = 0;$$

(ii) If $\lim_{n\to\infty} \frac{m_n}{n} = 0$, then

$$\lim_{n\to\infty} d[B_n^{(m_n)}(f), f] = 0.$$

Moreover, for any fixed $q \in \mathbb{N}$, the following estimates hold

$$\|B_n^{(m)}(f) - f\|_q \le \frac{m}{n} \sum_{k=2}^{\infty} |c_k| k(k-1) r_q^k,$$

and

$$d[B_n^{(m)}(f), f] \le \frac{m}{n} \sum_{k=2}^{\infty} |c_k| k(k-1) r_q^k + \frac{1}{2^q},$$

where $\sum_{k=2}^{\infty} |c_k| k(k-1) r_q^k < \infty$.
(iii) If $\lim_{n\to\infty} \frac{m_n}{n} = \infty$, then

$$\lim_{n\to\infty} d[B_n^{(m_n)}(f), B_1(f)] = 0;$$

(iv) If $\lim_{n\to\infty} \frac{m_n}{n} = t \in (0,\infty)$, *then*

$$\lim_{n\to\infty} d[B_n^{(m_n)}(f), T(t)(f)] = 0,$$

where $L(f)(z) = (1-z)f(0) + zf(1), z \in \mathbb{D}_R$,

$$T(t)(f)(z) = L(f)(z) + z(1-z)\int_0^1 G_t(z,y)[f(y) - L(f)(y)]dy, z \in \mathbb{D}_R,$$

$$G_t(z,y) = \sum_{k=2}^{\infty} \frac{k(2k-1)}{k-1} e^{-k(k-1)t/2} P_{k-2}^{(1,1)}(2z-1) P_{k-2}^{(1,1)}(2y-1),$$

$z \in \mathbb{D}_R, y \in [0,1]$, *and* $P_{k-2}^{(1,1)}(z), |z| < R$, *are the Jacobi polynomials normalized to be* $k-1$ *at* $z = 1$.

Proof. (i) Since from Karlin-Ziegler [114] it is known that if $m \to \infty$, then $B_n^{(m)}(f)(x) \to B_1(f)(x)$, uniformly on the interval $[0,1]$, according to the classical Vitali's result (see e.g. Kohr-Mocanu [118], p. 112, Theorem 3.2.10], it suffices to show that for any fixed $n \in \mathbb{N}$, the iterate sequence of polynomials $B_n^{(m)}(f)(z)$, $m = 1, 2, ...$, is uniformly bounded with respect to $m \in \mathbb{N}$ in each $\overline{\mathbb{D}}_r$ with $1 \le r < R$.

Let $n \in \mathbb{N}$ be fixed. According to He [107], p. 580, relationship (7), we can write

$$B_n(e_k)(z) = \sum_{j=1}^{k} S(k,j) \frac{n(n-1)...[n-(j-1)]}{n^k} z^j,$$

where $S(k,j)$ are the Stirling numbers of second kind. It is well-known that these numbers satisfy $S(k,j) \ge 0$, for all $j, k \in \mathbb{N}$ and

$$\sum_{j=1}^{k} S(k,j)n(n-1)...[n-(j-1)] = n^k, \text{ for } k, n \in \mathbb{N}.$$

Let $|z| \le r$ with $r \ge 1$. Since $S(k,j)n(n-1)...[n-(j-1)] \ge 0$, for all $k, n, j \in \mathbb{N}$ with $1 \le j \le k$, it follows

$$|B_n(e_k)(z)| \le \sum_{j=1}^{k} S(k,j) \frac{n(n-1)...[n-(j-1)]}{n^k}|e_j(z)|$$

$$\le \sum_{j=1}^{k} S(k,j) \frac{n(n-1)...[n-(j-1)]}{n^k} r^j$$

$$\le \sum_{j=1}^{k} S(k,j) \frac{n(n-1)...[n-(j-1)]}{n^k} r^k = r^k.$$

Applying now B_n to the above equality, we obtain

$$B_n^{(2)}(e_k)(z) = \sum_{j=1}^{k} S(k,j) \frac{n(n-1)...[n-(j-1)]}{n^k} B_n(e_j)(z),$$

which from the last inequality implies

$$|B_n^{(2)}(e_k)(z)| = \sum_{j=1}^{k} S(k,j) \frac{n(n-1)...[n-(j-1)]}{n^k} |B_n(e_j)(z)|$$

$$\leq \sum_{j=1}^{k} S(k,j) \frac{n(n-1)...[n-(j-1)]}{n^k} r^j \leq r^k.$$

Reasoning by recurrence, we easily get

$$|B_n^{(m)}(e_k)(z)| \leq r^k,$$

for all $k, n, m \in \mathbb{N}$ and $z \in \overline{\mathbb{D}}_r$.

This implies

$$|B_n^{(m)}(f)(z)| \leq \sum_{k=0}^{\infty} |c_k| r^k < +\infty,$$

for all $m, n \in \mathbb{N}$ and $z \in \overline{\mathbb{D}}_r$, which proves (i).

(ii) Since from the last inequality in (i), for each $r \in [1, R)$ the sequence $B_n^{(m)}(f)(z)$, $m, n = 1, 2, ...$, is in fact uniformly bounded in $\overline{\mathbb{D}}_r$ with respect to both $m, n \in \mathbb{N}$, and since by Kelisky-Rivlin [115], we have $B_n^{(m_n)}(f)(x) \to f(x)$ as $n \to \infty$, uniformly for $x \in [0,1]$, it follows that the Vitali's convergence theorem implies the first convergence in (ii).

Since $B_n^{(m)}(f)(z) = \sum_{k=0}^{\infty} c_k B_n^{(m)}(e_k)(z)$, with $e_k(z) = z^k$, we get

$$|B_n^{(m)}(f)(z) - f(z)| \leq \sum_{k=2}^{\infty} |c_k| \cdot |B_n^{(m)}(e_k)(z) - e_k(z)|.$$

But according to He [107], we have

$$B_n(e_k)(z) - e_k(z) = \sum_{j=1}^{k} S(k,j) \frac{n(n-1)...[n-(j-1)]}{n^k} e_j(z) - e_k(z).$$

Therefore,

$$B_n(e_k)(z) - e_k(z) = \sum_{j=1}^{k-1} S(k,j) \frac{n(n-1)...[n-(j-1)]}{n^k} e_j(z)$$
$$+ [(1-1/n)...(1-(k-1)/n) - 1] e_k(z),$$

which immediately implies

$$B_n^{(p)}[B_n(e_k)(z) - e_k(z)] = \sum_{j=1}^{k-1} S(k,j) \frac{n(n-1)...[n-(j-1)]}{n^k} B_n^{(p)}(e_j)(z)$$
$$+ [(1-1/n)...(1-(k-1)/n) - 1] B_n^{(p)}(e_k)(z).$$

Taking into account that by the proof of the above point (i), we have $|B_n^{(p)}(e_j)(z)| \leq r^j$, for all $p, n, j \in \mathbb{N}$ and $|z| \leq r$, it follows

$$|B_n^{(p)}[B_n(e_k) - e_k](z)| \leq \sum_{j=1}^{k-1} S(k,j) \frac{n(n-1)...[n-(j-1)]}{n^k} |B_n^{(p)}(e_j)(z)|$$

$$+[1 - (1-1/n)...(1-(k-1)/n)]|B_n^{(p)}(e_k)(z)|$$

$$\leq \sum_{j=1}^{k-1} S(k,j) \frac{n(n-1)...[n-(j-1)]}{n^k} r^j$$

$$+[1 - (1-1/n)...(1-(k-1)/n)]r^k$$

$$\leq 2[1 - (1-1/n)...(1-(k-1)/n)]r^k.$$

But

$$B_n^{(m)}(e_k)(z) - e_k(z) = \sum_{p=0}^{m-1} B_n^{(p)}[B_n(e_k)(z) - e_k(z)],$$

which implies for all $|z| \leq r$ that

$$|B_n^{(m)}(e_k) - e_k| \leq \sum_{p=0}^{m-1} |B_n^{(p)}[B_n(e_k) - e_k](z)|$$

$$\leq 2m[1 - (1-1/n)...(1-(k-1)/n)]r^k,$$

and finally

$$|B_n^{(m)}(f)(z) - f(z)| \leq 2m \sum_{k=0}^{\infty} |c_k| \cdot [1 - (1-1/n)...(1-(k-1)/n)]r^k.$$

Since $[1 - (1-1/n)...(1-(k-1)/n)] \leq \frac{1}{n}[1 + ... + (k-1)] = \frac{k(k-1)}{2n}$, for all $k \in \mathbb{N}$, $k \geq 2$, by choosing $r = r_q$, we get the first required inequality in the statement. Note that $\sum_{k=2}^{\infty} |c_k| k(k-1) r^{k-2} < \infty$, since we have $f''(z) = \sum_{k=2}^{\infty} c_k k(k-1) z^{k-2}$, for all $|z| \leq r$.

The second estimate in (ii) is a direct consequence of the inequality

$$d(f,g) = \sum_{j=1}^{q} \frac{1}{2^j} \cdot \frac{\|f-g\|_j}{1 + \|f-g\|_j} + \sum_{j=q+1}^{\infty} \frac{1}{2^j} \cdot \frac{\|f-g\|_j}{1 + \|f-g\|_j}$$

$$\leq \frac{\|f-g\|_q}{1 + \|f-g\|_q} \sum_{j=1}^{q} \frac{1}{2^j} + \sum_{j=q+1}^{\infty} \frac{1}{2^j}$$

$$\leq \frac{\|f-g\|_q}{1 + \|f-g\|_q} + \frac{1}{2^q} \leq \|f-g\|_q + \frac{1}{2^q}.$$

(iii) Since $B_n^{(m_n)}(f)(x) \to B_1(f)(x)$, as $n \to \infty$, uniformly for $x \in [0,1]$ (see Kelisky-Rivlin [115]) the proof is similar with that of (ii).

(iv) First we prove that for any $t > 0$, the complex series $G_t(z, y)$ is uniformly and absolutely convergent for $|z|, |y| \leq r$, with $r \geq 1$. Indeed, from the representation formula

$$P_n^{(1,1)}(z) = \frac{1}{2^n} \sum_{k=0}^{n} \binom{n+1}{k} \binom{n+1}{n-k} (z+1)^k (z-1)^{n-k}, |z| \leq r,$$

we get

$$|P_n^{(1,1)}(z)| \leq \left(\frac{r+1}{2}\right)^n \sum_{k=0}^{n} \binom{n+1}{k} \binom{n+1}{n-k} = \left(\frac{r+1}{2}\right)^n \binom{2n+2}{n}$$

$$\leq \left(\frac{r+1}{2}\right)^n \left(\frac{(2n+2)e}{n}\right)^n \leq \left(\frac{r+1}{2}\right)^n (4e)^n \leq [6(r+1)]^n.$$

We used above the Vandermond's equality $\sum_{k=0}^{j} \binom{n}{k} \binom{m}{j-k} = \binom{n+m}{j}$ and the inequality $\binom{n}{k} \leq \left(\frac{ne}{k}\right)^k$.

Now, since for $|z|, |y| \leq r$ we get $|2z - 1|, |2y - 1| \leq 2r + 1$, denoting $\rho = 2r + 1$, it follows

$$|G_t(z, y)| \leq \sum_{k=2}^{\infty} \frac{k(2k-1)}{k-1} e^{-k(k-1)t/2} |P_{k-2}^{(1,1)}(2z-1)| \cdot |P_{k-2}^{(1,1)}(2y-1)|$$

$$\leq \sum_{k=2}^{\infty} \frac{k(2k-1)}{k-1} e^{-k(k-1)t/2} [6\rho]^{2k-4}.$$

By applying the ratio test, the last series (of positive numbers) is convergent.

This shows that $T(t)(f)(z)$ is well defined for any $t > 0$ and all $|z| < R$.

In what follows, it suffices to prove that for any fixed $r \in [1, R)$ we have

$$\lim_{n \to \infty} \|B_n^{(m_n)}(f) - T(t)(f)\|_r = 0.$$

Since from Karlin-Ziegler [114], for $t > 0$ we have

$$\lim_{n \to \infty} B_n^{(m_n)}(f)(x) = L(f)(x) + x(1-x) \int_0^1 G_t(x, y)[f(y) - L(f)(y)]dy,$$

uniformly with respect to $x \in [0, 1]$, according to the Vitali's theorem, it suffices to prove that the sequence $(B_n^{(m_n)}(f)(z))_{n \in \mathbb{N}}$ is uniformly bounded in $\overline{\mathbb{D}_r}$. However this fact was proved by the last inequality at the above point (i) (see also the remark at the beginning of point (ii)). Therefore the theorem has been proved. □

Remarks. 1) The property (iv) in Theorem 1.2.1 suggests that for $f \in \mathbb{A}_R$, the limit of the iterates $B_n^{(m_n)}(f)(z)$ represents the semigroup of operators $T(t)(f)(z)$ defined on the locally convex space (Fréchet) $\mathbb{A}_R$.

2) The results in Theorem 1.2.1 extend some related results in the case of iterates of real Bernstein polynomials on $[0, 1]$ (see Karlin-Ziegler [114], Kelisky-Rivlin [115]).

In what follows we prove that the shape preserving properties for complex Bernstein polynomials in Theorem 1.1.7 hold for their iterates too.

In the proofs of these properties we need the following two auxiliary lemmas.

Lemma 1.2.2. (Gal [78]) *Let* $f \in \mathbb{A}_R$ *with* $R > 1$, *that is* $f(z) = \sum_{k=0}^{\infty} c_k z^k$, *for all* $z \in \mathbb{D}_R$. *For any* $m \in \mathbb{N}$, *we have*

$$\lim_{n \to \infty} d[B_n^{(m)}(f), f] = 0,$$

that is the sequence $(B_n^{(m)}(f))_{n \in \mathbb{N}}$ *uniformly converges to* f *on compacts disks in* $\mathbb{D}_R$.

Proof. Note that Lemma 1.2.2 is a particular case of Theorem 1.2.1, (ii), for the constant sequence $m_n \equiv m$. □

Lemma 1.2.3. (Gal [78]) *Let* $f \in \mathbb{A}_R$, $R > 1$, *be satisfying* $f(0) = 0$. *For all* $m, n \in \mathbb{N}$ *we have* $[B_n^{(m)}(f)]'(0) = nB_n^{(m-1)}(f)(1/n)$ *and* $\lim_{n \to \infty}[n \cdot B_n^{(j)}(f)(1/n)] = f'(0)$, *for any fixed* j, *where by convention,* $B_n^{(0)}(f) = f$ *and* $[B_n^{(m)}(f)]'(0)$ *denotes the first derivative of* $B_n^{(m)}(f)(z)$ *at* 0.

Proof. We have

$$B_n^{(m)}(f)(z) = B_n[B_n^{(m-1)}(f)](z),$$

which by $B_n'(f)(0) = nf(1/n)$ implies

$$[B_n^{(m)}(f)]'(0) = nB_n^{(m-1)}(f)(1/n).$$

For the second part of lemma, let us observe that it suffices to prove it for real functions $f(x)$, $x \in [0, 1]$, in $C^2[0, 1]$. Indeed, by $f(z) = U(x, y) + iV(x, y)$, $z = x + iy$, where U and V have partial derivatives of any order, we get $f(x) = U(x, 0) + iV(x, 0) := g(x) + ih(x)$, for all $x \in [0, 1]$, where g, h are continuously differentiable of any order. Also, we take into account that $B_n^{(m)}(\cdot)$ is a linear operator on $C[0, 1]$, for any $n, m \in \mathbb{N}$.

We obviously have $nf(1/n) = \frac{f(1/n) - f(0)}{(1/n)} \to f'(0)$, as $n \to \infty$.

In what follows, we will use the well-known pointwise estimate for Bernstein polynomials when $f \in C^2[0, 1]$, given by

$$|B_n(f)(x) - f(x)| \leq C\|f''\|\frac{x(1-x)}{n}, \text{ for all } x \in [0, 1], n \in \mathbb{N},$$

where $\| \cdot \|$ denotes the uniform norm in $C[0, 1]$.

We get

$$nB_n(f)(1/n) = \frac{B_n(f)(1/n) - B_n(f)(0)}{(1/n)}$$

$$= \frac{B_n(f)(1/n) - f(1/n)}{(1/n)} + \frac{f(1/n) - f(0)}{(1/n)} \to f'(0),$$

as $n \to \infty$, since by the above estimate we have

$$\left|\frac{B_n(f)(1/n) - f(1/n)}{(1/n)}\right| \leq \frac{C\|f''\|}{n} \to 0, \text{ for } n \to \infty.$$

Then, we get

$$nB_n^{(2)}(f)(1/n)$$
$$= \frac{B_n[B_n(f)](1/n) - B_n(f)(1/n)}{(1/n)} + \frac{B_n(f)(1/n) - B_n(f)(0)}{(1/n)} \to f'(0),$$

as $n \to \infty$. Indeed, by applying again the above pointwise estimate for f replaced by $B_n(f)$ and taking into account the inequality $\|B_n''(f)\| \le \|f''\|$ (which easily follows from e.g. Lorentz [125], p. 12, relation (2)), we obtain

$$\left| \frac{B_n[B_n(f)](1/n) - B_n(f)(1/n)}{(1/n)} \right| \le \frac{C\|B_n''(f)\|}{n} \le \frac{C\|f''\|}{n} \to 0,$$

as $n \to \infty$.

But $\lim_{n\to\infty}[n \cdot B_n^{(j)}(f)(1/n)] = f'(0)$, for any j, easily follows by mathematical induction, which proves the lemma. $\qquad\square$

The main result is the following.

Theorem 1.2.4. (Gal [78]) *Let us suppose that $G \subset \mathbb{C}$ is open, so that $\overline{\mathbb{D}}_1 \subset G$ and $f : G \to \mathbb{C}$ be analytic in G. Also, let $m \in \mathbb{N}$ be fixed.*

(i) If f is univalent in $\overline{\mathbb{D}}_1$, then there exists an index n_0 depending on f and m, so that the mth iterates $B_n^{(m)}(f)(z)$, be univalent in $\overline{\mathbb{D}}_1$, for all $n \ge n_0$.

(ii) If $f(0) = f'(0) - 1 = 0$ and f is starlike in $\overline{\mathbb{D}}_1$, that is

$$Re\left(\frac{zf'(z)}{f(z)}\right) > 0, \text{ for all } z \in \overline{\mathbb{D}}_1,$$

then there exists an index n_0 depending on f and m, so that the mth iterates $B_n^{(m)}(f)(z)$, be starlike in $\overline{\mathbb{D}}_1$, for all $n \ge n_0$.

If $f(0) = f'(0) - 1 = 0$ and f is starlike only in $\mathbb{D}_1$, then for any disk of radius $0 < r < 1$ and center 0 denoted by $\mathbb{D}_r$, there exists an index $n_0 = n_0(f, m, \mathbb{D}_r)$, so that the mth iterates $B_n^{(m)}(f)(z)$, be starlike in $\overline{\mathbb{D}}_r$ for all $n \ge n_0$, that is,

$$Re\left(\frac{z[B_n^{(m)}]'(f)(z)}{B_n^{(m)}(f)(z)}\right) > 0, \text{ for all } z \in \overline{\mathbb{D}}_r.$$

(Here $[B_n^{(m)}]'(f)(z)$ denotes the first derivative of $B_n^{(m)}(f)(z)$.)

(iii) If $f(0) = f'(0) - 1 = 0$ and f is convex in $\overline{\mathbb{D}}_1$, that is

$$Re\left(\frac{zf''(z)}{f'(z)}\right) + 1 > 0, \text{ for all } z \in \overline{\mathbb{D}}_1,$$

then there exists an index n_0 depending on f and m, so that the mth iterates $B_n^{(m)}(f)(z)$, be convex in $\overline{\mathbb{D}}_1$, for all $n \ge n_0$.

If $f(0) = f'(0) - 1 = 0$ and f is convex only in $\mathbb{D}_1$, then for any disk of radius $0 < r < 1$ and center 0 denoted by $\mathbb{D}_r$, there exists an index $n_0 = n_0(f, m, \mathbb{D}_r)$, so that for all $n \ge n_0$, the mth iterates $B_n^{(m)}(f)(z)$ be convex in $\overline{\mathbb{D}}_r$, that is,

$$Re\left(\frac{z[B_n^{(m)}]''(f)(z)}{[B_n^{(m)}]'(f)(z)}\right) + 1 > 0, \text{ for all } z \in \overline{\mathbb{D}}_r.$$

(iv) If $f(0) = f'(0) - 1 = 0$, $f(z) \neq 0$, for all $z \in \overline{\mathbb{D}}_1 \setminus \{0\}$ and f is spirallike of type $\gamma \in (-\pi/2, \pi/2)$ in $\overline{\mathbb{D}}_1$, that is

$$Re \left(e^{i\gamma} \frac{z f'(z)}{f(z)} \right) > 0, \text{ for all } z \in \overline{\mathbb{D}}_1,$$

then there exists an index n_0 depending on f, m and γ, so that the mth iterates $B_n^{(m)}(f)(z) \neq 0$, for all $z \in \overline{\mathbb{D}}_1 \setminus \{0\}$, and $B_n^{(m)}(f)(z)$ be spirallike of type γ in $\overline{\mathbb{D}}_1$, for all $n \geq n_0$.

If $f(0) = f'(0) - 1 = 0$, $f(z) \neq 0$ for all $z \in \mathbb{D}_1 \setminus \{0\}$ and f is spirallike of type γ only in $\mathbb{D}_1$, then for any disk of radius $0 < r < 1$ and center 0 denoted by $\mathbb{D}_r$, there exists an index $n_0 = n_0(f, m, \mathbb{D}_r, \gamma)$, so that for all $n \geq n_0$, the mth iterates $B_n^{(m)}(f)(z) \neq 0$ for all $z \in \overline{\mathbb{D}}_r \setminus \{0\}$ be spirallike of type γ in $\overline{\mathbb{D}}_r$, that is,

$$Re \left(e^{i\gamma} \frac{z [B_n^{(m)}]'(f)(z)}{B_n^{(m)}(f)(z)} \right) > 0, \text{ for all } z \in \overline{\mathbb{D}}_r.$$

Proof. (i) It is immediate from the uniform convergence in Lemma 1.2.2 and a well-known result concerning sequences of analytic functions converging locally uniformly to an univalent function (see e.g. Kohr-Mocanu [118], p. 130, Theorem 4.1.17).

For the proofs of the next points (ii), (iii) and (iv), let us make some general useful considerations. According to Lemma 1.2.2, combined with the Weierstrass's well-known result, it follows that as $n \to \infty$, uniformly in $\overline{\mathbb{D}}_1$ we have

$$B_n^{(m)}(f)(z) \to f(z), [B_n^{(m)}]'(f)(z) \to f'(z) \text{ and } [B_n^{(m)}]''(f)(z) \to f''(z).$$

In what follows we denote $C_{n,m} = [B_n^{(m)}(f)]'(0)$ and $P_n^{(m)}(f)(z) = \frac{B_n^{(m)}(f)(z)}{C_{n,m}}$.

Note here that $f'(0) = 1$ combined with Lemma 1.2.3, implies that for any $m \in \mathbb{N}$, there exists $n(m, f)$ so that $C_{n,m} > 0$, for all $n \geq n(m, f)$, (in fact we have $\lim_{n \to \infty} C_{n,m} = 1$).

From $f(0) = 0$ we get $B_n^{(m)}(f)(0) = 0$ and $P_n^{(m)}(f)(0) = 0$, while from the definition of $P_n^{(m)}$ we obtain $[P_n^{(m)}]'(f)(0) = 1$.

By combining all of these facts with Lemma 1.2.2, we obtain that for $n \to \infty$, we have $P_n^{(m)}(f)(z) \to f(z)$, $[P_n^{(m)}]'(f)(z) \to f'(z)$, $[P_n^{(m)}]''(f)(z) \to f''(z)$, uniformly in $\overline{\mathbb{D}}_1$.

(ii) By hypothesis we get $|f(z)| > 0$ for all $z \in \overline{\mathbb{D}}_1$ with $z \neq 0$, which from the univalence of f in $\mathbb{D}_1$, implies that we can write $f(z) = z g(z)$, with $g(z) \neq 0$, for all $z \in \overline{\mathbb{D}}_1$, where g is analytic in $\mathbb{D}_1$ and continuous in $\overline{\mathbb{D}}_1$.

By writing $P_n^{(m)}(f)(z)$ in the form $P_n^{(m)}(f)(z) = z Q_{n,m}(f)(z)$, it is obvious that $Q_{n,m}(f)(z)$ is a polynomial of degree $\leq n - 1$.

Let $|z| = 1$. We have

$$|f(z) - P_n^{(m)}(f)(z)| = |z| \cdot |g(z) - Q_{n,m}(f)(z)| = |g(z) - Q_{n,m}(f)(z)|,$$

which by the uniform convergence in $\overline{\mathbb{D}}_1$ of $P_n^{(m)}(f)$ to f and by the maximum modulus principle implies the uniform convergence in $\overline{\mathbb{D}}_1$ of $Q_{n,m}(f)(z)$ to $g(z)$.

Since g is continuous in $\overline{\mathbb{D}}_1$ and $|g(z)| > 0$ for all $z \in \overline{\mathbb{D}}_1$, there exist an index $n_1 \in \mathbb{N}$ and $a > 0$ depending on g, so that $|Q_{n,m}(f)(z)| > a > 0$, for all $z \in \overline{\mathbb{D}}_1$ and all $n \geq n_0$.

Also, for all $|z| = 1$, we have

$$
\begin{aligned}
|f'(z) - [P_n^{(m)}]'(f)(z)| &= |z[g'(z) - Q'_{n,m}(f)(z)] + [g(z) - Q_{n,m}(f)(z)]| \\
&\geq |\ |z||g'(z) - Q'_{n,m}(f)(z)| - |g(z) - Q_{n,m}(f)(z)|\ | \\
&= |\ |g'(z) - Q'_{n,m}(f)(z)| - |g(z) - Q_{n,m}(f)(z)|\ |
\end{aligned}
$$

which from the maximum modulus principle, the uniform convergence of $[P_n^{(m)}]'(f)$ to f' and of $Q_{n,m}(f)$ to g, evidently implies the uniform convergence of $Q'_{n,m}(f)$ to g'.

Then, for $|z| = 1$, we get

$$
\begin{aligned}
\frac{z[P_n^{(m)}]'(f)(z)}{P_n^{(m)}(f)} &= \frac{z[zQ'_{n,m}(f)(z) + Q_{n,m}(f)(z)]}{zQ_{n,m}(f)(z)} \\
&= \frac{zQ'_{n,m}(f)(z) + Q_{n,m}(f)(z)}{Q_{n,m}(f)(z)} \rightarrow \frac{zg'(z) + g(z)}{g(z)} \\
&= \frac{f'(z)}{g(z)} = \frac{zf'(z)}{f(z)},
\end{aligned}
$$

which again from the maximum modulus principle, implies

$$
\frac{z[P_n^{(m)}]'(f)(z)}{P_n^{(m)}(f)} \rightarrow \frac{zf'(z)}{f(z)}, \text{ uniformly in } \overline{\mathbb{D}}_1.
$$

Since $Re\left(\frac{zf'(z)}{f(z)}\right)$ is continuous in $\overline{\mathbb{D}}_1$, there exists $\alpha \in (0,1)$, so that

$$
Re\left(\frac{zf'(z)}{f(z)}\right) \geq \alpha, \text{ for all } z \in \overline{\mathbb{D}}_1.
$$

Therefore,

$$
Re\left[\frac{z[P_n^{(m)}]'(f)(z)}{P_n^{(m)}(f)(z)}\right] \rightarrow Re\left[\frac{zf'(z)}{f(z)}\right] \geq \alpha > 0
$$

uniformly on $\overline{\mathbb{D}}_1$, i.e. for any $0 < \beta < \alpha$, there is n_0 so that for all $n \geq n_0$ we have

$$
Re\left[\frac{z[P_n^{(m)}]'(f)(z)}{P_n^{(m)}(f)(z)}\right] > \beta > 0, \text{ for all } z \in \overline{\mathbb{D}}_1.
$$

Since $P_n^{(m)}(f)(z)$ differs from $B_n^{(m)}(f)(z)$ only by a constant, this proves the first part in (ii).

For the second part, the proof is identical with the first part, with the only difference that instead of $\overline{\mathbb{D}}_1$, we reason for $\overline{\mathbb{D}}_r$.

(iv) Obviously we have

$$
Re\left[e^{i\gamma}\frac{z[P_n^{(m)}]'(f)(z)}{P_n^{(m)}(f)(z)}\right] \rightarrow Re\left[e^{i\gamma}\frac{zf'(z)}{f(z)}\right],
$$

uniformly in $\overline{\mathbb{D}}_1$.

We also note that since f is univalent in $\overline{\mathbb{D}}_1$, according to Lemma 1.2.2, there exists $n_1(m, f)$ so that $B_n^{(m)}(f)(z)$ be univalent in $\overline{\mathbb{D}}_1$ for all $n \geq n_1(m, f)$. Therefore, $B_n^{(m)}(f)(0) = 0$ implies $B_n^{(m)}(f)(z) \neq 0$, for all $z \in \overline{\mathbb{D}}_1 \setminus \{0\}$, $n \geq n_1(m, f)$. For the rest, the proof is identical with that from the above point (ii).

(iii) For the first part, by hypothesis there is $\alpha \in (0, 1)$, so that

$$Re \left[\frac{zf''(z)}{f'(z)} \right] + 1 \geq \alpha > 0,$$

uniformly in $\overline{\mathbb{D}}_1$. It is not difficult to show that this is equivalent with the fact that for any $\beta \in (0, \alpha)$, the function $zf'(z)$ is starlike of order β in $\overline{\mathbb{D}}_1$ (see e.g. Mocanu-Bulboacă-Sălăgean [138], p. 77), which implies that $f'(z) \neq 0$, for all $z \in \overline{\mathbb{D}}_1$, i.e. $|f'(z)| > 0$, for all $z \in \overline{\mathbb{D}}_1$. Also, by using the same type of reasonings as those mentioned in the above point (ii), we get

$$Re \left[\frac{z[P_n^{(m)}]''(f)(z)}{[P_n^{(m)}]'(f)(z)} \right] + 1 \rightarrow Re \left[\frac{zf''(z)}{f'(z)} \right] + 1 \geq \alpha > 0,$$

uniformly in $\overline{\mathbb{D}}_1$. As a conclusion, for any $0 < \beta < \alpha$, there is n_0 depending on f, so that for all $n \geq n_0$ we have

$$Re \left[\frac{z[P_n^{(m)}]''(f)(z)}{[P_n^{(m)}]'(f)(z)} \right] + 1 > \beta > 0, \text{ for all } z \in \overline{\mathbb{D}}_1.$$

The proof of second part in (iii) is similar, which proves the theorem. □

1.3 Generalized Voronovskaja Theorems for Bernstein Polynomials

It is well-known the fact that the classical Voronovskaja's theorem for real variable was generalized by Bernstein [42] as follows.

Theorem 1.3.1. (see e.g. Lorentz [125], p. 22-23) *If f is defined and bounded on $[0, 1]$ and the derivative $f^{(2p)}(x)$ exists at x, then we can write*

$$B_n(f)(x) = f(x) + \sum_{j=1}^{2p} \frac{f^{(j)}(x)}{j!} n^{-j} T_{n,j}(x) + \frac{\varepsilon_n}{n^p},$$

where $B_n(f)(x) = \sum_{k=0}^{n} \binom{n}{k} x^k (1 - x)^k f(k/n)$ denotes the Bernstein polynomials, $T_{n,j}(x) = \sum_{i=0}^{n} (i - nx)^j \binom{n}{i} x^i (1 - x)^{n-i}$ and $\varepsilon_n \rightarrow 0$ as $n \rightarrow \infty$.

Remark. The classical Voronovskaja's theorem is recaptured for $k = 1$.

The goal of this section is to obtain a similar result to Theorem 1.3.1 for the complex Bernstein polynomials attached to analytic functions in compact disks with the centers in origin and radii ≥ 1. For the particular case $p = 1$ one recapture Theorem 1.1.3. Moreover, the analyticity of f will imply exact orders of approximation in the generalized Voronovskaja's theorems.

The first main result of this section is the following.

Theorem 1.3.2. (Gal [89]) *Let $R > 1$ and let $f : \mathbb{D}_R \to \mathbb{C}$ be an analytic function, that is $f(z) = \sum_{k=0}^{\infty} c_k z^k$. Then for any $1 \leq r < R$ and any natural number p there exists a constant $C_p > 0$ such that the following estimate*

$$\left| B_n(f)(z) - f(z) - \sum_{j=1}^{2p} \frac{f^{(j)}(z)}{j!} n^{-j} T_{n,j}(z) \right| \leq \frac{C_{p,r}(f)}{n^{p+1}}$$

holds for all $z \in \mathbb{C}$ with $|z| \leq r$ and for all $n \in \mathbb{N}$, where

$$C_{p,r}(f) = C_p \cdot \sum_{k=2p+1}^{\infty} |c_k| \frac{k! \cdot (k-2p)}{(k-2p-1)!} r^k < \infty.$$

Remark. Theorem 1.3.2 is the complex analogue of the Bernstein's result, with the quantitative estimate $\varepsilon_n \leq 1/n$.

Proof of Theorem 1.3.2. Let e_k be the function defined by $e_k(z) = z^k$. Let us put $\pi_{k,n} = B_n(e_k)$. By Lemma 2.2 in Pop [155] one has

$$\pi_{k,n}(z) = \sum_{j=0}^{k} \frac{1}{j! n^j} T_{n,j}(z) \cdot \left(z^k\right)^{(j)}, \qquad (1.1)$$

where $\left(z^k\right)^{(j)}$ is the j-th derivative of the function $e_k(z) = z^k$. Note that

$$\left(z^k\right)^{(j)} = k(k-1)\dots(k-j+1) z^{k-j}. \qquad (1.2)$$

Let us introduce the polynomial $E_{k,n,p}$ by defining

$$E_{k,n,p}(z) = \pi_{k,n}(z) - e_k(z) - \sum_{j=1}^{2p} \frac{1}{j! n^j} T_{n,j}(z) \cdot \left(z^k\right)^{(j)}. \qquad (1.3)$$

Since $T_{n,0}(z) = 1$ we see by (1.1) and (1.2) that for $k \geq 2p+1$ we have

$$E_{k,n,p}(z) = \sum_{j=2p+1}^{k} \frac{1}{j! n^j} T_{n,j}(z) \cdot \left(z^k\right)^{(j)} = \sum_{j=2p+1}^{k} \frac{1}{n^j} \binom{k}{j} z^{k-j} T_{n,j}(z). \qquad (1.4)$$

First we need the following auxiliary result.

Lemma 1.3.3. (Gal [89]) *For given $p \in \mathbb{N}$ and $r \geq 1$ there exists a constant $C_p > 0$ such that for all $k \geq 2p+1$, $|z| \leq r$ and $n \in \mathbb{N}$ the following estimate*

$$|E_{k,n,p}(z)| \leq C_p r^k \frac{k!}{n^{p+1}(k-2p)!}(k-2p)^2$$

holds.

Proof of Lemma 1.3.3. We shall use mathematical induction over p. For $p = 1$ this inequality has been proved in the proof of Theorem 1.1.3, (ii). Now suppose that the inequality is valid for p, and we want to prove it for $p + 1$. At first we observe from (1.4) that $E_{k,n,p+1}(z)$ is equal to

$$E_{k,n,p}(z)\binom{k}{2p+1}\frac{z^{k-(2p+1)}}{n^{2p+1}}T_{n,2p+1}(z)\binom{k}{2p+2}\frac{z^{k-(2p+2)}}{n^{2p+2}}T_{n,2p+2}(z). \quad (1.5)$$

Let us define

$$R_{k,n,p}(z) = E_{k,n,p+1}(z) - zE_{k-1,n,p+1}(z) - \frac{z(1-z)}{n}E'_{k-1,n,p+1}(z). \quad (1.6)$$

Using (1.4), a simple computation shows that $R_{k,n,p}(z)$ is equal to

$$\sum_{j=2p+3}^{k}\frac{1}{n^j}\binom{k}{j}z^{k-j}T_{n,j}(z) - z\sum_{j=2p+3}^{k-1}\frac{1}{n^j}\binom{k-1}{j}z^{k-1-j}T_{n,j}(z) \quad (1.7)$$

$$-\frac{z(1-z)}{n}\sum_{j=2p+3}^{k-1}\frac{1}{n^j}\binom{k-1}{j}\frac{d}{dz}\left[z^{k-1-j}T_{n,j}(z)\right]. \quad (1.8)$$

Now let us rewrite (1.1) in the following trivial way

$$\sum_{j=2p+3}^{k}\frac{1}{n^j}\binom{k}{j}z^{k-j}T_{n,j}(z) = \pi_{k,n}(z) - \sum_{j=0}^{2p+2}\frac{1}{n^j}\binom{k}{j}z^{k-j}T_{n,j}(z) \quad (1.9)$$

and replace the summands in (1.7) and (1.8) by the corresponding expressions induced by (1.9). Then

$$R_{k,n,p}(z) = \pi_{k,n}(z) - \sum_{j=0}^{2p+2}\frac{1}{n^j}\binom{k}{j}z^{k-j}T_{n,j}(z) - z\pi_{k-1,n}(z)$$

$$+ \sum_{j=0}^{2p+2}\frac{1}{n^j}\binom{k-1}{j}z^{k-j}T_{n,j}(z) - \frac{z(1-z)}{n}\pi'_{k-1,n}(z)$$

$$+\frac{z(1-z)}{n}\sum_{j=0}^{2p+2}\frac{1}{n^j}\binom{k-1}{j}\frac{d}{dz}\left[z^{k-1-j}T_{n,j}(z)\right]'$$

$$:= S_1 - zS_2 - \frac{z(1-z)}{n}S_3.$$

Note that D. Andrica [24] has proved (see also the proof of Theorem 1.1.2, (i)) that

$$\pi_{k,n}(z) = z\pi_{k-1,n}(z) + \frac{z(1-z)}{n}\pi'_{k-1,n}(z),$$

which simplifies the above expression for $R_{k,n,p}(z)$ in an obvious way.

We want to deduce from the above formula that

$$|R_{k,n,p}(z)| \leq C_p^*\frac{1}{n^{p+2}}r^k k(k-1)\ldots(k-2p-2), \quad (1.10)$$

for all $|z| \leq r$, $n \in \mathbb{N}$ and $k \geq 2p+3$.

For this purpose, we observe that in the above expression of $R_{k,n,p}(z)$ (with respect to S_1, S_2, S_3), the expression S_1 is equal to the left-hand side in (1.9), S_2 is equal to the left-hand side in (1.9) written for $k-1$ and S_3 is equal to the derivative with respect to z of the left-hand side in (1.9) written for $k-1$.

But since by Lorentz [125], p. 14, Theorem 1.5.1, $T_{n,j}(z)$ is a polynomial of degree $[j/2]$ with respect to n, it is clear from its form that the left-hand side in (1.9) contains only terms having at denominator $n^{j-[j/2]}$, $j \geq 2p+3$, that is only terms having at the denominator n^j with $j \geq p+2$. This immediately implies that $R_{k,n,p}(z)$ contains only terms having at denominator n^j with $j \geq p+2$. Now, since by the same Lorentz [125], p. 14, Theorem 1.5.1, $T_{n,j}(z)$ is a polynomial of degree j with respect to z, (by using again (1.9)) this immediately implies that for all $|z| \leq r$ we have an estimate of the form $|R_{n,k,p}(z)| \leq C_{p,k} r^k \frac{1}{n^{p+2}}$.

Therefore it remains to find out the form of the constant $C_{p,k}$. By the recurrence formula for $T_{n,j}(z)$ in Lorentz [125], p. 14, relation (3), it follows that $T_{n,j}(z)$ is a polynomial in n of degree $\leq [j/2]$ with at most $[j/2]$ terms containing the powers of n (where $j \leq 2p+2$), satisfying the estimate $|T_{n,j}(z)| \leq r^j A_j n^{[j/2]}$, for all $|z| \leq r$. Combining with the estimates $\binom{k}{j} \leq k(k-1)...(k-(2p+2)+1) \leq k(k-1)...(k-2p-2)$, $\binom{k-1}{j} \leq (k-1)(k-2)...(k-(2p+2)) \leq k(k-1)...(k-2p-2)$, $j = 1,,, 2p+2$, it easily follows that the modulus of all the nominators of the terms having at the denominators n^j with $j \geq p+2$, can be bounded by $C_p r^k k(k-1)...(k-2p-2)$, with a suitable chosen constant C_p depending only on p.

Now we shall estimate $E_{k,n,p+1}(z)$ by using (1.10). Indeed, by (1.6) we have

$$E_{k,n,p+1}(z) = z E_{k-1,n,p+1}(z) + \frac{z(1-z)}{n} E'_{k-1,n,p+1}(z) + R_{k,n,p}(z).$$

Let us denote $\|f\|_r = \sup_{|z| \leq r} |f(z)|$ and let us recall Bernstein's inequality

$$\|P'_j\|_r \leq \frac{j}{r} \|P_j\|_r,$$

valid for any polynomial P_j of degree $\leq j$. Since $|z(1-z)| \leq r(1+r) \leq 2r^2$ for all $|z| \leq r$ (recall that $1 \leq r$) and $E_{k-1,n,p+1}(z)$ is a polynomial of degree $\leq k$ we conclude that for $|z| \leq r$

$$|E_{k,n,p+1}(z)| \leq r|E_{k-1,n,p+1}(z)| + \frac{2r}{n} k |E_{k-1,n,p+1}(z)| + |R_{k,n,p}(z)|. \quad (1.11)$$

By equation (1.5) (applied to $k-1$ instead of k) one obtains

$$|E_{k-1,n,p+1}(z)| \leq |E_{k-1,n,p}(z)| + \binom{k-1}{2p+1} \frac{1}{n^{2p+1}} \left| z^{k-1-(2p+1)} T_{n,2p+1}(z) \right|$$

$$+ \binom{k-1}{2p+2} \frac{1}{n^{2p+2}} \left| z^{k-1-(2p+2)} T_{n,2p+2}(z) \right|.$$

We use the last inequality in order to estimate the middle term in (1.11) in the following way:

$$\frac{2r}{n} k |E_{k-1,n,p+1}(z)| \leq \frac{2rk}{n} \left[C_p r^{k-1} \frac{(k-1)...(k-2p)(k-2p-1)^2}{n^{p+1}} \right.$$

$$+ A_{2p+1} r^{k-1} \frac{(k-1)...(k-(2p+1))}{n^{p+1}} + A_{2p} r^{k-1} \frac{(k-1)...(k-(2p+2))}{n^{p+2}} \Bigg].$$

From this we conclude that

$$|E_{k,n,p+1}(z)| \leq r |E_{k-1,n,p+1}(z)| + C_p r^k \frac{1}{n^{p+2}} k(k-1)...(k-2p-2).$$

Since by Lemma 2.2 in Pop [155] we have $E_{2p+2,n,p+1}(z) = 0$, from the above inequality by an inductive argument applied for $k = 2p+3, ...$ we finally obtain

$$|E_{k,n,p+1}(z)| \leq C_p r^k \frac{1}{n^{p+2}} k(k-1)...(k-2p-1)(k-2p-2)^2,$$

which proves the lemma. $\qquad \square$

Now we are in position to prove Theorem 1.3.2. Indeed, taking into account that by the estimate in Lemma 1.3.3 we have $E_{0,n,p}(z) = E_{1,n,p}(z) = ... = E_{2p,n,p}(z) = 0$, it follows

$$\left| B_n(f)(z) - f(z) - \sum_{j=1}^{2p} \frac{f^{(j)}(z)}{j!} n^{-j} T_{n,j}(z) \right|$$

$$\leq \sum_{k=2p+1}^{\infty} |c_k| \cdot |E_{k,n,p}(z)|$$

$$\leq \frac{C_p \cdot \sum_{k=2p+1}^{\infty} |c_k| k(k-1)...(k-2p+1)(k-2p)^2 r^k}{n^{p+1}},$$

which proves the theorem. $\qquad \square$

Remark. Analysing the proof of Lemma 1.3.3 and the reasonings for the proof of Theorem 1.3.2, in a similar way we can prove that in the case when $r = 1$ the pointwise estimate

$$\left| B_n(f)(z) - f(z) - \sum_{j=1}^{2p} \frac{f^{(j)}(z)}{j!} n^{-j} T_{n,j}(z) \right| \leq \frac{|z| \cdot |1 - z| C_{p,1}(f)}{n^{p+1}}, \forall |z| \leq 1,$$

holds, where $C_{p,1}(f) = C_p \cdot \sum_{k=2p+1}^{\infty} |c_k| k(k-1)...(k-2p+1)(k-2p)^2 < \infty$.

Unlike the real case, for complex analytic functions the order of approximation in Theorem 1.3.2 is exactly $\frac{1}{n^{p+1}}$. More exactly, the second main result of this paper is the following.

Corollary 1.3.4. (Gal [89]) *Let $R > 1$ and let $f : \mathbb{D}_R \to \mathbb{C}$ be an analytic function, say $f(z) = \sum_{k=0}^{\infty} c_k z^k$. If f is not a polynomial of degree $\leq 2p$ then for any $1 \leq r < R$ and any natural number p we have*

$$\left\| B_n(f) - f - \sum_{j=1}^{2p} \frac{f^{(j)}}{j!} n^{-j} T_{n,j} \right\|_r \sim \frac{1}{n^{p+1}}, \ n \in \mathbb{N},$$

where the constants in the equivalence depend only on f, r and p and are independent of n. Here $\|f\|_r = \sup_{|z| \leq r} |f(z)|$.

Proof.   Taking into account Theorem 1.3.2, it remains to obtain the lower estimate for the quantity in the statement of Corollary 1.3.4. Thus, suppose that f is not a polynomial of degree $\leq 2p$. Keeping the notations in the previous section, since $f^{(s)}(z) = \sum_{k=s}^{\infty} c_k k(k-1)...(k-s+1)z^{k-s}$, by using (1.4) and simple calculations we easily obtain the identity

$$B_n(f)(z) - f(z) - \sum_{j=1}^{2p} \frac{f^{(j)}(z)}{j!} n^{-j} T_{n,j}(z)$$

$$= \sum_{k=2p+1}^{\infty} c_k E_{k,n,p}(z)$$

$$= \frac{1}{n^{p+1}} \left\{ \sum_{k=2p+1}^{\infty} c_k \left[\sum_{j=2p+1}^{k} \binom{k}{j} z^{k-j} n^{p+1-j} T_{n,j}(z) \right] \right\}$$

$$= \frac{1}{n^{p+1}} \left\{ \frac{T_{n,2p+1}(z)}{n^p (2p+1)!} f^{(2p+1)}(z) + \frac{T_{n,2p+2}(z)}{n^{p+1}(2p+2)!} f^{(2p+2)}(z) \right.$$

$$\left. + \frac{1}{n} \left[n^{p+2} \sum_{k=2p+3}^{\infty} c_k E_{k,n,p+1}(z) \right] \right\}.$$

By the recurrence formula for $T_{n,j}(z)$ in Lorentz [125], p. 14, relation (3), it follows that $T_{n,j}(z)$ is a polynomial in n of degree $\leq [j/2]$ with at most $[j/2]$ terms containing the powers of n (where $j \leq 2p+2$). Also, by Lorentz [125], p. 14, Theorem 1.5.1, the coefficient of n^{p+1} in $T_{n,2p+2}(z)$ is $\frac{(2p+2)!}{2^{p+1}(p+1)!} \cdot [x(1-x)]^{p+1}$, while from the recurrence formula in Lorentz [125], p. 14, relation (3), it easily follows that the coefficient of n^p in $T_{n,2p+1}(z)$ is of the form $a_p(1-2z)[z(1-z)]^p$ with the constant $a_p > 0$ (a_p depends only on p). Therefore, it is easy to see that the sum $\frac{T_{n,2p+1}(z)}{n^p(2p+1)!} f^{(2p+1)}(z) + \frac{1}{n^{p+1}} \cdot \frac{T_{n,2p+2}(z)}{(2p+2)!} f^{(2p+2)}(z)$ can be written in the form

$$\frac{T_{n,2p+1}(z)}{n^p (2p+1)!} f^{(2p+1)}(z) + \frac{1}{n^{p+1}} \cdot \frac{T_{n,2p+2}(z)}{(2p+2)!} f^{(2p+2)}(z)$$

$$= \frac{a_p}{(2p+1)!}(1-2z)[z(1-z)]^p f^{(2p+1)}(z)$$

$$+ \frac{[z(1-z)]^{p+1}}{2^{p+1}(p+1)!} f^{(2p+2)}(z)$$

$$+ \frac{1}{n} F(z) f^{(2p+1)}(z) + \frac{1}{n} G(z) f^{(2p+2)}(z),$$

where the polynomials $F(z) := P_1(z) + \frac{1}{n} P_2(z) + ... + \frac{1}{n^{p-1}} P_p(z)$ and $G(z) := Q_1(z) + \frac{1}{n} Q_2(z) + ... + \frac{1}{n^p} Q_{p+1}(z)$ are bounded in any closed disk $|z| \leq r$ by constants depending on r and p but independent of n.

Replacing this form in the above identity and taking into account the inequalities

$$\|h + g\|_r \geq |\, \|h\|_r - \|g\|_r \,| \geq \|h\|_r - \|g\|_r,$$

we obtain

$$\left\| B_n(f) - f - \sum_{j=1}^{2p} \frac{f^{(j)}}{j!} n^{-j} T_{n,j} \right\|_r$$

$$\geq \frac{1}{n^{p+1}} \left\{ \left\| \frac{a_p}{(2p+1)!}(1-2e_1)[e_1(1-e_1)]^p f^{(2p+1)} + \frac{[e_1(1-e_1)]^{p+1}}{2^{p+1}(p+1)!} f^{(2p+2)} \right\|_r \right.$$

$$\left. - \frac{1}{n} \left[\left\| n^{p+2} \sum_{k=2p+3}^{\infty} c_k E_{k,n,p+1} + F f^{(2p+1)} + G f^{(2p+2)} \right\|_r \right] \right\}$$

$$:= \frac{1}{n^{p+1}} \left\{ \|U\|_r - \frac{1}{n} [\|V\|_r] \right\} \geq \frac{1}{n^{p+1}} \cdot \frac{1}{2} \|U\|_r,$$

for all $n \geq n_0$ (n_0 depends on f, p and r), under the conditions that $\|U\|_r > 0$ and if $\|V\|_r$ is bounded by a constant depending only on f, p and r. But this is exactly what happens, because from Theorem 1.3.2 (written for $p+1$) and from the above considerations on F and G it is immediate that $\|V\|_r$ is bounded by a constant depending only on f, p and r while by the fact that f is not a polynomial of degree $\leq 2p$ it follows $\|U\|_r > 0$. Indeed, for the last fact let us suppose the contrary. It follows that f must satisfy the differential equation (here recall that $a_p > 0$)

$$\frac{a_p}{(2p+1)!}(1-2z)[z(1-z)]^p f^{(2p+1)}(z) + \frac{[z(1-z)]^{p+1}}{2^{p+1}(p+1)!} f^{(2p+2)}(z) = 0, |z| \leq r.$$

Making the substitution $f^{(2p+1)}(z) := y(z)$ it follows that $y(z)$ necessarily is analytic in $\mathbb{D}_R$ (since f is supposed analytic there) and is solution of the first order differential equation

$$\frac{a_p}{(2p+1)!}(1-2z)[z(1-z)]^p y(z) + \frac{[z(1-z)]^{p+1}}{2^{p+1}(p+1)!} y'(z) = 0, |z| \leq r.$$

After simplification with $[z(1-z)]^p$, we get that $y(z)$ is an analytic function in $\mathbb{D}_R$ satisfying the differential equation

$$\frac{a_p}{(2p+1)!}(1-2z)y(z) + \frac{z(1-z)}{2^{p+1}(p+1)!} y'(z) = 0, |z| \leq r, z \neq 0, z \neq 1.$$

Writing $y(z)$ in the form $y(z) = \sum_{k=0}^{\infty} b_k z^k$, by comparison of coefficients, we easily obtain that $b_k = 0$, for all $k = 0, 1, ...,$ which implies that $y(z)$ is identical zero in $\overline{\mathbb{D}}_r \setminus \{0,1\}$. Since y is analytic, it is continuous and therefore $y(0) = y(1) = 0$, which implies that $y(z) = 0$, for all $|z| \leq r$. But from the identity's theorem of analytic functions, it necessarily follows that $y(z) = 0$ for all $|z| < R$, obviously in contradiction with the hypothesis that f is not a polynomial of degree $\leq 2p$ in $\mathbb{D}_R$.

For $n \in \{1, ..., n_0 - 1\}$ we obviously have

$$\left\| B_n(f) - f - \sum_{j=1}^{2p} \frac{f^{(j)}}{j!} n^{-j} T_{n,j} \right\|_r \geq \frac{M_{r,n}(f)}{n^{p+1}},$$

with $M_{r,n}(f) = n^{p+1} \cdot \left\| B_n(f) - f - \sum_{j=1}^{2p} \frac{f^{(j)}}{j!} n^{-j} T_{n,j} \right\|_r > 0$, which finally implies

$$\left\| B_n(f) - f - \sum_{j=1}^{2p} \frac{f^{(j)}}{j!} n^{-j} T_{n,j} \right\|_r \geq \frac{C_{p,r}(f)}{n^{p+1}}, \text{ for all } n \in \mathbb{N},$$

where $C_{p,r}(f) = \min\{M_{r,1}(f), ..., M_{r,n_0-1}(f), \frac{1}{2}\|U\|_r\}$. This completes the proof.$\square$

1.4   Butzer's Linear Combination of Bernstein Polynomials

In the paper of Butzer [51], were inductively introduced the operators $L_n^{[q]}(x)$ of real variable $x \in [0, 1]$ by setting $L_n^{[0]}(f)(x) := B_n(f)(x)$ and

$$(2^q - 1) L_n^{[2q]}(f)(x) = 2^q L_{2n}^{[2q-2]}(f)(x) - L_n^{[2q-2]}(f)(x),$$

for $q \in \mathbb{N}$. For example, for $q = 1$ one obtains

$$L_n^{[2]}(f)(x) := 2L_{2n}^{[0]}(f)(x) - L_n^{[0]}(f)(x) = 2B_{2n}(f)(x) - B_n(f)(x).$$

In the same paper of Butzer [51], by using the generalized Voronovskaja's theorem of Bernstein [42] (that is Theorem 1.3.1), he proved that

$$|L_n^{[2q-2]}(f)(x) - f(x)| = O(n^{-q}).$$

The first main result of this section is the extension of Butzer's result to the case of complex variable and can be stated as follows.

Theorem 1.4.1. (Gal [89]) *For any analytic function $f : \mathbb{D}_R \to \mathbb{C}$ with $R > 1$, for each $1 \leq r < R$ and given natural number q there exists a constant $d_{q,r}(f) > 0$ such that the following estimate*

$$\left| L_n^{[2q-2]}(f)(z) - f(z) \right| \leq \frac{d_{q,r}(f)}{n^q},$$

is valid for all $|z| \leq r$ and $n \in \mathbb{N}$.

Proof.   The proof of Theorem 1.4.1 is simple. Indeed, let us consider Butzer's linear combination of complex Bernstein polynomials defined by the recurrence

$$L_n^{[0]}(f)(z) = B_n(f)(z), (2^q - 1)L_n^{[2q]}(f)(z) = 2^q L_{2n}^{[2q-2]}(f)(z) - L_n^{[2q-2]}(f)(z),$$

where $z \in \mathbb{C}, q = 1, 2, ...$

Analysing the proofs of Lemma 1, Lemma 2 and Theorem 1 in Butzer [51] and taking into account Theorem 1.3.2, it is easy to see that their reasonings can analogously be used for the above linear combinations of complex Bernstein polynomials, so that finally we get the statement of Theorem 1.4.1. $\square$

Remarks. 1) By the Remark after the proof of Lemma 1.3.3, it easily follows that in a similar way we get the pointwise estimate

$$|L_n^{[2q-2]}(f)(z) - f(z)| \leq d_{q,1}(f) \cdot \frac{|z| \cdot |1 - z|}{n^q}, \text{ for all } |z| \leq 1.$$

2) For $q = 2$, the estimate in Theorem 1.4.1 can easily be obtained with an explicit constant $d_{q,r}(f)$, by using the following estimate in Theorem 1.1.3, (ii) :

$$|B_n(f)(z) - f(z) - \frac{z(1-z)}{2n}f''(z)| \leq \frac{5M_r(f)(1+r)^2}{2n^2}, r \geq 1, |z| \leq r, n \in \mathbb{N},$$

where $M_r(f) = \sum_{k=3}^{\infty} |c_k|k(k-1)(k-2)r^{k-2}$.

Indeed, taking into account that $L_n^{[2]}(f)(z) = B_{2n}(f)(z) - B_n(f)(z)$ and that we can write the identity

$$L_n^{[2]}(f)(z) - f(z) = 2\left[B_{2n}(f)(z) - f(z) - \frac{z(1-z)}{4n}f''(z)\right]$$
$$+ \left[f(z) - B_n(f)(z) + \frac{z(1-z)}{2n}f''(z)\right],$$

we immediately obtain

$$|L_n^{[2]}(f)(z) - f(z)|$$
$$\leq 2\left|B_{2n}(f)(z) - f(z) - \frac{z(1-z)}{4n}f''(z)\right| + \left|B_n(f)(z) - f(z) - \frac{z(1-z)}{2n}f''(z)\right|$$
$$\leq 2\frac{5M_r(f)(1+r)^2}{8n^2} + \frac{5M_r(f)(1+r)^2}{2n^2} = \frac{15(1+r)^2M_r(f)}{4n^2}.$$

3) For $q = 3$, in Theorem 1.4.1 similar reasonings can be applied. Indeed, first it easily follows that $L_n^{[4]}(f)(z) = \frac{8}{3}B_{4n}(f)(z) - 2B_{2n}(f)(z) + \frac{1}{3}B_n(f)(z)$ and that we can write the identity

$$L_n^{[4]}(f)(z) - f(z)$$
$$= \frac{8}{3}\left[B_{4n}(f)(z) - f(z) - \sum_{j=1}^{4}\frac{f^{(j)}(z)}{j!}(4n)^{-j}T_{4n,j}(z)\right]$$
$$-2\left[B_{2n}(f)(z) - f(z) - \sum_{j=1}^{4}\frac{f^{(j)}(z)}{j!}(2n)^{-j}T_{2n,j}(z)\right]$$
$$+\frac{1}{3}\left[B_n(f)(z) - f(z) - \sum_{j=1}^{4}\frac{f^{(j)}(z)}{j!}(n)^{-j}T_{n,j}(z)\right]$$
$$+\frac{f^{(4)}(z)[1 - 6z(1-z)]z(1-z)}{4! \cdot 8 \cdot n^3}.$$

This identity follows from the identities $T_{n,1}(z) = 0$ and

$$\frac{f''(z)}{2!}\left[-\frac{8}{3} \cdot \frac{T_{4n,2}(z)}{16n^2} + 2 \cdot \frac{T_{2n,2}(z)}{4n^2} - \frac{1}{3} \cdot \frac{T_{n,2}(z)}{n^2}\right] = 0,$$

$$\frac{f'''(z)}{3!}\left[-\frac{8}{3} \cdot \frac{T_{4n,3}(z)}{64n^3} + 2 \cdot \frac{T_{2n,3}(z)}{8n^3} - \frac{1}{3} \cdot \frac{T_{n,3}(z)}{n^3}\right] = 0,$$

$$\frac{f^{(4)}(z)}{4!}\left[-\frac{8}{3} \cdot \frac{T_{4n,4}(z)}{256n^4} + 2 \cdot \frac{T_{2n,4}(z)}{16n^4} - \frac{1}{3} \cdot \frac{T_{n,4}(z)}{n^4}\right]$$

$$= -\frac{f^{(4)}(z)nz(1-z)[1-6z(1-z)]}{4! \cdot 8 \cdot n^4}.$$

As a conclusion, an estimate with explicit constant in Theorem 1.3.2 for $p = 2$ (which can be obtained by following the reasonings in the proof of Lemma 1.3.3 for $p = 2$, but with explicit constants), will immediately give an explicit constant d for the estimate $|L_n^{[4]}(f)(z) - f(z)| \le d/n^3$.

4) A nice consequence of Corollary 1.3.4 and of Theorem 1.4.1 is that if f is not a polynomial of degree $\le q$ then the order of approximation in Theorem 1.4.1 is exactly $\frac{1}{n^q}$. For simplicity first we illustrate the particular cases $q = 1, 2, 3$. Indeed, the case $q = 1$ is contained in Corollary 1.1.5. In the $q = 2$ case, taking into account the identity from the above Remark 2 and applying the reasonings in the proof of Corollary 1.3.4 for the case $p = 1$, it follows that $L_n^{[2]}(f)(z) - f(z)$ can be written in the form

$$L_n^{[2]}(f)(z) - f(z)$$

$$= 2\frac{1}{(2n)^2}\left[\frac{1}{2n}A_{1,n}(f)(z) + \frac{T_{2n,3}(z)}{3!(2n)}f^{(3)}(z) + \frac{T_{2n,4}(z)}{4!(2n)^2}f^{(4)}(z)\right]$$

$$- \frac{1}{n^2}\left[\frac{1}{n}A_{2,n}(f)(z) + \frac{T_{n,3}(z)}{3!n}f^{(3)}(z) + \frac{T_{n,4}(z)}{4!n^2}f^{(4)}(z)\right]$$

$$= \frac{1}{n^2}\left[\frac{1}{4n}A_{1,n}(f)(z) - \frac{1}{n}A_{2,n}(f)(z) + \frac{1}{2} \cdot \frac{T_{2n,3}(z)}{3!(2n)}f^{(3)}(z)\right.$$

$$\left. + \frac{1}{2} \cdot \frac{T_{2n,4}(z)}{4!(2n)^2}f^{(4)}(z) - \frac{T_{n,3}(z)}{3!n}f^{(3)}(z) - \frac{T_{n,4}(z)}{4!n^2}f^{(4)}(z)\right]$$

$$= \frac{1}{n^2}\left[\frac{1}{n}A_{3,n}(f)(z) - \frac{(1-2z)z(1-z)f^{(3)}(z)}{2 \cdot 3!} + \frac{1}{2} \cdot \frac{3z^2(1-z)^2f^{(4)}(z)}{2 \cdot 4!}\right],$$

where for all $n \in \mathbb{N}$

$$\|A_{1,n}(f)\|_r, \|A_{2,n}(f)\|_r, \|A_{3,n}(f)\|_r \le C_r(f),$$

($C_r(f)$ is independent of n) and we used the formulas in Lorentz [125], p. 14, $T_{n,3}(z) = n(1-2z)z(1-z)$, $T_{n,4}(z) = nz(1-z)[3nz(1-z) + (1-6z(1-z))]$.

By using for

$$-\frac{(1-2z)z(1-z)f^{(3)}(z)}{12} + \frac{z^2(1-z)^2f^{(4)}(z)}{32}$$

similar reasonings with those in the proof of Corollary 1.3.4, case $p = 1$, finally we easily obtain

$$\|L_n^{[2]}(f)(z) - f(z)\|_r \sim \frac{1}{n^2}, \ n \in \mathbb{N},$$

where the constants in the equivalence depend only on f and r.

5) For the $q = 3$ case we can apply similar reasonings. Indeed, applying to the three expressions between the brackets in the formula for $L_n^{[4]}(f)(z) - f(z)$ from the

above Remark 3, the reasonings in the proof of Corollary 1.3.4, the case $p = 2$, we obtain

$$L_n^{[4]}(f)(z) - f(z)$$

$$= \frac{1}{n^3}\left[\frac{1}{n}A_{1,n}(f)(z) + \frac{8}{3 \cdot 4^3} \cdot \frac{T_{4n,5}(z)}{5! \cdot (4n)^2}f^{(5)}(z) + \frac{8}{3 \cdot 4^3} \cdot \frac{T_{4n,6}(z)}{6! \cdot (4n)^3}f^{(6)}(z)\right.$$

$$- 2 \cdot \frac{1}{2^3} \cdot \frac{T_{2n,5}(z)}{5! \cdot (2n)^2}f^{(5)}(z) - 2 \cdot \frac{1}{2^3} \cdot \frac{T_{2n,6}(z)}{6! \cdot (2n)^3}f^{(6)}(z)$$

$$\left.+ \frac{1}{3} \cdot \frac{T_{n,5}(z)}{5! \cdot n^2}f^{(5)}(z) + \frac{1}{3} \cdot \frac{T_{n,6}(z)}{6! \cdot n^3}f^{(6)}(z)\right] + \frac{f^{(4)}(z)[1 - 6z(1-z)]z(1-z)}{4! \cdot 8 \cdot n^3}$$

$$= \frac{1}{n^3}\left[\frac{1}{n}A_n(f)(z) + \frac{10z^2(1-z)^2(1-2z)f^{(5)}(z)}{8 \cdot 5!} + \frac{15z^3(1-z)^3f^{(6)}(z)}{8 \cdot 6!}\right.$$

$$\left.+ \frac{z(1-z)(1 - 6z(1-z))f^{(4)}(z)}{8 \cdot 4!}\right],$$

where $\|A_{1,n}(f)\|_r, \|A_n(f)\|_r \leq C_r(f)$ for all $n \in \mathbb{N}$ ($C_r(f)$ is independent of n) and we used the formula in Lorentz [125], p. 14, $T_{n,5}(z) = (1 - 2z)[10n^2z^2(1-z)^2 + n(z(1-z) - 12z^2(1-z)^2)]$ and the Theorem 1.5.1 in Lorentz [125], p. 14, for the formula of n^3 in the polynomial $T_{n,6}(z)$.

Now, by using for

$$G(z) = \frac{10z^2(1-z)^2(1-2z)f^{(5)}(z)}{8 \cdot 5!} + \frac{15z^3(1-z)^3f^{(6)}(z)}{8 \cdot 6!}$$

$$+ \frac{z(1-z)(1 - 6z(1-z))f^{(4)}(z)}{8 \cdot 4!}$$

similar reasonings with those for $U(z)$ in the proof of Corollary 1.3.4, finally we easily obtain

$$\|L_n^{[4]}(f)(z) - f(z)\|_r \sim \frac{1}{n^3},$$

where the constants in the equivalence depend only on f and r.

In what follows we present a proof of the equivalence result in approximation by $L_n^{[2q-2]}(f)(z)$ for general $q \geq 3$. Taking into account Theorem 1.4.1, therefore it suffices to prove the following lower estimate for general q.

Theorem 1.4.2. *Let $\mathbb{D}_R = \{z \in \mathbb{C}; |z| < R\}$ be with $R > 1$, $f : \mathbb{D}_R \to \mathbb{C}$ analytic in $\mathbb{D}_R$, i.e. $f(z) = \sum_{j=0}^{\infty} c_k z^k$, $z \in \mathbb{D}_R$, $1 \leq r < R$ and $q \geq 3$.*

If f is not a polynomial of degree $\leq q$ then for all $n \in \mathbb{N}$ we have

$$\|L_n^{[2q-2]}(f) - f\|_r \geq \frac{C}{n^q},$$

where the constant C is independent of n and depends on f, r and q.

The proof of Theorem 1.4.2 requires the following two auxiliary results.

Lemma 1.4.3. *Let $\mathbb{D}_R = \{z \in \mathbb{C}; |z| < R\}$ be with $R > 1$, $f : \mathbb{D}_R \to \mathbb{C}$ analytic in $\mathbb{D}_R$, and $1 \leq r < R$. Also, let $L_n^{[q]}(f)(z)$ be the Butzer's polynomials defined by the recurrence in the previous section, $q \geq 2$.*

(i) We have the representation formula

$$L_n^{[2q-2]}(f)(z) = \sum_{i=0}^{q-1} \alpha_{i,q-1} B_{2^i n}(f)(z),$$

where $\alpha_{i,q-1} = (-1)^{q-1-i} S_{i,q-1} / \Pi_{k=1}^{q-1}(2^k - 1)$, $S_{0,q-1} = 1$ by convention and

$$S_{i,q-1} = \sum_{j_1,j_2,...,j_i=1, \ j_1<...<j_i} 2^{j_1} 2^{j_2} ... 2^{j_i}, \text{ for all } i \in \{1,...,q-1\}.$$

(ii) The sums $S_{i,q}$ satisfy the recurrence formula

$$S_{i,q} = S_{i,q-1} + 2^q S_{i-1,q-1}, \text{ for all } i = 1,...,q-1.$$

(iii) The numbers $\alpha_{i,q-1}$ satisfy $\sum_{i=0}^{q-1} \alpha_{i,q-1} = 1$ and the recurrence

$$\alpha_{i,q} = \frac{2^q}{2^q - 1} \alpha_{i-1,q-1} - \frac{1}{2^q - 1} \alpha_{i,q-1}, \text{ for all } i = 1, q-1,$$

where $\alpha_{0,q-1} = \frac{(-1)^{q-1}}{\Pi_{k=1}^{q-1}(2^k-1)}$.

(iv) For all $i = 0,...,q$ we have

$$\alpha_{i,q} = (-1)^{q-i} \frac{2^{i(i+1)/2}[2^{\binom{q}{i}} - 1]}{\displaystyle\prod_{j=1}^{q}(2^j - 1)}.$$

Proof.   The proofs of (i) and (ii) can easily be done by mathematical induction after q. Also, (iii) is an immediate consequence of (ii) and (iv) is an immediate consequence of (iii).   □

Lemma 1.4.4. *Let $q \geq 3$.*
(i) For all $j = 1,...,q$, $z \in \mathbb{C}$ and $n \in \mathbb{N}$ we have

$$\sum_{i=0}^{q-1} \alpha_{i,q-1}(2^i)^{-j} T_{2^i n,j}(z) = 0.$$

(ii) For all $j = q+1,...,2q-2$, $z \in \mathbb{C}$ and $n \in \mathbb{N}$ we have

$$\sum_{i=0}^{q-1} \alpha_{i,q-1}(2^i)^{-j} T_{2^i n,j}(z) = n B_{q-1,j}(z),$$

where $B_{q-1,j}(z)$ is a polynomial of degree $\leq j$ in z.

Proof.   By using the recurrence formulas in Lemma 1.4.3, (iii), the proofs of (i) and (ii) are immediate by mathematical induction with respect to q. The degree of the polynomials $B_{q-1,j}(z)$ with respect to z follows from Lorentz [125], p. 14, Theorem 1.5.1.   □

Proof of Theorem 1.4.2. Taking into account the representation formula in Lemma 1.4.3 (i), we can write

$$L_n^{[2q-2]}(f)(z) - f(z) = \sum_{i=0}^{q-1} \alpha_{i,q-1}[B_{2^i n}(f)(z) - f(z)] =$$

$$\sum_{i=0}^{q-1} \alpha_{i,q-1}\left[B_{2^i n}(f)(z) - f(z) - \sum_{j=1}^{2q-2} \frac{f^{(j)}(z)}{j!}(2^i)^{-j}T_{2^i n,j}(z)\right] + R_{n,q-1}(f)(z),$$

where

$$R_{n,q-1}(f)(z) = \sum_{i=0}^{q-1} \alpha_{i,q-1}\left[\sum_{j=1}^{2q-2} \frac{f^{(j)}(z)}{j!}(2^i n)^{-j}T_{2^i n,j}(z)\right]$$

$$= \sum_{j=1}^{2q-2} \frac{f^{(j)}(z)}{j!}n^{-j}\left[\sum_{i=0}^{q-1} \alpha_{i,q-1}(2^i)^{-j}T_{2^i n,j}(z)\right].$$

Taking into account Lemma 1.4.4, (i) we have

$$R_{n,q-1}(f)(z) = \sum_{j=q+1}^{2q-2} \frac{f^{(j)}(z)}{j!}n^{-j}\left[\sum_{i=0}^{q-1} \alpha_{i,q-1}(2^i)^{-j}T_{2^i n,j}(z)\right],$$

which combined with Lemma 1.4.4, (ii) immediately implies

$$R_{n,q-1}(f)(z) = \frac{1}{n^q}\left[\sum_{j=q+1}^{2q-2} \frac{f^{(j)}(z)}{j!} \cdot \frac{nB_{q-1,j}(z)}{n^{j-q}}\right]$$

$$= \frac{1}{n^q}\left[\frac{f^{(q+1)}(z)}{(q+1)!}B_{q-1,q+1}(z) + \frac{1}{n}M_q(f)(z)\right],$$

where $\|M_q(f)\|_r \leq C$, with C independent of n.

Also, since from Theorem 1.5.1, p. 14 in Lorentz [125] we have that each $T_{2^i n,j}(z)$ is a polynomial in $z(1-z)$, this implies that $B_{q-1,q+1}(z)$ is of the form $B_{q-1,q+1}(z) = z(1-z)P_q(z)$, with degree $(P_q(z)) \leq q-1$ and $P_q(0) \neq 0$.

On the other hand, reasoning exactly as in the proof of Corollary 1.3.4, case $q = p - 1$, we get

$$B_{2^i n}(f)(z) - f(z) - \sum_{j=1}^{2(q-1)} \frac{f^{(j)}(z)}{j!}(2^i n)^{-j}T_{2^i n,j}(z)$$

$$= \frac{1}{(2^i n)^q}\left\{\frac{T_{2^i n,2q-1}(z)}{(2^i n)^{q-1}(2q-1)!}f^{(2q-1)}(z) + \frac{T_{2^i n,2q}(z)}{(2q)!}f^{(2q)}(z)\right.$$

$$\left. + \frac{1}{n}\left[(2^i n)^{q+1}\sum_{k=2q+1}^{\infty} c_k E_{k,2^i n,q}(z)\right]\right\}.$$

Clearly hat the expressions $(2^i n)^{q+1}\sum_{k=2q+1}^{\infty} c_k E_{k,2^i n,q}(z)$ are bounded in $\mathbb{D}_r$ with bounds independent of n (see Corollary 1.3.4).

Also, since by Theorem 1.5.1, p. 14 in Lorentz [125] the coefficient of $(2^i n)^q$ in $T_{2^i n, 2q}(z)$ is $\frac{(2q)!}{2^q \cdot q!}[z(1-z)]^q$, while from the recurrence formula in Lorentz [125], p. 14, relation (3), it easily follows that the coefficient of $(2^i n)^{q-1}$ in $T_{2^i n, 2q-1}(z)$ is of the form $a_{q-1}(1-2z)[z(1-z)]^{q-1}$ with $a_{q-1} > 0$. Therefore, reasoning exactly as in the proof of Corollary 1.3.4 we can write

$$\frac{T_{2^i n, 2q-1}(z)}{(2^i n)^{q-1}(2q-1)!} f^{(2q-1)}(z) + \frac{T_{2^i n, 2q}(z)}{(2q)!} f^{(2q)}(z)$$

$$= \frac{a_{q-1}}{(2q-1)!}(1-2z)[z(1-z)]^{q-1} f^{(2q-1)}(z) + \frac{[z(1-z)]^q}{2^q(q)!} f^{(2q)}(z)$$

$$+ \frac{1}{n} F_i(z) f^{(2q-1)}(z) + \frac{1}{n} G_i(z) f^{(2q)}(z),$$

where $F_i(z)$ and $G_i(z)$ are polynomials bounded in $\mathbb{D}_r$ by constants independent of n.

Collecting all the above considerations in conclusion we obtain

$$L_n^{[2q-2]}(f)(z) - f(z)$$

$$= \frac{1}{n^q} \left[\frac{f^{(q+1)}(z)}{(q+1)!} z(1-z) P_q(z) + \frac{2^q - 1}{2^{q-1}} a_{q-1}(1-2z)[z(1-z)]^{q-1} f^{(2q-1)}(z) \right.$$

$$\left. + \frac{2^q - 1}{2^{q-1}} \cdot \frac{[z(1-z)]^q}{2^q q!} f^{(2q)}(z) + \frac{1}{n} K_q(f)(z) \right],$$

where $\|K_q(f)\|_r \leq C$ with C independent of n.

Denoting

$$H_q(f)(z) = \frac{f^{(q+1)}(z)}{(q+1)!} z(1-z) P_q(z)$$

$$+ \frac{2^q - 1}{2^{q-1}} a_{q-1}(1-2z)[z(1-z)]^{q-1} f^{(2q-1)}(z)$$

$$+ \frac{2^q - 1}{2^{q-1}} \cdot \frac{[z(1-z)]^q}{2^q q!} f^{(2q)}(z),$$

if $\|H_q(f)\|_r > 0$ then the expected lower estimate follows from the inequalities

$$\|L_n^{[2q-2]}(f) - f\|_r$$

$$\geq \frac{1}{n^q} \left| \|H_q(f)\|_r - \frac{1}{n} \|K_q(f)\|_r \right|$$

$$\geq \frac{1}{n^q} \left[\|H_q(f)\|_r - \frac{1}{n} \|K_q(f)\|_r \right] \geq \frac{\|H_q(f)\|_r}{2n^q},$$

for all $n \geq n_0$. From the reasonings in the same Corollary 1.3.4 we get the lower estimate of the same order for all $n \in \mathbb{N}$.

Now, to finish the proof it will be enough to show that if f is not a polynomial of degree $\leq q$ then we have $\|H_q(f)\|_r > 0$. For this purpose, it will be enough to prove that the differential equation $H_q(f)(z) = 0$ for $z \in \mathbb{D}_r$ implies that f is a polynomial of degree $\leq q$. Making the substitution $f^{(q+1)}(z) = y(z)$ it will be enough to prove that the differential equation in $z \in \mathbb{D}_r$

$$y(z)z(1-z)P_q(z) + A_q(1-2z)[z(1-z)]^{q-1}y^{(q-2)}(z) + B_q[z(1-z)]^q y^{(q-1)}(z) = 0,$$

has as analytic solution only $y(z) = 0$. Here $A_q, B_q > 0$ and $P_q(0) \neq 0$.

Simplifying with $z(1-z)$ supposed to be $\neq 0$, it follows the differential equation in $z \in \mathbb{D}_{r_1} \setminus \{0\}$

$$y(z)P_q(z) + A_q(1-2z)[z(1-z)]^{q-2}y^{(q-2)}(z) + B_q[z(1-z)]^{q-1}y^{(q-1)}(z) = 0,$$

with $r_1 < 1$. Passing now with $z \to 0$ in this equation we immediately obtain $y(0) = 0$. This means that we can write $y(z) = zh(z)$, with h analytic in $\mathbb{D}_{r_1}$. Calculating $y'(z) = h(z) + zh'(z)$ and $y''(z) = 2h'(z) + zh''(z)$, replacing in the above differential equation, simplifying with z^2 and then passing to limit in the simplified differential equation with $z \to 0$, we immediately obtain $h(0) = 0$. Therefore we can write $h(z) = zu(z)$ and $y(z) = z^2 u(z)$, that is $y'(0) = 0$. Repeating the same reasonings for $y(z)$ written in this form, we arrive at the form $y(z) = z^3 v(z)$, that is $y''(0) = 0$. Step by step by this kind of reasoning we will obtain $y^{(k)}(0) = 0$ for all $k = 0, 1, 2, ...,$. In conclusion we obtain $y(z) = 0$ for all $z \in \mathbb{D}_{r_1}$, which proves the Theorem 1.4.2. $\qquad\square$

Remarks. 1) Let us suppose that $f \in C[0,1]$. By taking $\lambda = 1$ in Guo-Li-Liu [106], Theorem 2, we immediately obtain the following upper estimate valid for all $n \in \mathbb{N}$

$$\|L_n^{[2q-2]}(f) - f\| \leq C\left[\frac{\|f\|}{n^q} + \omega_{2q}^\varphi(f; 1/\sqrt{n})\right],$$

where $\|\cdot\|$ denotes the uniform norm on $C[0,1]$, $C > 0$ is an absolute constant and $\omega_{2q}^\varphi(f; 1/\delta)$ denotes the Ditzian-Totik modulus of smoothness of order $2q$ with respect to the weight $\varphi(x) = \sqrt{x(1-x)}$.

Then, the equivalence results in the above Remarks 4 and 5 and Theorem 1.4.1 and 1.4.2 suggest the following open question : there exists an absolute constant $C' > 0$ such that for all $n \in \mathbb{N}$ we have

$$C'\left[\frac{\|f\|}{n^q} + \omega_{2q}^\varphi(f; 1/\sqrt{n})\right] \leq \|L_n^{[2q-2]}(f) - f\|.$$

2) Analogously, Corollary 1.3.4 suggests equivalence result with respect to some suitable expressions involving Ditzian-Totik moduli of smoothness for the generalized Voronovskaja's theorem in the case of functions of real variable.

For $G \subset \mathbb{C}$ a compact set and $q \in \mathbb{N}$, define $\mathcal{L}_n^{[q]}(z)$, $z \in G$ by setting $\mathcal{L}_n^{[0]}(f; \overline{G})(z) := \mathcal{B}_n(f; \overline{G})(z)$ and

$$(2^q - 1)\mathcal{L}_n^{[2q]}(f; \overline{G})(z) = 2^q \mathcal{L}_{2n}^{[2q-2]}(f; \overline{G})(z) - \mathcal{L}_n^{[2q-2]}(f; \overline{G})(z).$$

Taking into account that these Butzer kind polynomials are linear combinations of Bernstein polynomials attached to G, by the above Theorems 1.4.1 and 1.4.2 and by similar reasonings with those in the proof of Theorem 1.1.9 (since T and T^{-1} are linear), we immediately obtain the following result.

Theorem 1.4.5. *Let G be a compact Faber set such that $\tilde{\mathbb{C}} \setminus G$ is simply connected. If f is analytic on G, that is there exists $R > 1$ such that f is analytic in G_R and if is not a polynomial of degree $\leq q$ then for any $1 < r < R$ we have*

$$\|\mathcal{L}_n^{[2q-2]}(f; \overline{G}) - f\|_{\overline{G}_r} \sim \frac{C}{n^q}, \text{ for all } n, q \in \mathbb{N},$$

where the constants in the equivalence depend on f, q, r and G_r but are independent of n. Here $\|f\|_{\overline{G}_r} = \sup_{z \in \overline{G}_r} |f(z)|$.

1.5 q-Bernstein Polynomials

In this section we present the approximation and shape preserving properties of the complex q-Bernstein polynomials. First we present upper estimates in approximation and we prove the Voronovskaja's convergence theorem in compact disks in $\mathbb{C}$, centered at origin, with quantitative estimate of this convergence. These results alow us to obtain the exact degrees in simultaneous approximation by complex q-Bernstein polynomials and their derivatives. Then we study the approximation properties of their iterates and finally we prove that the complex q-Bernstein polynomials preserve in the unit disk (beginning with an index) the starlikeness, convexity and spirallikeness. For $q = 1$, all these results become those proved for complex Bernstein polynomials in Sections 1.1 and 1.2.

Let $q > 0$. For any $n = 1, 2, ...$, define the q-integer $[n]_q := 1 + q + ... + q^{n-1}$, $[0]_q := 0$ and the q-factorial $[n]_q! := [1]_q[2]_q...[n]_q$, $[0]_q! := 1$. For $q = 1$ we obviously get $[n]_q = n$.

For integers $0 \leq k \leq n$, define

$$\binom{n}{k}_q := \frac{[n]_q!}{[k]_q![n-k]_q!}.$$

Evidently, for $q = 1$ we get $[n]_1 = n$, $[n]_1! = n!$ and $\binom{n}{k}_1 = \binom{n}{k}$.

Now, for $f : [0,1] \to \mathbb{C}$, the complex q-Bernstein polynomials are defined simply replacing z by x in the Phillips definition in [149], that is

$$B_{n,q}(f)(z) = \sum_{k=0}^{n} f\left(\frac{[k]_q}{[n]_q}\right) \binom{n}{k}_q z^k \prod_{s=0}^{n-1-k} (1 - q^s z), n \in \mathbb{N}, z \in \mathbb{C}.$$

Here conventionally, the empty product is equal to 1. Also, note that for $q = 1$ we recapture the classical complex Bernstein polynomials.

First let us briefly expose the present situation of the main approximation results for the complex q-Bernstein polynomials. Thus, concerning the estimates in the convergence of complex q-Bernstein polynomials attached to analytic functions, by the next theorem and remarks we mention the following known results.

Theorem 1.5.1. (Ostrovska [146], Gal [87]) *Let $q > 0$, $R > 1$, $\mathbb{D}_R = \{z \in \mathbb{C}; |z| \leq R\}$ and let us suppose that $f : \mathbb{D}_R \to \mathbb{C}$ is analytic in $\mathbb{D}_R$, that is we can write $f(z) = \sum_{k=0}^{\infty} c_k z^k$, for all $z \in \mathbb{D}_R$.*

Then for the complex q-Bernstein polynomials we have the estimate

$$|B_{n,q}(f)(z) - f(z)| \leq \frac{M_{r,q}(f)}{[n]_q}, \text{ for all } n \in \mathbb{N},$$

valid for all $n \in \mathbb{N}$ and $|z| \leq r$, with $1 \leq r < R$, where

$$0 < M_{r,q}(f) = 2 \sum_{k=2}^{\infty} (k-1)[k-1]_q |c_k| r^k.$$

Moreover, $M_{r,q}(f) \leq 2 \sum_{k=2}^{\infty} (k-1)k|c_k|r^k := M_r(f) < \infty$, for all $r \in [1, R)$ and $q \in (0, 1]$, while if $q > 1$, then $M_{r,q}(f) < \infty$, for all $q < R$ and $r \in [1, \frac{R}{q})$.

Proof. Since only the case $q \geq 1$ (and $|z| \leq 1$) was stated explicitly in Ostrovska [146], let us indicate below the proof in its full generality by using some results already proved in Ostrovska [146]. Indeed, first one easily observe that Lemma 3 in Ostrovska [146] is valid for all $q > 0$. Also, analysing the proof of Theorem 5 in Ostrovska [146] (which use Lemma 3), again it easily follows that its estimate is valid for any $q > 0$, and not only for $q \geq 1$. That is, for any $q > 0$, denoting $e_k(z) = z^k$, $k = 1, 2...$, $z \in \mathbb{C}$, for all $k, n \in \mathbb{N}$ and $|z| \leq r$, we get the kind of estimate in Theorem 5 in Ostrovska [146], i.e.

$$|B_{n,q}(e_k)(z) - e_k(z)| \leq 2r^k \frac{(k-1)[k-1]_q}{[n]_q},$$

which by $B_{n,q}(f)(z) - f(z) = \sum_{k=0}^{\infty} c_k (B_{n,q}(e_k)(z) - e_k(z))$, immediately implies the estimate.

Now, if $0 < q \leq 1$, since $[k-1]_q \leq k$, it is immediate that $M_{r,q}(f) = 2 \sum_{k=2}^{\infty} (k-1)[k-1]_q |c_k| r^k \leq 2 \sum_{k=2}^{\infty} (k-1)k|c_k|r^k < \infty$.

If $q > 1$, then by the estimates $[k-1]_q \leq [k]_q \leq \frac{q^k}{q-1}$ and

$$M_{r,q}(f) = 2 \sum_{k=2}^{\infty} (k-1)[k-1]_q |c_k| r^k \leq \frac{2}{q-1} \sum_{k=2}^{\infty} (k-1)|c_k|r^k q^k,$$

it follows that $M_{r,q}(f) < \infty$ for $rq < R$, which proves the theorem. $\square$

Remarks. 1) Let $0 < q \leq 1$ be fixed. Since we have $\frac{1}{[n]_q} \to 1 - q$ as $n \to \infty$, by passing to limit with $n \to \infty$ in the estimate in Theorem 1.5.1 we don't obtain convergence of $B_n(f)(z)$ to $f(z)$. But this situation can be improved by choosing $0 < q_n < 1$ with $q_n \nearrow 1$ as $n \to \infty$. Indeed, since in this case $\frac{1}{[n]_{q_n}} \to 0$ as $n \to \infty$ (see Videnskii [194], formula (2.7)), from Theorem 1.5.1 we get that $B_{n,q_n}(f)(z) \to f(z)$, uniformly for $|z| \leq r$, for any $1 \leq r < R$.

2) If $q > 1$, since $\frac{1}{[n]_q} \leq \frac{1}{n}$, then by Theorem 1.5.1 it follows that for any $r \geq 1$ with $rq < R$, we have $B_{n,q}(f)(z) \to f(z)$ as $n \to \infty$, uniformly for $|z| \leq r$. In

fact, in this case by Theorem 6 in Ostrovska [146] (for upper estimate) and by Corollary 1 in Wang-Wu [198] we know much more : if f is not a linear function then $\|B_{n,q}(f) - f\|_r \sim q^{-n}$, for any $0 < r < R/q$. Here $\|f\|_r = \sup\{|f(z)|; |z| \leq r\}$.

3) It is worth mentioning other two interesting papers in the topic : approximation by complex q-Bernstein polynomials of the Cauchy kernel $1/(z-a)$ (see Ostrovska [147]) and of the logarithmic function (see Ostrovska [148]).

Now, concerning Voronovskaja-type results and approximation by iterates we can mention the following known results.

If $q \geq 1$ then qualitative Voronovskaja-type and saturation-type results for complex q-Bernstein polynomials were obtained in Wang-Wu [198].

If $0 < q < 1$ then for the real q-Bernstein polynomials, qualitative Voronovskaja-type and saturation results (see Wang [197]) and quantitative Voronovskaja's result (see Videnskii [195]) were recently obtained.

In this section we fulfil this gap for the complex case and obtain a Voronovskaja-type result with quantitative estimate for complex q-Bernstein polynomials with $0 < q < 1$. Compared with the quantitative result proved for the real q-Bernstein polynomials in Videnskii [195], our result is essentially better. Also, as an application of our quantitative Voronovskaja's result, the exact order in approximation by complex q_n-Bernstein polynomials with $0 < q_n \leq 1$ and $\lim_{n\to\infty} q_n = 1$ is obtained.

Taking into account that until present only iterates for real q-Bernstein polynomial were studied (see Ostrovska [146], Xiang-He-Yang [202]), also we fulfil this gap for the complex case, by obtaining approximation results for the iterates of complex q-Bernstein polynomials with $q > 0$.

Also, suggested by the fact that for n sufficiently large the classical complex Bernstein polynomial $B_n(f)(z)$ preserves in the unit disk the starlikeness, convexity and spirallikeness, we will extend these kind of results to complex q_n-Bernstein polynomials, $B_{n,q_n}(f)(z), n \in \mathbb{N}$, with $0 < q_n < 1$ and $q_n \to 1$.

First we present the Voronovskaja-type formula.

Theorem 1.5.2. (Gal [87]) *Let* $0 < q < 1$, $R > 1$, $\mathbb{D}_R = \{z \in \mathbb{C}; |z| < R\}$ *and let us suppose that* $f : \mathbb{D}_R \to \mathbb{C}$ *is analytic in* $\mathbb{D}_R$, *that is we can write* $f(z) = \sum_{k=0}^{\infty} c_k z^k$, *for all* $z \in \mathbb{D}_R$.

(i) The following estimate holds :

$$\left| B_{n,q}(f)(z) - f(z) - \frac{z(1-z)}{2[n]_q} f''(z) \right| \leq \frac{|z(1-z)|}{2[n]_q} \cdot \frac{9M(f)}{[n]_q},$$

for all $n \in \mathbb{N}, z \in \overline{\mathbb{D}}_1$, *where* $0 < M(f) = \sum_{k=3}^{\infty} |c_k| k(k-1)(k-2)^2 < \infty$.

(ii) Let $r \in [1, R)$. *Then*

$$\left| B_{n,q}(f)(z) - f(z) - \frac{z(1-z)}{2[n]_q} f''(z) \right| \leq \frac{(1+r)}{2[n]_q} \cdot \frac{9K_r(f)}{[n]_q},$$

for all $n \in \mathbb{N}, |z| \leq r$, *where* $K_r(f) = \sum_{k=3}^{\infty} |c_k| k(k-1)(k-2)^2 r^k < \infty$.

Proof. (i) Denoting $e_k(z) = z^k$, $k = 0, 1, ...$, and $\pi_{k,n,q}(z) = B_{n,q}(e_k)(z)$, we can write $B_{n,q}(f)(z) = \sum_{k=0}^{\infty} c_k \pi_{k,n,q}(z)$, which immediately implies

$$\left| B_{n,q}(f)(z) - f(z) - \frac{z(1-z)}{2[n]_q} f''(z) \right|$$

$$\leq \sum_{k=2}^{\infty} |c_k| \cdot \left| \pi_{k,n,q}(z) - e_k(z) - \frac{z^{k-1}(1-z)k(k-1)}{2[n]_q} \right|,$$

for all $z \in \overline{\mathbb{D}}_1$, $n \in \mathbb{N}$.

Denote $D_q(f)(z) = \frac{f(z) - f(qz)}{z - qz}$, $q \neq 1$. In what follows, we prove the recurrence formula

$$\pi_{k+1,n,q}(z) = \frac{z(1-z)}{[n]_q} D_q[\pi_{k,n,q}](z) + z\pi_{k,n,q}(z),$$

for all $n \in \mathbb{N}$, $z \in \mathbb{C}$ and $k = 0, 1, 2, ...$

For $z = 0$ and $z = 1$, this recurrence is obviously satisfied. Therefore let us suppose $z \neq 0$ and $z \neq 1$.

Denoting

$$S_{k,n,q}(z) = \sum_{j=0}^{n} [j]_q^k \binom{n}{j}_q z^j \Pi_{s=0}^{n-1-j}(1 - q^s z),$$

and taking into account the formulas $q^j \frac{1-q^{n-j}}{1-q} = [n]_q - [j]_q$ and

$$D_q[f \cdot g](z) = g(z)D_q(f)(z) + f(qz)D_q(g)(z), \quad D_q(z^j)(z) = [j]_q z^{j-1},$$

we obtain

$$D_q[S_{k,n,q}](z)$$

$$= \sum_{j=0}^{n} [j]_q^k \binom{n}{j}_q \left\{ D_q(z^j)(z) \prod_{s=0}^{n-1-j}(1 - q^s z) + q^j z^j D_q(\Pi_{s=0}^{n-1-j}(1 - q^s z)) \right\}$$

$$= \frac{S_{k+1,n,q}(z)}{z} - \sum_{j=0}^{n} [j]_q^k \binom{n}{j}_q z^j \prod_{s=0}^{n-1-j}(1 - q^s z) \frac{q^j}{1-z} \cdot \frac{1-q^{n-j}}{1-q}$$

$$= \frac{S_{k+1,n,q}(z)}{z} - \frac{[n]_q}{1-z} S_{k,n,q}(z) + \frac{S_{k+1,n,q}(z)}{1-z} = \frac{S_{k+1,n,q}(z)}{z(1-z)} - \frac{[n]_q}{1-z} S_{k,n,q}(z).$$

Dividing now by $[n]_q^{k+1}$, the recurrence formula for $\pi_{k,n,q}(z)$ is immediate. Note that from this recurrence we easily obtain that degree $(\pi_{k,n,q}(z)) = k$.

Now, let us denote $E_{k,n,q}(z) = \pi_{k,n,q}(z) - e_k(z) - \frac{z^{k-1}(1-z)k(k-1)}{2[n]_q}$.

For all $k \geq 2$, $n \in \mathbb{N}$ and $z \in \overline{\mathbb{D}}_1$, the above recurrence leads us to

$$E_{k,n,q}(z) = \frac{z(1-z)}{[n]_q} D_q[E_{k-1,n,q}](z) + z E_{k-1,n,q}(z) + G_{k,n,q}(z),$$

where

$$G_{k,n,q}(z) = \frac{z^{k-2}(1-z)}{2[n]_q} \cdot \frac{(k-1)(k-2)[k-2]_q}{[n]_q}$$

$$- \frac{z^{k-2}(1-z)}{2[n]_q} \cdot z \left(\frac{(k-1)[k-1]_q(k-2)}{[n]_q} + 2(k-1) - 2[k-1]_q \right),$$

Taking into account that by the mean value theorem in complex analysis we have $|D_q(f)(z)| \leq \|f'\|_1$, where $\|\cdot\|_1$ denotes the uniform norm in $C(\overline{\mathbb{D}}_1)$, and by using the relationships $0 < 1 - q < \frac{1}{[n]_q}$ and

$$
\begin{aligned}
(k-1) - [k-1]_q &= (1-q) + ... + (1-q^{k-2}) \\
&= (1-q)[1 + (1+q) + ... + (1+q+...q^{k-3})] \\
&\leq \frac{1}{[n]_q}(1 + 2 + ... + k - 2) = \frac{(k-2)(k-1)}{2[n]_q},
\end{aligned}
$$

we obtain, for all $|z| \leq 1$, $k \geq 2$, $n \in \mathbb{N}$,

$$
|E_{k,n,q}(z)| \leq \frac{|z| \cdot |1-z|}{2[n]_q}[2\|E'_{k-1,n,q}\|_1] + |E_{k-1,n,q}(z)| + \frac{|z| \cdot |1-z| \cdot |z|^{k-3}}{2[n]_q}
$$

$$
\cdot \left[\frac{(k-1)(k-2)[k-2]_q}{[n]_q} + \frac{(k-1)(k-2)[k-1]_q}{[n]_q} + 2(k-1) - 2[k-1]_q\right]
$$

$$
\leq |E_{k-1,n,q}(z)| + \frac{|z| \cdot |1-z|}{2[n]_q}
$$

$$
\cdot \left[2\|E'_{k-1,n,q}\|_1 + \frac{2(k-1)(k-2)[k-1]_q}{[n]_q} + \frac{(k-2)(k-1)}{[n]_q}\right]
$$

$$
\leq |E_{k-1,n,q}(z)| + \frac{|z| \cdot |1-z|}{2[n]_q}
$$

$$
\cdot \left[2(k-1)\|E_{k-1,n,q}\|_1 + \frac{2(k-1)(k-2)[k-1]_q}{[n]_q} + \frac{(k-2)(k-1)}{[n]_q}\right]
$$

$$
\leq |E_{k-1,n,q}(z)| + \frac{|z| \cdot |1-z|}{2[n]_q}\left[2(k-1)\|\pi_{k-1,n,q} - e_{k-1}\|_1\right.
$$

$$
+ 2(k-1)^2(k-2)\frac{\|e_{k-2}(z) - e_{k-1}(z)\|_1}{2[n]_q} + \frac{2(k-1)(k-2)[k-1]_q}{[n]_q} +
$$

$$
+ \left.\frac{(k-2)(k-1)}{[n]_q}\right] \leq \text{ (by the proof of Theorem 1.5.1)}
$$

$$
\leq |E_{k-1,n,q}(z)| + \frac{|z| \cdot |1-z|}{2[n]_q}\left[2(k-1)\frac{2(k-2)[k-2]_q}{[n]_q}\right.
$$

$$
+ \left.\frac{2(k-1)^2(k-2)}{[n]_q} + \frac{2(k-1)(k-2)[k-1]_q}{[n]_q} + \frac{(k-2)(k-1)}{[n]_q}\right]
$$

$$
\leq |E_{k-1,n,q}(z)| + \frac{9|z| \cdot |1-z|k(k-1)(k-2)}{2[n]_q^2}.
$$

Therefore, we have obtained

$$
|E_{k,n,q}(z)| \leq |E_{k-1,n,q}(z)| + \frac{9|z| \cdot |1-z|k(k-1)(k-2)}{2[n]_q^2}.
$$

The last inequality is trivial for $k = 1, 2$, since $E_{1,n,q}(z) = E_{2,n,q}(z) = 0$, for any $z \in \mathbb{C}$.

Writing now the last inequality for $k = 3, 4, ...$, step by step we easily obtain

$$|E_{k,n,q}(z)| \leq \frac{|z| \cdot |1-z|}{2[n]_q} \cdot \frac{9}{[n]_q} \cdot \sum_{j=3}^{k} j(j-1)(j-2)$$

$$\leq \frac{|z| \cdot |1-z|}{2[n]_q} \cdot \frac{9}{[n]_q} \cdot k(k-1)(k-2)^2.$$

In conclusion,

$$\left| B_{n,q}(f)(z) - f(z) - \frac{z(1-z)}{2[n]_q} f''(z) \right| \leq \sum_{k=3}^{\infty} |c_k| \cdot |E_{k,n,q}(z)|$$

$$\leq \frac{|z| \cdot |1-z|}{2[n]_q} \cdot \frac{9}{[n]_q} \cdot \sum_{k=3}^{\infty} |c_k| k(k-1)(k-2)^2.$$

Note that since $f^{(4)}(z) = \sum_{k=4}^{\infty} c_k k(k-1)(k-2)(k-3)z^{k-4}$ and the series is absolutely convergent in $\overline{\mathbb{D}}_1$, it easily follows that $\sum_{k=3}^{\infty} |c_k| k(k-1)(k-2)^2 < \infty$, which proves (i).

(ii) First we use the relationship in the proof of Theorem 1.5.1

$$|\pi_{k,n,q}(z) - e_k(z)| \leq 2r^k \frac{(k-1)[k-1]_q}{[n]_q},$$

for all $k, n \in \mathbb{N}$, $|z| \leq r$, with $1 \leq r < R$.

Denoting with $\| \cdot \|_r$ the norm in $C(\overline{\mathbb{D}}_r)$, where $\overline{\mathbb{D}}_r = \{z \in \mathbb{C}; |z| \leq r\}$, one observes that by a linear transformation, the Bernstein's inequality in the closed unit disk becomes $|P_k'(z)| \leq \frac{k}{r} \|P_k\|_r$, for all $|z| \leq r$, $r \geq 1$, which combined with the mean value theorem in complex analysis, implies $|D_q(P_k)(z)| \leq \|P_k'\|_r \leq \frac{k}{r} \|P_k\|_r$, for all $|z| \leq r$, where $P_k(z)$ is a complex polynomial of degree $\leq k$.

Now, taking into account the formula proved at the above point (i), given by

$$E_{k,n,q}(z) = \frac{z(1-z)}{[n]_q} D_q[E_{k-1,n,q}](z) + z E_{k-1,n,q}(z)$$

$$+ \frac{z^{k-2}(1-z)}{2[n]_q} \left[\frac{(k-1)(k-2)[k-2]_q}{[n]_q} \right.$$

$$\left. - z \left(\frac{(k-1)[k-1]_q(k-2)}{[n]_q} + 2(k-1) - [k-1]_q \right) \right],$$

it follows for all $k, n \in \mathbb{N}$, $k \geq 2$ and $|z| \leq r$,

$$|E_{k,n,q}(z)| \leq \frac{r(1+r)}{[n]_q} |D_q[E_{k-1,n,q}](z)| + r|E_{k-1,n,q}(z)| + \frac{(1+r)r^{k-2}}{2[n]_q} \cdot$$

$$\left[\frac{(k-1)(k-2)[k-2]_q}{[n]_q} + r \left(\frac{(k-1)(k-2)[k-1]_q}{[n]_q} + \frac{(k-2)(k-1)}{[n]_q} \right) \right]$$

$$\leq r|E_{k-1,n,q}(z)| + \frac{r(1+r)}{[n]_q} |D_q[E_{k-1,n,q}](z)| + \frac{3(1+r)r^{k-1}k(k-1)(k-2)}{2[n]_q^2}.$$

Now, we will estimate $|D_q[E_{k-1,n,q}](z)|$, for $k \geq 3$. Taking into account that $E_{k-1,n,q}(z)$ is a polynomial of degree $\leq (k-1)$, we obtain

$$
\begin{aligned}
|D_q[E_{k-1,n,q}](z)| &\leq \frac{k-1}{r} \|E_{k-1,n,q}(z)\|_r \\
&\leq \frac{k-1}{r} \left[\|\pi_{k-1,n,q} - e_{k-1}\|_r + \| \frac{(k-1)(k-2)(e_{k-1} - e_{k-2})}{2[n]_q} \|_r \right] \\
&\leq \frac{k-1}{r} \left[\frac{2(k-2)[k-2]_q r^k}{[n]_q} + \frac{2r^{k-1}(k-1)(k-2)}{2[n]_q} \right] \\
&\leq \frac{3r^{k-1}k(k-1)(k-2)}{[n]_q}.
\end{aligned}
$$

This implies

$$
\frac{r(1+r)}{[n]_q} |D_q[E_{k-1,n,q}](z)| \leq \frac{3(1+r)k(k-1)(k-2)r^k}{[n]_q^2},
$$

and

$$
\begin{aligned}
|E_{k,n,q}(z)| &\leq r|E_{k-1,n,q}(z)| \\
&\quad + \frac{3(1+r)k(k-1)(k-2)r^k}{[n]_q^2} + \frac{3(1+r)k(k-1)(k-2)r^{k-1}}{2[n]_q^2} \\
&\leq r|E_{k-1,n,q}(z)| + \frac{9(1+r)k(k-1)(k-2)r^k}{2[n]_q^2}.
\end{aligned}
$$

But $E_{0,n}(z) = E_{1,n}(z) = E_{2,n}(z) = 0$, for any $z \in \mathbb{C}$. Writing now the last inequality for $k = 3, 4, ...$, step by step we easily obtain

$$
\begin{aligned}
|E_{k,n,q}(z)| &\leq \frac{9(1+r)r^k}{2[n]_q^2} \left[\sum_{j=3}^{k} j(j-1)(j-2) \right] \\
&\leq \frac{9(1+r)k(k-1)(k-2)^2 r^k}{2[n]_q^2}.
\end{aligned}
$$

As a conclusion, we obtain

$$
\begin{aligned}
\left| B_{n,q}(f)(z) - f(z) - \frac{z(1-z)}{2[n]_q} f''(z) \right| &\leq \sum_{k=3}^{\infty} |c_k| \cdot |E_{k,n}(z)| \\
&\leq \frac{9(1+r)}{2[n]_q^2} \sum_{k=3}^{\infty} |c_k| k(k-1)(k-2)^2 r^k.
\end{aligned}
$$

Note that since $f^{(4)}(z) = \sum_{k=4}^{\infty} c_k k(k-1)(k-2)(k-3) z^{k-4}$ and the series is absolutely convergent in $|z| \leq r$, it easily follows that $\sum_{k=3}^{\infty} |c_k| k(k-1)(k-2)^2 r^k < \infty$, which proves (ii). $\qquad \square$

Remarks. 1) In the hypothesis on f in Theorem 1.5.2 and choosing $0 < q_n < 1$ with $q_n \nearrow 1$ as $n \to \infty$, it follows that

$$
\lim_{n \to \infty} [n]_{q_n} [B_{n,q_n}(f)(z) - f(z)] = \frac{z(1-z)f''(z)}{2},
$$

uniformly in any compact disk included in the open disk of center 0 and radius R.

2) In Videnskii [194], Theorem 5.1, estimate (5.7), for the real q-Bernstein polynomials it is proved that for $f \in C^2[0,1]$, $x \in [0,1]$, and $0 < q_n < 1$ with $\lim_{n \to \infty} q_n = 1$, we have

$$\left| B_{n,q_n}(f)(x) - f(x) - \frac{f''(x)}{2} \cdot \frac{x(1-x)}{[n]_{q_n}} \right| \leq \frac{Kx(1-x)}{[n]_{q_n}} \omega_1(f''; [n]_{q_n}^{-1/2}),$$

where ω_1 denotes the modulus of continuity. Obviously that the best order of approximation that can be obtained from this estimate is $O(1/[n]_{q_n}^{-3/2})$ (for $f \in C^m[0,1]$ with $m \geq 3$), while the order given by our Theorem 1.5.2 is $O(1/[n]_{q_n}^{-2})$, which is essentially better taking into account that as $n \to \infty$ we have $[n]_{q_n} \to \infty$.

In what follows we obtain the exact orders in approximation by complex q-Bernstein polynomials and their derivatives on compact disks.

In this sense, the first result is the following.

Theorem 1.5.3. (Gal [87]) *Let $0 < q_n \leq 1$ be with $\lim_{n \to \infty} q_n = 1$, $R > 1$, $\mathbb{D}_R = \{z \in \mathbb{C}; |z| < R\}$ and let us suppose that $f : \mathbb{D}_R \to \mathbb{C}$ is analytic in $\mathbb{D}_R$, that is we can write $f(z) = \sum_{k=0}^{\infty} c_k z^k$, for all $z \in \mathbb{D}_R$. If f is not a polynomial of degree ≤ 1, then for any $r \in [1, R)$ we have*

$$\|B_{n,q_n}(f) - f\|_r \geq \frac{C_r(f)}{[n]_{q_n}}, n \in \mathbb{N},$$

where $\|f\|_r = \max\{|f(z)|; |z| \leq r\}$ and the constant $C_r(f) > 0$ depends on f, r and on the sequence $(q_n)_{n \in \mathbb{N}}$ but it is independent of n.

Proof. For all $z \in \mathbb{D}_R$ and $n \in \mathbb{N}$ we have

$$B_{n,q_n}(f)(z) - f(z) = \frac{1}{[n]_{q_n}} \left\{ \frac{z(1-z)}{2} f''(z) \right.$$
$$\left. + \frac{1}{[n]_{q_n}} \left[[n]_{q_n}^2 \left(B_{n,q_n}(f)(z) - f(z) - \frac{z(1-z)}{2[n]_{q_n}} f''(z) \right) \right] \right\}.$$

We will apply to this identity the following obvious property :

$$\|F + G\|_r \geq |\|F\|_r - \|G\|_r| \geq \|F\|_r - \|G\|_r.$$

Denoting $e_1(z) = z$ it follows

$$\|B_{n,q_n}(f) - f\|_r \geq$$

$$\frac{1}{[n]_{q_n}} \left\{ \left\| \frac{e_1(1 - e_1)}{2} f'' \right\|_r - \frac{1}{[n]_{q_n}} \left[[n]_{q_n}^2 \left\| B_{n,q_n}(f) - f - \frac{e_1(1 - e_1)}{2[n]_{q_n}} f'' \right\|_r \right] \right\}.$$

Taking into account that by hypothesis f is not a polynomial of degree ≤ 1 in $\mathbb{D}_R$, we get $\left\| \frac{e_1(1-e_1)}{2} f'' \right\|_r > 0$. Indeed, supposing the contrary it follows that $\frac{z(1-z)}{2} f''(z) = 0$ for all $z \in \overline{\mathbb{D}}_r$, which implies $f''(z) = 0$ for all $z \in \overline{\mathbb{D}}_r \setminus \{0, 1\}$. Since f is supposed to be analytic, from the identity's theorem of analytic (holomorphic) functions this necessarily implies that $f''(z) = 0$, for all $z \in \mathbb{D}_R$, i.e. that f is a polynomial of degree ≤ 1, which is a contradiction.

But by Theorem 1.5.2 we have

$$
([n]_{q_n})^2 \left\| B_{n,q_n}(f) - f - \frac{e_1(1 - e_1)}{2[n]_{q_n}} f'' \right\|_r \leq ([n]_{q_n})^2 \frac{9K_r(f)(1 + r)}{2([n]_{q_n})^2}
$$

$$
= \frac{9K_r(f)(1 + r)}{2}.
$$

Since by the Remark after the proof of Theorem 1.5.1 we have $\frac{1}{[n]_{q_n}} \to 0$ as $n \to \infty$, it follows that there exists an index n_0 depending only on f, r and on the sequence $(q_n)_n$, such that for all $n \geq n_0$ we have

$$
\left\| \frac{e_1(1 - e_1)}{2} f'' \right\|_r - \frac{1}{[n]_{q_n}} \left[([n]_{q_n})^2 \left\| B_{n,q_n}(f) - f - \frac{e_1(1 - e_1)}{2[n]_{q_n}} f'' \right\|_r \right]
$$

$$
\geq \left\| \frac{e_1(1 - e_1)}{2} f'' \right\|_r > 0,
$$

which immediately implies

$$
\| B_{n,q_n}(f) - f \|_r \geq \frac{1}{[n]_{q_n}} \cdot \left\| \frac{e_1(1 - e_1)}{4} f'' \right\|_r, \forall n \geq n_0.
$$

For $1 \leq n \leq n_0 - 1$ we obviously have

$$
\| B_{n,q_n}(f) - f \|_r \geq \frac{M_{r,n}(f)}{[n]_{q_n}},
$$

with $M_{r,n}(f) = [n]_{q_n} \cdot \| B_{n,q_n}(f) - f \|_r > 0$, which finally implies

$$
\| B_{n,q_n}(f) - f \|_r \geq \frac{C_r(f)}{[n]_{q_n}}, \text{ for all } n \in \mathbb{N},
$$

where $C_r(f) = \min \left\{ M_{r,1}, M_{r,2}(f), ..., M_{r,n_0-1}(f), \left\| \frac{e_1(1-e_1)}{4} f'' \right\|_r \right\}$. $\square$

Combining Theorem 1.5.3 with Theorem 1.5.1 we get the following.

Corollary 1.5.4. (Gal [87]) *Let Let $0 < q_n \leq 1$ be with $\lim_{n \to \infty} q_n = 1$, $R > 1$, $\mathbb{D}_R = \{z \in \mathbb{C}; |z| < R\}$ and let us suppose that $f : \mathbb{D}_R \to \mathbb{C}$ is analytic in $\mathbb{D}_R$. If f is not a polynomial of degree ≤ 1, then for any $r \in [1, R)$ we have*

$$
\| B_{n,q_n}(f) - f \|_r \sim \frac{1}{[n]_{q_n}}, n \in \mathbb{N},
$$

where the constants in the equivalence depend on f, r and on the sequence $(q_n)_n$ but are independent of n.

Proof.   Since $q_n \leq 1$ for all $n \in \mathbb{N}$, by Theorem 1.5.1 it follows the upper estimate with the constant depending only on f and r (independent of the sequence $(q_n)_n$). Theorem 1.5.3 assures the lower estimate with the constant depending on f, r and on the sequence $(q_n)_n$, but independent of n. $\square$

Remark. Theorem 1.5.3 and Corollary 1.5.4 in the case when $q_n = 1$ for all $n \in \mathbb{N}$ were obtained by Theorem 1.1.4 and Corollary 1.1.5.

In the case of approximation by the derivatives of complex q-Bernstein polynomials we can present the following new result which appears for the first time here.

Theorem 1.5.5. *Let $0 < q_n \le 1$ be with $\lim_{n \to \infty} q_n = 1$, $R > 1$, $\mathbb{D}_R = \{z \in \mathbb{C}; |z| < R\}$ and let us suppose that $f : \mathbb{D}_R \to \mathbb{C}$ is analytic in $\mathbb{D}_R$, i.e. $f(z) = \sum_{k=0}^{\infty} c_k z^k$, for all $z \in \mathbb{D}_R$. Also, let $1 \le r < r_1 < R$ and $p \in \mathbb{N}$ be fixed. If f is not a polynomial of degree $\le \max\{1, p-1\}$, then we have*

$$\|B_{n,q_n}^{(p)}(f) - f^{(p)}\|_r \sim \frac{1}{[n]_{q_n}},$$

where the constants in the equivalence depend on f, r, r_1, p and on the sequence $(q_n)_n$, but are independent of n.

Proof. Denoting by Γ the circle of radius $r_1 >$ and center 0 (where $r_1 > r \ge 1$), by the Cauchy's formulas it follows that for all $|z| \le r$ and $n \in \mathbb{N}$ we have

$$B_{n,q_n}^{(p)}(f)(z) - f^{(p)}(z) = \frac{p!}{2\pi i} \int_{\Gamma} \frac{B_{n,q_n}(f)(v) - f(v)}{(v - z)^{p+1}} dv,$$

which by Theorem 1.5.1 and by the inequality $|v - z| \ge r_1 - r$ valid for all $|z| \le r$ and $v \in \Gamma$, immediately implies

$$\|B_{n,q_n}^{(p)}(f) - f^{(p)}\|_r \le \frac{p!}{2\pi} \cdot \frac{2\pi r_1}{(r_1 - r)^{p+1}} \|B_{n,q_n}(f) - f\|_{r_1}$$

$$\le M_{r_1}(f) \frac{p! r_1}{[n]_{q_n}(r_1 - r)^{p+1}}.$$

It remains to prove the lower estimate for $\|B_{n,q_n}^{(p)}(f) - f^{(p)}\|_r$. For this purpose, as in the proof of Theorem 1.5.3, for all $v \in \Gamma$ and $n \in \mathbb{N}$ we have

$$B_{n,q_n}(f)(v) - f(v) = \frac{1}{[n]_{q_n}} \left\{ \frac{v(1-v)}{2} f''(v) \right.$$

$$+ \left. \frac{1}{[n]_{q_n}} \left[([n]_{q_n})^2 \left(B_{n,q_n}(f)(v) - f(v) - \frac{v(1-v)}{2[n]_{q_n}} f''(v) \right) \right] \right\},$$

which replaced in the above Cauchy's formula implies

$$B_{n,q_n}^{(p)}(f)(z) - f^{(p)}(z) = \frac{1}{[n]_{q_n}} \left\{ \frac{p!}{2\pi i} \int_{\Gamma} \frac{v(1-v)f''(v)}{2(v - z)^{p+1}} dv \right.$$

$$+ \frac{1}{[n]_{q_n}} \cdot \frac{p!}{2\pi i} \int_{\Gamma} \frac{n^2 \left(B_{n,q_n}(f)(v) - f(v) - \frac{v(1-v)}{2[n]_{q_n}} f''(v) \right)}{(v - z)^{p+1}} dv \right\}$$

$$= \frac{1}{[n]_{q_n}} \left\{ \left[\frac{z(1-z)}{2} f''(z) \right]^{(p)} \right.$$

$$+ \frac{1}{[n]_{q_n}} \cdot \frac{p!}{2\pi i} \int_{\Gamma} \frac{([n]_{q_n})^2 \left(B_{n,q_n}(f)(v) - f(v) - \frac{v(1-v)}{2[n]_{q_n}} f''(v) \right)}{(v - z)^{p+1}} dv \right\}.$$

Passing now to $\| \cdot \|_r$ it follows

$$\left\| B_{n,q_n}^{(p)}(f) - f^{(p)} \right\|_r \geq \frac{1}{[n]_{q_n}} \left\{ \left\| \left[\frac{e_1(1-e_1)}{2} f'' \right]^{(p)} \right\|_r \right.$$

$$\left. - \frac{1}{[n]_{q_n}} \left\| \frac{p!}{2\pi} \int_\Gamma \frac{([n]_{q_n})^2 \left(B_{n,q_n}(f)(v) - f(v) - \frac{v(1-v)}{2[n]_{q_n}} f''(v) \right)}{(v - e_1)^{p+1}} dv \right\|_r \right\},$$

where by using Theorem 1.5.2 and denoting $e_1(z) = z$ we get

$$\left\| \frac{p!}{2\pi} \int_\Gamma \frac{([n]_{q_n})^2 \left(B_{n,q_n}(f)(v) - f(v) - \frac{v(1-v)}{2[n]_{q_n}} f''(v) \right)}{(v - e_1)^{p+1}} dv \right\|_r$$

$$\leq \frac{p!}{2\pi} \cdot \frac{2\pi r_1 ([n]_{q_n})^2}{(r_1 - r)^{p+1}} \left\| B_{n,q_n}(f) - f - \frac{e_1(1-e_1)}{2[n]_{q_n}} f'' \right\|_{r_1}$$

$$\leq \frac{5 K_{r_1}(f)(1+r_1)^2}{2} \cdot \frac{p! r_1}{(r_1 - r)^{p+1}}.$$

But by hypothesis on f we have $\left\| \left[\frac{e_1(1-e_1)}{2} f'' \right]^{(p)} \right\|_r > 0$. Indeed, supposing the contrary it follows that $\frac{z(1-z)}{2} f''(z)$ is a polynomial of degree $\leq p - 1$. Now, if $p = 1$ and $p = 2$ then the analyticity of f obviously implies that f necessarily is a polynomial of degree $\leq 1 = \max\{1, p - 1\}$, which contradicts the hypothesis. If $p > 2$ then the analyticity of f obviously implies that f necessarily is a polynomial of degree $\leq p - 1 = \max\{1, p - 1\}$, which again contradicts the hypothesis.

In continuation reasoning exactly as in the proof of Theorem 1.5.3, we immediately get the desired conclusion. $\square$

Remark. Theorem 1.5.5 in the case when $q_n = 1$ for all $n \in \mathbb{N}$ was obtained by Theorem 1.1.6.

In what follows we consider the approximation properties for iterates. First we recall some considerations in Section 1.2. For $R > 1$ let us define by $\mathbb{A}_R$ the space of all functions defined and analytic in the open disk of center 0 and radius R denoted by $\mathbb{D}_R$. Denoting $r_j = R - \frac{R-1}{j}$, $j \in \mathbb{N}$ and for $f \in \mathbb{A}_R$, $\|f\|_j = max\{|f(z)|; |z| \leq r_j\}$, since $r_1 = 1$ and $r_j \nearrow R$ as $j \to \infty$, it is well-known that $\{\| \cdot \|_j, j \in \mathbb{N}\}$ it is a countable family of increasing semi-norms on $\mathbb{A}_R$ and that $\mathbb{A}_R$ becomes a metrizable complete locally convex space (Fréchet space), with respect to the metric

$$d(f, g) = \sum_{j=1}^{\infty} \frac{1}{2^j} \cdot \frac{\|f - g\|_j}{1 + \|f - g\|_j}, f, g \in \mathbb{A}_R.$$

It is well-known that $\lim_{n \to \infty} d(f_n, f) = 0$ is equivalent to the fact that the sequence $(f_n)_{n \in \mathbb{N}}$ converges to f uniformly on compacts in $\mathbb{D}_R$.

Now, for $f \in \mathbb{A}_R$, that is of the form $f(z) = \sum_{k=0}^{\infty} c_k z^k$, for all $z \in \mathbb{D}_R$, let us define the iterates of complex q-Bernstein polynomial $B_{n,q}(f)(z)$, by $B_{n,q}^{(1)}(f)(z) =$

$B_{n,q}(f)(z)$ and $B_{n,q}^{(m)}(f)(z) = B_{n,q}[B_{n,q}^{(m-1)}(f)](z)$, for any $m \in \mathbb{N}$, $m \geq 2$. Since we have $B_{n,q}(f)(z) = \sum_{k=0}^{\infty} c_k B_{n,q}(e_k)(z)$, by recurrence for all $m \geq 1$, it easily follows $B_{n,q}^{(m)}(f)(z) = \sum_{k=0}^{\infty} c_k B_{n,q}^{(m)}(e_k)(z)$, with $e_k(z) = z^k$.

The main result is the following.

Theorem 1.5.6. (Gal [87]) *Let $f \in \mathbb{A}_R$, that is $f(z) = \sum_{k=0}^{\infty} c_k z^k$, for all $z \in \mathbb{D}_R$.*

(i) Let $q \in (0, 1)$. If $\lim_{n \to \infty} m_n = +\infty$, then

$$\lim_{n \to \infty} d[B_{n,q}^{(m_n)}(f), L_1(f)] = 0.$$

(ii) If $q \in (0, 1]$ then for any fixed $s \in \mathbb{N}$, the following estimates hold

$$\|B_{n,q}^{(m)}(f) - f\|_s \leq \frac{2m}{[n]_q} \sum_{k=2}^{\infty} |c_k| k(k-1) r_s^k,$$

and

$$d[B_{n,q}^{(m)}(f), f] \leq \frac{2m}{[n]_q} \sum_{k=2}^{\infty} |c_k| k(k-1) r_s^k + \frac{1}{2^s},$$

where $\sum_{k=2}^{\infty} |c_k| k(k-1) r_s^k < \infty$.

(iii) Let $q \in (1, \infty)$. If $\lim_{n \to \infty} \frac{m_n}{[n]_q} = 0$, then

$$\lim_{n \to \infty} d[B_{n,q}^{(m_n)}(P), P] = 0,$$

for any polynomial P.

(iv) Let $q \in (1, \infty)$. If $1 \leq r < R$, then the following estimate holds for all $|z| \leq r$

$$|B_{n,q}^{(m)}(f)(z) - f(z)| \leq \frac{2m}{n} \sum_{k=2}^{\infty} |c_k| \left[\frac{q^k - 1 - k(q-1)}{(q-1)^2} + k(k-1) \right] r^k.$$

If, in addition, $q < R$, then since $\sum_{k=2}^{\infty} |c_k| \left[\frac{q^k - 1 - k(q-1)}{(q-1)^2} + k(k-1) \right] < \infty$, we obtain $B_{n,q}^{(m_n)}(f)(z) \to f(z)$, uniformly in $\overline{\mathbb{D}_1}$, for $\frac{m_n}{n} \to 0$ as $n \to \infty$.

(v) Let $q \in (1, \infty)$. If $\lim_{n \to \infty} \frac{m_n}{[n]_q} = \infty$, then

$$\lim_{n \to \infty} d[B_{n,q}^{(m_n)}(P), L_1(P)] = 0,$$

for any polynomial P.

Proof. (i) Let $0 < q < 1$ and $n \in \mathbb{N}$ be fixed. Denoting $s = min\{n, k\}$, by Lemma 3 in Ostrovska [146] we can write $B_{n,q}(e_k)(z) = \sum_{j=1}^{s} \alpha_{j,k,n,q} z^j$, where $\alpha_{(j}, k, n, q) \geq 0$, for all $j, k, n \in \mathbb{N}$ and $\sum_{j=1}^{s} \alpha_{j,k,n,q} = 1$, for $k, n \in \mathbb{N}$.

Let $|z| \leq r$ with $r \geq 1$. It follows

$$|B_{n,q}(e_k)(z)| \leq \sum_{j=1}^{s} \alpha_{j,k,n,q} |e_j(z)| \leq \sum_{j=1}^{s} \alpha_{j,k,n,q} r^j \leq \sum_{j=1}^{s} \alpha_{j,k,n,q} r^s = r^s \leq r^k.$$

Applying now above $B_{n,q}$, we obtain

$$B_{n,q}^{(2)}(e_k)(z) = \sum_{j=1}^{s} \alpha_{j,k,n,q} B_{n,q}(e_j)(z),$$

which from the last inequality implies

$$|B_{n,q}^{(2)}(e_k)(z)| \leq \sum_{j=1}^{s} \alpha_{j,k,n,q} |B_{n,q}(e_j)(z)|$$

$$\leq \sum_{j=1}^{s} \alpha_{j,k,n,q} r^j \leq \sum_{j=1}^{s} \alpha_{j,k,n,q} r^s = r^s \leq r^k.$$

Reasoning by recurrence, we easily get $|B_{n,q}^{(m)}(e_k)(z)| \leq r^k$ for all $k, n, m \in \mathbb{N}$ and $z \in \overline{\mathbb{D}_r}$.

This implies

$$|B_{n,q}^{(m)}(f)(z)| = |\sum_{k=0}^{\infty} c_k B_{n,q}^{(m)}(e_k)(z)| \leq \sum_{k=0}^{\infty} |c_k| r^k < +\infty,$$

for all $m, n \in \mathbb{N}$ and $z \in \overline{\mathbb{D}_r}$, that is for each $r \in [1, R)$, the sequence $B_{n,q}^{(m)}(f)(z)$, $m, n = 1, 2, \ldots$, is uniformly bounded in $\overline{\mathbb{D}_r}$ with respect to both $m, n \in \mathbb{N}$.

Therefore, the sequence $B_{n,q}^{(m_n)}(f)(z)$, $n = 1, 2, \ldots$, is uniformly bounded in $\overline{\mathbb{D}_r}$ with respect to $n \in \mathbb{N}$.

Since by Theorem 8 in Ostrovska [146], for each $q \in (0,1)$ and for $n \to \infty$ we have $B_{n,q}^{(m_n)}(f)(x) \to L_1(f)(x)$, for $x \in [0,1]$, (even uniformly), the classical Vitali's convergence theorem implies the uniform convergence (as $n \to \infty$) on compacts in $\mathbb{D}_R$ of the sequence $B_{n,q}^{(m_n)}(f)(z)$ to $L_1(f)(z)$. Taking into account that the uniform convergence on compacts is equivalent to the convergence with respect to the metric d, (i) is proved.

(ii) Since $B_{n,q}^{(m)}(f)(z) = \sum_{k=0}^{\infty} c_k B_{n,q}^{(m)}(e_k)(z)$, with $e_k(z) = z^k$, we get

$$|B_{n,q}^{(m)}(f)(z) - f(z)| \leq \sum_{k=2}^{\infty} |c_k| \cdot |B_{n,q}^{(m)}(e_k)(z) - e_k(z)|.$$

To estimate $|B_{n,q}^{(m)}(e_k)(z) - e_k(z)|$, we have two possibilities : 1) $0 \leq k \leq n$; 2) $k > n \geq 1$.

Case 1). According to Lemma 3 in Ostrovska [146], we have

$$B_{n,q}(e_k)(z) - e_k(z) = \sum_{j=1}^{k} \alpha_{j,k,n,q} e_j(z) - e_k(z).$$

Therefore,

$$B_{n,q}(e_k)(z) - e_k(z) = \sum_{j=1}^{k-1} \alpha_{j,k,n,q} e_j(z) + [\alpha_{k,k,n,q} - 1] e_k(z),$$

which immediately implies

$$B_{n,q}^{(p)}[B_{n,q}(e_k)(z) - e_k(z)] = \sum_{j=1}^{k-1} \alpha_{j,k,n,q} B_{n,q}^{(p)}(e_j)(z) + [\alpha_{k,k,n,q} - 1]B_{n,q}^{(p)}(e_k)(z).$$

Taking into account that by the proof of (i) we have $|B_{n,q}^{(p)}(e_j)(z)| \leq r^j$, for all $p, n, j \in \mathbb{N}$ and $|z| \leq r$ with $r \geq 1$, it follows

$$|B_{n,q}^{(p)}[B_{n,q}(e_k) - e_k](z)|$$

$$\leq \sum_{j=1}^{k-1} \alpha_{j,k,n,q}|B_{n,q}^{(p)}(e_j)(z)| + |1 - \alpha_{k,k,n,q}| \cdot |B_{n,q}^{(p)}(e_k)(z)|$$

$$\leq \sum_{j=1}^{k-1} \alpha_{j,k,n,q} r^j + |1 - \alpha_{k,k,n,q}| r^k \leq 2|1 - \alpha_{k,k,n,q}| r^k.$$

But

$$B_{n,q}^{(m)}(e_k)(z) - e_k(z) = \sum_{p=0}^{m-1} B_{n,q}^{(p)}[B_{n,q}(e_k)(z) - e_k(z)],$$

which implies, for all $|z| \leq r$

$$|B_{n,q}^{(m)}(e_k) - e_k| \leq \sum_{p=0}^{m-1} |B_{n,q}^{(p)}[B_{n,q}(e_k) - e_k](z)| \leq 2m|1 - \alpha_{k,k,n,q}| r^k.$$

Since by the same Lemma 3 in Ostrovska [146], we have

$$\alpha_{k,k,n,q} = \left(1 - \frac{1}{[n]_q}\right) \cdots \left(1 - \frac{[k-1]_q}{[n]_q}\right),$$

by using the inequality

$$1 - \prod_{j=1}^{k} x_j \leq \sum_{j=1}^{k}(1 - x_j), 0 \leq x_j \leq 1, j = 1, ..., k,$$

it follows

$$|1 - \alpha_{k,k,n,q}| \leq \sum_{j=1}^{k-1} \frac{[j]_q}{[n]_q} = \frac{1}{[n]_q} \sum_{j=1}^{k-1}[j]_q$$

$$= \frac{1}{[n]_q} \sum_{j=0}^{k-2} q^j[k - (j+1)] \leq \frac{1}{[n]_q} \sum_{j=0}^{k-2}[k - (j+1)]$$

$$= \frac{1}{[n]_q}[k(k-1) - k(k-1)/2] = \frac{k(k-1)}{2[n]_q}.$$

Therefore

$$|B_{n,q}^{(m)}(e_k) - e_k| \leq \frac{m}{[n]_q} k(k-1) r^k.$$

Case 2). For $1 \leq r < R$, $|z| \leq r$ and $k > n \geq 1$, we get

$$|B_{n,q}^{(m)}(e_k)(z) - e_k(z)| \leq |B_{n,q}^{(m)}(e_k)(z)| + |e_k(z)|$$

$$\leq 2r^k \leq 2\frac{k(k-1)}{n}r^k \leq 2\frac{k(k-1)}{[n]_q}r^k,$$

since for $q \in (0, 1]$ we have $[n]_q \leq n$.

As a conclusion, for both Cases 1) and 2), for $|z| \leq r$ we obtain

$$|B_{n,q}^{(m)}(f)(z) - f(z)| \leq \sum_{k=1}^{\infty} |c_k| \cdot |B_{n,q}^{(m)}(e_k)(z) - e_k(z)|$$

$$= \sum_{k=1}^{n} |c_k| \cdot |B_{n,q}^{(m)}(e_k)(z) - e_k(z)|$$

$$+ \sum_{k=n+1}^{\infty} |c_k| \cdot |B_{n,q}^{(m)}(e_k)(z) - e_k(z)|$$

$$\leq \frac{2m}{[n]_q} \sum_{k=2}^{\infty} |c_k| k(k-1) r^k.$$

Now, by choosing $r = r_s$, we get the first required inequality in the statement. Note that $\sum_{k=2}^{\infty} |c_k| k(k-1) r^{k-2} < \infty$, since we have $f''(z) = \sum_{k=2}^{\infty} c_k k(k-1) z^{k-2}$, for all $|z| \leq r$.

The second estimate in (ii) is a direct consequence of the inequality

$$d(f, g) = \sum_{j=1}^{s} \frac{1}{2^j} \cdot \frac{\|f - g\|_j}{1 + \|f - g\|_j} + \sum_{j=s+1}^{\infty} \frac{1}{2^j} \cdot \frac{\|f - g\|_j}{1 + \|f - g\|_j}$$

$$\leq \frac{\|f - g\|_s}{1 + \|f - g\|_s} \sum_{j=1}^{s} \frac{1}{2^j} + \sum_{j=s+1}^{\infty} \frac{1}{2^j}$$

$$\leq \frac{\|f - g\|_s}{1 + \|f - g\|_s} + \frac{1}{2^s} \leq \|f - g\|_s + \frac{1}{2^s}.$$

This proves (ii).

(iii) The proof is similar with that of the point (i), by taking into account that from Theorem 10 in Ostrovska [146], for each $q \in (1, \infty)$, any polynomial P and for $\lim_{n \to \infty} \frac{m_n}{[n]_q} = 0$ we have $\lim_{n \to \infty} B_{n,q}^{(m_n)}(P)(x) = P(x)$, $x \in [0, 1]$.

(iv) Let us suppose that $q \in (1, \infty)$ and $|z| \leq r$, with $1 \leq r < R$. We reason exactly as in the proof of the point (ii). First, to estimate $|B_{n,q}^{(m)}(e_k)(z) - e_k(z)|$, again we have two possibilities : 1) $0 \leq k \leq n$; 2) $k > n \geq 1$.

Case 1). Reasoning exactly as at the proof of the above point (ii), Case 1), by simple calculation we obtain

$$|B_{n,q}^{(m)}(e_k)(z) - e_k(z)| \leq \frac{2m}{[n]_q} \sum_{j=0}^{k-2} q^j [k - (j+1)] r^k = \frac{2m}{[n]_q} \cdot \frac{q^k - 1 - k(q-1)}{(q-1)^2} r^k.$$

Since for $q \in (1, \infty)$ we have $n \leq [n]_q$, it follows

$$|B_{n,q}^{(m)}(e_k)(z) - e_k(z)| \leq \frac{2m}{n} \cdot \frac{q^k - 1 - k(q-1)}{(q-1)^2} r^k.$$

Case 2). Reasoning exactly as the proof of the above point (ii), Case 2, we get

$$|B_{n,q}^{(m)}(e_k)(z) - e_k(z)| \leq 2\frac{k(k-1)}{n} r^k.$$

As a conclusion, collecting the estimates in the Cases 1) and 2), we get

$$|B_{n,q}^{(m)}(f)(z) - f(z)| \leq \sum_{k=1}^{\infty} |c_k| \cdot |B_{n,q}^{(m)}(e_k)(z) - e_k(z)|$$

$$= \sum_{k=2}^{n} |c_k| \cdot |B_{n,q}^{(m)}(e_k)(z) - e_k(z)|$$

$$+ \sum_{k=n+1}^{\infty} |c_k| \cdot |B_{n,q}^{(m)}(e_k)(z) - e_k(z)|$$

$$\leq \frac{2m}{n} \sum_{k=2}^{n} |c_k| \cdot \frac{q^k - 1 - k(q-1)}{(q-1)^2} r^k$$

$$+ \frac{2}{n} \sum_{k=n+1}^{\infty} |c_k| \cdot k(k-1) r^k$$

$$\leq \frac{2m}{n} \sum_{k=2}^{\infty} |c_k| \left[\frac{q^k - 1 - k(q-1)}{(q-1)^2} + k(k-1) \right] r^k.$$

If, in addition we take $r = 1$ and $q < R$, then obviously $\sum_{k=2}^{\infty} |c_k| q^k < \infty$ and $\sum_{k=2}^{\infty} |c_k| k(k-1) < \infty$, which for $\frac{m_n}{n} \to 0$ as $n \to \infty$, by the above inequality implies that $B_{n,q}^{(m_n)}(f)(z) \to f(z)$, uniformly in $\overline{\mathbb{D}_1}$.

(v) Let $q \in (1, \infty)$ and $1 \leq r < R$ be arbitrary. Taking into account that for any polynomial P, by Theorem 10 in Ostrovska [146], for $\lim_{n\to\infty} \frac{m_n}{[n]_q} = \infty$ as $n \to \infty$ we have $B_{n,q}^{(m_n)}(P)(x) \to L_1(P)(x)$, uniformly for $x \in [0,1]$ and since by the above point (i), it is immediate that $B_{n,q}^{(m_n)}(P)(z)$, $n = 1, 2, ...$, is uniformly bounded in $\overline{\mathbb{D}_r}$ with respect to $n \in \mathbb{N}$, by Vitali's theorem it follows the uniform convergence on $\overline{\mathbb{D}_r}$ to $L_1(P)(z)$ of $B_{n,q}^{(m_n)}(P)(z)$ (as $n \to \infty$). Because r is arbitrary, it follows the convergence in d too (see the considerations at the beginning of this section). Note that for the Vitali's convergence result, the uniform convergence on $[0,1]$ is not necessary, it suffices to have only pointwise convergence there. The theorem is proved. $\square$

Finally we present the geometric properties for the complex q-Bernstein polynomials $B_{n,q_n}(f)(z)$, with $0 < q_n < 1$, and $q_n \to 1$ as $n \to \infty$.

Theorem 1.5.7. (Gal [87]) *Let us suppose that $G \subset \mathbb{C}$ is open, such that $\overline{\mathbb{D}_1} \subset G$ and $f : G \to \mathbb{C}$ is analytic in G. Also, let us consider $B_{n,q_n}(f)(z)$, with $0 < q_n < 1$, and $q_n \to 1$ as $n \to \infty$.*

If $f(0) = f'(0) - 1 = 0$ and f is starlike (convex, spirallike of type η, respectively) in $\overline{\mathbb{D}}_1$, that is for all $z \in \overline{\mathbb{D}}_1$ (see e.g. Mocanu-Bulboacă-Sălăgean [138])

$$Re\left(\frac{zf'(z)}{f(z)}\right) > 0 \left(Re\left(\frac{zf''(z)}{f'(z)}\right) + 1 > 0, Re\left(e^{i\eta}\frac{zf'(z)}{f(z)}\right) > 0, resp.\right),$$

then there exists an index n_0 depending on f (and on η for spirallikeness), such that for all $n \geq n_0$, $B_{n,q_n}(f)(z)$, are starlike (convex, spirallike of type η, respectively) in $\overline{\mathbb{D}}_1$.

If $f(0) = f'(0) - 1 = 0$ and f is starlike (convex, spirallike of type η, respectively) only in $\mathbb{D}_1$ (that is the corresponding inequalities hold only in $\mathbb{D}_1$), then for any disk of radius $0 < r < 1$ and center 0 denoted by $\mathbb{D}_r$, there exists an index $n_0 = n_0(f, \mathbb{D}_r)$ (n_0 depends on η too in the case of spirallikeness), such that for all $n \geq n_0$, $B_{n,q_n}(f)(z)$, are starlike (convex, spirallike of type η, respectively) in $\overline{\mathbb{D}}_r$ (that is, the corresponding inequalities hold in $\overline{\mathbb{D}}_r$).

Proof. By Theorem 2 in Phillips [149] and by the classical Vitali's theorem, it follows that we have $B_{n,q_n}(f)(z) \to f(z)$, uniformly for $|z| \leq 1$, which by the well-known Weierstrass's theorem implies $[B_{n,q_n}(f)]'(z) \to f'(z)$ and $[B_{n,q_n}(f)]''(z) \to f''(z)$, for $n \to \infty$, uniformly in $\overline{\mathbb{D}}_1$. In all what follows, denote $P_n(f)(z) = \frac{B_{n,q_n}(f)(z)}{[B_{n,q_n}(f)]'(0)}$, well defined for sufficiently large n. We easily get $P_n(f)(0) = 0$, $P'_n(f)(0) = 1$ for sufficiently large n, and $P_n(f)(z) \to f(z)$, $P'_n(f)(z) \to f'(z)$ and $P''_n(f)(z) \to f''(z)$, uniformly in $\overline{\mathbb{D}}_1$.

Suppose first that f is starlike in $\overline{\mathbb{D}}_1$. Then, by hypothesis we get $|f(z)| > 0$ for all $z \in \overline{\mathbb{D}}_1$ with $z \neq 0$, which from the univalence of f in $\mathbb{D}_1$, implies that we can write $f(z) = zg(z)$, with $g(z) \neq 0$, for all $z \in \overline{\mathbb{D}_1}$, where g is analytic in $\mathbb{D}_1$ and continuous in $\overline{\mathbb{D}}_1$.

Writing $P_n(f)(z)$ in the form $P_n(f)(z) = zQ_n(f)(z)$, obviously $Q_n(f)(z)$ is a polynomial of degree $\leq n - 1$. Also, for $|z| = 1$ we have

$$|f(z) - P_n(f)(z)| = |z| \cdot |g(z) - Q_n(f)(z)| = |g(z) - Q_n(f)(z)|,$$

which by the uniform convergence in $\overline{\mathbb{D}}_1$ of $P_n(f)$ to f and by the maximum modulus principle, implies the uniform convergence in $\overline{\mathbb{D}}_1$ of $Q_n(f)(z)$ to $g(z)$.

Since g is continuous in $\overline{\mathbb{D}}_1$ and $|g(z)| > 0$ for all $z \in \overline{\mathbb{D}}_1$, there exist an index $n_1 \in \mathbb{N}$ and $a > 0$ depending on g, such that $|Q_n(f)(z)| > a > 0$, for all $z \in \overline{\mathbb{D}}_1$ and all $n \geq n_0$. Also, for all $|z| = 1$, we have

$$\begin{aligned}|f'(z) - P'_n(f)(z)| &= |z[g'(z) - Q'_n(f)(z)] + [g(z) - Q_n(f)(z)]| \\ &\geq | \ \ |z| \cdot |g'(z) - Q'_n(f)(z)| - |g(z) - Q_n(f)(z)| \ \ | \\ &= | \ \ |g'(z) - Q'_n(f)(z)| - |g(z) - Q_n(f)(z)| \ \ |,\end{aligned}$$

which from the maximum modulus principle, the uniform convergence of $P'_n(f)$ to f' and of $Q_n(f)$ to g, evidently implies the uniform convergence of $Q'_n(f)$ to g'.

Then, for $|z| = 1$, we get

$$
\begin{aligned}
\frac{zP_n'(f)(z)}{P_n(f)} &= \frac{z[zQ_n'(f)(z) + Q_n(f)(z)]}{zQ_n(f)(z)} \\
&= \frac{zQ_n'(f)(z) + Q_n(f)(z)}{Q_n(f)(z)} \to \frac{zg'(z) + g(z)}{g(z)} \\
&= \frac{f'(z)}{g(z)} = \frac{zf'(z)}{f(z)},
\end{aligned}
$$

which again from the maximum modulus principle, implies

$$
\frac{zP_n'(f)(z)}{P_n(f)} \to \frac{zf'(z)}{f(z)}, \text{ uniformly in } \overline{\mathbb{D}}_1.
$$

Since $Re\left(\frac{zf'(z)}{f(z)}\right)$ is continuous in $\overline{\mathbb{D}}_1$, there exists $\varepsilon \in (0,1)$, such that

$$
Re\left(\frac{zf'(z)}{f(z)}\right) \geq \varepsilon, \text{ for all } z \in \overline{\mathbb{D}}_1.
$$

Therefore

$$
Re\left[\frac{zP_n'(f)(z)}{P_n(f)(z)}\right] \to Re\left[\frac{zf'(z)}{f(z)}\right] \geq \varepsilon > 0
$$

uniformly on $\overline{\mathbb{D}}_1$, i.e. for any $0 < \rho < \varepsilon$, there is n_0 such that for all $n \geq n_0$ we have

$$
Re\left[\frac{zP_n'(f)(z)}{P_n(f)(z)}\right] > \rho > 0, \text{ for all } z \in \overline{\mathbb{D}}_1.
$$

Since $P_n(f)(z)$ differs from $B_{n,q_n}(f)(z)$ only by a constant, this proves the starlikeness of $B_{n,q_n}(f)(z)$, for sufficiently large n.

If f is supposed to be starlike only in $\mathbb{D}_1$, the proof is identical, with the only difference that instead of $\overline{\mathbb{D}}_1$, we reason for $\overline{\mathbb{D}}_r$.

The proofs in the cases when f is convex or spirallike of order η are similar and follow from the following uniform convergency (on $\overline{\mathbb{D}}_1$ or on $\overline{\mathbb{D}}_r$)

$$
Re\left[\frac{zP_n''(f)(z)}{P_n'(f)(z)}\right] + 1 \to Re\left[\frac{zf''(z)}{f'(z)}\right] + 1
$$

and

$$
Re\left[e^{i\eta}\frac{zP_n'(f)(z)}{P_n(f)(z)}\right] \to Re\left[e^{i\eta}\frac{zf'(z)}{f(z)}\right].
$$

The theorem is proved. $\qquad\square$

1.6 Bernstein-Stancu Polynomials

In this section for two kinds of complex Bernstein-Stancu polynomials we study similar properties with those for the classical complex Bernstein polynomials and

q-Bernstein polynomials in Sections 1.1, 1.2 and 1.5. More exactly we consider the following two kinds of polynomials :

$$S_n^{(\alpha,\beta)}(f)(z) = \sum_{k=0}^{n} \binom{n}{k} z^k (1-z)^{n-k} f[(k+\alpha)/(n+\beta)], z \in \mathbb{C},$$

where $0 \le \alpha \le \beta$ are independent of n, (introduced and studied for the case of real variable in Stancu [173]) and

$$S_n^{<\gamma>}(f)(z) = \sum_{k=0}^{n} p_{n,k}^{<\gamma>}(z) f(k/n), z \in \mathbb{C},$$

(introduced and studied for the case of real variable in Stancu [174]) where $\gamma \ge 0$ may to depend on n and

$$p_{n,k}^{<\gamma>}(z)$$
$$= \binom{n}{k} \frac{z(z+\gamma)...(z+(k-1)\gamma)(1-z)(1-z+\gamma)...(1-z+(n-k-1)\gamma)}{(1+\gamma)(1+2\gamma)...(1+(n-1)\gamma)}.$$

Note that $S_n^{(0,0)}(f)(z) = S_n^{<0>}(f)(z) = B_n(f)(z)$.

We begin our study with $S_n^{(\alpha,\beta)}(f)(z)$. First we present upper estimates in simultaneous approximation.

Theorem 1.6.1. (Gal [80]) *Let $\mathbb{D}_R = \{z \in \mathbb{C}; |z| < R\}$ be with $R > 1$ and let us suppose that $f : \mathbb{D}_R \to \mathbb{C}$ is analytic in $\mathbb{D}_R$, i.e. $f(z) = \sum_{k=0}^{\infty} c_k z^k$, for all $z \in \mathbb{D}_R$. Suppose $0 \le \alpha \le \beta$ and $1 \le r < R$ are arbitrary fixed. For all $|z| \le r$ and $n \in \mathbb{N}$, we have*

$$|S_n^{(\alpha,\beta)}(f)(z) - f(z)| \le \frac{M_{2,r}^{(\beta)}(f)}{n+\beta},$$

where $0 < M_{2,r}^{(\beta)}(f) = 2r^2 \sum_{j=2}^{\infty} j(j-1)|c_j|r^{j-2} + 2\beta r \sum_{j=1}^{\infty} j|c_j|r^{j-1} < \infty$.

Also, if $1 \le r < r_1 < R$, then for all $|z| \le r$ and $n, p \in \mathbb{N}$, we have

$$|[S_n^{(\alpha,\beta)}(f)]^{(p)}(z) - f^{(p)}(z)| \le \frac{M_{2,r_1}^{(\beta)}(f)p! r_1}{(n+\beta)(r_1-r)^{p+1}}.$$

Proof.   Denoting $e_k(z) = z^k$, we get $S_n^{(\alpha,\beta)}(f)(z) = \sum_{k=0}^{\infty} c_k S_n^{(\alpha,\beta)}(e_k)(z)$ and

$$|S_n^{(\alpha,\beta)}(f)(z) - f(z)| \le \sum_{k=0}^{\infty} |c_k| \cdot |S_n^{(\alpha,\beta)}(e_k)(z) - e_k(z)|.$$

To estimate $|S_n^{(\alpha,\beta)}(e_k)(z) - e_k(z)|$ for fixed $n \in \mathbb{N}$, we consider two possible cases :
1) $0 \le k \le n$; 2) $k > n$.

Denoting by Δ^k the finite difference of order k, we will use the representation formula (see Stancu [173])

$$S_n^{(\alpha,\beta)}(f)(z) = \sum_{p=0}^{n} \binom{n}{p} \Delta_{1/(n+\beta)}^p f(\alpha/(n+\beta)) e_p(z).$$

Case 1). If $k = 0$, then obviously we have $S_n^{(\alpha,\beta)}(e_k)(z) - e_k(z) = 0$. Therefore, let us suppose that $1 \leq k \leq n$ and denote

$$C_{n,p,k}^{(\alpha,\beta)} = \binom{n}{p} \Delta_{1/(n+\beta)}^p e_k(\alpha/(n+\beta))$$

$$= \binom{n}{p}[\alpha/(n+\beta), (\alpha+1)/(n+\beta), ..., (\alpha+p)/(n+\beta); e_k](p!)/(n+\beta)^p.$$

Since e_k is convex of any order, it follows $C_{n,p,k} \geq 0$ and since $S_n^{(\alpha,\beta)}(f)(1) = f[(n+\alpha)/(n+\beta)]$, we get $\sum_{p=0}^{n} C_{n,p,k}^{(\alpha,\beta)} = \frac{(n+\alpha)^k}{(n+\beta)^k} \leq 1$.

For any $|z| \leq r$ with $1 \leq r < R$, we can write

$$|S_n^{(\alpha,\beta)}(e_k)(z) - e_k(z)|$$

$$= |\sum_{p=0}^{k} C_{n,p,k}^{(\alpha,\beta)} e_p(z) - e_k(z)| = |[C_{n,k,k}^{(\alpha,\beta)} - 1]e_k(z) + \sum_{p=0}^{k-1} C_{n,p,k}^{(\alpha,\beta)} e_p(z)|$$

$$\leq \left[1 - \frac{n(n-1)...(n-(k-1))}{(n+\beta)^k}\right] r^k$$

$$+ \left[\frac{(n+\alpha)^k}{(n+\beta)^k} - \frac{n(n-1)...(n-(k-1))}{(n+\beta)^k}\right] r^k$$

$$= 2\left[1 - \frac{n(n-1)...(n-(k-1))}{(n+\beta)^k}\right] r^k + \left[\frac{(n+\alpha)^k}{(n+\beta)^k} - 1\right] r^k$$

$$\leq 2\left[1 - \frac{n(n-1)...(n-(k-1))}{(n+\beta)^k}\right] r^k \leq \frac{1}{n+\beta}[2\beta k + k(k-1)] r^k.$$

Here we have applied the formula (easily proved by mathematical induction)

$$1 - \Pi_{j=1}^{k} x_j \leq \sum_{j=1}^{k}(1 - x_j), 0 \leq x_j \leq 1, j = 1, ..., k.$$

Case 2). For $1 \leq r < R$, $|z| \leq r$ and $k > n \geq 1$, we get

$$|S_n^{(\alpha,\beta)}(e_k)(z) - e_k(z)| \leq |S_n^{(\alpha,\beta)}(e_k)(z)| + r^k$$

$$\leq \sum_{p=0}^{n} C_{n,p,k}^{(\alpha,\beta)} r^p + r^k \leq r^n + r^k \leq 2r^k$$

$$\leq 2nr^k \leq 2(k-1) \cdot \frac{k+\beta}{n+\beta} r^k$$

$$= \frac{2k(k-1) + 2\beta(k-1)}{n+\beta} r^k \leq \frac{2k(k-1) + 2\beta k}{n+\beta} r^k.$$

Combining it with the above Case 1, we get the desired inequality.

For the simultaneous approximation, denoting by Γ the circle of radius $r_1 > r$ and center 0, since for any $|z| \leq r$ and $v \in \Gamma$, we have $|v - z| \geq r_1 - r$, by the

Cauchy's formulas it follows that for all $|z| \le r$ and $n \in \mathbb{N}$, we have

$$|[S_n^{(\alpha,\beta)}(f)]^{(p)}(z) - f^{(p)}(z)| = \frac{p!}{2\pi} \left| \int_\Gamma \frac{S_n^{(\alpha,\beta)}(f)(v) - f(v)}{(v-z)^{p+1}} dv \right|$$

$$\le \frac{M_{2,r_1}^{(\beta)}(f)}{n+\beta} \frac{p!}{2\pi} \frac{2\pi r_1}{(r_1-r)^{p+1}}$$

$$= \frac{M_{2,r_1}^{(\beta)}(f)}{n+\beta} \frac{p! r_1}{(r_1-r)^{p+1}}.$$

Finally, since by hypothesis, $f(z) = \sum_k^\infty c_k z^k$ is absolutely and uniformly convergent in $|z| \le r$, for any $1 \le r < R$, it is clear that $M_{2,r}^{(\beta)}(f) < \infty$. $\square$

Remark. For $\alpha = \beta = 0$ in Theorem 1.6.1 the estimates in Theorem 1.1.2 for classical Bernstein polynomials are obtained.

A quantitative Voronovskaja-type formula follows.

Theorem 1.6.2. (Gal [80]) *Let $\mathbb{D}_R = \{z \in \mathbb{C}; |z| < R\}$ be with $R > 1$ and let us suppose that $f : \mathbb{D}_R \to \mathbb{C}$ is analytic in $\mathbb{D}_R$, that is we can write $f(z) = \sum_{k=0}^\infty c_k z^k$, for all $z \in \mathbb{D}_R$. Let $0 \le \alpha \le \beta$. For all $|z| \le 1$ and $n \in \mathbb{N}$, we have*

$$\left| S_n^{(\alpha,\beta)}(f)(z) - f(z) + \frac{\beta z - \alpha}{n+\beta} f'(z) - \frac{nz(1-z)}{2(n+\beta)^2} f''(z) \right|$$

$$\le \frac{|z| \cdot |1-z|}{(n+\beta)^2} M_1^{(\alpha,\beta)}(f) + \frac{M_2^{(\alpha,\beta)}(f)}{(n+\beta)^2},$$

where $0 < M_1^{(\alpha,\beta)}(f)$,

$$M_1^{(\alpha,\beta)}(f) = \sum_{k=2}^\infty |c_k| \left[\frac{9(k-1)^3(k-2)}{2} + \frac{(k-1)^2(k-2)^2}{2} + 4\beta(k-1)^3 \right.$$

$$\left. + \frac{3\beta(k-1)^2(k-2)}{2} + \frac{3\alpha(k-1)^2(k-2)}{2} + \beta k(k-1)^2(k-2) \right] < \infty,$$

$$0 < M_2^{(\alpha,\beta)}(f) = (\alpha+\beta)^2 \sum_{k=2}^\infty |c_k| \frac{k(k-1)}{2} < \infty.$$

Proof. Denoting $e_k(z) = z^k$ and $\pi_{n,k}(z) = S_n^{(\alpha,\beta)}(e_k)(z)$, we obtain

$$\left| S_n^{(\alpha,\beta)}(f)(z) - f(z) + \frac{\beta z - \alpha}{n+\beta} f'(z) - \frac{nz(1-z)}{2(n+\beta)^2} f''(z) \right|$$

$$\le \sum_{k=1}^\infty |c_k| \left| \pi_{n,k}(z) - e_k(z) + \frac{z^{k-1}(\beta z - \alpha)k}{n+\beta} - \frac{nz^{k-1}(1-z)k(k-1)}{2(n+\beta)^2} \right|.$$

Differentiating the sum $s_{n,k}(z) = \sum_{j=0}^{n}(j+\alpha)^k\binom{n}{j}z^j(1-z)^{n-j}$ and then dividing the formula by $(n+\beta)^{k+1}$, by simple calculation we get the recurrence formula

$$\pi_{n,k+1}(z) = \frac{z(1-z)}{n+\beta}\pi'_{n,k}(z) + \frac{\alpha+nz}{n+\beta}\pi_{n,k}(z), z \in \mathbb{C}.$$

Denoting $G_{n,k}(z) = \pi_{n,k}(z) - e_k(z) + \frac{z^{k-1}(\beta z - \alpha)k}{n+\beta} - \frac{nz^{k-1}(1-z)k(k-1)}{2(n+\beta)^2}$, the above recurrence for $\pi_{n,k}(z)$ implies

$$G_{n,k}(z) = \frac{z(1-z)}{n+\beta}\pi'_{n,k-1}(z) + \frac{\alpha+nz}{n+\beta}\pi_{n,k-1}(z) - e_k(z)$$
$$+ \frac{z^{k-1}(\beta z - \alpha)k}{n+\beta} - \frac{nz^{k-1}(1-z)k(k-1)}{2(n+\beta)^2},$$

which by simple calculation implies the following recurrence formula for $G_{n,k}(z)$ (valid for all $k \geq 2$ since $G_{n,0}(z) = G_{n,1}(z) = 0$)

$$G_{n,k}(z) = \frac{z(1-z)}{n+\beta}G'_{n,k-1}(z) + \frac{\alpha+nz}{n+\beta}G_{n,k-1}(z) + A,$$

where

$$A := \frac{z^{k-1}(1-z)(k-1)}{n+\beta} - \frac{z^{k-2}(1-z)(k-1)[(k-2)(\beta z - \alpha)+\beta z]}{(n+\beta)^2}$$
$$+ \frac{z^{k-2}(1-z)(k-1)(k-2)[(k-2)-(k-1)z]}{2(n+\beta)}\left(\frac{n+\beta}{(n+\beta)^2} - \frac{\beta}{(n+\beta)^2}\right)$$
$$+ \frac{\alpha+nz}{n+\beta}z^{k-1} - \frac{\alpha+nz}{(n+\beta)^2}z^{k-2}(\beta z - \alpha)(k-1)$$
$$+ \frac{\alpha+nz}{2(n+\beta)}z^{k-2}(1-z)(k-1)(k-2)\left[\frac{n+\beta}{(n+\beta)^2} - \frac{\beta}{(n+\beta)^2}\right] - \frac{n+\beta}{n+\beta}z^k$$
$$+ \frac{z^{k-1}(\beta z - \alpha)k}{n+\beta} - \frac{z^{k-1}(1-z)k(k-1)}{2(n+\beta)} + \frac{\beta z^{k-1}(1-z)k(k-1)}{2(n+\beta)^2}.$$

In what follows, we will write the expression A in the form $A := T_1 + T_2 + T_3$, where T_1 is the sum of all the terms containing $(n+\beta)$ at the denominator, T_2 is the sum of all the terms containing $(n+\beta)^2$ at the denominator and T_3 is the sum of all the terms containing $(n+\beta)^3$ at the denominator. Therefore, by writing (for T_2) $\alpha + nz = (\alpha - \beta z) + (n+\beta)z$, we obtain

$$T_1 = \frac{z^{k-1}}{n+\beta}[(1-z)(k-1)+(\alpha+nz)-z(n+\beta)+k(\beta z - \alpha)$$
$$- k(k-1)(1-z)/2] = \frac{z^{k-1}(k-1)}{n+\beta}[z(k/2+\beta-1)+1-\alpha-k/2],$$

$$T_2$$
$$= \frac{z^{k-2}(k-1)}{(n+\beta)^2}\left[\frac{(1-z)(k-2)z(n+\beta)}{2} + \frac{(1-z)(k-2)z(-2\beta-(k-1))}{2}\right.$$

$$\left. + \frac{(1-z)(k-2)[3\alpha + (k-2)]}{2} - (\beta z - \alpha)(n+\beta)z + (\beta z - \alpha)^2 \right]$$

$$= -T_1 + \frac{z^{k-2}(k-1)}{2(n+\beta)^2}\big[(1-z)z(k-2)[-2\beta - (k-1)]$$

$$+ (1-z)(k-2)[3\alpha + (k-2)] + 2(\beta z - \alpha)^2\big]$$

$$= -T_1 + \frac{z^{k-1}(1-z)(k-1)(k-2)}{2(n+\beta)^2}[-2\beta - (k-1)]$$

$$+ \frac{z^{k-2}(1-z)(k-1)(k-2)[3\alpha + (k-2)]}{2(n+\beta)^2} + \frac{z^{k-2}(k-1)}{(n+\beta)^2}(\beta z - \alpha)^2,$$

$$T_3 = -\frac{\beta z^{k-2}(1-z)(k-1)(k-2)[(k-2) - (k-1)z]}{2(n+\beta)^3}$$

$$- \frac{\beta(\alpha + nz)z^{k-2}(1-z)(k-1)(k-2)}{2(n+\beta)^2}$$

$$= -\frac{\beta z^{k-2}(1-z)(k-1)(k-2)}{2(n+\beta)^3}[z(n-(k-1)) + \alpha + (k-2)].$$

First, we observe that in the sum $T_1 + T_2 + T_3$, the terms containing $(n+\beta)$ at the denominator cancel. Now, we will estimate $|A| = |T_1 + T_2 + T_3|$ for $|z| \le 1$.

For all $k \ge 2$ we obtain

$$|A| \le \frac{|z| \cdot |1-z|(k-1)(k-2)[2\beta + (k-1)]}{2(n+\beta)^2}$$

$$+ \frac{|z| \cdot |1-z|(k-1)(k-2)[3\alpha + (k-2)]}{2(n+\beta)^2} + \frac{(k-1)(\beta + \alpha)^2}{(n+\beta)^2}$$

$$+ \frac{|z| \cdot |1-z|\beta(k-1)(k-2)[n + \alpha + 2k - 3]}{2(n+\beta)^3}$$

$$= |z| \cdot |1-z| \left\{ \frac{(k-1)(k-2)[2\beta + (k-1)]}{2(n+\beta)^2} + \frac{(k-1)(k-2)[3\alpha + k - 2]}{2(n+\beta)^2} \right.$$

$$\left. + \frac{\beta(k-1)(k-2)[n + \alpha + 2k - 3]}{2(n+\beta)^3} \right\} + \frac{(k-1)(\beta + \alpha)^2}{(n+\beta)^2}$$

$$\le \frac{|z| \cdot |1-z|}{(n+\beta)^2} \left\{ (k-1)^2(k-2)/2 + 2\beta(k-1)(k-2)/2 \right.$$

$$+ (k-1)\left[(k-2)^2/2 + \frac{3\alpha(k-2)}{2}\right] + \frac{\beta(k-1)(k-2)}{2}$$

$$\left. + \frac{\beta(\alpha - \beta)(k-1)(k-2)}{2(n+\beta)} + \frac{2\beta k(k-1)(k-2)}{2(n+\beta)} \right\} + \frac{(k-1)(\beta + \alpha)^2}{(n+\beta)^2}$$

$$\leq \frac{|z| \cdot |1-z|}{(n+\beta)^2} \left\{ (k-1)^2(k-2)/2 + \beta(k-1)(k-2) + (k-1)(k-2)^2/2 \right.$$

$$\left. + 3\alpha(k-1)(k-2)/2 + \frac{\beta(k-1)(k-2)}{2} + \beta k(k-1)(k-2) \right\}$$

$$+ \frac{(k-1)(\alpha+\beta)^2}{(n+\beta)^2} = \frac{|z| \cdot |1-z|}{(n+\beta)^2} \left\{ (k-1)^2(k-2)/2 + \frac{3\beta(k-1)(k-2)}{2} \right.$$

$$\left. + (k-1)(k-2)^2/2 + 3\alpha(k-1)(k-2)/2 + \beta k(k-1)(k-2) \right\}$$

$$+ \frac{(k-1)(\alpha+\beta)^2}{(n+\beta)^2}.$$

Therefore, denoting by $\|\cdot\|$ the uniform norm in the closed unit disk and estimating for $|z| \leq 1$ the absolute value of $G_{n,k}(z)$ in the formula of recurrence, also by using the Bernstein's inequality (since $G_{n,k}(z)$ is a polynomial of degree k) and the estimate for $\|\pi_{n,k} - e_k\|$ in the proof of Theorem 1.6.1, we obtain

$$|G_{n,k}(z)| \leq \frac{|z| \cdot |1-z|}{n+\beta} \|G'_{n,k-1}\| + \frac{n+\alpha}{n+\beta} |G_{n,k-1}(z)| + |A|$$

$$\leq |G_{n,k-1}(z)| + \frac{|z| \cdot |1-z|(k-1)}{n+\beta} \|G_{n,k-1}\| + |A|$$

$$\leq |G_{n,k-1}(z)| + \frac{|z| \cdot |1-z|(k-1)}{n+\beta} [\|\pi_{n,k-1} - e_{k-1}\|$$

$$+ \frac{(k-1)(\alpha+\beta)}{n+\beta} + \frac{n(k-1)(k-2)}{(n+\beta)^2}] + |A| \leq |G_{n,k-1}(z)|$$

$$+ \frac{|z| \cdot |1-z|(k-1)}{n+\beta} \left[\frac{2(k-1)(k-2) + 2\beta(k-1)}{n+\beta} \right.$$

$$+ \frac{(k-1)(\alpha+\beta)}{n+\beta} + \frac{(k-1)(k-2)}{n+\beta} \Big] + |A| \leq |G_{n,k-1}(z)|$$

$$+ \frac{|z| \cdot |1-z|(k-1)^2}{n+\beta} \cdot \frac{2(k-2) + 2\beta + (\alpha+\beta) + (k-2)}{n+\beta}$$

$$+ |A| \leq |G_{n,k-1}(z)| + \frac{4|z| \cdot |1-z|(k-1)^2}{(n+\beta)^2}(k-2+\beta) + |A|$$

$$\leq |G_{n,k-1}(z)| + \frac{4|z| \cdot |1-z|(k-1)^2}{(n+\beta)^2}(k-2+\beta)$$

$$+ \frac{|z| \cdot |1-z|}{(n+\beta)^2} \left\{ (k-1)^2(k-2)/2 + \frac{3\beta(k-1)(k-2)}{2} \right.$$

$$\left. + (k-1)(k-2)^2/2 + 3\alpha(k-1)(k-2)/2 + \beta k(k-1)(k-2) \right\}$$

$$+ \frac{(k-1)(\alpha+\beta)^2}{(n+\beta)^2} := |G_{n,k-1}(z)| + A(n,k,\alpha,\beta)(z).$$

That is, in conclusion for all $|z| \leq 1$ we can write

$$|G_{n,k}(z)| \leq |G_{n,k-1}(z)| + A(n,k,\alpha,\beta)(z).$$

Since $G_{n,1}(z) = 0$, reasoning by recurrence for $k = 2, 3, \ldots$, we immediately obtain

$$|G_{n,k}(z)|$$

$$\leq \frac{|z| \cdot |1-z|}{(n+\beta)^2} \sum_{j=2}^{k} \Big[9(j-1)^2(j-2)/2 + 4\beta(j-1)^2 + (j-1)(j-2)^2/2$$

$$+ \frac{3\beta(j-1)(j-2)}{2} + 3\alpha(j-1)(j-2)/2 + \beta j(j-1)(j-2) \Big]$$

$$+ \frac{(\alpha+\beta)^2}{(n+\beta)^2} \sum_{j=2}^{k} (j-1)$$

$$\leq \frac{|z| \cdot |1-z|}{(n+\beta)^2} \Big[9(k-1)^3(k-2)/2 + (k-1)^2(k-2)^2/2 + 4\beta(k-1)^3$$

$$+ \frac{3\beta(k-1)^2(k-2)}{2} + 3\alpha(k-1)^2(k-2)/2 + \beta k(k-1)^2(k-2) \Big]$$

$$+ \frac{k(k-1)(\alpha+\beta)^2}{2(n+\beta)^2}.$$

As a conclusion, the desired estimate is immediate from

$$\left| S_n^{(\alpha,\beta)}(f)(z) - f(z) + \frac{\beta z - \alpha}{n+\beta} f'(z) - \frac{nz(1-z)}{2(n+\beta)^2} f''(z) \right| \leq \sum_{k=1}^{\infty} |c_k| \cdot |G_{n,k}(z)|. \qquad \square$$

Remarks. 1) Taking now $\alpha = \beta = 0$ in Theorem 1.6.2 we get the Voronovskaja's theorem with an upper estimate for the classical Bernstein polynomials in Theorem 1.1.3.

2) Following exactly the lines in the proof of Theorem 1.6.2 it is immediate that in fact for any $1 \leq r < R$ we have an upper estimate of the form

$$\left\| S_n^{(\alpha,\beta)}(f) - f + \frac{\beta e_1 - \alpha}{n+\beta} f' - \frac{ne_1(1-e_1)}{2(n+\beta)^2} f'' \right\|_r \leq \frac{M_r^{(\alpha,\beta)}(f)}{(n+\beta)^2},$$

where the constant $M_r^{(\alpha,\beta)}(f) > 0$ is independent of n and depends on f, r, α and β.

In what follows we prove that the degrees of approximation in Theorem 1.6.1 in fact are exact. Since the particular case $\alpha = \beta = 0$ (that is the case of classical Bernstein polynomials) was already considered by Theorem 1.1.4, Corollary 1.1.5 and Theorem 1.1.6, in the next Theorem 1.6.3, Corollary 1.6.4 and Theorem 1.6.5 it will be excluded.

First we present :

Theorem 1.6.3. (Gal [81]) *Let $R > 1$, $0 \leq \alpha \leq \beta$ with $\alpha + \beta > 0$, $\mathbb{D}_R = \{z \in \mathbb{C}; |z| < R\}$ and let us suppose that $f : \mathbb{D}_R \to \mathbb{C}$ is analytic in $\mathbb{D}_R$, that is we can write $f(z) = \sum_{k=0}^{\infty} c_k z^k$, for all $z \in \mathbb{D}_R$. If f is not a polynomial of degree 0 and $1 \leq r < R$, then we have*

$$\|S_n^{(\alpha,\beta)}(f) - f\|_r \geq \frac{C_r^{(\alpha,\beta)}(f)}{n+\beta}, n \in \mathbb{N},$$

where the constant $C_r^{(\alpha,\beta)}(f)$ depends only on f, r, α and β.

Proof. For all $z \in \mathbb{D}_R$ and $n \in \mathbb{N}$ we have

$$S_n^{(\alpha,\beta)}(f)(z) - f(z) = \frac{1}{n+\beta}\left\{-(\beta z - \alpha)f'(z) + \frac{z(1-z)}{2}f''(z)\right.$$

$$+\frac{1}{n+\beta}\left[(n+\beta)^2\left(S_n^{(\alpha,\beta)}(f)(z) - f(z) + \frac{\beta z - \alpha}{n+\beta}f'(z) - \frac{nz(1-z)}{2(n+\beta)^2}f''(z)\right)\right.$$

$$\left.\left.-\frac{\beta z(1-z)}{2}f''(z)\right]\right\}.$$

Note that in the case $\alpha = \beta = 0$ in Corollary 1.1.5, necessarily f was supposed to be not a polynomial of degree ≤ 1.

In what follows we will apply to the above identity the following obvious property :

$$\|F + G\|_r \geq |\,\|F\|_r - \|G\|_r\,| \geq \|F\|_r - \|G\|_r.$$

It follows

$$\|S_n^{(\alpha,\beta)}(f) - f\|_r \geq \frac{1}{n+\beta}\left\{\left\|-(\beta e_1 - \alpha)f' + \frac{e_1(1-e_1)}{2}f''\right\|_r\right.$$

$$-\frac{1}{n+\beta}\cdot\left[\left\|(n+\beta)^2\left(S_n^{(\alpha,\beta)}(f) - f + \frac{\beta e_1 - \alpha}{n+\beta}f' - \frac{ne_1(1-e_1)}{2(n+\beta)^2}f''\right)\right.\right.$$

$$\left.\left.\left.-\frac{\beta e_1(1-e_1)}{2}f''\right\|_r\right]\right\}.$$

Since by Remark 2 after the proof of Theorem 1.6.2 we have

$$\left\|(n+\beta)^2\left(S_n^{(\alpha,\beta)}(f) - f + \frac{\beta e_1 - \alpha}{n+\beta}f' - \frac{ne_1(1-e_1)}{2(n+\beta)^2}f''\right) - \frac{\beta e_1(1-e_1)}{2}f''\right\|_r$$

$$\leq M_r^{(\alpha,\beta)}(f) + \beta\|f''\|_r,$$

and denoting $H(z) = -(\beta z - \alpha)f'(z) + \frac{z(1-z)}{2}f''(z)$, if we prove that $\|H\|_r > 0$, then it is clear that there exists an index n_0 depending only on f, α and β, such that

$$\|S_n^{(\alpha,\beta)}(f) - f\|_r \geq \frac{1}{n+\beta}\cdot\frac{\|H\|_r}{2}, \forall n \geq n_0.$$

For $n \in \{1, 2, ..., n_0 - 1\}$ we have $\|S_n^{(\alpha,\beta)}(f) - f\|_r \geq \frac{A_{n,r}^{(\alpha,\beta)}(f)}{n+\beta}$ with $A_{n,r}^{(\alpha,\beta)}(f) = (n+\beta)\cdot\|S_n^{(\alpha,\beta)}(f) - f\|_r > 0$, which finally implies $\|S_n^{(\alpha,\beta)}(f) - f\|_r \geq \frac{C_r^{(\alpha,\beta)}(f)}{n+\beta}$ for all $n \in \mathbb{N}$, with $C_r^{(\alpha,\beta)}(f) = \min\left\{A_{1,r}^{(\alpha,\beta)}, A_{2,r}^{(\alpha,\beta)}(f), ..., A_{n_0-1,r}^{(\alpha,\beta)}(f), \frac{\|H\|_r}{2}\right\}$.

Therefore it remains to show that $\|H\|_r > 0$. Indeed, suppose that $\|H\|_r = 0$. We have two possibilities : 1) $0 = \alpha < \beta$ or 2) $0 < \alpha \leq \beta$.

Case 1). We obtain $H(z) = -\beta z f'(z) + \frac{z(1-z)}{2} f''(z) = 0$, for all $|z| \le r$ and denoting $y(z) = f'(z)$, it follows that $y(z)$ is an analytic function in $\mathbb{D}_R$, solution of the differential equation $-\beta z y(z) + \frac{z(1-z)}{2} y'(z) = 0, |z| \le r$, which after simplification with $z \ne 0$ becomes $-\beta y(z) + \frac{(1-z)}{2} y'(z) = 0, |z| \le r$. Now, seeking $y(z)$ in the form $y(z) = \sum_{k=0}^{\infty} b_k z^k$ and replacing it in the differential equation, by the identification of the coefficients we easily obtain $b_k = 0$ for all $k = 0, 1, ...,$. Therefore $y(z) = 0$ for all $|z| \le r$, which by the identity's theorem on analytic (holomorphic) functions implies $y(z) = 0$ for all $z \in \mathbb{D}_R$ and the contradiction that f is a polynomial of degree ≤ 0.

Case 2). Denoting $y(z) = f'(z)$ by hypothesis it follows that $y(z)$ is an analytic function in $\mathbb{D}_R$ solution of the differential equation $(-\beta z + \alpha) y(z) + \frac{z(1-z)}{2} y'(z) = 0, |z| \le r$.

Taking $z = 0$ it follows $\alpha y(0) = 0$, which means $y(0) = 0$. Seeking $y(z)$ in the form $y(z) = \sum_{k=1}^{\infty} b_k z^k$ and replacing it in the differential equation, by the identification of the coefficients we easily obtain $b_k = 0$ for all $k = 1, 2, ...,$, which finally leads to the contradiction that f is a constant. $\square$

Combining now Theorem 1.6.3 with Theorem 1.6.1 we immediately get the following.

Corollary 1.6.4. (Gal [81]) *Let $R > 1$, $0 \le \alpha \le \beta$ with $\alpha + \beta > 0$, $\mathbb{D}_R = \{z \in \mathbb{C}; |z| < R\}$ and let us suppose that $f : \mathbb{D}_R \to \mathbb{C}$ is analytic in $\mathbb{D}_R$. If f is not a polynomial of degree 0 and $1 \le r < R$, then we have*

$$\|S_n^{(\alpha,\beta)}(f) - f\|_r \sim \frac{1}{n+\beta}, n \in \mathbb{N},$$

where the constants in the equivalence depend on f, r, α and β.

In the case of simultaneous approximation we present :

Theorem 1.6.5. (Gal [81]) *Let $\mathbb{D}_R = \{z \in \mathbb{C}; |z| < R\}$ be with $R > 1$, $0 \le \alpha \le \beta$ with $\alpha + \beta > 0$ and let us suppose that $f : \mathbb{D}_R \to \mathbb{C}$ is analytic in $\mathbb{D}_R$, i.e. $f(z) = \sum_{k=0}^{\infty} c_k z^k$, for all $z \in \mathbb{D}_R$. Also, let $1 \le r < r_1 < R$ and $p \in \mathbb{N}$ be fixed. If f is not a polynomial of degree $\le p - 1$, then we have*

$$\|[S_n^{(\alpha,\beta)}(f)]^{(p)} - f^{(p)}\|_r \sim \frac{1}{n+\beta},$$

where the constants in the equivalence depend on f, α, β, r, r_1 and p.

Proof. Taking into account Theorem 1.6.1, it remains to prove the lower estimate for $\|[S_n^{(\alpha,\beta)}(f)]^{(p)} - f^{(p)}\|_r$ only. Denoting by Γ the circle of radius $r_1 > r$ (with $r \ge 1$) and center 0, by the Cauchy's formulas it follows that for all $|z| \le r$ and $n \in \mathbb{N}$ we have

$$[S_n^{(\alpha,\beta)}(f)]^{(p)}(z) - f^{(p)}(z) = \frac{p!}{2\pi i} \int_{\Gamma} \frac{S_n^{(\alpha,\beta)}(f)(v) - f(v)}{(v-z)^{p+1}} dv,$$

where we have the inequality $|v - z| \ge r_1 - r$ valid for all $|z| \le r$ and $v \in \Gamma$.

As in the proof of Theorem 1.6.3 (keeping the notation for H), for all $v \in \Gamma$ and $n \in \mathbb{N}$ we have

$$
S_n^{(\alpha,\beta)}(f)(v) - f(v) = \frac{1}{n+\beta} \left\{ H(v) \right.
$$

$$
+ \frac{1}{n+\beta} \left[(n+\beta)^2 \left(S_n^{(\alpha,\beta)}(f)(v) - f(v) + \frac{\beta v - \alpha}{n+\beta} f'(v) - \frac{nv(1-v)}{2(n+\beta)^2} f''(v) \right) \right.
$$

$$
\left. \left. - \frac{\beta v(1-v)}{2} f''(v) \right] \right\},
$$

which replaced in the above Cauchy's formula implies

$$
[S_n^{(\alpha,\beta)}(f)]^{(p)}(z) - f^{(p)}(z) = \frac{1}{n+\beta} \left\{ H^{(p)}(z) + \frac{1}{n+\beta} \cdot \right.
$$

$$
\left[\frac{p!}{2\pi i} \int_\Gamma \frac{(n+\beta)^2 \left(S_n^{(\alpha,\beta)}(f)(v) - f(v) + \frac{\beta v - \alpha}{n+\beta} f'(v) - \frac{nv(1-v)}{2(n+\beta)^2} f''(v) \right)}{(v-z)^{p+1}} dv \right.
$$

$$
\left. \left. - \frac{p!}{2\pi i} \int_\Gamma \frac{\beta v(1-v)}{2(v-z)^{p+1}} f''(v) dv \right] \right\}.
$$

Passing now to absolute value, for all $|z| \le r$ and $n \in \mathbb{N}$ it follows

$$
|[S_n^{(\alpha,\beta)}(f)]^{(p)}(z) - f^{(p)}(z)| \ge \frac{1}{n+\beta} \left\{ |H^{(p)}(z)| - \frac{1}{n+\beta} \cdot \right.
$$

$$
\left[\left| \frac{p!}{2\pi i} \int_\Gamma \frac{(n+\beta)^2 \left(S_n^{(\alpha,\beta)}(f)(v) - f(v) + \frac{\beta v - \alpha}{n+\beta} f'(v) - \frac{nv(1-v)}{2(n+\beta)^2} f''(v) \right)}{(v-z)^{p+1}} dv \right. \right.
$$

$$
\left. \left. \left. - \frac{p!}{2\pi i} \int_\Gamma \frac{\beta v(1-v)}{2(v-z)^{p+1}} f''(v) dv \right| \right] \right\},
$$

where by using the Remark 2 after the proof of Theorem 1.6.2, for all $|z| \le r$ and $n \in \mathbb{N}$ we get

$$
\left| \frac{p!}{2\pi i} \int_\Gamma \frac{(n+\beta)^2 \left(S_n^{(\alpha,\beta)}(f)(v) - f(v) + \frac{\beta v - \alpha}{n+\beta} f'(v) - \frac{nv(1-v)}{2(n+\beta)^2} f''(v) \right)}{(v-z)^{p+1}} dv - \right.
$$

$$
\left. \frac{p!}{2\pi i} \int_\Gamma \frac{\beta v(1-v)}{2(v-z)^{p+1}} f''(v) dv \right| \le \frac{p!}{2\pi} \cdot \frac{2\pi r_1 M_{r_1}^{(\alpha,\beta)}}{(r_1-r)^{p+1}} + \frac{p!}{2\pi} \cdot \frac{2\pi r_1 \beta r_1(1+r_1)\|f''\|_{r_1}}{2(r_1-r)^{p+1}}.
$$

Denoting now $F_p(z) = H^{(p)}(z)$, we prove that $\|F_p\|_r > 0$. Indeed, if we suppose that $\|F_p\|_r = 0$ then it follows that f satisfies the differential equation

$$
-\beta z f'(z) + \frac{z(1-z)}{2} f''(z) = Q_{p-1}(z), \forall |z| \le r,
$$

where $Q_{p-1}(z)$ is a polynomial of degree $\le p - 1$. Simplifying with z, making the substitution $y(z) = f'(z)$, searching $y(z)$ in the form $y(z) = \sum_{k=0}^{\infty} b_k z^k$ and then replacing in the differential equation, by simple calculations we easily obtain that

$b_k = 0$ for all $k \geq p - 1$, that is $y(z)$ is a polynomial of degree $\leq p - 2$. This implies the contradiction that f is a polynomial of degree $\leq p - 1$.

Continuing exactly as in the proof of Theorem 1.6.3 (with $\|S_n^{(\alpha,\beta)}(f) - f\|_r$ replaced by $\|[S_n^{(\alpha,\beta)}(f)]^{(p)} - f^{(p)}\|_r$), finally there exists an index $n_0 \in \mathbb{N}$ depending on f, r, r_1 and p, such that for all $n \geq n_0$ we have

$$\|[S_n^{(\alpha,\beta)}(f)]^{(p)} - f^{(p)}\|_r \geq \frac{1}{n} \cdot \frac{C_0}{2}.$$

Also, the cases when $n \in \{1, 2, ..., n_0 - 1\}$ are similar with those in the proof of Theorem 1.6.3. $\qquad \square$

Now defining the m-th iterates by ${}^m S_n^{(\alpha,\beta)}(f)(z)$, first we prove the following

Theorem 1.6.6. (Gal [80]) *Let* $\mathbb{D}_R = \{z \in \mathbb{C}; |z| < R\}$ *be with* $R > 1$ *and let us suppose that* $f : \mathbb{D}_R \to \mathbb{C}$ *is analytic in* $\mathbb{D}_R$, *that is we can write* $f(z) = \sum_{k=0}^{\infty} c_k z^k$, *for all* $z \in \mathbb{D}_R$. *Let* $0 \leq \alpha \leq \beta$ *and* $1 \leq r < R$. *Then, uniformly in* $|z| \leq r$, $\forall n \in \mathbb{N}$, *we have*

$$\lim_{m \to \infty} {}^m S_n^{(0,\beta)}(f)(z) = f(0), \quad \lim_{m \to \infty} {}^m S_n^{(\alpha,\alpha)}(f)(z) = f(1),$$

$\lim_{m \to \infty} {}^m S_n^{(\alpha,\beta)}(f)(z) = b_0$, *where* b_0 *is of the form*

$$b_0 = \sum_{j=1}^{n} d_j f((j + \alpha)/(n + \beta)), \quad \text{with } d_j \geq 0, j = 0, ..., n, \sum_{j=1}^{n} d_j = 1,$$

all the values $d_j, j = 0, ..., n$ *being independent of* f.

Proof.   By Theorem 2 in Gonska-Piţul-Raşa [103], for any $n \in \mathbb{N}$ we have

$$\lim_{m \to \infty} {}^m S_n^{(0,\beta)}(f)(x) = f(0), \quad \lim_{m \to \infty} {}^m S_n^{(\alpha,\alpha)}(f)(x) = f(1),$$

$$\lim_{m \to \infty} {}^m S_n^{(\alpha,\beta)}(f)(x) = \sum_{j=1}^{n} d_j f[(j + \alpha)/(n + \beta)],$$

uniformly with respect to $x \in [0, 1]$. From the classical Vitali's result, it suffices to show that for any fixed $n \in \mathbb{N}$, the sequence $({}^m S_n^{(\alpha,\beta)}(f)(z))_{m \in \mathbb{N}}$ is uniformly bounded for $|z| \leq r$.

We obviously have ${}^m S_n^{(\alpha,\beta)}(f)(z) = \sum_{k=0}^{\infty} c_k \cdot {}^m S_n^{(\alpha,\beta)}(e_k)(z)$.

But from the proof of Theorem 1.6.1 (both Cases 1) and 2)) it easily follows that $|S_n^{(\alpha,\beta)}(e_k)(z)| \leq r^k$, for all $k, n \in \mathbb{N}$, $|z| \leq r$, which implies

$$|{}^2 S_n^{(\alpha,\beta)}(e_k)(z)| = \left| \sum_{p=0}^{min\{n,k\}} C_{n,p,k}^{(\alpha,\beta)} S_n^{(\alpha,\beta)}(e_p)(z) \right| \leq r^k,$$

and by recurrence it easily follows $|{}^m S_n^{(\alpha,\beta)}(e_k)(z)| \leq r^k$, for all $m, k, n \in \mathbb{N}$.

This implies that

$$|{}^m S_n^{(\alpha,\beta)}(f)(z)| \leq \sum_{k=0}^{\infty} |c_k| \cdot |{}^m S_n^{(\alpha,\beta)}(e_k)(z)| \leq \sum_{k=0}^{\infty} |c_k| r^k < \infty,$$

for all $m, n \in \mathbb{N}$, which proves the theorem. $\qquad \square$

Also, the following quantitative result holds.

Theorem 1.6.7. (Gal [80]) *Let $\mathbb{D}_R = \{z \in \mathbb{C}; |z| < R\}$ be with $R > 1$ and let us suppose that $f : \mathbb{D}_R \to \mathbb{C}$ is analytic in $\mathbb{D}_R$, that is we can write $f(z) = \sum_{k=0}^{\infty} c_k z^k$, for all $z \in \mathbb{D}_R$. Let $0 \leq \alpha \leq \beta$ and $1 \leq r < R$. Then, for all $|z| \leq r$, we have*

$$|{}^m S_n^{(\alpha,\beta)}(f)(z) - f(z)| \leq \frac{2m}{n+\beta} \sum_{k=1}^{\infty} |c_k| \cdot [\beta k + k(k-1)] r^k.$$

Proof. We easily see that

$$|{}^m S_n^{(\alpha,\beta)}(f)(z) - f(z)| \leq \sum_{k=1}^{\infty} |c_k| \cdot |{}^m S_n^{(\alpha,\beta)}(e_k)(z) - e_k(z)|.$$

We have two possibilities : 1) $0 \leq k \leq n$; 2) $k > n$.

Case 1). We successively get

$$S_n^{(\alpha,\beta)}(e_k)(z) - e_k(z) = \sum_{j=1}^{k} C_{n,j,k}^{(\alpha,\beta)} e_j(z) - e_k(z)$$

$$= [C_{n,k,k}^{(\alpha,\beta)} - 1]e_k(z) + \sum_{j=1}^{k-1} C_{n,j,k}^{(\alpha,\beta)} e_j(z),$$

$${}^p S_n^{(\alpha,\beta)}[S_n^{(\alpha,\beta)}(e_k)(z) - e_k(z)] = [C_{n,k,k}^{(\alpha,\beta)} - 1] \cdot {}^p S_n^{(\alpha,\beta)}(e_k)(z)$$

$$+ \sum_{j=0}^{k-1} C_{n,j,k}^{(\alpha,\beta)} \cdot {}^p S_n^{(\alpha,\beta)}(e_j)(z),$$

$$|{}^p S_n^{(\alpha,\beta)}[S_n^{(\alpha,\beta)}(e_k)(z) - e_k(z)]| \leq |1 - C_{n,k,k}^{(\alpha,\beta)}| \cdot |{}^p S_n^{(\alpha,\beta)}(e_k)(z)|$$

$$+ |1 - C_{n,k,k}^{(\alpha,\beta)}| \cdot \max_{j=0,\ldots,k-1} \{|{}^p S_n^{(\alpha,\beta)}(e_j)(z)|\}.$$

But by the proofs of Theorems 1.6.1 and 1.6.6, for all $p, n, k \in \mathbb{N}$ we have

$$|1 - C_{n,k,k}^{(\alpha,\beta)}| \leq \frac{1}{n+\beta}\left[\frac{k(k-1)}{2} + \beta k\right], |{}^p S_n^{(\alpha,\beta)}(e_k)(z)| \leq r^k,$$

which implies

$$|{}^p S_n^{(\alpha,\beta)}[S_n^{(\alpha,\beta)}(e_k)(z) - e_k(z)]| \leq 2|1 - C_{n,k,k}^{(\beta)}| r^k$$

$$= \frac{1}{n+\beta}[2\beta k + k(k-1)] r^k,$$

and

$$|{}^m S_n^{(\alpha,\beta)}(e_k)(z) - e_k(z)| = \left|\sum_{p=0}^{m-1} {}^p S_n^{(\alpha,\beta)}[S_n^{(\alpha,\beta)}(e_k)(z) - e_k(z)]\right|$$

$$\leq \frac{m}{n+\beta}[2\beta k + k(k-1)] r^k.$$

Case 2). As in the proof of Theorem 1.6.1, Case 2), for all $k > n$ we get

$$|^m S_n^{(\alpha,\beta)}(e_k)(z) - e_k(z)| \leq 2r^k \leq \frac{2k(k-1) + 2\beta k}{n + \beta} r^k.$$

As a conclusion, from both Cases 1) and 2), we obtain

$$\begin{aligned}
|^m S_n^{(\alpha,\beta)}(f)(z) - f(z)| &\leq \sum_{k=1}^{\infty} |c_k| \cdot |^m S_n^{(\alpha,\beta)}(e_k)(z) - e_k(z)| \\
&= \sum_{k=1}^{n} |c_k| \cdot |^m S_n^{(\alpha,\beta)}(e_k)(z) - e_k(z)| \\
&\quad + \sum_{k=n+1}^{\infty} |c_k| \cdot |^m S_n^{(\alpha,\beta)}(e_k)(z) - e_k(z)| \\
&\leq \sum_{k=1}^{n} |c_k| \frac{m}{n+\beta} [2\beta k + k(k-1)] r^k \\
&\quad + \sum_{k=n+1}^{\infty} |c_k| \frac{2\beta k + 2k(k-1)}{n+\beta} r^k \\
&\leq \frac{2m}{n+\beta} \sum_{k=1}^{\infty} |c_k| \cdot [\beta k + k(k-1)] r^k,
\end{aligned}$$

which proves the theorem. $\qquad\qquad\square$

Corollary 1.6.8. (Gal [80]) *Suppose $\frac{m_n}{n} \to 0$ when $n \to \infty$. Then*

$$^{m_n} S_n^{(\alpha,\beta)}(f)(z) \to f(z),$$

uniformly with respect to $|z| \leq r$, for any $1 \leq r < R$.

Proof. It is immediate by passing to limit with $n \to \infty$ in Theorem 1.6.7. $\quad\square$

Remark. Theorem 1.6.7 and Corollary 1.6.8 are new even for the case of real functions of one real variable, since they are not covered by Gonska-Kacsó-Piţul [101], Gonska-Piţul-Raşa [103].

In what follows we present some similar properties for the complex Bernstein-Stancu polynomials $S_n^{<\gamma>}(f)(z)$. The first results concern the approximation properties.

Theorem 1.6.9. (Gal [82]) *Let $\mathbb{D}_R = \{z \in \mathbb{C}; |z| < R\}$ be with $R > 1$ and let us suppose that $f : \mathbb{D}_R \to \mathbb{C}$ is analytic in $\mathbb{D}_R$, that is we can write $f(z) = \sum_{k=0}^{\infty} c_k z^k$, for all $z \in \mathbb{D}_R$.*

Let $0 \leq \gamma$ which can be dependent on n and $1 \leq r < R$. Then, for all $|z| \leq r$ and $n \in \mathbb{N}$, we have

$$|S_n^{<\gamma>}(f)(z) - f(z)| \leq M_{2,r,n}^{<\gamma>}(f),$$

where $0 < M_{2,r,n}^{<\gamma>}(f) = \frac{2}{n}\sum_{j=2}^{\infty} j(j-1)|c_j|r^j + \frac{\gamma(r+1)}{6r}\sum_{j=2}^{\infty} j(j-1)(2j-1)|c_j|r^j <$
∞.

Also, if $1 \le r < r_1 < R$, *then for all* $|z| \le r$ *and* $n, p \in \mathbb{N}$, *we have*

$$|[S_n^{<\gamma>}(f)]^{(p)}(z) - f^{(p)}(z)| \le \frac{M_{2,r_1,n}^{<\gamma>}(f)p!r_1}{(r_1-r)^{p+1}}.$$

Proof. Since $S_n^{<\gamma>}(f)(z) = \sum_{k=0}^{\infty} c_k S_n^{<\gamma>}(e_k)(z)$, we get

$$|S_n^{<\gamma>}(f)(z) - f(z)| \le \sum_{k=0}^{\infty} |c_k| \cdot |S_n^{<\gamma>}(e_k)(z) - e_k(z)|.$$

To estimate $|S_n^{<\gamma>}(e_k)(z) - e_k(z)|$ for any fixed $n \in \mathbb{N}$, we will consider two possible cases : 1) $0 \le k \le n$; 2) $k > n$.

We will use the well-known representation (see Stancu [174])

$$S_n^{<\gamma>}(f)(z) = \sum_{p=0}^{n} \binom{n}{p} \frac{z(z+\gamma)...(z+(p-1)\gamma)}{(1+\gamma)...(1+(p-1)\gamma)} \Delta_{1/n}^p f(0).$$

Denoting

$$D_{n,p,k} = \binom{n}{p}\Delta_{1/n}^p e_k(0) = \binom{n}{p}[0, 1/n, ..., p/n; e_k](p!)/n^p,$$

since e_k is convex of any order, it follows that all $D_{n,p,k} \ge 0$ and

$$S_n^{<\gamma>}(e_k)(z) = \sum_{p=0}^{min\{n,k\}} D_{n,p,k} \frac{z(z+\gamma)...(z+(p-1)\gamma)}{(1+\gamma)...(1+(p-1)\gamma)}.$$

Also, since $S_n^{<\gamma>}(f)(1) = f(1)$, we get $\sum_{p=0}^{n} D_{n,p,k} = \sum_{p=0}^{min\{n,k\}} D_{n,p,k} = 1$.

Note that since for any $j = 0, 1, ...,$ we have $\frac{r+j\gamma}{1+j\gamma} \le r$, for all $0 \le p \le min\{n,k\} \le k$ and $|z| \le r$ we obtain

$$\frac{|z(z+\gamma)...(z+(p-1)\gamma)|}{(1+\gamma)...(1+(p-1)\gamma)} \le r\frac{r+\gamma}{1+\gamma}\cdot...\frac{r+(p-1)\gamma}{1+(p-1)\gamma} \le r^p \le r^k,$$

which for all $|z| \le r$ and $n, k \in \mathbb{N}$, immediately implies

$$|S_n^{<\gamma>}(e_k)(z)| \le r^k \sum_{p=0}^{min\{n,k\}} D_{n,p,k} = r^k.$$

Case 1). If $k = 0$, then obviously we have $S_n^{<\gamma>}(e_k)(z) - e_k(z) = 0$. Therefore, let us suppose that $1 \le k \le n$. By using the representation in Stancu [174], we obtain

$$|S_n^{<\gamma>}(e_k)(z) - e_k(z)|$$
$$\le \left| \frac{n(n-1)...(n-(k-1))}{n^k} \cdot \frac{z(z+\gamma)...(z+(k-1)\gamma)}{(1+\gamma)...(1+(k-1)\gamma)} - z^k \right|$$
$$+ \sum_{p=0}^{k-1} D_{n,p,k} \left| \frac{z(z+\gamma)...(z+(p-1)\gamma)}{(1+\gamma)...(1+(p-1)\gamma)} \right|$$
$$:= E_{n,k}^{<\gamma>}(z) + F_{n,k}^{<\gamma>}(z).$$

For $|z| \leq r$ it follows

$$F_{n,k}^{<\gamma>}(z) \leq r^k \sum_{p=0}^{k-1} D_{n,p,k} = r^k[1 - D_{n,k,k}]$$

$$= r^k[1 - \frac{n(n-1)...(n-(k-1))}{n^k}] \leq r^k \frac{k(k-1)}{2n}.$$

Here we have applied the inequality $1 - \Pi x_i \leq \sum(1 - x_i)$, with all $0 \leq x_i \leq 1$.

Also,

$$E_{n,k}^{<\gamma>}(z) \leq \left| \frac{n(n-1)...(n-(k-1))}{n^k} \frac{z(z+\gamma)...(z+(k-1)\gamma)}{(1+\gamma)...(1+(k-1)\gamma)} \right.$$

$$\left. - \frac{z(z+\gamma)...(z+(k-1)\gamma)}{(1+\gamma)...(1+(k-1)\gamma)} \right| + \left| \frac{z(z+\gamma)...(z+(k-1)\gamma)}{(1+\gamma)...(1+(k-1)\gamma)} - z^k \right|$$

$$\leq \left| \frac{z(z+\gamma)...(z+(k-1)\gamma)}{(1+\gamma)...(1+(k-1)\gamma)} \right| \cdot \left| 1 - \frac{n(n-1)...(n-(k-1))}{n^k} \right|$$

$$+ \left| \frac{z(z+\gamma)...(z+(k-1)\gamma)}{(1+\gamma)...(1+(k-1)\gamma)} - z^k \right|$$

$$\leq r^k \frac{k(k-1)}{2n} + \left| \frac{z(z+\gamma)...(z+(k-1)\gamma)}{(1+\gamma)...(1+(k-1)\gamma)} - z^k \right|.$$

For any fixed $|z| \leq r$, let us denote $g_k(\alpha)(z) = \frac{z(z+\alpha)...(z+(k-1)\alpha)}{(1+\alpha)...(1+(k-1)\alpha)}$, where $\alpha \geq 0$. Then, by applying the mean value theorem, there is $\xi \in [0, \gamma]$ such that

$$\left| \frac{z(z+\gamma)...(z+(k-1)\gamma)}{(1+\gamma)...(1+(k-1)\gamma)} - z^k \right| = |g_k(\gamma)(z) - g_k(0)(z)| \leq \gamma \cdot \max \left| \frac{dg_k(\xi)(z)}{d\alpha} \right|.$$

But denoting $u_j(\alpha)(z) = \frac{z+j\alpha}{1+j\alpha}$, we have $g_k(\alpha)(z) = z\Pi_{j=1}^{k-1} u_j(\alpha)(z)$ and

$$\frac{dg_k(\alpha)(z)}{d\alpha} = z \sum_{j=1}^{k-1} \left(\frac{z+j\alpha}{1+j\alpha} \right)'_\alpha \cdot \prod_{i=1, i\neq j}^{k-1} \frac{z+i\alpha}{1+i\alpha}$$

$$= z \sum_{j=1}^{k-1} \frac{j(1-z)}{(1+j\alpha)^2} \prod_{i=1, i\neq j}^{k-1} \frac{z+i\alpha}{1+i\alpha}.$$

Since $\frac{j}{(1+j\xi)^2} \leq j^2$, passing to modulus (for $0 \leq \xi \leq \gamma$ and $|z| \leq r$), we obtain

$$\left| \frac{dg_k(\xi)(z)}{d\alpha} \right| \leq r(r+1) \sum_{j=1}^{k-1} j^2 r^{k-2} = (r+1)r^{k-1} \frac{k(k-1)(2k-1)}{6}.$$

It follows

$$E_{n,k}^{<\gamma>}(z) \leq r^k \frac{k(k-1)}{2n} + \gamma(r+1)r^{k-1} \frac{k(k-1)(2k-1)}{6}.$$

Collecting all the above estimates, we get for all $|z| \leq r$

$$|S_n^{<\gamma>}(e_k)(z) - e_k(z)| \leq r^k \frac{k(k-1)}{2n} + r^k \frac{k(k-1)}{2n}$$

$$+ \gamma(r+1)r^{k-1} \frac{k(k-1)(2k-1)}{6}$$

$$= r^k \left[\frac{k(k-1)}{n} + \gamma \cdot \frac{r+1}{r} \cdot \frac{k(k-1)(2k-1)}{6} \right].$$

Case 2). We have

$$|S_n^{<\gamma>}(e_k)(z) - e_k(z)| \leq |S_n^{<\gamma>}(e_k)(z)| + |e_k(z)|$$

$$\leq \sum_{p=0}^{n} D_{n,p,k} \left| \frac{z(z+\gamma)...(z+(p-1)\gamma)}{(1+\gamma)...(1+(p-1)\gamma)} \right| + |e_k(z)|.$$

Reasoning as in the above Case 1), we get

$$|S_n^{<\gamma>}(e_k)(z) - e_k(z)| \leq r^n + r^k \leq 2r^k \leq \frac{2(k-1)k}{n} r^k.$$

Collecting all the results in the Cases 1) and 2), we immediately obtain for all $|z| < r$ and $k = 0, 1, 2, ...,$

$$|S_n^{<\gamma>}(e_k)(z) - e_k(z)| \leq r^k \left[\frac{2k(k-1)}{n} + \gamma \cdot \frac{r+1}{r} \cdot \frac{k(k-1)(2k-1)}{6} \right],$$

which implies the corresponding estimate in statement.

For the simultaneous approximation, denoting by Γ the circle of radius $r_1 > r$ and center 0, since for any $|z| \leq r$ and $v \in \Gamma$, we have $|v - z| \geq r_1 - r$, by the Cauchy's formulas it follows that for all $|z| \leq r$ and $n \in \mathbb{N}$, we have

$$|[S_n^{<\gamma>}(f)]^{(p)}(z) - f^{(p)}(z)| = \frac{p!}{2\pi} \left| \int_\Gamma \frac{S_n^{<\gamma>}(f)(v) - f(v)}{(v-z)^{p+1}} dv \right|$$

$$\leq M_{2,r_1,n}^{<\gamma>}(f) \frac{p!}{2\pi} \frac{2\pi r_1}{(r_1 - r)^{p+1}}$$

$$= M_{2,r_1,n}^{<\gamma>}(f) \frac{p! r_1}{(r_1 - r)^{p+1}}. \qquad \square$$

Remark. For $\gamma = 0$ we get the results for the classical complex Bernstein polynomials in Theorem 1.1.2.

Now, defining the m-th iterates by $^m S_n^{<\gamma>}(f)(z)$, first we prove the following qualitative result.

Theorem 1.6.10. (Gal [82]) *Let* $\mathbb{D}_R = \{z \in \mathbb{C}; |z| < R\}$ *be with* $R > 1$ *and let us suppose that* $f : \mathbb{D}_R \to \mathbb{C}$ *is analytic in* $\mathbb{D}_R$, *that is we can write* $f(z) = \sum_{k=0}^{\infty} c_k z^k$, *for all* $z \in \mathbb{D}_R$. *Let* $0 \leq \gamma$. *Uniformly in* $|z| \leq r$, *where* $1 \leq r < R$, *we have*

$$\lim_{m \to \infty} {}^m S_n^{<\gamma>}(f)(z) = (1-z)f(0) + zf(1), \forall n \in \mathbb{N}.$$

Proof. From Agratini-Rus [4], Remark 2 after Theorem 9, p. 165, for any $n \in \mathbb{N}$, we have $\lim_{m \to \infty} {}^m S_n^{<\gamma>}(f)(x) = (1-x)f(0) + xf(1)$, uniformly with respect to $x \in [0,1]$. From the classical Vitali's result, it suffices to show that for any fixed $n \in \mathbb{N}$, the sequence $({}^m S_n^{<\gamma>}(f)(z))_{m \in \mathbb{N}}$ is uniformly bounded for $|z| \leq r$.

We have $^m S_n^{<\gamma>}(f)(z) = \sum_{k=0}^{\infty} c_k \cdot {}^m S_n^{<\gamma>}(e_k)(z)$. We will prove that for all $n, m, k \in \mathbb{N}$ and $|z| \leq r$, we have $|^m S_n^{<\gamma>}(e_k)(z)| \leq r^k$.

Indeed, for $m = 1$ it easily follows (also see the proof of Theorem 1.6.9) from the representation formula

$$S_n^{<\gamma>}(e_k)(z) = \sum_{j=0}^{n} D_{n,j,k} \frac{z(z+\gamma)...(z+(j-1)\gamma)}{(1+\gamma)...(1+(j-1)\gamma)}$$

$$= \sum_{j=0}^{min\{n,k\}} D_{n,j,k} \frac{z(z+\gamma)...(z+(j-1)\gamma)}{(1+\gamma)...(1+(j-1)\gamma)},$$

with $D_{n,j,k} \geq 0$ and $\sum_{j=0}^{n} D_{n,j,k} = \sum_{j=0}^{min\{n,k\}} D_{n,j,k} = 1$.

Denote $h_j(z) = z(z+\gamma)...(z+(j-1)\gamma) = \sum_{i=0}^{j} c_i^{(j)} e_i(z)$, where $c_i^{(j)} \geq 0$, $c_j^{(j)} = 1$ and $\sum_{i=0}^{j} c_i^{(j)} = h_j(1) = (1+\gamma)...(1+(j-1)\gamma)$.

By the linearity of $S_n^{<\gamma>}$, we get

$$|^2 S_n^{<\gamma>}(e_k)| = \left| \sum_{j=0}^{min\{n,k\}} D_{n,j,k} \frac{1}{(1+\gamma)...(1+(j-1)\gamma)} \cdot \sum_{i=0}^{j} c_i^{(j)} S_n^{<\gamma>}(e_i)(z) \right|$$

$$\leq \sum_{j=0}^{min\{n,k\}} D_{n,j,k} \frac{1}{(1+\gamma)...(1+(j-1)\gamma)} \cdot \sum_{i=0}^{j} c_i^{(j)} r^j \leq r^k,$$

and by mathematical induction it easily follows that for all $n, m, k \in \mathbb{N}$ we have

$$|^m S_n^{<\gamma>}(e_k)(z)| \leq r^k, \text{ for all } |z| \leq r.$$

This implies that

$$|^m S_n^{<\gamma>}(f)(z)| \leq \sum_{k=0}^{\infty} |c_k| \cdot |^m S_n^{<\gamma>}(e_k)(z)| \leq \sum_{k=0}^{\infty} |c_k| r^k < \infty,$$

for all $m, n \in \mathbb{N}$, which proves the theorem. $\qquad \square$

The following quantitative result is not correspondent to the above qualitative one.

Theorem 1.6.11. (Gal [82]) *Let* $\mathbb{D}_R = \{z \in \mathbb{C}; |z| < R\}$ *be with* $R > 1$ *and let us suppose that* $f : \mathbb{D}_R \to \mathbb{C}$ *is analytic in* $\mathbb{D}_R$, *that is we can write* $f(z) = \sum_{k=0}^{\infty} c_k z^k$, *for all* $z \in \mathbb{D}_R$.

Let $0 \leq \gamma$, $1 \leq r < R$ *and* $D_{n,k,k} = \frac{n(n-1)...(n-(k-1))}{n^k}$. *Then, for all* $|z| \leq r$ *we have*

$$|^m S_n^{<\gamma>}(f)(z) - f(z)|$$

$$\leq m \sum_{k=2}^{\infty} |c_k| \left[\frac{2k(k-1)}{n} + \left(1 - \frac{D_{n,k,k}}{(1+\gamma)...(1+(k-1)\gamma)} \right) + \gamma(k-1)^2 \right] r^k.$$

Proof. From the proof of Theorem 1.6.10, it follows that for all $n, m, k \in \mathbb{N}$ and $|z| \leq r$, we have $|^m S_n^{<\gamma>}(e_k)(z)| \leq r^k$. Also

$$|^m S_n^{<\gamma>}(f)(z) - f(z)| \leq \sum_{k=2}^{\infty} |c_k| \cdot |^m S_n^{<\gamma>}(e_k)(z) - e_k(z)|.$$

We have two possibilities : 1) $2 \leq k \leq n$; 2) $k > n$.

Case 1). With the notations for $g_j(\alpha)(z)$ in the proof of Theorem 1.6.9 and for $h_j(z), c_i^{(j)}$ in the proof of Theorem 1.6.10, we can write

$$|^m S_n^{<\gamma>}(e_k)(z) - e_k(z)|$$

$$= \left| \sum_{p=0}^{m-1} {}^p S_n^{<\gamma>}[S_n^{<\gamma>}(e_k)(z) - e_k(z)] \right|$$

$$= \left| \sum_{p=0}^{m-1} {}^p S_n^{<\gamma>} \left[\sum_{j=1}^{k} D_{n,j,k} \cdot g_j(\gamma)(z) - e_k(z) \right] \right|$$

$$= \left| \sum_{p=0}^{m-1} \left[\sum_{j=1}^{k} D_{n,j,k} \cdot {}^p S_n^{<\gamma>}(g_j(\gamma))(z) - {}^p S_n^{<\gamma>}(e_k)(z) \right] \right|$$

$$\leq \sum_{p=0}^{m-1} \sum_{j=1}^{k-1} D_{n,j,k} |^p S_n^{<\gamma>}(g_j(\gamma))(z)|$$

$$+ \sum_{p=0}^{m-1} |D_{n,k,k} \cdot {}^p S_n^{<\gamma>}(g_k(\gamma))(z) - {}^p S_n^{<\gamma>}(e_k)(z)|$$

$$= \sum_{p=0}^{m-1} \sum_{j=1}^{k-1} D_{n,j,k} |^p S_n^{<\gamma>}(g_j(\gamma))(z)|$$

$$+ \sum_{p=0}^{m-1} \left| \frac{D_{n,k,k}}{(1+\gamma)...(1+(k-1)\gamma)} \cdot {}^p S_n^{<\gamma>}[\sum_{i=0}^{k} c_i^{(k)} e_i(z)] - {}^p S_n^{<\gamma>}(e_k)(z) \right|$$

$$\leq \sum_{p=0}^{m-1} \sum_{j=1}^{k-1} D_{n,j,k} |^p S_n^{<\gamma>}(g_j(\gamma))(z)|$$

$$+ \sum_{p=0}^{m-1} \left| \frac{D_{n,k,k}}{(1+\gamma)...(1+(k-1)\gamma)} \cdot {}^p S_n^{<\gamma>}(e_k)(z) - {}^p S_n^{<\gamma>}(e_k)(z) \right|$$

$$+ \sum_{p=0}^{m-1} \frac{D_{n,k,k}}{(1+\gamma)...(1+(k-1)\gamma)} \left| \sum_{i=0}^{k-1} c_i^{(k)} \cdot {}^p S_n^{<\gamma>}(e_i)(z) \right|$$

$$:= T_1 + T_2 + T_3.$$

Reasoning exactly as in the proof of Theorem 1.6.10, we easily get for all j, p and $|z| \leq r$ that

$$|^p S_n^{<\gamma>}(g_j(\gamma))(z)| \leq r^j.$$

Taking into account the formula for $1 - D_{n,k,k}$ in the proof of Theorem 1.6.9, we get

$$T_1 \leq \sum_{p=0}^{m-1} \sum_{j=1}^{k-1} r^k D_{n,j,k} = mr^k[1 - D_{n,k,k}] \leq mr^k \frac{k(k-1)}{2n}.$$

Also,

$$T_2 = \sum_{p=0}^{m-1} |{}^p S_n^{<\gamma>}(e_k)(z)| \left[1 - \frac{D_{n,k,k}}{(1+\gamma)...(1+(k-1)\gamma)} \right]$$

$$\leq mr^k \left[1 - \frac{D_{n,k,k}}{(1+\gamma)...(1+(k-1)\gamma)} \right].$$

Finally,

$$T_3 \leq \sum_{p=0}^{m-1} \frac{D_{n,k,k}}{(1+\gamma)...(1+(k-1)\gamma)} [(1+\gamma)...(1+(k-1)\gamma) - 1]r^k$$

$$= mr^k D_{n,k,k} \left[1 - \frac{1}{(1+\gamma)...(1+(k-1)\gamma)} \right].$$

But, taking into account the inequalities $D_{n,k,k} \leq 1$ and

$$1 - \Pi_{j=1}^{k-1} x_j \leq \sum_{j=1}^{k-1} (1 - x_j), 0 \leq x_j \leq 1, j = 1, ..., k-1,$$

applied for $x_j = \frac{1}{1+j\gamma}$, we obtain

$$D_{n,k,k} \left[1 - \frac{1}{(1+\gamma)...(1+(k-1)\gamma)} \right] \leq \sum_{j=1}^{k-1} [1 - 1/(1+j\gamma)] = \sum_{j=1}^{k-1} \frac{j\gamma}{1+j\gamma}$$

$$\leq (k-1) \cdot \frac{\gamma(k-1)}{1+\gamma(k-1)} \leq \gamma(k-1)^2.$$

Collecting all these inequalities, we obtain

$$|{}^m S_n^{<\gamma>}(e_k)(z) - e_k(z)|$$

$$\leq mr^k \left[\frac{k(k-1)}{2n} + \left(1 - \frac{D_{n,k,k}}{(1+\gamma)...(1+(k-1)\gamma)} \right) + \gamma(k-1)^2 \right].$$

Case 2). We get

$$|{}^m S_n^{<\gamma>}(e_k)(z) - e_k(z)| \leq |{}^m S_n^{<\gamma>}(e_k)(z)| + |e_k(z)|$$

$$\leq 2r^k \leq \frac{2k(k-1)}{n} r^k.$$

As a conclusion, from both Cases 1) and 2), we obtain

$$|{}^m S_n^{<\gamma>}(f)(z) - f(z)|$$

$$\leq \sum_{k=2}^{\infty} |c_k| \cdot |{}^m S_n^{<\gamma>}(e_k)(z) - e_k(z)|$$

$$= \sum_{k=2}^{n} |c_k| \cdot |{}^m S_n^{<\gamma>}(e_k)(z) - e_k(z)| + \sum_{k=n+1}^{\infty} |c_k| \cdot |{}^m S_n^{<\gamma>}(e_k)(z) - e_k(z)|$$

$$\leq \sum_{k=2}^{n} |c_k| mr^k \left[\frac{k(k-1)}{2n} + \left(1 - \frac{D_{n,k,k}}{(1+\gamma)...(1+(k-1)\gamma)} \right) + \gamma(k-1)^2 \right]$$

$$+ \sum_{k=n+1}^{\infty} |c_k| r^k \frac{2k(k-1)}{n}$$

$$\leq m \sum_{k=2}^{\infty} |c_k| r^k \left[\frac{2k(k-1)}{n} + \left(1 - \frac{D_{n,k,k}}{(1+\gamma)...(1+(k-1)\gamma)} \right) + \gamma(k-1)^2 \right] r^k,$$

which proves the theorem. $\qquad\qquad\qquad\qquad\qquad\qquad\qquad\qquad\qquad\qquad$ $\square$

Remark. For $\gamma = 0$ we get some results for classical complex Bernstein polynomials in Section 1.2.

Corollary 1.6.12. (Gal [82]) *(i) Let $1 \leq r < R$. For $\gamma := \gamma_n = 1/n$ and $|z| \leq r$ we have the estimate*

$$|{}^m S_n^{<\gamma_n>}(f)(z) - f(z)| \leq \frac{m}{n} \sum_{k=2}^{\infty} |c_k| \left[2k(k-1) + 2(k-1)^3 + (k-1)^2 \right] r^k.$$

(ii) If $\gamma := \gamma_n = 1/n$ and $\frac{m_n}{n} \to 0$ as $n \to \infty$, then ${}^{m_n} S_n^{<\gamma_n>}(f)(z) \to f(z)$, uniformly with respect $|z| \leq r$.

Proof. (i) Taking $\gamma = 1/n$ we obtain for all $k \geq 2$

$$1 - \frac{D_{n,k,k}}{(1+\gamma)...(1+(k-1)\gamma)} = 1 - \Pi_{j=1}^{k-1} \frac{n-j}{n+j}$$

$$\leq \sum_{j=1}^{k-1} \left[1 - \frac{n-j}{n+j} \right] = 2 \sum_{j=1}^{k-1} \frac{j}{j+n}$$

$$\leq 2(k-1) \frac{k-1}{n+(k-1)} \leq 2 \frac{(k-1)^3}{n},$$

which replaced in Theorem 1.6.11 gives

$$|{}^m S_n^{<\gamma_n>}(f)(z) - f(z)| \leq \frac{m}{n} \sum_{k=2}^{\infty} |c_k| \left[2k(k-1) + 2(k-1)^3 + (k-1)^2 \right] r^k.$$

(ii) It is immediate by passing to limit with $n \to \infty$ in the estimate proved in (i). $\qquad\qquad\qquad\qquad\qquad\qquad\qquad\qquad\qquad\qquad\qquad\qquad$ $\square$

Remark. The results in Theorem 1.6.11 and Corollary 1.6.12 are new even for the case of real functions of one real variable, since they are not covered by those in Gonska-Kacsó-Piţul [101] or Gonska-Piţul-Raşa [103], whose estimates one refer to the difference $|{}^m L_n(f)(x) - B_1(f)(x)|$, with $B_1(f)(x) = f(0) + [f(1) - f(0)]x$ and ${}^m L_n(f)$ representing the mth iterate of the positive linear operator $L_n(f)$.

Finally we present the geometric properties of $S_n^{<\gamma>}(f)(z)$.

Theorem 1.6.13. (Gal [82]) *Let us suppose that $G \subset \mathbb{C}$ is open, such that $\overline{\mathbb{D}}_1 \subset G$ and $f : G \to \mathbb{C}$ is analytic in G. Also, let us consider $(S_n^{<\gamma(n)>}(f)(z))_{n\in\mathbb{N}}$, where we suppose that $\lim_{n\to\infty} \gamma(n) = 0$.*

If $f(0) = f'(0) - 1 = 0$ and f is starlike (convex, spirallike of type η, respectively) in $\overline{\mathbb{D}}_1$, that is for all $z \in \overline{\mathbb{D}}_1$ (see e.g. Mocanu, P. T., Bulboacă, T. and Sălăgean [138])

$$Re\left(\frac{zf'(z)}{f(z)} \right) > 0 \left(Re\left(\frac{zf''(z)}{f'(z)} \right) + 1 > 0, Re\left(e^{i\eta} \frac{zf'(z)}{f(z)} \right) > 0, resp. \right),$$

then there exists an index n_0 depending on f (and on η for spirallikeness), such that for all $n \geq n_0$, $S_n^{<\gamma(n)>}(f)(z)$, are starlike (convex, spirallike of type η, respectively) in $\overline{\mathbb{D}}_1$.

If $f(0) = f'(0) - 1 = 0$ and f is starlike (convex, spirallike of type η, respectively) only in $\mathbb{D}_1$ (that is the corresponding inequalities hold only in $\mathbb{D}_1$), then for any disk of radius $0 < r < 1$ and center 0 denoted by $\mathbb{D}_r$, there exists an index $n_0 = n_0(f, \mathbb{D}_r)$ (n_0 depends on η too in the case of spirallikeness), such that for all $n \geq n_0$, $S_n^{<\gamma(n)>}(f)(z)$, are starlike (convex, spirallike of type η, respectively) in $\overline{\mathbb{D}}_r$ (that is, the corresponding inequalities hold in $\overline{\mathbb{D}}_r$).

Proof.   By Theorem 1.6.9 it follows that we have $S_n^{<\gamma(n)>}(f)(z) \to f(z)$, uniformly for $|z| \leq 1$, which by the well-known Weierstrass's theorem implies $[S_n^{<\gamma(n)>}(f)]'(z) \to f'(z)$ and $[S_n^{<\gamma(n)>}(f)]''(z) \to f''(z)$, for $n \to \infty$, uniformly in $\overline{\mathbb{D}}_1$. In all what follows, denote $P_n(f)(z) = \frac{S_n^{<\gamma(n)>}(f)(z)}{[S_n^{<\gamma(n)>}(f)]'(0)}$, well defined for sufficiently large n. We easily get $P_n(f)(0) = 0$, $P_n'(f)(0) = 1$ for sufficiently large n, and $P_n(f)(z) \to f(z)$, $P_n'(f)(z) \to f'(z)$ and $P_n''(f)(z) \to f''(z)$, uniformly in $\overline{\mathbb{D}}_1$.

Suppose first that f is starlike in $\overline{\mathbb{D}}_1$. Then, by hypothesis we get $|f(z)| > 0$ for all $z \in \overline{\mathbb{D}}_1$ with $z \neq 0$, which from the univalence of f in $\mathbb{D}_1$, implies that we can write $f(z) = zg(z)$, with $g(z) \neq 0$, for all $z \in \overline{\mathbb{D}}_1$, where g is analytic in $\mathbb{D}_1$ and continuous in $\overline{\mathbb{D}}_1$.

Writing $P_n(f)(z)$ in the form $P_n(f)(z) = zQ_n(f)(z)$, obviously $Q_n(f)(z)$ is a polynomial of degree $\leq n - 1$. Also, for $|z| = 1$ we have

$$|f(z) - P_n(f)(z)| = |z| \cdot |g(z) - Q_n(f)(z)| = |g(z) - Q_n(f)(z)|,$$

which by the uniform convergence in $\overline{\mathbb{D}}_1$ of $P_n(f)$ to f and by the maximum modulus principle, implies the uniform convergence in $\overline{\mathbb{D}}_1$ of $Q_n(f)(z)$ to $g(z)$.

Since g is continuous in $\overline{\mathbb{D}}_1$ and $|g(z)| > 0$ for all $z \in \overline{\mathbb{D}}_1$, there exist an index $n_1 \in \mathbb{N}$ and $a > 0$ depending on g, such that $|Q_n(f)(z)| > a > 0$, for all $z \in \overline{\mathbb{D}}_1$ and all $n \geq n_0$. Also, for all $|z| = 1$, we have

$$
\begin{aligned}
|f'(z) - P_n'(f)(z)| &= |z[g'(z) - Q_n'(f)(z)] + [g(z) - Q_n(f)(z)]| \\
&\geq |\ |z| \cdot |g'(z) - Q_n'(f)(z)| - |g(z) - Q_n(f)(z)|\ | \\
&= |\ |g'(z) - Q_n'(f)(z)| - |g(z) - Q_n(f)(z)|\ |,
\end{aligned}
$$

which from the maximum modulus principle, the uniform convergence of $P_n'(f)$ to f' and of $Q_n(f)$ to g, evidently implies the uniform convergence of $Q_n'(f)$ to g'.

Then, for $|z| = 1$, we get

$$
\begin{aligned}
\frac{zP_n'(f)(z)}{P_n(f)} &= \frac{z[zQ_n'(f)(z) + Q_n(f)(z)]}{zQ_n(f)(z)} \\
&= \frac{zQ_n'(f)(z) + Q_n(f)(z)}{Q_n(f)(z)} \to \frac{zg'(z) + g(z)}{g(z)} \\
&= \frac{f'(z)}{g(z)} = \frac{zf'(z)}{f(z)},
\end{aligned}
$$

which again from the maximum modulus principle, implies

$$\frac{zP_n'(f)(z)}{P_n(f)} \to \frac{zf'(z)}{f(z)}, \text{ uniformly in } \overline{\mathbb{D}}_1.$$

Since $Re\left(\frac{zf'(z)}{f(z)}\right)$ is continuous in $\overline{\mathbb{D}}_1$, there exists $\varepsilon \in (0,1)$, such that

$$Re\left(\frac{zf'(z)}{f(z)}\right) \geq \varepsilon, \text{ for all } z \in \overline{\mathbb{D}}_1.$$

Therefore

$$Re\left[\frac{zP_n'(f)(z)}{P_n(f)(z)}\right] \to Re\left[\frac{zf'(z)}{f(z)}\right] \geq \varepsilon > 0$$

uniformly on $\overline{\mathbb{D}}_1$, i.e. for any $0 < \rho < \varepsilon$, there is n_0 such that for all $n \geq n_0$ we have

$$Re\left[\frac{zP_n'(f)(z)}{P_n(f)(z)}\right] > \rho > 0, \text{ for all } z \in \overline{\mathbb{D}}_1.$$

Since $P_n(f)(z)$ differs from $S_n^{<\gamma(n)>}(f)(z)$ only by a constant, this proves the star-likeness of $S_n^{<\gamma(n)>}(f)(z)$, for sufficiently large n.

If f is supposed to be starlike only in $\mathbb{D}_1$, the proof is identical, with the only difference that instead of $\overline{\mathbb{D}}_1$, we reason for $\overline{\mathbb{D}}_r$.

The proofs in the cases when f is convex or spirallike of order η are similar and follow from the following uniform convergences (on $\overline{\mathbb{D}}_1$ or on $\overline{\mathbb{D}}_r$)

$$Re\left[\frac{zP_n''(f)(z)}{P_n'(f)(z)}\right] + 1 \to Re\left[\frac{zf''(z)}{f'(z)}\right] + 1$$

and

$$Re\left[e^{i\eta}\frac{zP_n'(f)(z)}{P_n(f)(z)}\right] \to Re\left[e^{i\eta}\frac{zf'(z)}{f(z)}\right]. \qquad \square$$

Remark. If f is univalent in $\overline{\mathbb{D}}_1$, then from the uniform convergence in Theorem 1.6.9 and a well-known result in complex analysis, concerning sequences of analytic functions converging locally uniformly to an univalent function, it is immediate that for sufficiently large n, the complex polynomials $S_n^{<\gamma(n)>}(f)(z)$ (where $\gamma(n) \to 0$, for $n \to \infty$), must be univalent in $\overline{\mathbb{D}}_1$.

At the end of this section we will extend the Bernstein-Stancu polynomials $S_n^{(\alpha,\beta)}(f)(z)$ and some of their approximation results to compact subsets $G \subset \mathbb{C}$.

For this purpose, in what follows $G \subset \mathbb{C}$ we will be considered a compact set such that $\tilde{\mathbb{C}} \setminus G$ is connected. In this case, according to the Riemann Mapping Theorem, a unique conformal mapping Ψ of $\tilde{\mathbb{C}} \setminus \overline{\mathbb{D}}_1$ onto $\tilde{\mathbb{C}} \setminus G$ exists so that $\Psi(\infty) = \infty$ and $\Psi'(\infty) > 0$.

By using the Faber polynomials $F_p(z)$ attached to G (see Definition 1.0.10), for $f \in A(\overline{G})$ and $0 \leq \alpha \leq \beta$ we can introduce the Bernstein-Stancu-Faber polynomials given by the formula

$$\mathcal{S}_n^{(\alpha,\beta)}(f;\overline{G})(z) = \sum_{p=0}^{n} \binom{n}{p} \Delta_{1/(n+\beta)}^p F(\alpha/(n+\beta)) \cdot F_p(z), z \in G, \ n \in \mathbb{N},$$

where $0 \leq \alpha \leq \beta$, $F(w) = \frac{1}{2\pi i} \int_{|u|=1} \frac{f(\Psi(u))}{u-w} du$, $w \in \mathbb{D}_1$, and

$$\Delta_h^p F(\alpha/(n+\beta)) = \sum_{k=0}^{p} (-1)^{p-k} \binom{p}{k} F(\alpha/(n+\beta) + kh).$$

Here, in the case when $\alpha = \beta$ since $F(1)$ is involved in $\Delta_{1/(n+\beta)}^n F(\alpha/(n+\beta))$ and therefore in the definition of $\mathcal{S}_n^{(\alpha,\beta)}(f; G)(z)$ too, in addition we will suppose that F can be extended by continuity on the boundary $\partial \mathbb{D}_1$.

Remarks. 1) For $G = \overline{\mathbb{D}}_1$ it is easy to see that the above Bernstein-Stancu-Faber polynomials one reduce to the classical complex Bernstein-Stancu polynomials given by

$$S_n^{(\alpha,\beta)}(f)(z) = \sum_{p=0}^{n} \binom{n}{p} \Delta_{1/(n+\beta)}^p f(\alpha/(n+\beta)) z^p$$

$$= \sum_{p=0}^{n} \binom{n}{p} z^p (1-z)^{n-p} f[(p+\alpha)/(n+\beta)].$$

2) It is known that, for example, $\int_0^1 \frac{\omega_p(f \circ \Psi; u)_{\partial \mathbb{D}_1}}{u} du < \infty$ is a sufficient condition for the continuity on $\partial \mathbb{D}_1$ of F in the above definition of the Bernstein-Stancu-Faber polynomials (see e.g. Gaier [76], p. 52, Theorem 6). Here $p \in \mathbb{N}$ is arbitrary fixed.

3) In the case when $\alpha = \beta = 0$, $\mathcal{S}_n^{(\alpha,\beta)}(f; \overline{G})(z)$ becomes $\mathcal{B}_n(f; \overline{G})(z)$.

The first main result one refers to approximation on compact sets without any restriction on their boundaries and can be stated as follows.

Theorem 1.6.14. *Let G be a continuum (that is a connected compact subset of $\mathbb{C}$) and suppose that f is analytic in G, that is there exists $R > 1$ such that f is analytic in G_R. Here recall that G_R denotes the interior of the closed level curve Γ_R given by $\Gamma_R = \{z; |\Phi(z)| = R\} = \{\Psi(w); |w| = R\}$ (and that $G \subset \overline{G}_r$ for all $1 < r < R$). Also, we suppose that F given in the definition of Bernstein-Stancu-Faber polynomials can be extended by continuity on $\partial \mathbb{D}_1$.*

For any $1 < r < R$ the following estimate

$$|\mathcal{S}_n^{(\alpha,\beta)}(f; \overline{G})(z) - f(z)| \leq \frac{C}{n}, \ z \in \overline{G}_r, \ n \in \mathbb{N},$$

holds, where $C > 0$ depends on f, α, β, r and G_r but it is independent of n.

Proof. First we note that since G is a continuum then it follows that $\tilde{\mathbb{C}} \setminus G$ is simply connected. By the proof of Theorem 2, p. 52 in Suetin [186], for any fixed $1 < \eta < R$ we have $f(z) = \sum_{k=0}^{\infty} a_k(f) F_k(z)$ uniformly in $\overline{G}_\eta$, where $a_k(f)$ are the Faber coefficients and are given by $a_k(f) = \frac{1}{2\pi i} \int_{|u|=\eta} \frac{f[\Psi(u)]}{u^{k+1}} du$. Note here that $G \subset \overline{G}_\eta$.

First we will prove that

$$\mathcal{S}_n^{(\alpha,\beta)}(f; \overline{G})(z) = \sum_{k=0}^{\infty} a_k(f) \mathcal{S}_n^{(\alpha,\beta)}(F_k; \overline{G})(z),$$

for all $z \in G$. (Note here that by hypothesis we have $\overline{G} = G$). For this purpose, denote $f_m(z) = \sum_{k=0}^m a_k(f) F_k(z)$, $m \in \mathbb{N}$.

Since by the linearity of $\mathcal{S}_n(\alpha, \beta)$ we easily get

$$\mathcal{S}_n^{(\alpha,\beta)}(f_m; \overline{G})(z) = \sum_{k=0}^m a_k(f) \mathcal{S}_n^{(\alpha,\beta)}(F_k; \overline{G})(z), \text{ for all } z \in G,$$

it suffices to prove that $\lim_{m \to \infty} \mathcal{S}_n^{(\alpha,\beta)}(f_m; \overline{G})(z) = \mathcal{S}_n^{(\alpha,\beta)}(f; \overline{G})(z)$, for all $z \in G$ and $n \in \mathbb{N}$.

First we have

$$\mathcal{S}_n^{(\alpha,\beta)}(f_m; \overline{G})(z) = \sum_{p=0}^n \binom{n}{p} \Delta_{1/(n+\beta)}^p G_m(\alpha/(n+\beta)) F_k(z),$$

where $G_m(w) = \frac{1}{2\pi i} \int_{|u|=1} \frac{f_m(\Psi(u))}{u-w} du$ and $F(w) = \frac{1}{2\pi i} \int_{|u|=1} \frac{f(\Psi(u))}{u-w} du$.

Note here that since by Gaier [76], p. 48, first relation before (6.17), we have

$$\mathcal{F}_k(w) = \frac{1}{2\pi i} \int_{|u|=1} \frac{F_k(\Psi(u))}{u-w} du = w^k, \text{ for all } |w| < 1,$$

evidently that $\mathcal{F}_k(w)$ can be extended by continuity on $\partial \mathbb{D}_1$. This also immediately implies that $G_m(w) = \frac{1}{2\pi i} \int_{|u|=1} \frac{f_m(\Psi(u))}{u-w} du$ can be extended by continuity on $\partial \mathbb{D}_1$, which means that $\mathcal{S}_n^{(\alpha,\beta)}(F_k; G)(z)$ and $\mathcal{S}_n^{(\alpha,\beta)}(f_m; G)(z)$ are well defined.

Now, taking into account the Cauchy's theorem we also can write

$$G_m(w) = \frac{1}{2\pi i} \int_{|u|=\eta} \frac{f_m(\Psi(u))}{u-w} du \text{ and } F(w) = \frac{1}{2\pi i} \int_{|u|=\eta} \frac{f(\Psi(u))}{u-w} du.$$

For all $n, m \in \mathbb{N}$ and $z \in G$ it follows

$$|\mathcal{S}_n^{(\alpha,\beta)}(f_m; \overline{G})(z) - \mathcal{S}_n^{(\alpha,\beta)}(f; \overline{G})(z)|$$

$$\leq \sum_{p=0}^n \binom{n}{p} |\Delta_{1/(n+\beta)}^p (G_m - F)(\alpha/(n+\beta))| \cdot |F_k(z)|$$

$$\leq \sum_{p=0}^n \binom{n}{p} \sum_{j=0}^p \binom{p}{j} |(G_m - F)(\alpha/(n+\beta) + (p-j)/(n+\beta))| \cdot |F_k(z)|$$

$$\leq \sum_{p=0}^n \binom{n}{p} \sum_{j=0}^p \binom{p}{j} C_{j,p,\eta,\alpha,\beta} \|f_m - f\|_{\overline{G}_\eta} \cdot |F_k(z)|$$

$$\leq M_{n,p,\eta,\alpha,\beta,G_\eta} \|f_m - f\|_{\overline{G}_\eta},$$

which by $\lim_{m \to \infty} \|f_m - f\|_{\overline{G}_\eta} = 0$ (see e.g. the proof of Theorem 2, p. 52 in Suetin [186]) implies the desired conclusion. Here $\|f_m - f\|_{\overline{G}_\eta}$ denotes the uniform norm of $f_m - f$ on $\overline{G}_\eta$.

Consequently we obtain

$$|\mathcal{S}_n^{(\alpha,\beta)}(f;\overline{G})(z) - f(z)| \leq \sum_{k=0}^{\infty} |a_k(f)| \cdot |\mathcal{S}_n^{(\alpha,\beta)}(F_k;\overline{G})(z) - F_k(z)|$$

$$= \sum_{k=0}^{n} |a_k(f)| \cdot |\mathcal{S}_n^{(\alpha,\beta)}(F_k;\overline{G})(z) - F_k(z)|$$

$$+ \sum_{k=n+1}^{\infty} |a_k(f)| \cdot |\mathcal{S}_n^{(\alpha,\beta)}(F_k;\overline{G})(z) - F_k(z)|.$$

Therefore it remains to estimate $|a_k(f)| \cdot |\mathcal{S}_n^{(\alpha,\beta)}(F_k;\overline{G})(z) - F_k(z)|$, firstly for all $0 \leq k \leq n$ and secondly for $k \geq n+1$, where

$$\mathcal{S}_n^{(\alpha,\beta)}(F_k;\overline{G})(z) = \sum_{p=0}^{n} \binom{n}{p} [\Delta_{1/(n+\beta)}^p \mathcal{F}_k(\alpha/(n+\beta))] \cdot F_p(z).$$

First it is useful to observe that by Gaier [76], p. 48, combined with the Cauchy's theorem, for any fixed $1 < \eta < R$ we have

$$\mathcal{F}_k(w) := \frac{1}{2\pi i} \int_{|u|=\eta} \frac{F_k[\Psi(u)]}{u-w} du = w^k = e_k(w), \text{ for all } |w| < \eta.$$

Denote

$$D_{n,p,k}^{(\alpha,\beta)} = \binom{n}{p} \Delta_{1/(n+\beta)}^p e_k(\alpha/(n+\beta))$$

$$= \binom{n}{p} [\alpha/(n+\beta), (\alpha+1)/(n+\beta), ..., (\alpha+p)/(n+\beta); e_k] \cdot (p!)/(n+\beta)^p.$$

It follows

$$\mathcal{S}_n^{(\alpha,\beta)}(F_k;\overline{G})(z) = \sum_{p=0}^{n} D_{n,p,k}^{(\alpha,\beta)} \cdot F_p(z).$$

Since $S_n^{(\alpha,\beta)}(f)(1) = f[(n+\alpha)/(n+\beta)]$ and since each e_k is convex of any order, it follows $D_{n,p,k}^{(\alpha,\beta)} > 0$ and, by taking $f(z) = e_k(z)$ we get $\sum_{p=0}^n D_{n,p,k}^{(\alpha,\beta)} = \frac{(n+\alpha)^k}{(n+\beta)^k} \leq 1$, for all k and n.

In the estimation of $|a_k(f)| \cdot |\mathcal{S}_n^{(\alpha,\beta)}(F_k;\overline{G})(z) - F_k(z)|$ we distinguish two cases : 1) $0 \leq k \leq n$; 2) $k > n$.

Case 1. We have

$$|\mathcal{S}_n^{(\alpha,\beta)}(F_k;\overline{G})(z) - F_k(z)| \leq |F_k(z)| \cdot |1 - D_{n,k,k}^{(\alpha,\beta)}| + \sum_{p=0}^{k-1} D_{n,p,k}^{(\alpha,\beta)} \cdot |F_p(z)|.$$

Fix now $1 < r < \eta$. By the inequality (13), p. 44 in Suetin [186] we have

$$|F_p(z)| \leq C(r)r^p, \text{ for all } z \in \overline{G}_r, \ p \geq 0,$$

which immediately implies

$$|\mathcal{S}_n^{(\alpha,\beta)}(F_k;\overline{G})(z) - F_k(z)|$$

$$\leq \left[1 - \frac{n(n-1)...(n-(k-1))}{(n+\beta)^k}\right]C(r)r^k$$

$$+ \left[\frac{(n+\alpha)^k}{(n+\beta)^k} - \frac{n(n-1)...(n-(k-1))}{(n+\beta)^k}\right]C(r)r^k$$

$$= 2\left[1 - \frac{n(n-1)...(n-(k-1))}{(n+\beta)^k}\right]C(r)r^k + \left[\frac{(n+\alpha)^k}{(n+\beta)^k} - 1\right]C(r)r^k$$

$$\leq 2\left[1 - \frac{n(n-1)...(n-(k-1))}{(n+\beta)^k}\right]C(r)r^k \leq \frac{1}{n+\beta}\left[2\beta k + k(k-1)\right]C(r)r^k.$$

Here we used the inequality $1 - \Pi_{i=1}^k x_i \leq \sum_{i=1}^k (1-x_i)$, valid if all $x_i \in [0,1]$. Also by the above formula for $a_k(f)$ we easily obtain $|a_k(f)| \leq \frac{C(\eta,f)}{\eta^k}$, for all $k \geq 0$. Note that $C(r), C(\eta,f) > 0$ are constants independent of k.

For all $z \in \overline{G}_r$ and $k = 0,1,2,...n$ it follows

$$|a_k(f)| \cdot |\mathcal{S}_n^{(\alpha,\beta)}(F_k;\overline{G})(z) - F_k(z)| \leq \frac{C(r,\eta,f)}{n+\beta}\left[2\beta k + k(k-1)\right]\cdot\left[\frac{r}{\eta}\right]^k,$$

that is

$$\sum_{k=0}^n |a_k(f)| \cdot |\mathcal{S}_n^{(\alpha,\beta)}(F_k;\overline{G})(z) - F_k(z)| \leq \frac{C(r,\eta,f)}{n+\beta}\sum_{k=1}^n[2\beta k + k(k-1)]d^k,$$

for all $z \in \overline{G}_r$, where $0 < d = r/\eta < 1$.

Also, clearly we have $\sum_{k=1}^n[2\beta k + k(k-1)]d^k \leq \sum_{k=1}^\infty[2\beta k+]k(k-1)d^k < \infty$ which finally implies that

$$\sum_{k=0}^n |a_k(f)| \cdot |\mathcal{S}_n^{(\alpha,\beta)}(F_k;\overline{G})(z) - F_k(z)| \leq \frac{C^*(r,\eta,\beta,f)}{n}.$$

Case 2. We have

$$\sum_{k=n+1}^\infty |a_k(f)| \cdot |\mathcal{S}_n^{(\alpha,\beta)}(F_k;\overline{G})(z) - F_k(z)| \leq \sum_{k=n+1}^\infty |a_k(f)| \cdot |\mathcal{S}_n^{(\alpha,\beta)}(F_k;\overline{G})(z)|$$

$$+ \sum_{k=n+1}^\infty |a_k(f)| \cdot |F_k(z)|.$$

By the estimates mentioned in the case 1), we immediately get

$$\sum_{k=n+1}^\infty |a_k(f)| \cdot |F_k(z)| \leq C(r,\eta,f) \sum_{k=n+1}^\infty d^k, \text{ for all } z \in \overline{G}_r,$$

with $d = r/\beta$.

Also,

$$\sum_{k=n+1}^{\infty} |a_k(f)| \cdot |\mathcal{S}_n^{(\alpha,\beta)}(F_k; \overline{G})(z)| = \sum_{k=n+1}^{\infty} |a_k(f)| \cdot \left| \sum_{p=0}^{n} D_{n,p,k}^{(\alpha,\beta)} \cdot F_p(z) \right|$$

$$\leq \sum_{k=n+1}^{\infty} |a_k(f)| \cdot \sum_{p=0}^{n} D_{n,p,k}^{(\alpha,\beta)} \cdot |F_p(z)|.$$

But for $p \leq n < k$ and taking into account the estimates obtained in the Case 1) we get

$$|a_k(f)| \cdot |F_p(z)| \leq C(r,\eta,f) \frac{r^p}{\eta^k} \leq C(r,\beta,f) \frac{r^k}{\eta^k}, \text{ for all } z \in \overline{G}_r,$$

which implies

$$\sum_{k=n+1}^{\infty} |a_k(f)| \cdot |\mathcal{S}_n^{(\alpha,\beta)}(F_k; \overline{G})(z) - F_k(z)| \leq C(r,\eta,f) \sum_{k=n+1}^{\infty} \sum_{p=0}^{n} D_{n,p,k}^{(\alpha,\beta)} \left[\frac{r}{\beta} \right]^k$$

$$= C(r,\beta,f) \sum_{k=n+1}^{\infty} \left[\frac{r}{\beta} \right]^k$$

$$= C(r,\beta,f) \frac{d^{n+1}}{1-d},$$

with $d = r/\beta$.

In conclusion, collecting the estimates in the Cases 1) and 2) we obtain

$$|\mathcal{S}_n^{(\alpha,\beta)}(f; \overline{G})(z) - f(z)| \leq \frac{C_1}{n+\beta} + C_2 d^{n+1} \leq \frac{C}{n}, \; z \in \overline{G}_r, \; n \in \mathbb{N}.$$

This proves the theorem.   □

Remark. In the case when $\alpha = \beta = 0$ we recapture Theorem 1.1.8.

As a consequence of the Remarks 1 and 2 after the proof of Theorem 1.1.8, we can state the following result.

Theorem 1.6.15. *Let G be a compact Faber set such that $\tilde{\mathbb{C}} \setminus G$ is simply connected. If f is analytic on G, that is there exists $R > 1$ such that f is analytic in G_R and if f is not a polynomial of degree ≤ 1, then for any $1 < r < R$ we have*

$$\|\mathcal{S}_n^{(\alpha,\beta)}(f; \overline{G}) - f\|_{\overline{G}_r} \sim \frac{1}{n}, \; n \in \mathbb{N},$$

where the constants in the equivalence depend on f, α, β, r and G_r but are independent of n. Here $\|f\|_{\overline{G}_r} = \sup_{z \in \overline{G}_r} |f(z)|$.

Proof.   According to Remark 2 after the proof of Theorem 1.1.8, there exists g analytic in $\overline{\mathbb{D}}_r$ such that $f = T(g)$, that is $g = T^{-1}(f)$ (therefore F can be extended by continuity on $\partial \mathbb{D}_1$. By hypothesis on f it follows that f cannot be of the form $f(z) = c_0 F_0(z) + c_1 F_1(z)$ where F_0 and F_1 are the Faber polynomials of degree 0 and

1 respectively and $c_0, c_1 \in \mathbb{C}$. This immediately implies that g is not a polynomial of degree ≤ 1.

First we have $S_n^{(\alpha,\beta)}(T^{-1}(f)) = T^{-1}[\mathcal{S}_n^{(\alpha,\beta)}(f; \overline{G})]$. Indeed,

$$S_n^{(\alpha,\beta)}(T^{-1}(f))(z) = \sum_{p=0}^{n} \binom{n}{p} \Delta_{1/(n+\beta)}^p T^{-1}(f)(\alpha/(n+\beta)) z^p$$

$$= \sum_{p=0}^{n} \binom{n}{p} \Delta_{1/(n+\beta)}^p F(\alpha/(n+\beta)) z^p,$$

since $T^{-1}(f)(\xi) = \frac{1}{2\pi i} \int_{|w|=1} \frac{f[\Psi(w)]}{w-\xi} dw = F(\xi)$, and

$$T^{-1}[\mathcal{S}_n^{(\alpha,\beta)}(f; \overline{G})](z)$$

$$= \frac{1}{2\pi i} \int_{|w|=1} \frac{\mathcal{S}_n^{(\alpha,\beta)}(f; \overline{G})[\Psi(w)]}{w-z} dw$$

$$= \sum_{p=0}^{n} \binom{n}{p} \Delta_{1/(n+\beta)}^p F(\alpha/(n+\beta)) \frac{1}{2\pi i} \int_{|w|=1} \frac{F_p[\Psi(w)]}{w-z} dw$$

$$= \sum_{p=0}^{n} \binom{n}{p} \Delta_{1/(n+\beta)}^p F(\alpha/(n+\beta)) z^p,$$

since according to Gaier [76], p. 48, first relation before (6.17), we have

$$\frac{1}{2\pi i} \int_{|w|=1} \frac{F_p[\Psi(w)]}{w-z} dw = z^p.$$

Then by Theorem 1.6.3 (see also Corollary 1.6.4) and by the linearity and continuity of T^{-1} we get

$$\frac{C}{n} \leq \|S_n^{(\alpha,\beta)}(g) - g\|_r = \|S_n^{(\alpha,\beta)}(g) - T^{-1}(f)\|_r$$

$$= \|T^{-1}[\mathcal{S}_n^{(\alpha,\beta)}(f; \overline{G})] - T^{-1}(f)\|_r$$

$$\leq \|T^{-1}\| \cdot \|\mathcal{S}_n^{(\alpha,\beta)}(f; \overline{G}) - f\|_{\overline{G}_r}$$

$$\leq M\|\mathcal{S}_n^{(\alpha,\beta)}(f; \overline{G}) - f\|_{\overline{G}_r},$$

which proves the lower estimate.

On the other hand we have $T[S_n^{(\alpha,\beta)}(g)] = \mathcal{S}_n^{(\alpha,\beta)}(T(g); \overline{G})$. Indeed,

$$T[S_n^{(\alpha,\beta)}(g)](z) = \sum_{p=0}^{n} \Delta_{1/(n+\beta)}^p g(\alpha/(n+\beta)) F_p(z),$$

and

$$\mathcal{S}_n^{(\alpha,\beta)}(T(g); \overline{G})(z) = \sum_{p=0}^{n} \binom{n}{p} \Delta_{1/(n+\beta)}^p H(\alpha/(n+\beta)) F_p(z),$$

where according to Gaier [76], p. 49. relation (6.17') we have

$$H(w) = \frac{1}{2\pi i} \int_{|u|=1} \frac{T(g)[\Psi(u)]}{u-w} du = g(w).$$

Therefore by the same Corollary 1.6.4 and by the linearity and continuity of T we obtain

$$
\begin{aligned}
\|\mathcal{S}_n^{(\alpha,\beta)}(f;\overline{G}) - f\|_{\overline{G}_r} &= \|\mathcal{S}_n^{(\alpha,\beta)}(T(g);\overline{G}) - T(g)\|_{\overline{G}_r} \\
&= \|T[S_n^{(\alpha,\beta)}(g)] - T(g)\|_{\overline{G}_r} \\
&\leq \|\|T\|\| \cdot \|S_n^{(\alpha,\beta)}(g) - g\|_r \leq \frac{C}{n},
\end{aligned}
$$

which proves the upper estimate and the theorem. $\qquad\square$

Remark. For $\alpha = \beta = 0$, we recapture Theorem 1.1.9.

1.7   Bernstein-Kantorovich Type Polynomials

In this section we extend some results in the previous sections to the following complex Kantorovich variants of these polynomials defined (for the case of real variable) by Kantorovich [112]

$$
K_n(f)(z) = (n+1)\sum_{k=0}^{n} p_{n,k}(z) \int_{k/(n+1)}^{(k+1)/(n+1)} f(t)dt,
$$

and (for the case of real variable) Bărbosu [36]

$$
K_n^{(\alpha,\beta)}(f)(z) = (n+1+\beta)\sum_{k=0}^{n} p_{n,k}(z) \int_{(k+\alpha)/(n+1+\beta)}^{(k+1+\alpha)/(n+1+\beta)} f(t)dt.
$$

For our purpose will be useful the results expressed by the following.

Theorem 1.7.1. *Let $F(z) = \int_0^z f(t)dt$.*

(i) (see e.g. Lorentz [125], p. 30) Denoting by $B_n(f)(z)$ the Bernstein polynomials, we have

$$
K_n(f)(z) = B'_{n+1}(F)(z), z \in \mathbb{C}.
$$

(ii) (Gal [91]) Denoting $S_n^{(\alpha,\beta)}(f)(z) = \sum_{k=0}^{n} \binom{n}{k} z^k (1-z)^{n-k} f[(k+\alpha)/(n+\beta)], z \in \mathbb{C}$, the Bernstein-Stancu polynomials studied in Section 1.6, where $0 \leq \alpha \leq \beta$ are independent of n, we have

$$
K_n^{(\alpha,\beta)}(f)(z) = \frac{n+1+\beta}{n+1}\left[S_{n+1}^{(\alpha,\beta)}(F)\right]'(z), z \in \mathbb{C}.
$$

Proof.   (ii) It is immediate by the formula

$$
\begin{aligned}
&[S_{n+1}^{(\alpha,\beta)}(F)]'(z) \\
&= (n+1+\beta)\sum_{k=0}^{n} p_{n,k}(z)\left[F\left(\frac{k+\alpha+1}{n+\beta+1}\right) - F\left(\frac{k+\alpha}{n+1+\beta}\right)\right] \\
&\quad -\beta\sum_{k=0}^{n} p_{n,k}(z)\left[F\left(\frac{k+\alpha+1}{n+\beta+1}\right) - F\left(\frac{k+\alpha}{n+1+\beta}\right)\right] \\
&= K_n^{(\alpha,\beta)}(f)(z) - \frac{\beta}{n+1+\beta}K_n^{(\alpha,\beta)}(f)(z).
\end{aligned}
$$

$\qquad\square$

Now, as a consequence of Theorem 1.7.1, (i) and Theorem 1.1.6, we immediately get the following.

Corollary 1.7.2. (Gal [91]) *Let $f : \mathbb{D}_R \to \mathbb{C}$ be analytic in $\mathbb{D}_R$ with $R > 1$ and $1 \leq r < R$.*

(i) If f is not a polynomial of degree ≤ 0 then for all $n \in \mathbb{N}$ we have

$$\|K_n(f) - f\|_r \sim \frac{1}{n},$$

where the constants in the equivalence depend only on f and r.

(ii) If f is not a polynomial of degree $\leq \max\{1, p-1\}$ then for all $p, n \in \mathbb{N}$ we have

$$\|K_n^{(p)}(f) - f^{(p)}\|_r \sim \frac{1}{n},$$

with the constants in the equivalence depending only on f, r and p.

Proof. We combine Theorem 1.7.1, (i) with Theorem 1.1.6.

(i) We get

$$\|K_n(f) - f\|_r = \|B'_{n+1}(F) - F'\|_r \sim \frac{1}{n+1},$$

if F is not a polynomial of degree $\leq \max\{1, 1\} = 1$, which ends the proof.

(ii) We obtain

$$\|K_n^{(p)}(f) - f^{(p)}\|_r = \|B_{n+1}^{(p+1)}(F) - F^{(p+1)}\|_r \sim \frac{1}{n+1},$$

if F is not a polynomial of degree $\leq \max\{1, p\} = p$, which ends the proof. $\square$

As a consequence of Theorems 1.7.1, (ii) and Theorem 1.6.5, we also get the following.

Corollary 1.7.3. (Gal [91]) *Let $f : \mathbb{D}_R \to \mathbb{C}$ be analytic in $\mathbb{D}_R$ with $R > 1$, $1 \leq r < R$ and $0 \leq \alpha \leq \beta$, $\alpha + \beta > 0$.*

(i) If f is not identical 0, then for all $n \in \mathbb{N}$ we have

$$\|K_n^{(\alpha,\beta)}(f) - f\|_r \sim \frac{1}{n+\beta},$$

where the constants in the equivalence depend only on f, r, α and β.

(ii) If f is not a polynomial of degree $\leq p - 1$ then for all $p, n \in \mathbb{N}$ we have

$$\|[K_n^{(\alpha,\beta)}(f)]^{(p)} - f^{(p)}\|_r \sim \frac{1}{n},$$

with the constants in the equivalence depending only on f, r, α, β and p.

Proof. We combine Theorem 1.7.1, (ii) with Theorem 1.6.5.

(i) We get

$$\|K_n^{(\alpha,\beta)}(f) - f\|_r = \|[S_{n+1}^{(\alpha,\beta)}(F)]' - F'\|_r \sim \frac{1}{n+\beta},$$

if F is not a polynomial of degree ≤ 0, which ends the proof.

(ii) We obtain

$$\|[K_n^{(\alpha,\beta)}(f)]^{(p)} - f^{(p)}\|_r = \|[S_{n+1}^{(\alpha,\beta)}(F)]^{(p+1)} - F^{(p+1)}\|_r \sim \frac{1}{n+\beta},$$

if F is not a polynomial of degree $\leq p$, which ends the proof. $\square$

Upper estimates with explicit constants in approximation by these kind of poly-nomials and in Voronovskaja-type formula can be derived as follows. First we consider the case of $K_n(f)(z)$ polynomials.

Theorem 1.7.4. (Gal [91]) *Let $f : \mathbb{D}_R \to \mathbb{C}$ be analytic in $\mathbb{D}_R = \{z \in \mathbb{C}; |z| < R\}$ with $R > 1$, i.e. $f(z) = \sum_{k=0}^{\infty} c_k z^k$, for all $z \in \mathbb{D}_R$. Suppose $1 \leq r < r_1 < R$. Then for all $|z| \leq r$ and $n, p \in \mathbb{N}$, we have :*
    (i)

$$|K_n^{(p)}(f)(z) - f^{(p)}(z)| \leq \frac{C_{2,r_1}(f)(p+1)! r_1}{(n+1)(r_1 - r)^{p+2}},$$

where $0 < C_{2,r_1}(f) = 2 \sum_{j=2}^{\infty} (j-1)|c_{j-1}| r_1^j < \infty$;
    (ii)

$$|K_n(f)(z) - f(z) - \frac{1-2z}{2(n+1)} \cdot f'(z) - \frac{z(1-z)}{2(n+1)} \cdot f''(z)|$$

$$\leq C_{r_1, n+1}(f) \cdot \frac{r_1}{(r_1 - r)^2},$$

where

$$C_{r_1,n}(f) = \frac{5(1+r_1)^2}{2n} \cdot \frac{\sum_{k=3}^{\infty} |c_{k-1}|(k-1)(k-2)^2 r_1^{k-2}}{n}.$$

Proof.   (i) Combining Theorem 1.7.1, (i) with Theorem 1.1.2, (ii), we obtain

$$|K_n^{(p)}(f)(z) - f^{(p)}(z)| = |B_{n+1}^{(p+1)}(F)(z) - F^{(p+1)}(z)| \leq \frac{M_{2,r_1}(F)(p+1)! r_1}{(n+1)(r_1 - r)^{p+2}},$$

where $0 < M_{2,r_1}(F) = 2 \sum_{j=2}^{\infty} j(j-1)|C_j| r_1^j < \infty$ and $F(z) = \sum_{k=0}^{\infty} C_k z^k$, $z \in \mathbb{D}_R$. But we also get

$$F(z) = \int_0^z [\sum_{k=0}^{\infty} c_k t^k] dt = \sum_{k=0}^{\infty} \frac{c_k}{k+1} z^{k+1} = \sum_{k=1}^{\infty} \frac{c_{k-1}}{k} z^k,$$

which implies $C_k = \frac{c_{k-1}}{k}$ and $C_{2,r_1}(f) = 2 \sum_{j=2}^{\infty} (j-1)|c_{j-1}| r_1^j$.
    (ii) Replacing in Theorem 1.1.3, (ii), n by $n+1$, r by r_1 and f by F, for all $|z| \leq r_1$ and $n \in \mathbb{N}$, we obtain

$$\left| B_{n+1}(F)(z) - F(z) - \frac{z(1-z)}{2(n+1)} F''(z) \right| \leq \frac{5(1+r_1)^2}{2(n+1)} \cdot \frac{M_{r_1}(F)}{n+1},$$

where

$$M_{r_1}(F) = \sum_{k=3}^{\infty} |C_k| k(k-1)(k-2)^2 r_1^{k-2} = \sum_{k=3}^{\infty} |c_{k-1}|(k-1)(k-2)^2 r_1^{k-2}$$
$$:= A_{r_1}(f).$$

Here again we wrote $F(z) = \sum_{k=0}^{\infty} C_k z^k$, for all $z \in \mathbb{D}_R$.

Now, denoting $C_{r_1,n}(f) = \frac{5(1+r_1)^2}{2n} \cdot \frac{A_{r_1}(f)}{n}$, by Γ the circle of radius $r_1 > r$ and center 0, and $E_n(F)(z) = B_{n+1}(F)(z) - F(z) - \frac{z(1-z)}{2(n+1)}F''(z)$, since for any $|z| \le r$ and $v \in \Gamma$, we have $|v - z| \ge r_1 - r$, by the Cauchy's formula it follows that for all $|z| \le r$ and $n \in \mathbb{N}$, we obtain

$$|E_n'(F)(z)| = \frac{1}{2\pi}\left|\int_\Gamma \frac{E_n(f)(z)}{(v-z)^2}dv\right| \le C_{r_1,n+1}(f)\frac{1}{2\pi}\frac{2\pi r_1}{(r_1-1)^2}$$

$$= C_{r_1,n+1}(f) \cdot \frac{r_1}{(r_1-r)^2}.$$

But by Theorem 1.7.1, (i) we obtain

$$E_n'(F)(z) = K_n(f)(z) - f(z) - \frac{1-2z}{2(n+1)} \cdot f'(z) - \frac{z(1-z)}{2(n+1)} \cdot f''(z),$$

which proves the theorem.   □

In the case of $K_n^{(\alpha,\beta)}(f)(z)$ polynomials we have the following.

Theorem 1.7.5. (Gal [91]) *Let $f : \mathbb{D}_R \to \mathbb{C}$ be analytic in $\mathbb{D}_R = \{z \in \mathbb{C}; |z| < R\}$ with $R > 1$, i.e. $f(z) = \sum_{k=0}^\infty c_k z^k$, for all $z \in \mathbb{D}_R$. Suppose $1 \le r < r_1 < R$. Then for all $|z| \le r$ and $n, p \in \mathbb{N}$, we have :*
(i)

$$|[K_n^{(\alpha,\beta)}(f)]^{(p)}(z) - f^{(p)}(z)| \le \frac{C_{2,r_1}^{(\beta)}(f)(p+1)!r_1}{(n+1)(r_1-r)^{p+2}} + \frac{\beta}{n+1}\|f\|_r,$$

where $0 < C_{2,r_1}^{(\beta)}(f) = 2\sum_{j=2}^\infty (j-1)|c_{j-1}|r_1^j + 2\beta\sum_{j=1}^\infty |c_{j-1}|r_1^j < \infty$;
(ii)

$$\left|K_n^{(\alpha,\beta)}(f)(z) - f(z) + \left(\frac{\beta z - \alpha}{n+1} - \frac{1-2z}{2(n+\beta+1)}\right)f'(z)\right.$$

$$\left. - \frac{z(1-z)}{2(n+\beta+1)}f''(z)\right| \le \frac{C(f,r_1,\alpha,\beta)}{(n+1)(n+\beta+1)} \cdot \frac{r_1}{(r_1-r)^2},$$

where $C(f,r_1,\alpha,\beta)$ is a positive constant depending only on f, r_1, α and β.

Proof.   (i) Combining Theorem 1.7.1, (ii) with Theorem 1.6.1, for all $|z| \le r$ we obtain

$$|[K_n^{(\alpha,\beta)}(f)]^{(p)}(z) - f^{(p)}(z)| = \left|\frac{n+1+\beta}{n+1}[S_{n+1}^{(\alpha,\beta)}(F)]^{(p+1)}(z) - F^{(p+1)}(z)\right|$$

$$\le \left|\frac{n+1+\beta}{n+1}|[S_{n+1}^{(\alpha,\beta)}(F)]^{(p+1)}(z) - F^{(p+1)}(z)| + \frac{\beta}{n+1}|F^{(p+1)}(z)|\right|$$

$$\le \frac{n+1+\beta}{n+1} \cdot \frac{M_{2,r_1}^{(\beta)}(F)(p+1)!r_1}{(n+\beta+1)(r_1-r)^{p+2}} + \frac{\beta}{n+1} \cdot |f^{(p)}(z)|$$

$$\leq \frac{M_{2,r_1}^{(\beta)}(F)(p+1)!r_1}{(n+1)(r_1-r)^{p+2}} + \frac{\beta}{n+1} \cdot \|f^{(p)}\|_r,$$

where $\|f^{(p)}\|_r = \sup\{|f^{(p)}(z)|; |z| \leq r\}$ and reasoning exactly as in the proof of Theorem 1.7.4, (i), we get

$$M_{2,r_1}^{(\beta)}(F) = 2\sum_{j=2}^{\infty} j(j-1)|C_j|r_1^j + 2\beta\sum_{j=1}^{\infty} j|C_j|r_1^j$$

$$= 2\sum_{j=2}^{\infty} (j-1)|c_{j-1}|r_1^j + 2\beta\sum_{j=1}^{\infty} |c_{j-1}|r_1^j := C_{2,r_1}^{(\beta)}(f).$$

(ii) Replacing in Remark 2 after Theorem 1.6.2 n by $n+1$, r by r_1 and f by F, for all $|z| \leq r_1$ and $n \in \mathbb{N}$, we obtain

$$\left| S_{n+1}^{(\alpha,\beta)}(F)(z) - F(z) + \frac{\beta z - \alpha}{n+\beta+1}F'(z) - \frac{(n+1)z(1-z)}{2(n+\beta+1)^2}F''(z) \right|$$

$$\leq \frac{C(f,r_1,\alpha,\beta)}{(n+\beta+1)^2},$$

where the positive constant $C(f,r_1,\alpha,\beta)$ depends only on f,r,α and β. Let us denote

$$E_n(F)(z) = S_{n+1}^{(\alpha,\beta)}(F)(z) - F(z) + \frac{\beta z - \alpha}{n+\beta+1}F'(z) - \frac{(n+1)z(1-z)}{2(n+\beta+1)^2}F''(z).$$

If Γ is the circle of radius $r_1 > r$ and center 0, and since for any $|z| \leq r$ and $v \in \Gamma$, we have $|v - z| \geq r_1 - r$, by the Cauchy's formula it follows that for all $|z| \leq r$ and $n \in \mathbb{N}$, we obtain as in the proof of Theorem 1.7.4, (ii)

$$|E_n'(F)(z)| \leq C(f,r_1,\alpha,\beta) \cdot \frac{r_1}{(r_1-r)^2} \cdot \frac{1}{(n+\beta+1)^2}.$$

But by Theorem 1.7.1, (ii) we obtain

$$E_n'(F)(z) = \frac{n+1}{n+1+\beta}K_n^{(\alpha,\beta)}(f)(z) - f(z) + \frac{1}{n+\beta+1}[(\beta z - \alpha)f(z)]'$$

$$- \frac{n+1}{2(n+\beta+1)^2}[(z-z^2)f'(z)]' = \frac{n+1}{n+\beta+1}$$

$$\cdot \left[K_n^{(\alpha,\beta)}(f)(z) - f(z) + f'(z)\left(\frac{\beta z - \alpha}{n+1} - \frac{1-2z}{2(n+\beta+1)} \right)f'(z) \right.$$

$$\left. - \frac{z(1-z)}{2(n+\beta+1)}f''(z) \right],$$

which immediately proves the theorem. $\qquad\square$

In what follows we will prove an equivalence result for approximation in Voronovskaja's theorem in the case of complex Bernstein-Kantorovich polynomials $K_n(f)(z)$, analogous to that for complex Bernstein polynomials contained by Corollary 1.3.4.

Theorem 1.7.6. *Let $R > 1$ and let $f : \mathbb{D}_R \to \mathbb{C}$ be an analytic function, say $f(z) = \sum_{k=0}^{\infty} c_k z^k$. If f is not a polynomial of degree $\leq 2p-1$ then for any $1 \leq r < R$ and any natural number p we have*

$$\left\| K_n(f) - f - \sum_{j=1}^{2p} \frac{(n+1)^{-j}}{j!} [f^{(j)} T_{n+1,j} + f^{(j-1)} T'_{n+1,j}] \right\|_r \sim \frac{1}{n^{p+1}}, \ n \in \mathbb{N},$$

where the constants in the equivalence depend only on f, r and p and are independent of n. Recall here that $T_{n+1,j}(z) = \sum_{i=0}^{n+1} (i - nz)^j \binom{n+1}{i} z^i (1-z)^{n+1-i}$.

Proof. Denoting $F(z) = \int_0^z f(t) dt$, by Theorem 1.3.2 we get

$$\left\| B_{n+1}(F) - F - \sum_{j=1}^{2p} \frac{F^{(j)}}{j!} (n+1)^{-j} T_{n+1,j} \right\|_r \leq \frac{C_{p,r}(f)}{(n+1)^{p+1}}.$$

Let $1 \leq r < r_1 < R$ and denote by γ the circle of radius $r_1 > r$ and center 0 and

$$H(z) = B_{n+1}(F)(z) - F(z) - \sum_{j=1}^{2p} \frac{F^{(j)}(z)}{j!} (n+1)^{-j} T_{n+1,j}(z).$$

Since for any $|z| \leq r$ and $v \in \gamma$ we have $|v - z| \geq r_1 - r$, by the Cauchy's formula we get $H'(z) = \frac{1}{2\pi i} \int_\gamma \frac{H(v)}{(v-z)^2} dv$, which implies

$$\|H'\|_r \leq \frac{\|H\|_{r_1}}{2\pi} \cdot \frac{2\pi r_1}{(r_1 - r)^2} \leq \frac{C_{p,r}(F) r_1}{(r_1 - r)^2} \cdot \frac{1}{(n+1)^{p+1}} \leq \frac{C_{p,r,r_1}(F)}{n^{p+1}},$$

which is exactly the upper estimate in the statement of Theorem 1.7.6. Note that if $f(z) = \sum_{k=0}^{\infty} c_k z^k$ then

$$F(z) = \sum_{k=0}^{\infty} \frac{c_k}{k+1} z^{k+1} = \sum_{j=1}^{\infty} \frac{c_{j-1}}{j} z^j := \sum_{j=1}^{\infty} C_j^* z^j,$$

where $C_j^* = \frac{c_{j-1}}{j}$, for all $j = 1, 2, \ldots,$.

So it remains to prove the lower estimate. In the proof of Corollary 1.3.4, write the first identity for $B_{n+1}(F)$ and F and then take the first derivative. It follows

$$K_n(f)(z) - f(z) - \sum_{j=1}^{2p} \frac{(n+1)^{-j}}{j!} [f^{(j)}(z) T_{n+1,j} + f^{j-1}(z) T'_{n+1,j}(z)]$$

$$= \frac{1}{(n+1)^{p+1}} \left\{ \frac{T'_{n+1,2p+1}(z)}{(n+1)^p (2p+1)!} F^{(2p+1)}(z) + \frac{T_{n+1,2p+1}(z)}{(n+1)^p (2p+1)!} F^{(2p+2)}(z) \right.$$

$$+ \frac{T'_{n+1,2p+2}(z)}{(n+1)^{p+1} (2p+2)!} F^{(2p+2)}(z) + \frac{T_{n+1,2p+2}(z)}{(n+1)^{p+1} (2p+2)!} F^{(2p+3)}(z)$$

$$+ \frac{1}{n+1} \left[(n+1)^{p+2} \sum_{k=2p+3}^{\infty} C_k^* E'_{k,n+1,p+1}(z) \right] \right\}.$$

Here $E_{k,n+1,p}(z)$ does not depend on F and it is a polynomial of degree $\leq k$ in z. By the Bernstein's inequality and by Lemma 1.3.3 it follows

$$\|E'_{k,n+1,p+1}\|_r \leq \frac{k}{r}\|E_{k,n+1,p+1}\|_r \leq C_p r^{k-1}\frac{(k+1)!(k-2(p+1))^2}{(n+1)^{p+2}(k-2(p+1))!}.$$

By the same Lemma 1.3.3 we get that $(n+1)^{p+2}\sum_{k=2p+3}^{\infty} C_k^* E'_{k,n+1,p+1}(z)$ is bounded in $\mathbb{D}_r$ by a constant independent of n.

It remains to deal with the expression

$$A := \frac{T'_{n+1,2p+1}(z)}{(n+1)^p(2p+1)!}F^{(2p+1)}(z) + \frac{T_{n+1,2p+1}(z)}{(n+1)^p(2p+1)!}F^{(2p+2)}(z)$$

$$+ \frac{T'_{n+1,2p+2}(z)}{(n+1)^{p+1}(2p+2)!}F^{(2p+2)}(z) + \frac{T_{n+1,2p+2}(z)}{(n+1)^{p+1}(2p+2)!}F^{(2p+3)}(z).$$

By using the recurrence formula in Lorentz [125], p. 14, relation (3)

$$T_{n+1,j+1}(z) = z(1-z)[T'_{n+1,j}(z) + (n+1)jT_{n+1,j-1}(z)],$$

for $j = 2p+1$ and $j = 2p+2$, we obtain

$$T'_{n+1,2p+1}(z) = \frac{T_{n+1,2p+2}(z)}{z(1-z)} - (n+1)(2p+1)T_{n+1,2p}(z),$$

and

$$T'_{n+1,2p+2}(z) = \frac{T_{n+1,2p+3}(z)}{z(1-z)} - (n+1)(2p+2)T_{n+1,2p+1}(z).$$

Replacing these in A, exactly as in the proof of Corollary 1.3.4 it follows that in the expression of A only the terms independent of n matter for the lower estimate. Simple calculation shows that these terms are given by

$$G(f)(z) = \frac{a_p[(1-2z)(z(1-z))^p]'f^{(2p)}(z)}{(2p+1)!}$$

$$+ \frac{a_{p+1}(1-2z)[z(1-z)]^{p+1}}{z(1-z)(2p+2)!}f^{(2p+1)}(z)$$

$$+ \frac{[z(1-z)]^{p+1}}{2^{p+1}(p+1)!}f^{(2p+2)}(z),$$

where $a_p, a_{p+1} > 0$. In order to obtain the lower estimate of order $\frac{1}{n^{p+1}}$, reasoning as in the proof of Corollary 1.3.4 it suffices to prove that if f is not a polynomial of degree $\leq 2p-1$ then $\|G(f)\|_r > 0$. Making the substitution $f^{(2p)}(z) = y(z)$ it follows that it suffices to prove that if $y(z)$ is not identical zero then $\|G(y)\|_r > 0$. For this purpose we reason as in the proof of Corollary 1.3.4. Thus we can show that the only solution of the differential equation

$$\frac{a_p[(1-2z)(z(1-z))^p]'y(z)}{(2p+1)!} + \frac{a_{p+1}(1-2z)[z(1-z)]^{p+1}}{z(1-z)(2p+2)!}y'(z)$$

$$+ \frac{[z(1-z)]^{p+1}}{2^{p+1}(p+1)!}y''(z) = 0$$

is $y(z) = 0$ for all $z \in \mathbb{D}_r$. Writing $y(z) = \sum_{k=0}^{\infty} b_k z^k$ and reasoning as in the proof of Corollary 1.3.4 , we easily obtain step by step that $b_0 = 0$, $b_1 = 0$, so on, $b_k = 0$ for all k. We omit here the calculation details which are simple. The theorem is proved. $\square$

At the end of this section, concerning the mth iterates ${}^{m}K_{n}^{(\alpha,\beta)}(f)(z)$, we can prove the following result.

Theorem 1.7.7. (Gal [91]) *Let* $f : \mathbb{D}_R \to \mathbb{C}$ *be analytic in* $\mathbb{D}_R = \{z \in \mathbb{C}; |z| < R\}$ *with* $R > 1$, *i.e.* $f(z) = \sum_{k=0}^{\infty} c_k z^k$, *for all* $z \in \mathbb{D}_R$. *Suppose* $1 \leq r < r_1 < R$. *Then for all* $|z| \leq r$ *and* $n, p \in \mathbb{N}$, *we have*

$$|[{}^{m}K_{n}^{(\alpha,\beta)}(f)]^{(p)}(z) - f^{(p)}(z)|$$
$$\leq \frac{2m}{n+1+\beta} \sum_{k=1}^{\infty} |c_{k-1}| \cdot |\beta + (k-1)| r^k \cdot \frac{(p+1)! r_1}{(r_1 - r)^{p+1}}.$$

Proof. First we easily observe that

$$ {}^{m}K_{n}^{(\alpha,\beta)}(f)(z) = \frac{d}{dz}[{}^{m}S_{n+1}^{(\alpha,\beta)}(F)](z),$$

where $F(z) = \int_0^z f(t)dt = \sum_{k=0}^{\infty} C_k z^k$. Taking into account Theorem 1.6.7, the Cauchy's formula and reasoning exactly as in the proofs of Theorem 1.7.4, (i) and 1.7.5, (i), it follows

$$|[{}^{m}K_{n}^{(\alpha,\beta)}(f)]^{(p)}(z) - f^{(p)}(z)|$$
$$= |[{}^{m}S_{n+1}^{(\alpha,\beta)}(F)]^{(p+1)}(z) - F^{(p+1)}(z)|$$
$$\leq \frac{2m}{n+1+\beta} \sum_{k=1}^{\infty} |C_k| \cdot |\beta k + k(k-1)| r^k \cdot \frac{(p+1)! r_1}{(r_1 - r)^{p+1}}$$
$$= \frac{2m}{n+1+\beta} \sum_{k=1}^{\infty} |c_{k-1}| \cdot |\beta + (k-1)| r^k \cdot \frac{(p+1)! r_1}{(r_1 - r)^{p+1}},$$

which proves the theorem. $\square$

Remark. 1) For $\beta = 0$ in Theorem 1.7.7 we get corresponding results for the iterates of classical complex Kantorovich polynomials. Note that in the real case, some asymptotic results for iterates of Kantorovich polynomials were obtained in Nagel [143].

2) If $\frac{m_n}{n} \to 0$ when $n \to \infty$, then by Theorem 1.7.7 it is immediate that

$$[{}^{m_n}K_{n}^{(\alpha,\beta)}(f)]^{(p)}(z) \to f^{(p)}(z),$$

uniformly with respect to $|z| \leq 1$, for any $1 \leq r < R$.

1.8 Favard-Szász-Mirakjan Operators

In this section we obtain quantitative estimates of the convergence and of the Voronovskaja's theorem in compact disks, for complex Favard-Szász-Mirakjan operators attached to an analytic function in a disk of radius $R > 1$ and center 0. The section is divided in two parts. In the first part of it, the analytic function satisfies some suitable exponential-type growth condition, while in the second part it does not satisfy such of conditions. But in the second case, the price paid is that the uniform convergence and the estimates hold in closed disks of radii $< \frac{R}{2}$ only.

Also, we will prove that beginning with an index, these operators preserve the starlikeness, convexity and spirallikeness in the unit disk.

If $f : [0, \infty) \to \mathbb{R}$ then it is well-known that the Favard-Szász-Mirakjan operators are given by (see Favard [72], Szász [188], Mirakjan [137]) $S_n(f)(x) = e^{-nx} \sum_{j=0}^{\infty} \frac{(nx)^j}{j!} f(j/n)$, $x \in [0, \infty)$, where for the convergence of $S_n(f)(x)$ to $f(x)$, usually f is supposed to be of exponential growth, that is $|f(x)| \leq Ce^{Bx}$, for all $x \in [0, +\infty)$, with $C, B > 0$ (see Favard [72]). Also, concerning quantitative estimates in approximation of $f(x)$ by $S_n(f)(x)$, in e.g. Totik [191], it is proved that under some additional assumptions on f, we actually have $|S_n(f)(x) - f(x)| \leq \frac{C}{n}$, for all $x \in \mathbb{R}_+$, $n \in \mathbb{N}$.

The complex Favard-Szász-Mirakjan operator is obtained from the real version, simply replacing the real variable x by the complex one z, that is

$$S_n(f)(z) = e^{-nz} \sum_{j=0}^{\infty} \frac{(nz)^j}{j!} f(j/n).$$

Let us note that in our results, the domain of definition of the approximated function $f : \overline{\mathbb{D}}_R \bigcup [R, \infty) \to \mathbb{C}$ seem to be rather strange. However, the analyticity of f on $\mathbb{R}$ on $\mathbb{D}_R$ assures the representation $f(z) = \sum_{k=0}^{\infty} c_k z^k$, which is essential in the proof of quantitative estimates in any $\overline{\mathbb{D}}_r$ with $1 \leq r < R$ (while on $[0, \infty)$ the well known estimates in the case of real variable can be used).

Probably a more natural domain of definition for f would be a strip around the OX-axis, but in this case the representation $f(z) = \sum_{k=0}^{\infty} c_k z^k$ fails, fact which produces the failure of the methods of proofs in this case.

In this first part, supposing that $f : [0, +\infty) \to \mathbb{C}$ of exponential growth, can be prolonged to an analytic function in an open disk (with center in origin) by keeping exponential growth, we obtain quantitative estimates in closed disks with center in origin, similar in form with that in the real case in Totik [191] mentioned above.

Also, we recall that the first result concerning the convergence of complex $S_n(f)(z)$ to $f(z)$ belonging to a class of analytic functions satisfying a suitable exponential-type growth condition in a parabolic domain, was proved in Dressel-Gergen-Purcell [65], but without any estimate of the approximation error.

The first main result of this section can be summarized by the following.

Theorem 1.8.1. (Gal [83]) *Let $\mathbb{D}_R = \{z \in \mathbb{C}; |z| < R\}$ be with $1 < R < +\infty$ and suppose that $f : [R, +\infty) \cup \overline{\mathbb{D}}_R \to \mathbb{C}$ is continuous in $[R, +\infty) \cup \overline{\mathbb{D}}_R$, analytic in $\mathbb{D}_R$, i.e. $f(z) = \sum_{k=0}^{\infty} c_k z^k$, for all $z \in \mathbb{D}_R$, and that there exist $M, C, B > 0$ and $A \in (\frac{1}{R}, 1)$, with the property $|c_k| \leq M \frac{A^k}{k!}$, for all $k = 0, 1, ...,$ (which implies $|f(z)| \leq Me^{A|z|}$ for all $z \in \mathbb{D}_R$) and $|f(x)| \leq Ce^{Bx}$, for all $x \in [R, +\infty)$.*

(i) Let $1 \leq r < \frac{1}{A}$ be arbitrary fixed. For all $|z| \leq r$ and $n \in \mathbb{N}$, we have

$$|S_n(f)(z) - f(z)| \leq \frac{C_{r,A}}{n},$$

where $C_{r,A} = \frac{M}{2r} \sum_{k=2}^{\infty} (k+1)(rA)^k < \infty$.

(ii) If $1 \le r < r_1 < \frac{1}{A}$ are arbitrary fixed, then for all $|z| \le r$ and $n, p \in \mathbb{N}$,

$$|S_n^{(p)}(f)(z) - f^{(p)}(z)| \le \frac{p! r_1 C_{r_1, A}}{n(r_1 - r)^{p+1}},$$

where $C_{r_1, A}$ is given as at the above point (i).

Proof. (i) According to Theorem 2 in Lupaş [127], we can write

$$S_n(f)(z) = \sum_{j=0}^{\infty} [0, 1/n, ..., j/n; f] z^j,$$

where $[0, 1/n, ..., j/n; f]$ denotes the divided difference of f on the knots $0, 1/n, ..., j/n$. Note that the above formula was proved in Lupaş [127] for functions of real variable, but the formula holds in complex setting too, since only algebraic calculations were used (see the proof of Theorem 2 in Lupaş [127]).

Taking in this representation formula $e_k(z) = z^k$, we get that $T_{n,k}(z) := S_n(e_k)(z)$ is a polynomial of degree $\le k$, $k = 0, 1, 2, ...,$ and $T_{n,0}(z) = 1, T_{n,1}(z) = z$, for all $z \in \mathbb{C}$. Also, differentiating $T_{n,k}(z)$ with respect to $z \ne 0$, we get

$$T'_{n,k}(z) = \sum_{j=0}^{\infty} \frac{j^k}{n^k} \left[-n e^{-nz} \frac{(nz)^j}{j!} + e^{-nz} jn \frac{(nz)^{j-1}}{j!} \right]$$

$$= -n T_{n,k}(z) + \sum_{j=0}^{\infty} \frac{j^{k+1}}{n^{k+1}} e^{-nz} \frac{n}{z} \frac{(nz)^j}{j!}$$

$$= -n T_{n,k}(z) + \frac{n}{z} T_{n,k+1}(z),$$

which implies

$$T_{n,k+1}(z) = \frac{z}{n} T'_{n,k}(z) + z T_{n,k}(z),$$

for all $z \in \mathbb{C}$, $k \in \{0, 1, 2, ..., \}$, $n \in \mathbb{N}$. From this it is immediate the recurrence formula

$$T_{n,k}(z) - z^k = \frac{z}{n}[T_{n,k-1}(z) - z^{k-1}]' + z[T_{n,k-1}(z) - z^{k-1}] + \frac{k-1}{n} z^{k-1},$$

for all $z \in \mathbb{C}$, $k, n \in \mathbb{N}$.

Now, let $1 \le r < R$. Denoting with $\| \cdot \|_r$ the norm in $C(\overline{\mathbb{D}}_r)$, where $\overline{\mathbb{D}}_r = \{z \in \mathbb{C}; |z| \le r\}$, by a linear transformation, the Bernstein's inequality in the closed unit disk becomes $|P'_k(z)| \le \frac{k}{r} \|P_k\|_r$, for all $|z| \le r$, where $P_k(z)$ is a polynomial of degree $\le k$. Therefore, from the above recurrence formula, we get

$$\|T_{n,k} - e_k\|_r \le \frac{r}{n} \cdot \|T_{n,k-1} - e_{k-1}\|_r \frac{k-1}{r}$$

$$+ r\|T_{n,k-1} - e_{k-1}\|_r + \frac{r^{k-1}(k-1)}{n},$$

which implies the recurrence

$$\|T_{n,k} - e_k\|_r \leq \left(r + \frac{k-1}{n}\right) \|T_{n,k-1} - e_{k-1}\|_r + \frac{r^{k-1}(k-1)}{n}.$$

In what follows we prove by mathematical induction with respect to k (with $n \geq 1$ supposed to be fixed, arbitrary), that this recurrence implies

$$\|T_{n,k} - e_k\|_r \leq \frac{(k+1)!}{2n} r^{k-1}, \text{ for all } k \geq 2, n \geq 1.$$

Indeed, for $k = 2$ and $n \in \mathbb{N}$, the left-hand side is $\frac{r}{n}$ while the right-hand side is $\frac{3r}{n}$. Supposing now that it is true for k, the above recurrence implies

$$\|T_{n,k+1} - e_{k+1}\|_r \leq \left(r + \frac{k}{n}\right) \frac{(k+1)!}{2n} r^{k-1} + \frac{r^k k}{n}.$$

It remains to prove that

$$\left(r + \frac{k}{n}\right) \frac{(k+1)!}{2n} r^{k-1} + \frac{r^k k}{n} \leq \frac{(k+2)!}{2n} r^k,$$

or, after simplifications, equivalently to

$$\left(r + \frac{k}{n}\right)(k+1)! + 2rk \leq (k+2)!r.$$

It is easy to see that this last inequality holds true for all $k \geq 2$ and $n \in \mathbb{N}$.

Now, from the hypothesis on f (that is $|f(x)| \leq max\{M, C\}e^{max\{A,B\}x}$, for all $x \in \mathbb{R}_+$), it follows that (see e.g. Dressel-Gergen-Purcell [65], pp. 1171-1172 and p. 1178) $S_n(f)(z)$ is analytic in $\mathbb{D}_R$. Therefore, it is easy to see that we can write

$$S_n(f)(z) = \sum_{k=0}^{\infty} c_k S_n(e_k)(z) = \sum_{k=0}^{\infty} c_k T_{n,k}(z), \text{ for all } z \in \mathbb{D}_R,$$

which from the hypothesis on c_k, immediately implies for all $|z| \leq r$

$$|S_n(f)(z) - f(z)| \leq \sum_{k=2}^{\infty} |c_k| \cdot |T_{k,n}(z) - e_k(z)| \leq \sum_{k=2}^{\infty} M \frac{A^k}{k!} \frac{(k+1)!}{2n} r^{k-1}$$

$$= \frac{M}{2nr} \sum_{k=2}^{\infty} (k+1)(rA)^k = \frac{C_{r,A}}{n},$$

where $C_{r,A} = \frac{M}{2r} \sum_{k=2}^{\infty}(k+1)(rA)^k < \infty$, for all $1 \leq r < \frac{1}{A}$, taking into account that the series $\sum_{k=2}^{\infty} u^{k+1}$ and therefore its derivative $\sum_{k=2}^{\infty}(k+1)u^k$, are uniformly and absolutely convergent in any compact disk included in the open unit disk.

(ii) Denoting by γ the circle of radius $r_1 > r$ and center 0, since for any $|z| \leq r$ and $v \in \gamma$, we have $|v - z| \geq r_1 - r$, by the Cauchy's formulas it follows that for all $|z| \leq r$ and $n \in \mathbb{N}$, we have

$$|S_n^{(p)}(f)(z) - f^{(p)}(z)| = \frac{p!}{2\pi} \left|\int_\gamma \frac{S_n(f)(v) - f(v)}{(v-z)^{p+1}} dv\right| \leq \frac{C_{r_1,A}}{n} \frac{p!}{2\pi} \frac{2\pi r_1}{(r_1-r)^{p+1}}$$

$$= \frac{C_{r_1,A}}{n} \frac{p!r_1}{(r_1-r)^{p+1}},$$

which proves (ii) and the theorem.   □

Remark. Let us show in detail the relationship

$$S_n(f)(z) = \sum_{k=0}^{\infty} c_k S_n(e_k)(z),$$

used in the proof of Theorem 1.8.1, (i) as follows. For this purpose, for any $m \in \mathbb{N}$ let us define

$$f_m(z) = \sum_{j=0}^{m} c_j z^j \text{ if } |z| \le r \text{ and } f_m(x) = f(x) \text{ if } x \in (r, +\infty).$$

From the hypothesis on f it is clear that for any $m \in \mathbb{N}$ it follows $|f_m(x)| \le C_m e^{B_m x}$, for all $x \in [0, +\infty)$. This implies that for each fixed $m, n \in \mathbb{N}$ and z,

$$|S_n(f_m)(z)| \le |e^{-nz}| \sum_{j=0}^{\infty} \frac{(n|z|)^j}{j!} |f_m(j/n)|$$

$$\le C_m |e^{-nz}| \sum_{j=0}^{\infty} \frac{(n|z|)^j}{j!} e^{B_m j/n} < \infty,$$

since by the ratio criterium the last series is convergent. Therefore $S_n(f_m)(z)$ is well-defined.

Denoting

$$f_{m,k}(z) = c_k e_k(z) \text{ if } |z| \le r \text{ and } f_{m,k}(x) = \frac{f(x)}{m+1} \text{ if } x \in (r, \infty),$$

it is clear that each $f_{m,k}$ is of exponential growth on $[0, \infty)$ and that $f_m(z) = \sum_{k=0}^{m} f_{m,k}(z)$. Since from the linearity of S_n we have

$$S_n(f_m)(z) = \sum_{k=0}^{m} c_k S_n(e_k)(z), \text{ for all } |z| \le r,$$

it suffices to prove that $\lim_{m \to \infty} S_n(f_m)(z) = S_n(f)(z)$ for any fixed $n \in \mathbb{N}$ and $|z| \le r$. But this is immediate from $\lim_{m \to \infty} \|f_m - f\|_r = 0$, from $\|f_m - f\|_{B[0,+\infty)} \le \|f_m - f\|_r$ and from the inequality

$$|S_n(f_m)(z) - S_n(f)(z)| \le |e^{-nz}| \cdot e^{n|z|} \cdot \|f_m - f\|_{B[0,\infty)} \le M_{r,n} \|f_m - f\|_r,$$

valid for all $|z| \le r$. Here $\| \cdot \|_{B[0,+\infty)}$ denotes the uniform norm on $C[0, +\infty)$-the space of all real-valued bounded functions on $[0, +\infty)$.

In what follows we obtain the Voronovskaja-type formula with a quantitative estimate for the complex Favard-Szász-Mirakjan operator.

Theorem 1.8.2. (Gal [83]) *Suppose that the hypothesis on the function f and the constants R, M, C, B, A in the statement of Theorem 1.8.1 hold and let $1 \le r < \frac{1}{A}$ be arbitrary fixed.*

(i) The following upper estimate in the Voronovskaja-type formula holds

$$\left| S_n(f)(z) - f(z) - \frac{z}{2n} f''(z) \right| \le \frac{3MA|z|}{r^2 n^2} \sum_{k=2}^{\infty} (k+1)(rA)^{k-1},$$

for all $n \in \mathbb{N}, |z| \leq r$.

(ii) We have the following equivalence in the Voronovskaja's formula

$$\left\| S_n(f) - f - \frac{e_1}{2n} f'' \right\|_r \sim \frac{1}{n^2},$$

where the constants in the equivalence depend on f and r but are independent of n.

Proof. (i) Denoting $e_k(z) = z^k$, $k = 0, 1, ...$, and $T_{n,k}(z) = S_n(e_k)(z)$, by the proof of Theorem 1.8.1, (i), we can write $S_n(f)(z) = \sum_{k=0}^{\infty} c_k T_{n,k}(z)$, which immediately implies

$$\left| S_n(f)(z) - f(z) - \frac{z}{2n} f''(z) \right| \leq \sum_{k=2}^{\infty} |c_k| \cdot \left| T_{n,k}(z) - e_k(z) - \frac{z^{k-1} k(k-1)}{2n} \right|$$

for all $z \in \mathbb{D}_R$, $n \in \mathbb{N}$.

By the recurrence relationship in the proof of Theorem 1.8.1, (i), satisfied by $T_{n,k}(z)$, denoting $E_{k,n}(z) = T_{n,k}(z) - e_k(z) - \frac{z^{k-1} k(k-1)}{2n}$, we immediately get the new recurrence

$$E_{k,n}(z) = \frac{z}{n} E'_{k-1,n}(z) + z E_{k-1,n}(z) + \frac{z^{k-2}(k-1)(k-2)^2}{2n^2},$$

for all $k \geq 2$, $n \in \mathbb{N}$ and $z \in \mathbb{D}_R$.

This implies, for all $|z| \leq r$, $k \geq 2$, $n \in \mathbb{N}$,

$$|E_{k,n}(z)|$$
$$\leq \frac{|z|}{2n} [2\|E'_{k-1,n}\|_r] + |z| \cdot |E_{k-1,n}(z)| + \frac{|z|}{2n} \cdot \frac{r^{k-3}(k-1)(k-2)^2}{n}$$
$$\leq r|E_{k-1,n}(z)| + \frac{|z|}{2n} [2\|E'_{k-1,n}\|_r + \frac{r^{k-3}(k-1)(k-2)^2}{n}]$$
$$\leq r|E_{k-1,n}(z)| + \frac{|z|}{2n} \left[\frac{2(k-1)}{r} \|E_{k-1,n}\|_r + \frac{r^{k-3}(k-1)(k-2)^2}{n} \right]$$
$$\leq r|E_{k-1,n}(z)| + \frac{|z|}{2n} \left[\frac{2(k-1)}{r} \|T_{n,k-1} - e_{k-1}\|_r \right.$$
$$\left. + \frac{2(k-1)}{r} \cdot \frac{r^{k-2}(k-1)(k-2)}{2n} + \frac{r^{k-3}(k-1)(k-2)^2}{n} \right] \leq r|E_{k-1,n}(z)|$$
$$+ \frac{|z|}{2n} \left[\frac{2(k-1)}{r} \cdot \frac{r^{k-2} k!}{2n} + \frac{2(k-1)}{r} \cdot \frac{r^{k-2}(k-1)(k-2)}{2n} \right.$$
$$\left. + \frac{r^{k-3}(k-1)(k-2)^2}{n} \right] \leq r|E_{k-1,n}(z)| + \frac{3|z| r^{k-3}}{2n^2} (k-1) k!$$
$$\leq r|E_{k-1,n}(z)| + \frac{3|z| r^{k-3}}{2n^2} (k+1)!,$$

that is

$$|E_{k,n}(z)| \leq r|E_{k-1,n}(z)| + \frac{3|z| r^{k-3}}{2n^2} (k+1)!, \text{ for all } |z| \leq r.$$

Taking $k = 2, 3, ...,$ in this last inequality, step by step we obtain

$$|E_{k,n}(z)| \le \frac{3|z|r^{k-3}}{2n^2} \sum_{j=3}^{k+1} j! \le \frac{3|z|r^{k-3}(k+1)!}{n^2},$$

which implies

$$\left| S_n(f)(z) - f(z) - \frac{z}{2n}f''(z) \right| \le \sum_{k=2}^{\infty} |c_k| \cdot |E_{k,n}(z)||$$

$$\le \frac{3M|z|}{n^2} \sum_{k=2}^{\infty} \frac{A^k}{k!}(k+1)!r^{k-3}$$

$$\le \frac{3MA|z|}{r^2n^2} \sum_{k=2}^{\infty} (k+1)(rA)^{k-1},$$

for all $|z| \le r$, where for $rA < 1$ we obviously have $\sum_{k=2}^{\infty}(k+1)(rA)^{k-1} < \infty$.

(ii) Taking into account the above point (i) it will be enough to prove the lower estimate. For this purpose we will use the ideas in the proof of Corollary 1.3.4. More exactly, let us consider the expression which appear in the generalized Voronovskaja's formula for the complex Favard-Szász-Mirakjan operators, that is

$$Q_{n,p}(f)(z) = S_n(f)(z) - f(z) - \sum_{j=1}^{2p} \frac{1}{j!n^j} A_{n,j}(z)f^{(j)}(z),$$

where

$$A_{n,j} = n^j S_n[(\cdot - z)^j](z) = e^{-nz} \sum_{k=0}^{\infty} \frac{(nz)^k}{k!}(k - nz)^j.$$

The idea in the proof of Corollary 1.3.4 is that in order to get the lower estimate $\|Q_{n,p}(f)\|_r \ge \frac{C}{n^{p+1}}$ we need first to prove the upper estimate $\|Q_{n,p+1}(f)\|_r \le \frac{C}{n^{p+2}}$. In what follows for simplicity we will consider above the case $p = 1$. Therefore, first we need an upper estimate for $|Q_{n,2}(f)(z)|$.

According to Lemma 1.2 in Pop [156], $A_{n,j}(z)$ is a polynomial of degree $[j/2]$. Also, from Lemmas 1.2 and 1.3 and Consequence 1.2 in Pop [156] we easily get $A_{n,0}(z) = 1$, $A_{n,1}(z) = 0$, $A_{n,2}(z) = nz$, $A_{n,3}(z) = nz$, $A_{n,4}(z) = 3n^2z^2 + nz$, which replaced in the expression of $Q_{n,2}(f)(z)$ will mean that we need to prove an upper estimate of the form (valid for all $|z| \le r$)

$$\left| S_n(f)(z) - f(z) - \frac{z}{2n}f''(z) - \frac{z}{6n^2}f^{(3)}(z) - \frac{3nz^2 + z}{24n^3}f^{(4)}(z) \right| \le \frac{C}{n^3}.$$

Since $S_n(f)(z) = \sum_{k=0}^{\infty} c_k T_{n,k}(z)$ and denoting

$$E_{k,n,2}(z) = T_{n,k}(z) - e_k(z) - \frac{k(k-1)z^{k-1}}{2n} - \frac{k(k-1)(k-2)z^{k-2}}{6n^2}$$
$$- \frac{k(k-1)(k-2)(k-2)}{24n^3}(3nz^2 + z)z^{k-4},$$

we can write

$$\left| S_n(f)(z) - f(z) - \frac{z}{2n} f''(z) - \frac{z}{6n^2} f^{(3)}(z) - \frac{3nz^2 + z}{24n^3} f^{(4)}(z) \right|$$

$$\leq \sum_{k=5}^{\infty} |c_k| \cdot |E_{k,n,2}(z)|,$$

by taking into account that by simple calculation we get $E_{k,n,2}(z) = 0$ for all $k = 0, 1, 2, 3, 4$. Also, clearly $E_{k,n,2}(z)$ is a polynomial of degree $\leq k$.

Now, if we denote $E_{k,n,1}(z) = T_{n,k}(z) - e_k - \frac{k(k-1)z^{k-1}}{2n}$, by the proof of the above point (i) we can write the recurrence formula

$$E_{k,n,1}(z) = \frac{z}{n} E'_{k-1,n,1}(z) + z E_{k-1,n,1}(z) + \frac{z^{k-2}(k-1)(k-2)^2}{2n^2}.$$

On the other hand, simple calculation lead us to the formula

$$E_{k,n,e}(z) = E_{k,n,1}(z) - P_{k,n}(z),$$

where

$$P_{k,n}(z) = \frac{k(k-1)(k-2)(3k-5)}{24n^2} z^{k-2} + \frac{k(k-1)(k-2)(k-3)}{24n^3} z^{k-3}.$$

The recurrence formula for $E_{k,n,1}(z)$ implies the following recurrence formula for $E_{k,n,2}(z)$

$$E_{k,n,2}(z) = \frac{z}{n} E'_{k-1,n,2}(z) + z E_{k-1,n,2}(z) + R_{k,n}(z),$$

where

$$R_{k,n}(z) = \frac{(k-1)(k-2)(k-3)(k-4)(3k-5)}{24n^3} z^{k-3}$$
$$+ \frac{(k-1)(k-2)(k-3)(k-4)^2}{24n^4} z^{k-4}.$$

From this recurrence, applying the Bernstein's inequality and the estimate for $\|E_{k-1,n,1}\|_r \leq \frac{3r^{k-3}k!}{n^2}$ from the above point (i), for all $|z| \leq r$, $n \in \mathbb{N}$ and $k \geq 5$ we obtain

$$\|E_{k,n,2}\|_r$$
$$\leq r\|E_{k-1,n,2}(z)\|_r + \frac{r}{n}\|E'_{k-1,n,2}\|_r + \|R_{k,n}\|_r \leq r\|E_{k-1,n,2}\|_r$$
$$+ \frac{r}{n} \cdot \frac{k-1}{r}\|E_{k-1,n,1}\|_r + \frac{r}{n} \cdot \frac{k-3}{r}\|P_{k,n}\|_r + \|R_{k,n}\|_r \leq r\|E_{k-1,n,2}\|_r$$
$$+ \frac{k-1}{n} \cdot \frac{3r^{k-3}k!}{n^2} + \frac{k-3}{n} \cdot \frac{r^{k-3}}{24n^2}(k-1)(k-2)(k-3)(3k-8)$$
$$+ \frac{k-3}{n} \cdot \frac{r^{k-4}}{24n^3}(k-1)(k-2)(k-3)(k-4)$$

$$+\frac{r^{k-3}}{24n^3}(k-1)(k-2)(k-3)(k-4)(3k-5)$$

$$+\frac{r^{k-4}}{24n^4}(k-1)(k-2)(k-3)(k-4)^2 \leq r\|E_{k-1,n,2}\|_r + \frac{3r^{k-3}(k+1)!}{n^3}$$

$$+\frac{3r^{k-3}(k+1)!}{24n^3}+\frac{r^{k-4}k!}{24n^4}+\frac{3r^{k-3}k!}{24n^3}+\frac{r^{k-4}k!}{24n^4}$$

$$\leq r\|E_{k-1,n,2}\|_r + \frac{3r^{k-3}(k+1)!}{n^3} + \frac{3r^{k-3}(k+1)!}{2n^3}$$

$$+\frac{r^{k-3}k!}{2n^3}+\frac{3r^{k-3}k!}{2n^3}+\frac{r^{k-3}k!}{2n^3}$$

$$= r\|E_{k-1,n,2}\|_r + \frac{7r^{k-3}(k+1)!}{n^3}.$$

Therefore we have obtained that for all $n \in \mathbb{N}$ and $k = 5, 6, \ldots$

$$\|E_{k,n,2}\|_r \leq r\|E_{k-1,n,2}\|_r + \frac{7r^{k-3}(k+1)!}{n^3}.$$

Taking here step by step $k = 5, 6, \ldots$ and taking into account that $E_{k,n,2}(z) = 0$ for $k = 0, 1, 2, 3, 4$, we easily obtain

$$\|E_{k,n,2}\|_r \leq \frac{7r^{k-3}}{n^3}\sum_{j=5}^{k}(j+1)! \leq \frac{7r^{k-3}(k+2)!}{n^3}, k = 5, 6, \ldots,$$

which implies

$$\left\|S_n(f) - f - \frac{e_1}{2n}f'' - \frac{e_1}{6n^2}f^{(3)} - \frac{3ne_2 + e_1}{24n^3}f^{(4)}\right\|_r \leq \sum_{k=5}^{\infty}|c_k| \cdot \|E_{k,n,2}\|_r$$

$$\leq \frac{C_r(f)}{n^3}.$$

Now, by similar reasonings with those in the case of Bernstein polynomials in Lemma 2.2 in Pop [155], we easily obtain

$$S_n(e_k)(z) = \sum_{j=0}^{k}\frac{1}{j!n^j}A_{n,j}(z)(z^k)^{(j)}.$$

Denoting for arbitrary $p \in \mathbb{N}$

$$E_{k,n,p}(z) = S_n(e_k)(z) - e_k(z) - \sum_{j=1}^{2p}\frac{1}{n^j j!}A_{n,j}(z)(z^k)^{(j)},$$

the above formula for $S_n(e_k)(z)$ implies that

$$E_{k,n,p}(z) = \sum_{j=2p+1}^{k}\frac{1}{n^j}\binom{k}{j}z^{k-j}A_{n,j}(z),$$

and by direct calculation we arrive at

$$S_n(f)(z) - f(z) - \sum_{j=1}^{2p} \frac{f^{(j)}(z)}{j!} n^{-j} A_{n,j}(z) = \sum_{k=2p+1}^{\infty} E_{k,n,p}(z)$$

$$= \frac{1}{n^{p+1}} \left\{ \sum_{k=2p+1}^{\infty} c_k \left[\sum_{j=2p+1}^{k} \binom{k}{j} z^{k-j} n^{p+1-j} A_{n,j}(z) \right] \right\}$$

$$= \frac{1}{n^{p+1}} \left\{ \frac{A_{n,2p+1}(z)}{n^p(2p+1)!} f^{(2p+1)}(z) + \frac{A_{n,2p+2}(z)}{n^{p+1}(2p+2)!} f^{(2p+2)}(z) \right.$$

$$\left. + \frac{1}{n} \left[n^{p+2} \sum_{k=2p+3}^{\infty} c_k E_{k,n,p+1}(z) \right] \right\}.$$

By Corollary 1.3 in Pop [156], $A_{n,j}(z)$ is a polynomial of degree $\leq [j/2]$ and by e.g. Agratini [3], p. 237, Lemma 3.9.4, is a polynomial of degree $\leq j$ in z. Therefore we can write

$$\frac{A_{n,2p+1}(z)}{n^p(2p+1)!} f^{(2p+1)}(z) + \frac{A_{n,2p+2}(z)}{n^{p+1}(2p+2)!} f^{(2p+2)}(z)$$

$$= P_{2p+1}(z) f^{(2p+1)}(z) + P_{2p+2}(z) f^{(2p+2)}(z) + \frac{1}{n} F(z) f^{(2p+1)}(z) + \frac{1}{n} G(z) f^{(2p+2)}(z),$$

where $F(z)$ and $G(z)$ are bounded polynomials on $\mathbb{D}_r$ by constants depending on r and p nut independent of n.

In what follows we will find the form of $P_{2p+1}(z)$ and $P_{2p+2}(z)$. First, by taking $p(z) = z$ in e.g. Lemma 1.3 in Pop [156] we get the recurrence formula

$$A_{n,j+1}(z) = z[A'_{n,j}(z) + nj A_{n,j-1}(z)],$$

which immediately implies $A_{n,0}(z) = 1$, $A_{n,1}(z) = 1$, $A_{n,2}(z) = nz$, $A_{n,3}(z) = nz$, $A_{n,4}(z) = 3n^2z^2 + nz$, $A_{n,5}(z) = 10n^2z^2 + nz$, $A_{n,6}(z) = 15n^3z^3 + 25n^2z^2$, $A_{n,7}(z) = 105n^3z^3 + 56n^2z^2$, and so on. By mathematical induction it easily follows that the coefficient of $n^{[(2p+1)/2]} = n^p$ in $A_{n,2p+1}(z)$ is of the form $c_p z^p$ with $c_p > 0$, while the coefficient of $n^{[(2p+2)/2]} = n^{p+1}$ in $A_{n,2p+2}(z)$ is of the form $d_p z^{p+1}$ with $d_p > 0$. Therefore we have $P_{2p+1}(z) = c_p z^p$ and $P_{2p+2}(z) = d_p z^{p+1}$.

Denoting now $U(f)(z) = P_{2p+1}(z) f^{(2p+1)}(z) + P_{2p+2}(z) f^{(2p+2)}(z)$, we will prove that if f is not a polynomial of degree $\leq 2p$ then $\|U(f)\|_r > 0$. Indeed, suppose that for such an f we have $\|U(f)\|_r = 0$, that is the following differential equation holds

$$P_{2p+1}(z) f^{(2p+1)}(z) + P_{2p+2}(z) f^{(2p+2)}(z) = 0, \ z \in \mathbb{D}_r.$$

Making the substitution $f^{(2p+1)}(z) = y(z)$ and replacing the above found form for $P_{2p+1}(z)$ and $P_{2p+2}(z)$, we obtain

$$c_p z^p y(z) + d_p z^{p+1} y'(z) = 0, \ z \in \mathbb{D}_r.$$

Simplifying with $z^p \neq 0$ we obtain

$$c_p y(z) + d_p z y'(z) = 0, \; z \in \mathbb{D}_r \setminus \{0\}.$$

Passing here with $z \to 0$ we get $y(0) = 0$. Writing $y(z) = \sum_{k=1}^{\infty} a_k z^k$ and replacing in the above differential, by the identification of the coefficients we easily obtain that $a_k = 0$ for all $k \geq 1$, that is $y(z) = 0$, for all $z \in \mathbb{D}_r$ and therefore f necessarily is a polynomial of degree $\leq 2p$, a contradiction.

Therefore for f a polynomial of degree $\leq 2p$, the supposition that $\|U(f)\|_r = 0$ is false, that is in this case we have $\|U(f)\|_r > 0$.

From this point, reasoning exactly as in the proof of Corollary 1.3.4 for $p = 1$ we obtain

$$\left\| S_n(f) - f - \sum_{j=1}^{2} \frac{f^{(j)}}{j!} n^{-j} A_{n,j} \right\|_r \geq \frac{C_r(f)}{n^2},$$

which proves the point (ii) and the theorem. □

The next result shows that the order of approximation in Theorem 1.8.1 is exactly $\frac{1}{n}$.

Corollary 1.8.3. (Gal [83]) *In the hypothesis of Theorem 1.8.1, if f is not a polynomial of degree ≤ 1 in the case (i) and if f is not a polynomial of degree $\leq p$, ($p \geq 1$) in the case (ii), then $\frac{1}{n}$ is in fact the exact order of approximation.*

Proof. Applying the norm $\| \cdot \|_r$ to the identity

$$S_n(f)(z) - f(z) = \frac{1}{n} \left\{ \frac{z}{2} f''(z) + \frac{1}{n} \left[n^2 \left(S_n(f)(z) - f(z) - \frac{z}{2n} f''(z) \right) \right] \right\},$$

it follows

$$\|S_n(f) - f\|_r \geq \frac{1}{n} \left\{ \left\| \frac{e_1}{2} f'' \right\|_r - \frac{1}{n} \left[n^2 \left\| S_n(f) - f - \frac{e_1}{2n} f'' \right\|_r \right] \right\}.$$

If f is not a polynomial of degree ≤ 1 then evidently $\left\| \frac{e_1}{2} f'' \right\|_r > 0$, which combined with the estimate in Theorem 1.8.2 immediately implies that $\|S_n(f) - f\|_r \geq \frac{C}{n}$, for all $n \geq n_0$, with $C > 0$ independent of n. Since for $n = 1, 2, ..., n_0 - 1$ the inequality $\|S_n(f) - f\|_r \geq \frac{C_1}{n}$ is trivial with a constant $C_1 > 0$ and taking into account the upper estimate in Theorem 1.8.1, (i), we get the desired conclusion.

Now, replacing $S_n(f)(z) - f(z)$ in the above identity, to the Cauchy formula in the proof of Theorem 1.8.1, (ii) and then applying the norm $\| \cdot \|_r$ to the integral identity, we get

$$\|S_n^{(p)}(f) - f^{(p)}\|_r$$

$$\geq \frac{1}{n} \left\{ \left\| \left[\frac{e_1}{2} f'' \right]^{(p)} \right\|_r - \frac{1}{n} \left\| \frac{p!}{2\pi} \int_{\Gamma} \frac{n^2 \left(S_n(f)(v) - f(v) - \frac{v}{2n} f''(v) \right)}{(v - e_1)^{p+1}} dv \right\|_r \right\},$$

which combined again with Theorem 1.8.2 and taking into account that $\left\| \left[\frac{e_1}{2} f'' \right]^{(p)} \right\|_r > 0$ (since f is not a a polynomial of degree $\leq p$), as above leads us to the same conclusion. □

In the second part of this section the growth conditions of exponential-type on f will be omitted. The only condition imposed to f is to be bounded on $[0, \infty)$, case when it is clear that the complex Favard-Szász-Mirakjan operators given by $S_n(f)(z) = e^{-nz} \sum_{j=0}^{\infty} \frac{(nz)^j}{j!} f(j/n)$ are well defined for all $z \in \mathbb{C}$.

In this sense, the first result is expressed by the following.

Theorem 1.8.4. (Gal [84]) *For $2 < R < +\infty$ let $f : [R, +\infty) \cup \overline{\mathbb{D}}_R \to \mathbb{C}$ be bounded on $[0, +\infty)$ and analytic in $\mathbb{D}_R$, that is $f(z) = \sum_{k=0}^{\infty} c_k z^k$, for all $z \in \mathbb{D}_R$.*

(i) If $1 \le r < \frac{R}{2}$ then for all $|z| \le r$ and $n \in \mathbb{N}$ it follows

$$|S_n(f)(z) - f(z)| \le \frac{C_{r,f}}{n},$$

with $C_{r,f} = 6 \sum_{k=2}^{\infty} |c_k|(k-1)(2r)^{k-1} < \infty$.

(ii) If $1 \le r < r_1 < \frac{R}{2}$ then for all $|z| \le r$ and $n, p \in \mathbb{N}$ it follows

$$|S_n^{(p)}(f)(z) - f^{(p)}(z)| \le \frac{p! r_1 C_{r_1, f}}{n(r_1 - r)^{p+1}},$$

where $C_{r_1, f}$ is as above.

Proof.   (i) From the proof of Theorem 1.8.1, (i) we can write

$$S_n(f)(z) = \sum_{j=0}^{\infty} [0, 1/n, ..., j/n; f] z^j,$$

where $[0, 1/n, ..., j/n; f]$ denotes the divided difference of f on the knots $0, 1/n, ..., j/n$.

For $f(z) = e_k(z) = z^k$ and applying the mean value theorem for divided differences, for all $|z| \le r$ and $k, n \in \mathbb{N}$ it follows

$$|S_n(e_k)(z)| \le \sum_{j=0}^{k} |[0, 1/n, ..., j/n; e_k]| r^j$$

$$= \sum_{j=1}^{k} \frac{k(k-1)...(k-j+1)}{j!} r^{k-j} r^j$$

$$\le r^k \sum_{j=0}^{k} \binom{k}{j} = (2r)^k.$$

Denoting $T_{n,k}(z) := S_n(e_k)(z)$ clearly that it is a polynomial of degree $\le k$, $k = 0, 1, 2, ..., T_{n,0}(z) = 1, T_{n,1}(z) = z$, for all $z \in \mathbb{C}$ and by the proof of Theorem 1.8.1, (i) the recurrence formula

$$T_{n,k}(z) - z^k = \frac{z}{n}[T_{n,k-1}(z) - z^{k-1}]' + z[T_{n,k-1}(z) - z^{k-1}] + \frac{k-1}{n} z^{k-1},$$

holds for all $z \in \mathbb{C}$, $k, n \in \mathbb{N}$.

Applying as in the proof of Theorem 1.8.1, (i) the Bernstein's inequality, from the above recurrence formula, for all $|z| \leq r$ we get

$$
\begin{aligned}
|T_{n,k} - e_k| &\leq \frac{r}{n} \cdot \|T_{n,k-1} - e_{k-1}\|_r \frac{k-1}{r} + r|T_{n,k-1} - e_{k-1}| + \frac{r^{k-1}(k-1)}{n} \\
&\leq \frac{k-1}{n}[\|T_{n,k-1}\|_r + r^{k-1}] + r|T_{n,k-1} - e_{k-1}| + \frac{r^{k-1}(k-1)}{n} \\
&\leq \frac{k-1}{n}[(2r)^{k-1} + r^{k-1}] + r|T_{n,k-1} - e_{k-1}| + \frac{r^{k-1}(k-1)}{n} \\
&\leq r|T_{n,k-1} - e_{k-1}| + \frac{3(k-1)(2r)^{k-1}}{n}.
\end{aligned}
$$

Since $T_{n,1}(z) = e_1(z)$, for $k = 2$ the above inequality implies $|T_{n,2}(z) - z^2| \leq \frac{3r}{n} 2^1$, for $k = 3$ it implies $|T_{n,3}(z) - z^3| \leq \frac{3r^2}{n}(1 \cdot 2^1 + 2 \cdot 2^2)$, and step by step for all $|z| \leq r$ we finally obtain

$$
\begin{aligned}
|T_{n,k}(z) - z^k| &\leq \frac{3r^{k-1}}{n}\left(\sum_{j=1}^{k-1} j 2^j\right) = \frac{3r^{k-1}}{n}\left[(k-2)2^k + 2\right] \\
&\leq \frac{6(k-1)}{n}(2r)^{k-1}.
\end{aligned}
$$

Here the formula $\sum_{j=1}^{k-1} j 2^j = (k-2)2^k + 2$ can easily be proved by mathematical induction.

Since $S_n(f)(z) = \sum_{k=0}^{\infty} c_k T_{n,k}(z)$, we get

$$
|S_n(f)(z) - f(z)| \leq \sum_{k=2}^{\infty} |c_k| \cdot |T_{n,k}(z) - z^k| \leq \frac{C_{r,f}}{n},
$$

which proves (i).

(ii) Denote by γ the circle of radius $r_1 > r$ and center 0. For any $|z| \leq r$ and $v \in \gamma$, we have $|v - z| \geq r_1 - r$ and by the Cauchy's formulas for all $|z| \leq r$ and $n \in \mathbb{N}$ it follows

$$
\begin{aligned}
|S_n^{(p)}(f)(z) - f^{(p)}(z)| &= \frac{p!}{2\pi}\left|\int_\gamma \frac{S_n(f)(v) - f(v)}{(v-z)^{p+1}} dv\right| \leq \frac{C_{r_1,f}}{n} \frac{p!}{2\pi} \frac{2\pi r_1}{(r_1 - r)^{p+1}} \\
&= \frac{C_{r_1,f}}{n} \frac{p! r_1}{(r_1 - r)^{p+1}},
\end{aligned}
$$

which proves (ii) and the theorem. $\quad\square$

Remark. A proof of the relationship $S_n(f)(z) = \sum_{k=0}^{\infty} c_k S_n(e_k)(z)$ used in the proof of Theorem 1.8.4, (i) is as follows. For any $m \in \mathbb{N}$ define

$$
f_m(z) = \sum_{j=0}^{m} c_j z^j \text{ if } |z| \leq r \text{ and } f_m(x) = f(x) \text{ if } x \in (r, +\infty).
$$

From the hypothesis on f it is clear that each f_m is bounded on $[0, +\infty)$, which implies that

$$|S_n(f_m)(z)| \leq |e^{-nz}| \sum_{j=0}^{\infty} \frac{(n|z|)^j}{j!} M(f_m) = |e^{-nz}| \cdot e^{n|z|} M(f_m) < \infty,$$

that is all $S_n(f_m)(z)$ are well-defined.

Denoting

$$f_{m,k}(z) = c_k e_k(z) \text{ if } |z| \leq r \text{ and } f_{m,k}(x) = \frac{f(x)}{m+1} \text{ if } x \in (r, \infty),$$

it is clear that each $f_{m,k}$ is bounded on $[0, \infty)$ and that $f_m(z) = \sum_{k=0}^{m} f_{m,k}(z)$. Since from the linearity of S_n we have

$$S_n(f_m)(z) = \sum_{k=0}^{m} c_k S_n(e_k)(z), \text{ for all } |z| \leq r,$$

it suffices to prove that $\lim_{m \to \infty} S_n(f_m)(z) = S_n(f)(z)$ for any fixed $n \in \mathbb{N}$ and $|z| \leq r$. But this is immediate from $\lim_{m \to \infty} \|f_m - f\|_r = 0$, from $\|f_m - f\|_{B[0,+\infty)} \leq \|f_m - f\|_r$ and from the inequality

$$|S_n(f_m)(z) - S_n(f)(z)| \leq |e^{-nz}| \cdot e^{n|z|} \cdot \|f_m - f\|_{B[0,\infty)} \leq M_{r,n} \|f_m - f\|_r,$$

valid for all $|z| \leq r$. Here $\| \cdot \|_{B[0,+\infty)}$ denotes the uniform norm on $C[0, +\infty)$-the space of all real-valued bounded functions on $[0, +\infty)$.

Also, the following Voronovskaja-type formula holds.

Theorem 1.8.5. (Gal [84]) *For $2 < R < +\infty$ let $f : [R, +\infty) \cup \overline{\mathbb{D}}_R \to \mathbb{C}$ be bounded on $[0, +\infty)$ and analytic in $\mathbb{D}_R$, that is $f(z) = \sum_{k=0}^{\infty} c_k z^k$, for all $z \in \mathbb{D}_R$. Also, let $1 \leq r < \frac{R}{2}$.*

(i) For all $|z| \leq r$ and $n \in \mathbb{N}$ it follows

$$\left| S_n(f)(z) - f(z) - \frac{z}{2n} f''(z) \right| \leq M_{r,f} \cdot \frac{|z|}{n^2},$$

with $M_{r,f} = 26 \sum_{k=3}^{\infty} |c_k|(k-1)^2(k-2)(2r)^{k-3} < \infty$.

(ii) For all $n \in \mathbb{N}$ we have

$$\left\| S_n(f) - f - \frac{e_1}{2n} f'' \right\|_r \sim \frac{1}{n^2},$$

where the constants in the equivalence depend on f and r but are independent of n.

Proof.   (i) Keeping the notations in the proof of Theorem 1.8.2 (i) we have

$$\left| S_n(f)(z) - f(z) - \frac{z}{2n} f''(z) \right| \leq \sum_{k=2}^{\infty} |c_k| \cdot |E_{k,n}(z)|,$$

where it is necessary to obtain a suitable estimate for $|E_{k,n}(z)|$. For this purpose we use the recurrence in the proof of Theorem 1.8.2 (i) given by

$$E_{k,n}(z) = \frac{z}{n} E'_{k-1,n}(z) + z E_{k-1,n}(z) + \frac{z^{k-2}(k-1)(k-2)^2}{2n^2},$$

for all $k \geq 2$, $n \in \mathbb{N}$ and $z \in \mathbb{D}_R$.

This implies, for all $|z| \leq r$, $k \geq 2$, $n \in \mathbb{N}$,

$$|E_{k,n}(z)|$$

$$\leq \frac{|z|}{2n}[2\|E'_{k-1,n}\|_r] + |z| \cdot |E_{k-1,n}(z)| + \frac{|z|}{2n} \cdot \frac{r^{k-3}(k-1)(k-2)^2}{n}$$

$$\leq r|E_{k-1,n}(z)| + \frac{|z|}{2n}\left[2\|E'_{k-1,n}\|_r + \frac{r^{k-3}(k-1)(k-2)^2}{n}\right] \leq r|E_{k-1,n}(z)|$$

$$+ \frac{|z|}{2n}\left[\frac{2(k-1)}{r}\|E_{k-1,n}\|_r + \frac{r^{k-3}(k-1)(k-2)^2}{n}\right] \leq r|E_{k-1,n}(z)|$$

$$+ \frac{|z|}{2n}\left[\frac{2(k-1)}{r}\|T_{n,k-1} - e_{k-1}\|_r + \frac{2(k-1)}{r} \cdot \frac{r^{k-2}(k-1)(k-2)}{2n}\right.$$

$$\left. + \frac{r^{k-3}(k-1)(k-2)^2}{n}\right] \leq r|E_{k-1,n}(z)|$$

$$+ \frac{|z|}{2n}\left[\frac{2(k-1)}{r} \cdot \frac{6(k-2)}{n} \cdot (2r)^{k-2} + \frac{2(k-1)}{r} \cdot \frac{r^{k-2}(k-1)(k-2)}{2n}\right.$$

$$\left. + \frac{r^{k-3}(k-1)(k-2)^2}{n}\right] \leq r|E_{k-1,n}(z)| + \frac{26|z|(k-1)^2(k-2)(2r)^{k-3}}{2n^2},$$

that is

$$|E_{k,n}(z)| \leq r|E_{k-1,n}(z)| + \frac{13|z|(k-1)^2(k-2)(2r)^{k-3}}{n^2}.$$

For $k = 1, 2$ we get $E_{k,n}(z) = 0$, for $k = 3$ in this last inequality we obtain $|E_{3,n}(z)| \leq \frac{13|z|r^0}{n^2}(3-1)^2(3-2)2^0$ and for $k = 4$ it follows

$$|E_{4,n}(z)| \leq \frac{13|z|r^1}{n^2}[(3-1)^2(3-2)2^0 + (4-1)^2(4-2)2^1].$$

Then step by step finally we arrive at

$$|E_{k,n}(z)| \leq \frac{13|z|r^{k-3}}{n^2}\left(\sum_{j=3}^{k}(j-1)^2(j-2)2^{j-3}\right)$$

$$\leq \frac{13|z|r^{k-3}}{n^2} \cdot (k-1)^2(k-2)\sum_{j=3}^{k}2^{j-3}$$

$$= \frac{13|z|r^{k-3}}{n^2} \cdot (k-1)^2(k-2)(2^{k-2}-1)$$

$$\leq \frac{26|z|(2r)^{k-3}}{n^2} \cdot (k-1)^2(k-2).$$

In conclusion it follows

$$\left|S_n(f)(z) - f(z) - \frac{z}{2n}f''(z)\right| \leq \sum_{k=2}^{\infty}|c_k| \cdot |E_{k,n}(z)|$$

$$\leq \frac{26|z|}{n^2} \sum_{k=3}^{\infty} |c_k|(k-1)^2(k-2)(2r)^{k-3} = M_{r,f} \cdot \frac{|z|}{n^2},$$

which proves the point (i).

(ii) The proof is similar to the proof of Theorem 1.8.2 (ii), with the only difference that for $|E_{k,n,1}(z)| := |E_{k,n}(z)|$ we use the upper estimate in the above point (i) and for the other terms appearing to be estimated we use upper estimates in accordance with this one. $\qquad \square$

Now we are in position to prove that the order of approximation in Theorem 1.8.4 is exactly $\frac{1}{n}$.

Thus we have :

Theorem 1.8.6. (Gal [84]) *Let* $2 < R < +\infty$, $1 \leq r < \frac{R}{2}$ *and* $f : [R, +\infty) \cup \overline{\mathbb{D}}_R \to \mathbb{C}$ *be bounded on* $[0, +\infty)$ *and analytic in* $\mathbb{D}_R$, *that is* $f(z) = \sum_{k=0}^{\infty} c_k z^k$, *for all* $z \in \mathbb{D}_R$.

If f *is not a polynomial of degree* ≤ 1, *then the estimate*

$$\|S_n(f) - f\|_r \geq \frac{C_r(f)}{n}, n \in \mathbb{N},$$

holds, where the constant $C_r(f)$ *depends on* f *and* r *but is independent of* n.

Proof.   For all $z \in \mathbb{D}_R$ and $n \in \mathbb{N}$ we get

$$S_n(f)(z) - f(z)$$

$$= \frac{1}{n}\left\{ \frac{z}{2}f''(z) + \frac{1}{n}\left[n^2\left(S_n(f)(z) - f(z) - \frac{z}{2n}f''(z) \right) \right] \right\}.$$

We apply to this identity the following property :

$$\|F + G\|_r \geq |\ \|F\|_r - \|G\|_r\ | \geq \|F\|_r - \|G\|_r.$$

We get

$$\|S_n(f) - f\|_r \geq \frac{1}{n}\cdot\left\{ \left\|\frac{e_1}{2}f''\right\|_r - \frac{1}{n}\left[n^2\left\|S_n(f) - f - \frac{e_1}{2n}f''\right\|_r \right] \right\}.$$

Because by hypothesis f is not a polynomial of degree ≤ 1 in $\mathbb{D}_R$, it follows $\left\|\frac{e_1}{2}f''\right\|_r > 0$. Indeed, supposing the contrary it follows that $\frac{z}{2}f''(z) = 0$ for all $z \in \overline{\mathbb{D}}_r$, that is $f''(z) = 0$ for all $z \in \overline{\mathbb{D}}_r \setminus \{0\}$. Since f is analytic, from the identity theorem on analytic functions this implies that $f''(z) = 0$, for all $z \in \mathbb{D}_R$, that is f is a polynomial of degree ≤ 1, which is contradiction with the hypothesis.

Now, by Theorem 1.8.5 we have

$$n^2\left\|S_n(f) - f - \frac{e_1}{2n}f''\right\|_r \leq 26r \sum_{k=3}^{\infty} |c_k|(k-1)^2(k-2)(2r)^{k-3} < \infty.$$

Consequently, there exists n_1 (depending only on f and r) such that for all $n \geq n_1$ we have

$$\left\|\frac{e_1}{2}f''\right\|_r - \frac{1}{n}\left[n^2\left\|S_n(f) - f - \frac{e_1}{2n}f''\right\|_r \right] \geq \frac{1}{2}\left\|\frac{e_1}{2}f''\right\|_r,$$

which implies

$$\|S_n(f) - f\|_r \geq \frac{1}{n} \cdot \frac{1}{2} \left\| \frac{e_1}{2} f'' \right\|_r, \forall n \geq n_1.$$

For $n \in \{1, ..., n_1 - 1\}$ we clearly have $\|S_n(f) - f\|_r \geq \frac{M_{r,n}(f)}{n}$ with $M_{r,n}(f) = n \cdot \|S_n(f) - f\|_r > 0$, which finally implies

$$\|S_n(f) - f\|_r \geq \frac{C_r(f)}{n},$$

for all n, with $C_r(f) = \min\{M_{r,1}(f), ..., M_{r,n_1-1}(f), \frac{1}{2} \left\| \frac{e_1}{2} f'' \right\|_r\}$. $\square$

From Theorem 1.8.6 and Theorem 1.8.4, (i) we immediately obtain the following consequence.

Corollary 1.8.7. (Gal [84]) *Let* $2 < R < +\infty$ *and* $f : [R, +\infty) \cup \overline{\mathbb{D}}_R \to \mathbb{C}$ *be bounded on* $[0, +\infty)$ *and analytic in* $\mathbb{D}_R$, *that is* $f(z) = \sum_{k=0}^{\infty} c_k z^k$, *for all* $z \in \mathbb{D}_R$.

If $1 \leq r < \frac{R}{2}$ *is arbitrary fixed and if* f *is not a polynomial of degree* ≤ 1, *then the estimate*

$$\|S_n(f) - f\|_r \sim \frac{1}{n}, n \in \mathbb{N},$$

holds, where the constants in the equivalence depend only on f *and* r.

Regarding the simultaneous approximation of the function and its derivatives we can present :

Theorem 1.8.8. (Gal [84]) *Let* $2 < R < +\infty$ *and* $f : [R, +\infty) \cup \overline{\mathbb{D}}_R \to \mathbb{C}$ *be bounded on* $[0, +\infty)$ *and analytic in* $\mathbb{D}_R$, *that is* $f(z) = \sum_{k=0}^{\infty} c_k z^k$, *for all* $z \in \mathbb{D}_R$.

If $1 \leq r < r_1 < \frac{R}{2}$, $p \in \mathbb{N}$ *and if* f *is not a polynomial of degree* $\leq p$, *then we have*

$$\|S_n^{(p)}(f) - f^{(p)}\|_r \sim \frac{1}{n},$$

where the constants in the equivalence depend only on f, r, r_1 *and* p.

Proof. The upper estimate is exactly Theorem 1.8.4, (ii), therefore it remains to prove the lower estimate. Denote by Γ the circle of radius r_1 and center 0 (where $\frac{R}{2} > r_1 > r \geq 1$). By the Cauchy's formulas for all $|z| \leq r$ and $n \in \mathbb{N}$ it follows

$$S_n^{(p)}(f)(z) - f^{(p)}(z) = \frac{p!}{2\pi i} \int_\Gamma \frac{S_n(f)(v) - f(v)}{(v - z)^{p+1}} dv,$$

where $|v - z| \geq r_1 - r$ for all $|z| \leq r$ and $v \in \Gamma$.

As in the proof of Theorem 1.8.6, for all $v \in \Gamma$ and $n \in \mathbb{N}$ we get

$$S_n(f)(v) - f(v)$$

$$= \frac{1}{n} \left\{ \frac{v}{2} f''(v) + \frac{1}{n} \left[n^2 \left(S_n(f)(v) - f(v) - \frac{v}{2n} f''(v) \right) \right] \right\},$$

which replaced in the Cauchy's formula implies

$$S_n^{(p)}(f)(z) - f^{(p)}(z) = \frac{1}{n} \left\{ \frac{p!}{2\pi i} \int_\Gamma \frac{v f''(v)}{2(v - z)^{p+1}} dv \right.$$

$$+ \frac{1}{n} \cdot \frac{p!}{2\pi i} \int_\Gamma \frac{n^2 \left(S_n(f)(v) - f(v) - \frac{v}{2n} f''(v) \right)}{(v-z)^{p+1}} dv \Bigg\}$$

$$= \frac{1}{n} \left\{ \left[\frac{z}{2} f''(z) \right]^{(p)} + \frac{1}{n} \cdot \frac{p!}{2\pi i} \int_\Gamma \frac{n^2 \left(S_n(f)(v) - f(v) - \frac{v}{2n} f''(v) \right)}{(v-z)^{p+1}} dv \right\}.$$

Passing to the norm $\| \cdot \|_r$, for all $n \in \mathbb{N}$ we obtain

$$\| S_n^{(p)}(f) - f^{(p)} \|_r$$

$$\geq \frac{1}{n} \left\{ \left\| \left[\frac{e_1}{2} f'' \right]^{(p)} \right\|_r - \frac{1}{n} \left\| \frac{p!}{2\pi} \int_\Gamma \frac{n^2 \left(S_n(f)(v) - f(v) - \frac{v}{2n} f''(v) \right)}{(v-z)^{p+1}} dv \right\|_r \right\},$$

where by Theorem 1.8.5, for all $n \in \mathbb{N}$ it follows

$$\left\| \frac{p!}{2\pi} \int_\Gamma \frac{n^2 \left(S_n(f)(v) - f(v) - \frac{v}{2n} f''(v) \right)}{(v-z)^{p+1}} dv \right\|_r$$

$$\leq \frac{p!}{2\pi} \cdot \frac{2\pi r_1 n^2}{(r_1 - r)^{p+1}} \left\| S_n(f) - f - \frac{e_1}{2n} f'' \right\|_{r_1}$$

$$\leq 26 r_1 \sum_{k=3}^\infty |c_k| (k-1)^2 (k-2)(2r_1)^{k-3} \cdot \frac{p! r_1}{(r_1 - r)^{p+1}}.$$

Now, by hypothesis on f we have $\left\| \left[\frac{e_1}{2} f'' \right]^{(p)} \right\|_r > 0$. Indeed, supposing the contrary it follows that $\frac{z}{2} f''(z)$ is a polynomial of degree $\leq p-1$, which by the analyticity of f obviously implies that f is a polynomial of degree $\leq p$, in contradiction with the hypothesis.

For the rest of the proof, reasoning exactly as in the proof of Theorem 1.8.6, we immediately get the required conclusion. □

Remarks. 1) Since the boundedness of f on $[0, \infty)$ in the Theorems 1.8.4, 1.8.5, 1.8.6, Corollary 1.8.7 and Theorem 1.8.8 is used only for the existence of the complex Favard-Szász-Mirakjan operator, taking into account the Remark after the proof of Theorem 1.8.1 it follows that in the above mentioned results it can replaced by the condition of exponential growth $|f(z)| \leq M e^{Bx}$, for all $x \in [0, \infty)$.

2) The domain of approximation $[R, +\infty) \cup \overline{\mathbb{D}}_R$ in the previous results of this section, seem less usual. More natural could be, for example, a strip of the form

$$T_R = \{ z = x + iy \in \mathbb{C}; x \in \mathbb{R} \text{ and } |y| \leq R \}.$$

In what follows we will obtain a weighted approximation result by the complex Favard-Szász-Mirakjan operator $S_n(f)(z)$ in such a strip T_R.

First we need the following.

Lemma 1.8.9. *For fixed arbitrary $z_0 \in \mathbb{C}$, let us denote $e_1(z) = z$ and*

$$T_{n,k,z_0}(z) := S_n((e_1 - z_0)^k)(z) = e^{-nz} \sum_{j=0}^\infty \frac{(nz)^j}{j!} \left(\frac{j}{n} - z_0 \right)^k.$$

(i) For all $n \in \mathbb{N}$, $k \in \mathbb{N} \bigcup \{0\}$ and $z \in \mathbb{C}$ we have

$$T_{n,k+1,z_0}(z) = \frac{z}{n} T'_{n,k,z_0}(z) + (z - z_0) T_{n,k,z_0}(z).$$

(ii) For all $n \in \mathbb{N}$, $k \in \mathbb{N}$ with $k \geq 2$ and $z \in \mathbb{C}$ we have

$$\begin{aligned}
T_{n,k,z_0}(z) - (z - z_0)^k &= \frac{z}{n} \left[T_{n,k-1,z_0}(z) - (z - z_0)^{k-1} \right]' \\
&\quad + (z - z_0)[T_{n,k-1,z_0}(z) - (z - z_0)^{k-1}] \\
&\quad + \frac{k-1}{n} z(z - z_0)^{k-2}.
\end{aligned}$$

(iii) For all $n, k \in \mathbb{N}$ and $z, z_0 \in \mathbb{C}$ with $|z - z_0| \leq r$ we have

$$|T_{n,k,z_0}(z)| \leq r^k (3 + 2|z_0|)^k.$$

Proof. (i) By differentiating T_{n,k,z_0} we easily get the required recurrence formula.

(ii) It is an immediate consequence of (i).

(iii) By the proof of Theorem 1.8.1 (i) we have the representation in Lupaş [127]

$$S_n(f)(z) = \sum_{j=0}^{\infty} [0, 1/n, ..., j/n; f] z^j,$$

which immediately implies

$$S_n((e_1 - z_0)^p)(z) = \sum_{j=0}^{p} [0, 1/n, ..., j/n; (e_1 - z_0)^p] z^j$$

$$= \sum_{j=0}^{p} \left[\sum_{k=j}^{p} [0, 1/n, ..., k/n; (e_1 - z_0)^p] \binom{k}{j} z_0^{k-j} \right] (z - z_0)^j.$$

For $|z - z_0| \leq r$ we obtain

$$|S_n((e_1 - z_0)^p)(z)| \leq \sum_{j=0}^{p} \left[\sum_{k=j}^{p} \binom{k}{j} \frac{p(p-1)...(p-k+1)}{k!} |z_0|^{k-j} r^{p-k} \right] r^j$$

$$\leq r^p \sum_{j=0}^{p} \left[\sum_{k=j}^{p} \binom{k}{j} \binom{p}{k} |z_0|^{k-j} \right]$$

$$\leq r^p \sum_{j=0}^{p} (1 + |z_0|)^{p-j} \left[\sum_{k=j}^{p} \binom{k}{j} \binom{p}{k} \right]$$

$$= r^p \sum_{j=0}^{p} (1 + |z_0|)^{p-j} \binom{p}{j} 2^{p-j} = r^p (3 + 2|z_0|)^p,$$

where we used the formula (see e.g. Tomescu [189], p. 11, Exercise 1.5, 2))

$$\sum_{k=j}^{p} \binom{k}{j} \binom{p}{k} = \binom{p}{j} 2^{p-j}.$$

$\square$

Corollary 1.8.10. *(i) For all $n, k \in \mathbb{N}$, $z_0 \in \mathbb{C}$ and $r \geq 1$ we have the estimate*

$$\|T_{n,k,z_0} - (e_1 - z_0)^k\|_{\overline{\mathbb{D}}(z_0,r)} \leq \frac{5}{6nr}(k-1)k(2k-1)[r(3+2|z_0|)]^k.$$

(ii) Let $r \geq 1$. Suppose that $f : T_r \to \mathbb{C}$ is analytic in the strip T_r and that f satisfies the conditions

$$|f^k(x_0)| \leq \frac{M}{k^3},$$

for all $k \in \mathbb{N}\bigcup\{0\}$ and all $x_0 \in \mathbb{R}$, where $M > 0$ is independent of x_0 and k. Denoting the weight $w_r(x) = e^{-r(3+2|x|)}$, $x \in \mathbb{R}$ and the weighted norm on T_r by

$$\|f\|_{T_r,w_r} = \sup_{x\in\mathbb{R}} w_r(x)\|f - S_n(f)\|_{\overline{\mathbb{D}}(x,r)},$$

we have

$$\|f - S_n(f)\|_{T_r,w_r} \leq \frac{15M}{6nr},$$

for all $n \in \mathbb{N}$.

Proof.   (i) First we estimate $|T_{n,k,z_0}(z) - (z - z_0)^k|$. By the recurrence in Lemma 1.8.9 (i) it easily follows that $T_{n,k,z_0}(z)$ is a polynomial in z of degree $\leq k$. We will use the following generalization of the Bernstein's inequality due to Pommerenke [151]

$$\|P_n'\|_K \leq \frac{en^2}{2cap(K)}\|P_n\|_K \leq \frac{2n^2}{cap(K)}\|P_n\|_K,$$

where $P_n(z)$ is a polynomial of degree n, K is a continuum in $\mathbb{C}$ and $cap(K)$ denotes the capacity of K.

By Lemma 1.8.9 (ii) and (iii) combined with the above Bernstein-type inequality applied for $K = \overline{\mathbb{D}}(z_0, r) = \{z \in \mathbb{C}; |z - z_0| \leq r\}$, since $cap(K) = r$ we obtain

$$
\begin{aligned}
&\|T_{n,k,z_0} - (e_1 - z_0)^k\|_{\overline{\mathbb{D}}(z_0,r)} \\
&\leq \frac{(r + |z_0|)}{n}\|T_{n,k-1,z_0} - (e_1 - z_0)^{k-1}\|_{\overline{\mathbb{D}}(z_0,R)} \cdot \frac{2(k-1)^2}{r} \\
&\quad + r\|T_{n,k-1,z_0} - (e_1 - z_0)^{k-1}\|_{\overline{\mathbb{D}}(z_0,r)} + \frac{k-1}{n}(r + |z_0|)r^{k-2} \\
&\leq \frac{r + |z_0|}{n}\left[\|T_{n,k-1,z_0}\|_{\overline{\mathbb{D}}(z_0,R)} + r^{k-1}\right]\frac{2(k-1)^2}{r} \\
&\quad + r\|T_{n,k-1,z_0} - (e_1 - z_0)^{k-1}\|_{\overline{\mathbb{D}}(z_0,r)} + \frac{k-1}{n}(r + |z_0|)r^{k-2} \\
&\leq \frac{r + |z_0|}{n}\left[r^{k-1}(3 + 2|z_0|)^{k-1} + r^{k-1}\right]\frac{2(k-1)^2}{r} \\
&\quad + r\|T_{n,k-1,z_0} - (e_1 - z_0)^{k-1}\|_{\overline{\mathbb{D}}(z_0,r)} + \frac{k-1}{n}(r + |z_0|)r^{k-2},
\end{aligned}
$$

which finally leads to the inequality

$$\|T_{n,k,z_0} - (e_1 - z_0)^k\|_{\overline{\mathbb{D}}(z_0,r)} \leq r\|T_{n,k-1,z_0} - (e_1 - z_0)^{k-1}\|_{\overline{\mathbb{D}}(z_0,r)}$$

$$+\frac{r(3 + 2|z_0|)2(k-1)^2}{n}[2r^{k-2}(3 + 2|z_0|)^{k-1}]$$

$$+\frac{k-1}{n}r^{k-2}r(3 + 2|z_0|)$$

$$\leq r\|T_{n,k-1,z_0} - (e_1 - z_0)^{k-1}\|_{\overline{\mathbb{D}}(z_0,r)} + \frac{5(k-1)^2[r(3 + 2|z_0|)]^k}{nr}.$$

For $k = 1$ this inequality obviously one reduces to $0 \leq 0$. Therefore, let $k \geq 2$.

For $k = 2$ we easily obtain

$$\|T_{n,2,z_0} - (e_1 - z_0)^2\|_{\overline{\mathbb{D}}(z_0,r)} \leq \frac{5}{nr}\{1^2[r(3 + 2|z_0|)]^2\}.$$

For $k = 3$ it follows

$$\|T_{n,3,z_0} - (e_1 - z_0)^3\|_{\overline{\mathbb{D}}(z_0,r)} \leq r \cdot \frac{5}{nr}\{1^2[r(3 + 2|z_0|)]^2\} + \frac{5 \cdot 2^2[r(3 + 2|z_0|)]^3}{nr}$$

$$\leq \frac{5}{nr}[r(3 + 2|z_0|)]^3(1^2 + 2^2),$$

and reasoning by recurrence finally we arrive at

$$\|T_{n,k,z_0} - (e_1 - z_0)^k\|_{\overline{\mathbb{D}}(z_0,r)} \leq \frac{5}{nr}[r(3 + 2|z_0|)]^k(1^2 + 2^2 + ... + (k-1)^2)$$

$$= \frac{5}{6nr}(k-1)k(2k-1)[r(3 + 2|z_0|)]^k,$$

which proves (i).

(ii) For arbitrary fixed $x \in \mathbb{R}$, since f is analytic in the strip T_r, we have the Taylor expansion

$$f(z) = \sum_{k=0}^{\infty}\frac{f^{(k)}(x)}{k!}(z - x)^k,$$

valid for all $z \in \mathbb{C}$ with $|z - x| \leq r$. First we prove that

$$S_n(f)(z) = \sum_{k=0}^{\infty}\frac{f^{(k)}(x)}{k!}S_n((e_1 - x)^k)(z),$$

for all $z \in \mathbb{C}$ with $|z-x| \leq r$. For this purpose let us define $f_m(z) = \sum_{k=0}^{m}\frac{f^{(k)}(x)}{k!}(z-x)^k$, if $|z - x| \leq r$ and $f_m(z) = f(z)$ if $z \in (-\infty, x - r]\bigcup[x + r, +\infty)$. Since by hypothesis f is bounded on $\mathbb{R}$, reasoning exactly as in the Remark after the proof of Theorem 1.8.4 we easily get the desired property.

Therefore, taking into account the above point (i), for all $|z - x| \leq r$ we can write

$$|f(z) - S_n(f)(z)| \leq \frac{5}{6nr} \cdot \sum_{k=2}^{\infty}\frac{|f^{(k)}(x)|}{k!}(k-1)k(2k-1)[r(3 + 2|x|)]^k$$

$$\leq \frac{5M}{6nr} \cdot \sum_{k=2}^{\infty}\frac{(k-1)k(2k-1)}{k^3} \cdot \frac{[r(3 + 2|x|)]^k}{k!}$$

$$\leq \frac{15M}{6nr} \cdot \sum_{k=2}^{\infty}\frac{[r(3 + 2|x|)]^k}{k!},$$

that is

$$\|f - S_n(f)\|_{\overline{\mathbb{D}(x,r)}} \leq \frac{15M}{6nr} \cdot \sum_{k=2}^{\infty} \frac{[r(3 + 2|x|)]^k}{k!}.$$

Multiplying this inequality by $w_r(x) = 1/\{\sum_{k=0}^{\infty} [r(3 + 2|x|)]^k/k!\}$ and passing to supremum with $x \in \mathbb{R}$ we easily obtain the weighted inequality in the statement. The corollary is proved. $\square$

1.9   Baskakov Operators

The aim of the present section is to extend some kinds of results in the previous sections to complex Baskakov operators.

For x real and ≥ 0, the original formula of the Baskakov operator is given by (see Baskakov [35])

$$Z_n(f)(x) = (1 + x)^{-n} \sum_{k=0}^{\infty} \binom{n + k - 1}{k} \left(\frac{x}{1 + x}\right)^k f(k/n).$$

Denoting

$$W_n(f)(x) = \sum_{j=0}^{\infty} \frac{n(n + 1)...(n + j - 1)}{n^j}[0, 1/n, ..., j/n; f]x^j, x \geq 0,$$

(where for $j = 0$ we take $n(n + 1)...(n + j - 1) = 1$), according to Lupaş [128], Theorem 2, $Z_n(f)(x) = W_n(f)(x)$, for all $x \geq 0$ (under the hypothesis on f that $Z_n(f)(x)$ is well defined). But if x is not positive then $W_n(f)(x)$ and $Z_n(f)(x)$ do not necessarily coincide. For example, if $x = -1/2$ then we easily get that for all $n \in \mathbb{N}$, $Z_n(f)(-1/2)$ represents the sum of a divergent series even for the simplest function $f(x) = 1$, for all x, while clearly $W_n(f)(-1/2) = 1$, for $f(x) = 1$ and all $n \in \mathbb{N}$.

Consequently, the complex versions of these two operators denoted by

$$Z_n(f)(z) = (1 + z)^{-n} \sum_{k=0}^{\infty} \binom{n + k - 1}{k} \left(\frac{z}{1 + z}\right)^k f(k/n),$$

and

$$W_n(f)(z) = \sum_{j=0}^{\infty} \frac{n(n + 1)...(n + j - 1)}{n^j}[0, 1/n, ..., j/n; f]z^j,$$

do not necessarily coincide for all $z \in \mathbb{C}$. Because of this reason in this section they will be studied separately, under different hypothesis on f and $z \in \mathbb{C}$.

Remarks. 1) A sufficient condition for the existence of the operator $W_n(f)(z)$, $z \in \mathbb{C}$, can be expressed by the fact that f has all its derivatives bounded in $[0, \infty)$

by the same constant $M > 0$. Indeed, in this case, for $r > 0$, $z \in \mathbb{C}$ and $|z| \leq r$, we get

$$|W_n(f)(z)| \leq M \sum_{k=0}^{\infty} \frac{n(n+1)...(n+k-1)r^k}{n^k k!}, \text{ for all } |z| \leq r.$$

Denoting $a_k(n,r) = \frac{n(n+1)...(n+k-1)r^k}{n^k k!}$, we have $\frac{a_{k+1}(n,r)}{a_k(n,r)} = \frac{r(n+k)}{n(1+k)}$. Since as $k \to \infty$ we have $\frac{a_{k+1}(n,r)}{a_k(n,r)} \searrow \frac{r}{n}$, then for a fixed $n_0 \in \mathbb{N}$ and $r < \frac{n_0}{2}$, there exists k_0 such that for all $k > k_0$ we have $\frac{a_{k+1}(n_0,r)}{a_k(n_0,r)} < \frac{2r}{n_0}$. By the ratio test and by the inequality $\frac{a_{k+1}(n,r)}{a_k(n,r)} < \frac{a_{k+1}(n_0,r)}{a_k(n_0,r)}$ for all $n > n_0$, we immediately get that $W_n(f)(z)$ is well-defined and analytic for all $n > n_0$ and $|z| \leq \frac{n_0}{2}$.

2) A sufficient condition for the existence of the complex operator $Z_n(f)(z)$ can be stated as follows. Suppose for example that $z \in \mathbb{C}$ satisfies $Re z \geq 0$, $|z| \leq r$ and that $|f(x)| \leq M$ for all $x \in [0, \infty)$. Then $1 + z \neq 0$ and for all $n \in \mathbb{N}$ it follows

$$|Z_n(f)(z)| \leq M|1 + z|^{-n} \sum_{k=0}^{\infty} \binom{n+k-1}{k} \left(\frac{|z|}{|1+z|} \right)^k$$

$$\leq M|1 + z|^{-n} \sum_{k=0}^{\infty} \binom{n+k-1}{k} \rho^k.$$

where $\rho = \sqrt{\frac{r^2}{1+r^2}} < 1$. Indeed, for $z = x + iy$ with $x \geq 0$ and $\sqrt{x^2 + y^2} \leq r$ first we get (since $h(t) = t/(1+t)$ is increasing for $t \geq 0$)

$$\left(\frac{|z|}{|1+z|} \right)^2 = \frac{x^2 + y^2}{1 + 2x + (x^2 + y^2)} \leq \frac{x^2 + y^2}{1 + (x^2 + y^2)} \leq \frac{r^2}{1+r^2}.$$

By the ratio test applied to the series $\sum_{k=0}^{\infty} \binom{n+k-1}{k} \rho^k := \sum_{k=0}^{\infty} a_k$, we get that for any fixed $n \in \mathbb{N}$ there exists k_0 such that $\frac{a_{k+1}}{a_k} = \rho \frac{k+n}{k+1} < \rho' < 1$, for all $k \geq k_0$. This implies that $|Z_n(f)(z)| < \infty$ and therefore $Z_n(f)(z)$ is analytic as function of z as above. Also, as we will see later, for $z \in \mathbb{C}$ as above the operator $Z_n(f)(z)$ exists under more general conditions on f.

3) As in the case of complex Favard-Szász-Mirakjan operators, for the complex Baskakov operators too we note that the domain of definition of the approximated function $f : \overline{\mathbb{D}}_R \bigcup [R, \infty) \to \mathbb{C}$ seems to be rather strange. But the analyticity of f on $\mathbb{D}_R$ assures the representation $f(z) = \sum_{k=0}^{\infty} c_k z^k$, which is essential in the proof of quantitative estimates in any $\overline{\mathbb{D}}_r$ with $1 \leq r < R$. On the other hand, on $[0, \infty)$ the well-known estimates in the case of real variable can be used.

A more natural domain of definition for f would be a strip around the OX-axis. Unfortunately, in this case the representation $f(z) = \sum_{k=0}^{\infty} c_k z^k$ fails and we cannot use the methods of proofs in this case.

Concerning the complex operator $W_n(f)(z)$, upper estimates in simultaneous approximation, Voronovskaja's result with a quantitative estimate and exact estimates in simultaneous approximation for these operators are obtained. The hypothesis on f in these cases consist in exponential growth in a compact disk and in the boundedness (by the same constant) on $[0, \infty)$ of the derivatives of all orders of f.

At the end of this section we present similar results for the complex operator $Z_n(f)(z)$, under different hypothesis on f and z. More exactly, the disks can be replaced by semidisks and the boundedness of all derivatives of f on $[0, \infty)$ can be replaced by the weaker condition that f is of exponential growth on $[0, \infty)$.

For future research would be of interest to find larger classes of functions for which similar results with those stated in the next Theorems 1.9.1-1.9.9 hold.

The first main result of this section can be summarized by the following.

Theorem 1.9.1. (Gal [90]) *For $n_0 \in \mathbb{N}$ and $R > 0$ with $3 \leq n_0 < 2R < +\infty$ let $f : [R, +\infty) \cup \overline{\mathbb{D}}_R \to \mathbb{C}$ be with all its derivatives bounded in $[0, \infty)$ by the same positive constant, analytic in $\mathbb{D}_R$, that is $f(z) = \sum_{k=0}^{\infty} c_k z^k$, for all $z \in \mathbb{D}_R$, and suppose that there exist $M > 0$ and $A \in (\frac{1}{R}, 1)$, with the property $|c_k| \leq M\frac{A^k}{k!}$, for all $k = 0, 1, ...,$ (this implies $|f(z)| \leq Me^{A|z|}$ for all $z \in \mathbb{D}_R$).*

(i) Let $1 \leq r < \min\{\frac{n_0}{2}, \frac{1}{A}\}$. For all $|z| \leq r$ and $n > n_0$ the estimate

$$|W_n(f)(z) - f(z)| \leq \frac{C_{r,A,M}}{n},$$

holds, with $C_{r,A,M} = 6M \sum_{k=2}^{\infty} (k+1)(k-1)(rA)^k < \infty$.

(ii) In the case of simultaneous approximation by complex Baskakov operators, we have : if $1 \leq r < r_1 < \min\{\frac{n_0}{2}, \frac{1}{A}\}$ are fixed, then for all $|z| \leq r$, $p \in \mathbb{N}$ and $n > n_0$ the estimate

$$|W_n^{(p)}(f)(z) - f^{(p)}(z)| \leq \frac{p!r_1 C_{r_1,A,M}}{n(r_1 - r)^{p+1}},$$

holds, with $C_{r_1,A,M}$ is given as above.

Proof. (i) Denoting $e_k(z) = z^k$, $T_{n,k}(z) := W_n(e_k)(z)$, clearly that $T_{n,k}(z)$ is a polynomial of degree $\leq k$, $k = 0, 1, 2, ...,$ and $T_{n,0}(z) = 1, T_{n,1}(z) = z$, for all $z \in \mathbb{C}$. Also, for all $z \in \mathbb{C}$ and $n, p \in \mathbb{N}$ the following recurrence holds :

$$T_{n,p+1}(z) = \frac{z(1+z)}{n} T'_{n,p}(z) + z T_{n,p}(z).$$

Indeed, simple calculation shows that this recurrence is equivalent to

$$[0, 1/n, ..., j/n; e_{p+1}] = \frac{j}{n}[0, 1/n, ..., j/n; e_p] + [0, 1/n, ..., (j-1)/n; e_p],$$

which is an immediate consequence of the well-known relation (see e.g. Stancu [172], p. 256, Exercise 4.9)

$$[x_0, ..., x_m; f \cdot g] = \sum_{i=0}^{m} [x_0, ..., x_i; f] \cdot [x_i, ..., x_m; g]$$

applied here for $m = j$, $f = e_p$, $g = e_1$ and $x_i = i/n$.

From the above recurrence we obtain

$$T_{n,p}(z) - z^p = \frac{z(1+z)}{n}[T_{n,p-1}(z) - z^{p-1}]'$$

$$+z[T_{n,p-1}(z) - z^{p-1}] + \frac{z^{p-1}(1+z)(p-1)}{n},$$

for all $z \in \mathbb{C}$, $p, n \in \mathbb{N}$.

In what follows we will use the Bernstein's inequality on $\overline{\mathbb{D}}_r = \{z \in \mathbb{C}; |z| \le r\}$.

Thus, passing to modulus for $|z| \le r$, $r \ge 1$, from the above recurrence formula, we obtain

$$|T_{n,p}(z) - e_p(z)| \le \frac{(p-1)(1+r)}{n}\|T_{n,p-1} - e_{p-1}\|_r + r|T_{n,p-1}(z) - e_{p-1}(z)|$$

$$+\frac{r^{p-1}(p-1)(1+r)}{n} \le r|T_{n,p-1}(z) - e_{p-1}(z)|$$

$$+\frac{(p-1)(1+r)}{n}[\|T_{n,p-1}\|_r + r^{p-1}] + \frac{r^{p-1}(p-1)(1+r)}{n}$$

$$\le r|T_{n,p-1}(z) - e_{p-1}(z)| + \frac{2(p-1)r}{n}[\|T_{n,p-1}\|_r + r^{p-1}]$$

$$+\frac{2r^p(p-1)}{n}.$$

Since for any $p \in \mathbb{N}$ we have

$$T_{n,p}(z) = \sum_{k=0}^{p} \frac{n(n+1)...(n+k-1)}{n^k}[0, 1/n, ..., k/n; e_p]z^k,$$

by using the mean value theorem in complex analysis we get

$$\|T_{n,p}(z)\|_r \le \sum_{k=1}^{p} \frac{n(n+1)...(n+k-1)}{n^k} \cdot \frac{p(p-1)...(p-k+1)}{k!}r^{p-k}r^k$$

$$= r^p \sum_{k=1}^{p} \binom{p}{k}\frac{n+1}{n} \cdot .... \cdot \frac{n+k-1}{n} \le r^p \sum_{k=1}^{p} \binom{p}{k}k!$$

$$= r^p \sum_{k=1}^{p} p(p-1)...(p-k+1) \le r^p p \cdot p! \le r^p(p+1)!.$$

Replacing in the above inequality, for all $|z| \le r$ it follows

$$|T_{n,p}(z) - e_p(z)| \le r|T_{n,p-1}(z) - e_{p-1}(z)| + \frac{2(p-1)r}{n} \cdot [r^{p-1}p! + r^{p-1}]$$

$$+\frac{2r^p(p-1)}{n} \le r|T_{n,p-1}(z) - e_{p-1}(z)| + \frac{6(p+1)!r^p}{n}.$$

Starting from $p = 2, 3, ...,$ and reasoning by mathematical induction with respect to p we get

$$\|T_{n,p} - e_p\|_r \le \frac{6r^p}{n}\left[\sum_{j=2}^{p}(p+1)!\right] \le \frac{6r^p(p+1)!(p-1)}{n}.$$

From the next Remark 1 after the present proof we have

$$W_n(f)(z) = \sum_{k=0}^{\infty} c_k W_n(e_k)(z) = \sum_{k=0}^{\infty} c_k T_{n,k}(z),$$

which from the hypothesis on c_k, implies that for all $|z| \le r$ and $n > n_0$, where $1 \le r < \min\{\frac{n_0}{2}, \frac{1}{A}\}$, we have

$$|W_n(f)(z) - f(z)| \le \sum_{k=2}^{\infty} |c_k| \cdot |T_{k,n}(z) - e_k(z)|$$

$$\le \sum_{k=2}^{\infty} M \frac{A^k}{k!} \cdot \frac{6r^k(k+1)!(k-1)}{n}$$

$$= \frac{6M}{n} \sum_{k=2}^{\infty} (k+1)(k-1)(rA)^k = \frac{C_{r,A,M}}{n},$$

where $C_{r,A,M} = 6M \sum_{k=2}^{\infty}(k+1)(k-1)(rA)^k < \infty$, for all $1 \le r < \frac{1}{A}$, because the series $\sum_{k=2}^{\infty} u^{k+1}$ and therefore its derivative $\sum_{k=2}^{\infty}(k+1)ku^{k-1}$ too, are uniformly and absolutely convergent in any compact disk included in the open unit disk.

(ii) Denote by γ the circle of radius $r_1 > r$ and center 0. Since for $|z| \le r$ and $v \in \gamma$, we have $|v - z| \ge r_1 - r$, by the Cauchy's formulas for all $|z| \le r$ and $n \in \mathbb{N}$, we obtain

$$|W_n^{(p)}(f)(z) - f^{(p)}(z)| = \frac{p!}{2\pi} \left| \int_{\gamma} \frac{W_n(f)(v) - f(v)}{(v-z)^{p+1}} dv \right|$$

$$\le \frac{C_{r_1,A,M}}{n} \frac{p!}{2\pi} \frac{2\pi r_1}{(r_1-r)^{p+1}} = \frac{C_{r_1,A,M}}{n} \frac{p! r_1}{(r_1-r)^{p+1}},$$

which proves the theorem. $\qquad\square$

Remarks. 1) The proof of the relation $W_n(f)(z) = \sum_{k=0}^{\infty} c_k W_n(e_k)(z)$ used in the proof of Theorem 1.9.1, (i) is indicated bellow. First we show that under the hypothesis in the statement of Theorem 1.9.1 we can write $f(x) = \sum_{k=0}^{\infty} c_k x^k$, for all $x \in [0, \infty)$, where the series is uniformly convergent in any compact interval $[0, b]$. Indeed, by hypothesis f is infinitely differentiable on $[0, \infty)$ with all the derivatives bounded by the same constant. Then, if we consider the Taylor series of f at 0 (here it is not important that $f : [0, \infty) \to \mathbb{C}$ since we can write the decomposition $f(x) = F(x) + iG(x)$ with $F, G : [0, \infty) \to \mathbb{R}$ and instead of f we reason on F and G), that is $\sum_{k=0}^{\infty} \frac{f^{(k)}(0)}{k!} x^k$, with $x \ge 0$, then by the Lagrange form of the remainder it follows that $f(x) = \sum_{k=0}^{\infty} \frac{f^{(k)}(0)}{k!} x^k$ for all $x \ge 0$, where the Taylor series uniformly converge on each compact interval $[0, b]$. Since for $z \in \mathbb{D}_R$ we have $f(z) = \sum_{k=0}^{\infty} c_k z^k$, obviously we obtain $f(x) = \sum_{k=0}^{\infty} c_k x^k$, for all $x \ge 0$.

For $m \in \mathbb{N}$ define

$$f_m(z) = \sum_{j=0}^{m} c_j z^j \text{ if } |z| \le r \text{ and } f_m(x) = \sum_{j=0}^{m} c_j x^j \text{ if } x \in (r, +\infty).$$

Clearly that on $[0, \infty)$, each f_m is infinitely differentiable with all its derivatives bounded by the same constant. Also, by the linearity of W_n we get

$$W_n(f_m)(z) = \sum_{k=0}^{m} c_k W_n(e_k)(z), \text{ for all } |z| \leq r.$$

For $n \in \mathbb{N}$, reasoning as in the proof of Theorem 1.9.1, (i) it follows that for all $|z| \leq r$ we have $|W_n(e_k)(z)| \leq r^k(k+1)!$, which implies

$$|W_n(f_m)(z)| \leq \sum_{k=0}^{m} |c_k| \cdot |W_n(e_k)(z)| \leq \sum_{k=0}^{\infty} |c_k| \cdot |W_n(e_k)(z)|$$

$$\leq M \sum_{k=0}^{\infty} (k+1)(rA)^k := M_r(f),$$

for all $m \in \mathbb{N}$, $|z| \leq r$, $n > n_0$ with $1 \leq r < \min\{\frac{n_0}{2}, \frac{1}{A}\}$.

From Vitali's result (Theorem 1.0.1) it suffices to show that for any $n > n_0 \geq 3$ there exists $0 < x_0 < 1$ (depending on n) such that

$$\lim_{m \to \infty} W_n(f_m)(x) = W_n(f)(x), \text{ for } x \in [0, x_0].$$

Indeed, for $x_0 \geq 0$ we have $W_n(f)(x_0) = Z_n(f)(x_0)$ and denoting $\rho_0 = \frac{x_0}{1+x_0} \in [0, 1)$, by the hypothesis on f we get

$$|W_n(f)(x_0) - W_n(f_m)(x_0)|$$

$$= \left| (1+x_0)^{-n} \sum_{k=0}^{\infty} \binom{n+k-1}{k} \left(\frac{x_0}{1+x_0} \right)^k \left[\sum_{j=m+1}^{\infty} c_j \left(\frac{k}{n} \right)^j \right] \right|$$

$$\leq (1+x_0)^{-n} \sum_{k=0}^{\infty} \binom{n+k-1}{k} \rho_0^k \sum_{j=m+1}^{\infty} |c_j| \frac{k^j}{n^j}$$

$$= (1+x_0)^{-n} \frac{1}{n^{m+1}} \sum_{k=0}^{\infty} \binom{n+k-1}{k} \rho_0^k \sum_{j=m+1}^{\infty} |c_j| \frac{k^j}{n^{j-m-1}}$$

$$\leq (1+x_0)^{-n} \frac{1}{n^{m+1}} \sum_{k=0}^{\infty} \binom{n+k-1}{k} \rho_0^k \sum_{j=m+1}^{\infty} |c_j| k^j$$

$$\leq M(1+x_0)^{-n} \frac{1}{n^{m+1}} \sum_{k=0}^{\infty} \binom{n+k-1}{k} \rho_0^k \sum_{j=m+1}^{\infty} \frac{(kA)^j}{j!}$$

$$= \frac{M(1+x_0)^{-n}}{n^{m+1}} \sum_{k=0}^{\infty} \binom{n+k-1}{k} \rho_0^k \left(e^{kA} - \sum_{j=0}^{m} \frac{(kA)^j}{j!} \right)$$

$$\leq \frac{M(1+x_0)^{-n}}{n^{m+1}} \sum_{k=0}^{\infty} \binom{n+k-1}{k} \rho_0^k \frac{e^{kA}(kA)^{m+1}}{(m+1)!}$$

$$\leq \frac{M(1+x_0)^{-n}}{n^{m+1}} \sum_{k=0}^{\infty} \binom{n+k-1}{k} \rho_0^k e^{2kA}$$

$$= \frac{M(1+x_0)^{-n}}{n^{m+1}} \sum_{k=0}^{\infty} \binom{n+k-1}{k} (\rho_0 e^{2A})^k.$$

Choose $x_0 > 0$ with $\rho_0 e^{2A} := \rho(x_0) < \frac{1}{n}$, where $n > n_0 \geq 3$. Because the function $h(t) = \frac{t}{1+t}$ is continuous and $h(0) = 0$, this is always possible. We get

$$(1 + x_0)^n |W_n(f)(x_0) - W_n(f_m)(x_0)| \leq \frac{M}{n^{m+1}} \sum_{k=0}^{\infty} \binom{n+k-1}{k} \rho(x_0)^k$$

$$\leq \frac{M}{n^{m+1}} \sum_{k=0}^{\infty} \binom{n+k-1}{k} \frac{1}{n^k}.$$

For $n > n_0 \geq 3$, by the ratio test we easily get that the series $\sum_{k=0}^{\infty} \binom{n+k-1}{k} \frac{1}{n^k}$ is convergent, which implies

$$\lim_{m \to \infty} W_n(f_m)(x_0) = W_n(f)(x_0).$$

But the above defined function $h(t)$ also is increasing on $[0, \infty)$, therefore for any $x \in [0, x_0]$ we get $\frac{x}{1+x} \leq \frac{x_0}{1+x_0}$ and $\rho(x) \leq \rho(x_0) < \frac{1}{n}$. Applying the above estimates for x instead of x_0 we get

$$\lim_{m \to \infty} W_n(f_m)(x) = W_n(f)(x),$$

which proves our assertion.

2) Simple examples of functions f satisfying the hypothesis in Theorem 1.9.1 are $f(z) = e^{-az}$, or $f(z) = sin(az)$ with $0 < a < 1$.

Now, let us recall that in the case of real Baskakov operators the following Voronovskaja-type formula is known.

Theorem 1.9.2. (Sikkema [163]) *If $f : [0, +\infty) \to \mathbb{R}$ is twice continuous differentiable on $[0, +\infty)$, then uniformly in any compact subinterval of $[0, \infty)$ we have*

$$\lim_{n \to \infty} n[W_n(f)(x) - f(x)] = \frac{x(1+x)}{2} f''(x).$$

In what follows we extend this result to the complex Baskakov operators obtaining, in addition, a quantitative estimate too.

We have :

Theorem 1.9.3. (Gal [90]) *Suppose that the hypothesis on the function f and on the constants n_0, R, M, A in the statement of Theorem 1.9.1 hold and let $1 \leq r < \min\{\frac{n_0}{2}, \frac{1}{A}\}$ be fixed. For all $n > n_0, |z| \leq r$ the following Voronovskaja-type result*

$$\left| W_n(f)(z) - f(z) - \frac{z(1+z)}{2n} f''(z) \right| \leq \frac{16M}{n^2} \sum_{k=3}^{\infty} (rA)^k (k-1)(k-2)^2,$$

holds, where for $rA < 1$ we have $\sum_{k=3}^{\infty} (rA)^k (k-1)(k-2) < \infty$.

Proof. Denote $e_k(z) = z^k$, $k = 0, 1, ...$, and $T_{n,k}(z) = W_n(e_k)(z)$. By the proof of Theorem 1.9.1, (i), we can write $W_n(f)(z) = \sum_{k=0}^{\infty} c_k T_{n,k}(z)$, which implies

$$\left| W_n(f)(z) - f(z) - \frac{z(1+z)}{2n} f''(z) \right|$$

$$\leq \sum_{k=2}^{\infty} |c_k| \cdot \left| T_{n,k}(z) - e_k(z) - \frac{z^{k-1}(1+z)k(k-1)}{2n} \right|.$$

By the recurrence in the proof of Theorem 1.9.1, (i), satisfied by $T_{n,k}(z)$, denoting $E_{k,n}(z) = T_{n,k}(z) - e_k(z) - \frac{z^{k-1}(1+z)k(k-1)}{2n}$, we obtain the new recurrence

$$E_{k,n}(z) =$$

$$\frac{z(1+z)}{n}E'_{k-1,n}(z) + zE_{k-1,n}(z) + \frac{z^{k-2}(1+z)(k-1)(k-2)}{2n^2}[(k-2)+z(k-1)],$$

for all $k \geq 2$, $n \in \mathbb{N}$.

Therefore, for all $|z| \leq r$, $k \geq 2$, $n \in \mathbb{N}$ we obtain

$$|E_{k,n}(z)| \leq \frac{r(1+r)}{n}|E'_{k-1,n}(z)| + r|E_{k-1,n}(z)|$$
$$+ \frac{(1+r)r^{k-2}(k-1)(k-2)}{2n^2}[(k-2)+r(k-1)].$$

Since $E_{k-1,n}(z)$ is a polynomial of degree $\leq (k-1)$, by applying the Bernstein's inequality we get

$$|E'_{k-1,n}(z)| \leq \frac{k-1}{r}\|E_{k-1,n}(z)\|_r$$
$$\leq \frac{k-1}{r}\left[\|T_{n,k-1} - e_{k-1}\|_r + \frac{r^{k-2}(r+1)(k-1)(k-2)}{2n}\right]$$
$$\leq \frac{k-1}{r}\left[\frac{6r^{k-1}k!(k-2)}{n} + \frac{r^{k-2}(r+1)(k-1)(k-2)}{2n}\right]$$
$$\leq \frac{7r^{k-2}k!(k-1)(k-2)}{n}.$$

Replacing above this inequality, for all $|z| \leq r$ we obtain

$$|E_{k,n}(z)| \leq r|E_{k-1,n}(z)| + \frac{14r^k k!(k-1)(k-2)}{n^2}$$
$$+ \frac{(1+r)r^{k-2}(k-1)(k-2)}{2n^2}[(k-2)+r(k-1)]$$
$$\leq r|E_{k-1,n}(z)| + \frac{16r^k k!(k-1)(k-2)}{n^2}.$$

Since $E_{0,n}(z) = E_{1,n}(z) = E_{2,n}(z) = 0$, taking $k = 3, 4, ...$, in the last inequality, step by step finally we arrive at

$$|E_{k,n}(z)| \leq \frac{16r^k}{n^2}\sum_{j=3}^{k}j!(j-1)(j-2) \leq \frac{16r^k k!(k-1)(k-2)^2}{n^2},$$

which implies

$$\left|W_n(f)(z) - f(z) - \frac{z(1+z)}{2n}f''(z)\right| \leq \sum_{k=3}^{\infty}|c_k| \cdot |E_{k,n}(z)|$$
$$\leq \frac{16M}{n^2}\sum_{k=3}^{\infty}(rA)^k(k-1)(k-2)^2.$$

Here for $rA < 1$ we obviously have $\sum_{k=3}^{\infty}(rA)^k(k-1)(k-2) < \infty$, which proves the theorem. $\qquad\square$

Remark. Under the hypothesis in Theorem 1.9.3 we have the equivalence

$$\left\|W_n(f) - f(z) - \frac{e_1(1+e_1)}{2n}f''\right\|_r \sim \frac{1}{n^2}.$$

For the proof of this equivalence we follow the ideas in the proof of Theorem 1.8.2 (ii), with the only differences that we use the upper estimate for $|E_{k,n,1}(z)| := |E_{k,n}(z)|$ in Theorem 1.9.3 and the recurrence formula satisfied in this case by $A_{n,j}(z) := W_n[(\cdot - z)^j](z)$, that is

$$A_{n,j+1}(z) = z(1+z)[A'_{n,j}(z) + njA_{n,j-1}(z)],$$

(the proof of this recurrence formula uses the same Lemma 1.3 in Pop [156]). Finally, everything one reduces to prove that the differential equation

$$P_{2p+1}(z)f^{(2p+1)}(z) + P_{2p+2}(z)f^{(2p+2)}(z) = 0, \ z \in \mathbb{D}_r,$$

has the only solutions for f a polynomial of degree $\leq 2p$. Here by the above recurrence formula we get $P_{2p+1}(z) = c_p[z(1+z)]^p(1+2z)$ and $P_{2p+2}(z) = d_p[z(1+z)]^{p+1}$, with $c_p, d_p > 0$. The proof is easy by writing $y(z)$ in the form $y(z) = \sum_{k=0}^{\infty} a_k z^k$.

In what follows we obtain the exact degree in approximation by $W_n(f)(z)$.

The first main result is the following.

Theorem 1.9.4. (Gal [90]) *Suppose that the hypothesis on the function f and on the constants n_0, R, M, A in the statement of Theorem 1.9.1 hold and let $1 \leq r < \min\{\frac{n_0}{2}, \frac{1}{A}\}$. If f is not a polynomial of degree ≤ 1, then for all $n > n_0$ the lower estimate*

$$\|W_n(f) - f\|_r \geq \frac{C_r(f)}{n}$$

holds, where the constant $C_r(f)$ depends only on f (that is on A, M) and r.

Proof. For all $|z| \leq r$ and $n > n_0$ we get

$$W_n(f)(z) - f(z)$$

$$= \frac{1}{n}\left\{\frac{z(1+z)}{2}f''(z) + \frac{1}{n}\left[n^2\left(W_n(f)(z) - f(z) - \frac{z(1+z)}{2n}f''(z)\right)\right]\right\}.$$

We apply to this identity the following simple property :

$$\|F + G\|_r \geq |\, \|F\|_r - \|G\|_r\,| \geq \|F\|_r - \|G\|_r.$$

We get

$$\|W_n(f) - f\|_r$$

$$\geq \frac{1}{n}\left\{\left\|\frac{e_1(1+e_1)}{2}f''\right\|_r - \frac{1}{n}\left[n^2\left\|W_n(f) - f - \frac{e_1(1+e_1)}{2n}f''\right\|_r\right]\right\}.$$

Since f is not a polynomial of degree ≤ 1 in $\mathbb{D}_R$, we get $\left\|\frac{e_1(1+e_1)}{2}f''\right\|_r > 0$. Indeed, supposing the contrary it follows that $\frac{z(1+z)}{2}f''(z) = 0$ for all $z \in \overline{\mathbb{D}}_r$, which clearly implies $f''(z) = 0$ for all $z \in \overline{\mathbb{D}}_r \setminus \{0, -1\}$. Since f is analytic, from the identity theorem on analytic functions this implies that $f''(z) = 0$, for all $z \in \mathbb{D}_R$, that is f is a polynomial of degree ≤ 1, in contradiction with the hypothesis.

By Theorem 1.9.3 we have

$$n^2 \left\| W_n(f) - f - \frac{e_1(1+e_1)}{2n}f'' \right\|_r \leq 16M \sum_{k=3}^{\infty} (rA)^k (k-1)(k-2)^2.$$

Therefore, there exists $n_1 > n_0$ depending only on f and r, such that for any $n \geq n_1$ we have

$$\left\| \frac{e_1(1+e_1)}{2}f'' \right\|_r - \frac{1}{n}\left[n^2 \left\| W_n(f) - f - \frac{e_1(1+e_1)}{2n}f'' \right\|_r \right]$$
$$\geq \frac{1}{2}\left\| \frac{e_1(1+e_1)}{2}f'' \right\|_r,$$

which implies

$$\|W_n(f) - f\|_r \geq \frac{1}{n} \cdot \frac{1}{2}\left\| \frac{e_1(1+e_1)}{2}f'' \right\|_r, \forall n \geq n_1.$$

For $n \in \{n_0+1, ..., n_1\}$ we get $\|W_n(f) - f\|_r \geq \frac{M_{r,n}(f)}{n}$ with $M_{r,n}(f) = n \cdot \|W_n(f) - f\|_r > 0$, which finally implies $\|W_n(f) - f\|_r \geq \frac{C_r(f)}{n}$ for all $n > n_0$, with $C_r(f) = \min\left\{ M_{r,n_0+1}(f), ..., M_{r,n_1-1}(f), \frac{1}{2}\left\| \frac{e_1(1+e_1)}{2}f'' \right\|_r \right\}$. This proves the theorem. $\square$

From Theorem 1.9.4 and Theorem 1.9.1, (i), we get the following consequence.

Corollary 1.9.5. (Gal [90]) *Suppose that the hypothesis on the function f and on the constants n_0, R, M, A in the statement of Theorem 1.9.1 hold and let $1 \leq r < \min\{\frac{n_0}{2}, \frac{1}{A}\}$ be arbitrary fixed. If f is not a polynomial of degree ≤ 1, then for all $n > n_0$ the estimate*

$$\|W_n(f) - f\|_r \sim \frac{1}{n},$$

holds, where the constants in the equivalence depend only on f (i.e. on A, M) and r.

Regarding the simultaneous approximation we have the following result.

Theorem 1.9.6. (Gal [90]) *Suppose that the hypothesis on the function f and on the constants n_0, R, M, A in the statement of Theorem 1.9.1 hold and let $1 \leq r < r_1 < \min\{\frac{n_0}{2}, \frac{1}{A}\}$, $p \in \mathbb{N}$. If f is not a polynomial of degree $\leq \max\{1, p-1\}$, then for all $n > n_0$ the estimate*

$$\|W_n^{(p)}(f) - f^{(p)}\|_r \sim \frac{1}{n},$$

holds, where the constants in the equivalence depend only on f (that is on A, M), r, r_1 and p.

Proof. By Theorem 1.9.1, (ii) we have the upper estimate for $\|W_n^{(p)}(f) - f^{(p)}\|_r$, therefore it remains to prove the lower estimate for $\|W_n^{(p)}(f) - f^{(p)}\|_r$. Denoting by Γ the circle of radius r_1 and center 0 (where $\min\{\frac{n_0}{2}, \frac{1}{A}\} > r_1 > r \geq 1$), we have the inequality $|v - z| \geq r_1 - r$ for all $|z| \leq r$ and $v \in \Gamma$. By the Cauchy's formula we obtain

$$W_n^{(p)}(f)(z) - f^{(p)}(z) = \frac{p!}{2\pi i} \int_\Gamma \frac{W_n(f)(v) - f(v)}{(v - z)^{p+1}} dv.$$

As in the proof of Theorem 1.9.1, (ii), for all $v \in \Gamma$ and $n > n_0$ we get

$$W_n(f)(v) - f(v)$$

$$= \frac{1}{n}\left\{ \frac{v(1 + v)}{2} f''(v) + \frac{1}{n}\left[n^2\left(W_n(f)(v) - f(v) - \frac{v(1 + v)}{2n} f''(v) \right)\right] \right\}.$$

Replaced in the above Cauchy's formula implies

$$W_n^{(p)}(f)(z) - f^{(p)}(z) = \frac{1}{n}\left\{ \frac{p!}{2\pi i} \int_\Gamma \frac{v(1 + v) f''(v)}{2(v - z)^{p+1}} dv \right.$$

$$\left. + \frac{1}{n} \cdot \frac{p!}{2\pi i} \int_\Gamma \frac{n^2\left(W_n(f)(v) - f(v) - \frac{v(1+v)}{2n} f''(v) \right)}{(v - z)^{p+1}} dv \right\}$$

$$= \frac{1}{n}\left\{ \left[\frac{z(1 + z)}{2} f''(z) \right]^{(p)} \right.$$

$$\left. + \frac{1}{n} \cdot \frac{p!}{2\pi i} \int_\Gamma \frac{n^2\left(W_n(f)(v) - f(v) - \frac{v(1+v)}{2n} f''(v) \right)}{(v - z)^{p+1}} dv \right\}$$

and passing to the norm $\|\cdot\|_r$, for all $n > n_0$ we obtain

$$\|W_n^{(p)}(f) - f^{(p)}\|_r \geq \frac{1}{n}\left\{ \left\| \left[\frac{e_1(1 + e_1)}{2} f'' \right]^{(p)} \right\|_r \right.$$

$$\left. - \frac{1}{n}\left\| \frac{p!}{2\pi} \int_\Gamma \frac{n^2\left(W_n(f)(v) - f(v) - \frac{v(1+v)}{2n} f''(v) \right)}{(v - z)^{p+1}} dv \right\|_r \right\},$$

where by Theorem 1.9.3, for all $n > n_0$ we have

$$\left\| \frac{p!}{2\pi} \int_\Gamma \frac{n^2\left(W_n(f)(v) - f(v) - \frac{v(1+v)}{2n} f''(v) \right)}{(v - z)^{p+1}} dv \right\|_r$$

$$\leq \frac{p!}{2\pi} \cdot \frac{2\pi r_1 n^2}{(r_1 - r)^{p+1}} \left\| W_n(f) - f - \frac{e_1(1 + e_1)}{2n} f'' \right\|_{r_1}$$

$$\leq 16M \sum_{k=3}^\infty (r_1 A)^k (k - 1)(k - 2)^2 \cdot \frac{p! r_1}{(r_1 - r)^{p+1}}.$$

But by hypothesis on f, we have $\left\| \left[\frac{e_1(1+e_1)}{2} f'' \right]^{(p)} \right\|_r > 0$. Indeed, supposing the contrary it follows that $\frac{z(1+z)}{2} f''(z)$ is a polynomial of degree $\leq p-1$. Now, if $p = 1$ and $p = 2$ then the analyticity of f implies that f is a polynomial of degree $\leq 1 = \max\{1, p-1\}$, in contradiction with the hypothesis. If $p > 2$ then the analyticity of f implies that f is a polynomial of degree $\leq p-1 = \max\{1, p-1\}$, which is again in contradiction with the hypothesis.

Finally, reasoning exactly as in the proof of Theorem 1.9.4 we immediately obtain the required conclusion. $\qquad \square$

At the end of this section we deal with the approximation properties of $Z_n(f)(z)$ given by

$$Z_n(f)(z) = (1+z)^{-n} \sum_{k=0}^{\infty} \binom{n+k-1}{k} \left(\frac{z}{1+z} \right)^k f(k/n),$$

mentioned and discussed at the beginning of this section. First we present some sufficient conditions on f for the analyticity of $Z_n(f)(z)$.

Theorem 1.9.7. (Gal [90]) *Suppose that $f : [0, \infty) \to \mathbb{C}$ is of exponential growth on $[0, \infty)$, that is there exists $M > 0$ and $A \geq 0$ such that $|f(x)| \leq Me^{Ax}$ for all $x \in [0, \infty)$.*

(i) For any $n \in \mathbb{N}$ there exists a $0 < \rho < 1$ (depending on n) such that $Z_n(f)(z)$ is well defined and analytic in the compact disk $\overline{\mathbb{D}}_{\rho/2}$.

(ii) Let $r \geq 1$ be fixed and denote $n_0 = \left[\frac{2A}{\ln(1+1/r^2)} \right] + 2$. For all $n \geq n_0$, $Z_n(f)(z)$ is analytic in the compact semi-disk $\overline{\mathbb{D}}_r^+ = \{z \in \mathbb{C}; |z| \leq r, Rez \geq 0\}$.

Proof. (i) It is clear that if $0 < \rho < 1$ then for $z \in \overline{\mathbb{D}}_{\rho/2}$ it follows $1 + z \neq 0$ and therefore $(1+z)^{-n}$ is analytic. We get

$$|Z_n(f)(z)| \leq M|1+z|^{-n} \sum_{k=0}^{\infty} \binom{n+k-1}{k} \left(\frac{|z|}{|1+z|} \right)^k e^{ak/n}.$$

For $z = x + iy$ with $|x| \leq \sqrt{x^2 + y^2} \leq \rho/2$ we obtain (since $h(t) = t/(1 - \rho + t)$ is increasing for $t \geq 0$)

$$\left(\frac{|z|}{|1+z|} \right)^2 = \frac{x^2 + y^2}{1 + 2x + (x^2 + y^2)} = \frac{x^2 + y^2}{1 - \rho + (\rho + 2x) + (x^2 + y^2)}$$

$$\leq \frac{x^2 + y^2}{1 - \rho + (x^2 + y^2)} \leq \frac{(\rho/2)^2}{1 - \rho + (\rho/2)^2}.$$

Denoting $\eta := \sqrt{\frac{(\rho/2)^2}{1 - \rho + (\rho/2)^2}} < 1$ we obtain

$$|Z_n(f)(z)| \leq M|1+z|^{-n} \sum_{k=0}^{\infty} \binom{n+k-1}{k} \eta^k e^{ak/n} := M|1+z|^{-n} \sum_{k=0}^{\infty} a_k.$$

We apply the ratio test to the last series. We have $\frac{a_{k+1}}{a_k} = \eta e^{a/n} \cdot \frac{k+n}{k+1}$. Since for $\rho \to 0$ we obtain $\eta \to 0$, for fixed n choose ρ so small that $\beta := \eta e^{a/n} n < 1$. By

$\frac{k+n}{k+1} \leq n$ for all $k \geq 0$ obviously that $\frac{a_{k+1}}{a_k} \leq \beta < 1$, for all $k \geq 0$, which proves that the series $\sum_{k=0}^{\infty} a_k$ is convergent and therefore $Z_n(f)(z)$ is analytic in $\overline{\mathbb{D}}_{\rho/2}$ for ρ sufficiently small (depending on n) chosen as above.

(ii) Clearly that for $z \in \overline{\mathbb{D}}_r^+$ we get $1 + z \neq 0$ and therefore $(1+z)^{-n}$ is analytic. Reasoning exactly as at the above point (i) and denoting $\eta = \frac{r}{\sqrt{1+r^2}} < 1$, for all $z \in \overline{\mathbb{D}}_r^+$ we get

$$|Z_n(f)(z)| \leq M|1+z|^{-n} \sum_{k=0}^{\infty} \binom{n+k-1}{k} \eta^k e^{Ak/n}$$

$$= M|1+z|^{-n} \sum_{k=0}^{\infty} \binom{n+k-1}{k} [\eta e^{A/n}]^k.$$

Denote $C = \frac{2A}{ln(1+1/r^2)}$, $n_1 = [C] + 1 \geq C$. We get that for $n \geq n_0 > n_1$ we have

$$\eta e^{A/n} \leq \eta e^{A/n_0} < \eta e^{A/n_1} \leq \eta e^{A/C} = 1.$$

Denoting $\gamma = \eta e^{A/n_0} < 1$, for all $n \geq n_0$ we obtain

$$|Z_n(f)(z)| \leq M|1+z|^{-n} \sum_{k=0}^{\infty} \binom{n+k-1}{k} [\eta e^{A/n}]^k$$

$$\leq M|1+z|^{-n} \sum_{k=0}^{\infty} \binom{n+k-1}{k} \gamma^k := M|1+z|^{-n} \sum_{k=0}^{\infty} a_k.$$

Let $n \geq n_0$. By the ratio test applied to the last series we have $\frac{a_{k+1}}{a_k} = \gamma \cdot \frac{k+n}{k+1}$. Since there exists a k_0 with $\frac{k+n}{k+1} < 2 - \gamma$ for all $k \geq k_0$, it follows that $\frac{a_{k+1}}{a_k} = \gamma \cdot \frac{k+n}{k+1} < \gamma(2 - \gamma) < 1$, for all $k \geq k_0$. This implies the convergence of the series $\sum_{k=0}^{\infty} a_k$ and therefore we obtain the analyticity of $Z_n(f)(z)$. □

In what follows we show that $Z_n(f)(z)$ can be written under the form of divided differences.

Corollary 1.9.8. (Gal [90]) *(i) Suppose that f satisfies the hypothesis in Theorem 1.9.7. Then for any $n \in \mathbb{N}$ there exists a sufficiently small $0 < \rho < 1$ (depending on n) such that for all $z \in \overline{\mathbb{D}}_{\rho/2}$ we have*

$$Z_n(f)(z) = \sum_{j=0}^{\infty} \frac{n(n+1)...(n+j-1)}{n^j} [0, 1/n, ..., j/n; f] z^j.$$

(ii) Let $p \in \mathbb{N} \cup \{0\}$ be fixed. For any $n \in \mathbb{N}$ there exists a sufficiently small $0 < \rho < 1$ such that for all $z \in \overline{\mathbb{D}}_{\rho/2}$ we have

$$Z_n(e_p)(z) = \sum_{k=0}^{p} \frac{n(n+1)...(n+k-1)}{n^k} [0, 1/n, ..., k/n; e_p] z^k.$$

(iii) Let $p \in \mathbb{N} \cup \{0\}$ and $r \geq 1$ be fixed and denote $n_0 = \left[\frac{2}{ln(1+1/r^2)}\right] + 2$. For all $n \geq n_0$ and $z \in \overline{\mathbb{D}}_r^+$ we have

$$Z_n(e_p)(z) = \sum_{k=0}^{p} \frac{n(n+1)...(n+k-1)}{n^k} [0, 1/n, ..., k/n; e_p] z^k.$$

Proof. (i) For fixed n let us define $g_{k,n}(z) = (1+z)^{-n-k}$, $k = 0, 1, 2, ...,$. Since $g_{k,n}$ is analytic in $\overline{\mathbb{D}}_{\rho/2}$, we get

$$g_{k,n}(z) = \sum_{p=0}^{\infty} \frac{g_{k,n}^{(p)}(0)}{p!} z^p = \sum_{p=0}^{\infty} (-1)^p \frac{(n+k-1+p)!}{(n+k-1)!p!} z^p,$$

which replaced in the expression of $Z_n(f)(z)$ implies

$$Z_n(f)(z) = \sum_{k=0}^{\infty} \binom{n+k-1}{k} z^k f(k/n) \sum_{p=0}^{\infty} (-1)^p \frac{(n+k-1+p)!}{(n+k-1)!p!} z^p$$

$$= \sum_{k=0}^{\infty} \sum_{p=0}^{\infty} z^{k+p} \left[f(k/n)(-1)^p \frac{(n+k-1+p)!}{(n-1)!k!p!} \right]$$

$$:= \sum_{k=0}^{\infty} \sum_{p=0}^{\infty} z^{k+p} A_{k,p} = \sum_{j=0}^{\infty} z^j B_j.$$

This implies the formulas

$$B_j = \sum_{\nu=0}^{j} A_{\nu,j-\nu} = \sum_{\nu=0}^{j} f(\nu/n)(-1)^{j-\nu} \frac{(n+j-1)!}{(n-1)!\nu!(j-\nu)!}$$

$$= \frac{(n+j-1)!}{(n-1)!n^j} \sum_{\nu=0}^{j} f(\nu/n)(-1)^{j-\nu} \frac{n^j}{\nu!(j-\nu)!}$$

$$= \frac{(n+j-1)!}{(n-1)!n^j} [0, 1/n, ..., j/n; f],$$

which proves the corollary.

(ii) It is immediate from the above point (i) since each e_p satisfies the hypothesis in Theorem 1.9.7 with $A = 1$.

(iii) By Theorem 1.9.7 (ii) we get that for all $n \geq n_0$, $Z_n(e_p)(z)$ is analytic in $\overline{\mathbb{D}}_r^+$. On the other hand, by the above point (ii) , $Z_n(e_p)(z)$ is a polynomial of degree $\leq p$ in a small disk $\overline{\mathbb{D}}_{\rho/2}$, that is its derivative of order $p+1$ is zero in $\overline{\mathbb{D}}_{\rho/2}$. By the identity theorem on analytic functions it is clear that the derivative of order $p+1$ of $Z_n(e_p)(z)$ also is zero in $\overline{\mathbb{D}}_r^+$, that is $Z_n(e_p)(z)$ is a polynomial of degree $\leq p$ in $\overline{\mathbb{D}}_r^+$. Since $Z_n(f)(z)$ analytically extends the values in $\overline{\mathbb{D}}_{\rho/2}$ to $\overline{\mathbb{D}}_r^+$, clearly that $Z_n(e_p)(z)$ must be of the same form

$$Z_n(e_p)(z) = \sum_{k=0}^{p} \frac{n(n+1)...(n+k-1)}{n^k} [0, 1/n, ..., k/n; e_p] z^k,$$

for all $z \in \overline{\mathbb{D}}_r^+$. $\qquad \square$

We are in position to prove the following result.

Theorem 1.9.9. (Gal [90]) *For $R > 1$ suppose that $f : [R, +\infty) \cup \overline{\mathbb{D}}_R \to \mathbb{C}$ is analytic in $\mathbb{D}_R$, that is $f(z) = \sum_{k=0}^{\infty} c_k z^k$, for all $z \in \mathbb{D}_R$, and that there exist*

$M > 0$ and $A \in (\frac{1}{R}, 1)$, with the property $|c_k| \leq M\frac{A^k}{k!}$, for all $k = 0, 1, ...,$ (this implies $|f(z)| \leq Me^{A|z|}$ for all $z \in \mathbb{D}_R^+$). In addition, let us suppose that f is of exponential growth on $[0, \infty)$ (for simplicity suppose that the exponent in the exponential growth also is A). Then the upper estimates in Theorems 1.9.1 (i) and 1.9.3 and the exact estimate in Corollary 1.9.5 hold for $Z_n(f)(z)$ in the compact semi-disk $\overline{\mathbb{D}}_r^+$, for any $1 \leq r < R$ and $n \geq n_0$ with $n_0 = \left[\frac{2A}{ln(1+1/r^2)} \right] + 2$.

Proof. By Corollary 1.9.8 (iii), the recurrence for $Z_n(e_p)(z)$ is similar to that in the proof of Theorem 1.9.1, taking place for z in the compact semi-disk $\overline{\mathbb{D}}_r^+ = \{z; |z| \leq r, \ Rez \geq 0\}$. The reasonings in the proofs of the corresponding results mentioned in the statement are similar. What remains to prove is the equality

$$Z_n(f)(z) = \sum_{k=0}^{\infty} c_k Z_n(e_k)(z),$$

for all $z \in \overline{\mathbb{D}}_r^+$ and $n \geq n_0$.

For this purpose, for any $m \in \mathbb{N}$ let us define

$$f_m(z) = \sum_{j=0}^{m} c_j z^j \text{ if } z \in \overline{\mathbb{D}}_r^+ \text{ and } f_m(x) = f(x) \text{ if } x \in (r, +\infty).$$

From the hypothesis on the coefficients of f it is clear that for any $m \in \mathbb{N}$ we have

$$|f_m(x)| \leq \sum_{j=0}^{m} |c_j| x^j \leq M \sum_{j=0}^{m} \frac{(Ax)^j}{j!} \leq Me^{Ax}, \text{ for all } x \in [0, r],$$

which from the hypothesis on f on $[0, \infty)$, immediately implies that for any $m \in \mathbb{N}$ we have

$$|f_m(x)| \leq Me^{Ax}, \text{ for all } x \in [0, \infty).$$

By Theorem 1.9.7 (ii) we get that $Z_n(f_m)(z)$, are well-defined and analytic in the compact semi-disk $\overline{\mathbb{D}}_r^+$, for all $m \in \mathbb{N}$ and $n \geq n_0$.

Denoting

$$f_{m,k}(z) = c_k e_k(z) \text{ if } z \in \overline{\mathbb{D}}_r^+ \text{ and } f_{m,k}(x) = \frac{f(x)}{m+1} \text{ if } x \in (r, \infty),$$

each $f_{m,k}$ is of exponential growth on $[0, \infty)$ (with the same exponent A) and $f_m(z) = \sum_{k=0}^{m} f_{m,k}(z)$. Since from the linearity of Z_n we have

$$Z_n(f_m)(z) = \sum_{k=0}^{m} c_k Z_n(e_k)(z), \text{ for all } z \in \overline{\mathbb{D}}_r^+,$$

it suffices to prove that $\lim_{m\to\infty} Z_n(f_m)(z) = Z_n(f)(z)$ for any fixed $n \geq n_0$ and $z \in \overline{\mathbb{D}}_r^+$. But this is immediate from $\lim_{m\to\infty} \|f_m - f\|_r = 0$, from $\|f_m - f\|_{B[0,+\infty)} \leq$

$\|f_m - f\|_r$ and from the inequality

$$|Z_n(f_m)(z) - Z_n(f)(z)|$$

$$\leq |1+z|^{-n} \sum_{j=0}^{\infty} \binom{n+j-1}{j} \left(\frac{|z|}{|1+z|}\right)^j \|f_m - f\|_{B[0,\infty)}$$

$$\leq |1+z|^{-n} \sum_{j=0}^{\infty} \binom{n+j-1}{j} \left(\frac{r}{\sqrt{1+r^2}}\right)^j \|f_m - f\|_{B[0,\infty)},$$

valid for all $z \in \overline{\mathbb{D}}_r^+$. Here, by using the ratio test it follows that the series $\sum_{j=0}^{\infty} \binom{n+j-1}{j} \left(\frac{r}{\sqrt{1+r^2}}\right)^j$ is convergent by using the ratio test. Also, $\|\cdot\|_{B[0,+\infty)}$ denotes the uniform norm on $C[0,+\infty)$-which represents the space of all real-valued bounded functions on $[0,+\infty)$. $\qquad\square$

1.10 Balázs-Szabados Operators

The goal of the present section is to obtain similar type of results for the rational complex Balázs-Szabados operators given by

$$R_n(f)(z) = \frac{1}{(1+a_n z)^n} \sum_{j=0}^{n} f(j/b_n) \binom{n}{j} (a_n z)^j,$$

where $a_n = n^{\beta-1}$, $b_n = n^\beta$, $0 < \beta \leq 2/3$, $n \in \mathbb{N}$ and $z \in \mathbb{C}, z \neq -\frac{1}{a_n}$.

The above complex form is obtained simply replacing x by z in the real form of rational operators introduced and studied in Balázs [33] and Balázs-Szabados [34]. Further studies on these operators in the case of real variable can be found in e.g. the paper Abel-Della Vecchia [1].

Remarks. 1) The complex operators $R_n(f)(z)$ are well-defined and analytic for all $n \geq n_0$ and $|z| \leq r < n_0^{1-\beta}$. Indeed, in this case we easily obtain that $z \neq -\frac{1}{a_n}$, for all $|z| \leq r < n_0^{1-\beta}$ and $n \geq n_0$, which implies that $\frac{1}{(1+a_n z)^n}$ is analytic.

2) There exists a close connection between $R_n(f)(z)$ and the classical complex Bernstein polynomials given by $B_n(f)(z) = \sum_{j=0}^{n} f(j/n) \binom{n}{j} z^j (1-z)^j$. Indeed, denoting $e_k(z) = z^k$, we easily get

$$R_n(e_k)(z) = n^{k(1-\beta)} B_n(e_k) \left(\frac{a_n z}{1+a_n z}\right),$$

valid for all $n \geq n_0$, $k \in \mathbb{N}$ and $|z| \leq r < n_0^{1-\beta}$. This connection will be essential in our reasonings.

First we will find some classes of analytic functions for which the uniform convergence of $R_n(f)(z)$ to $f(z)$ holds in some compact disks. As in the case of complex Favard-Szász-Mirakjan and complex Baskakov operators, for the complex Balázs-Szabados operators too let us note that in our results, the domain of definition of the approximated function $f : \mathbb{D}_R \bigcup [R, \infty) \to \mathbb{C}$ seem to be rather strange. However, the analyticity of f on $\mathbb{R}$ on $\mathbb{D}_R$ assures the representation $f(z) = \sum_{k=0}^{\infty} c_k z^k$, which is essential in the proof of quantitative estimates in any $\overline{\mathbb{D}}_r$ with $1 \leq r < R$ (while on $[0, \infty)$ the well known estimates in the case of real variable can be used).

Probably a more natural domain of definition for f would be a strip around the OX-axis, but in this case the representation $f(z) = \sum_{k=0}^{\infty} c_k z^k$ fails, fact which produces the failure of the proofs in this case.

Because of the Remark 2 from the beginning of this section first we need to deal with the estimate of $|B_n(e_k)(z)|$ for $|z| \leq r$ with $0 < r < 1$. Note that in the case when $r \geq 1$ the estimate of $|B_n(e_k)(z)|$ is completely different and it was found in the proof of Theorem 1.1.2 (i).

Lemma 1.10.1. *Denote $\pi_{k,n}(z) = B_n(e_k)(z)$, $\|f\|_r = \sup_{|z| \leq r}\{|f(z)|\}$ and consider $0 < r < 1$.*

(i) For all $|z| \leq r$, $k \in \mathbb{N} \bigcup \{0\}$ and $n \geq 1 + \frac{1}{r}$ we have

$$\|\pi_{k,n}\|_r \leq k!(1+r)r^k.$$

(ii) For all $|z| \leq r < 1$, $k = 0, 1, 2, \dots$ and $n \in \mathbb{N}$ we have $\|\pi_{k,n}\|_r \leq r^k + \frac{(1+r)k(k-1)}{2n}$.

Proof. (i) We consider the following recurrence formula for Bernstein polynomials (see the proof of Theorem 1.1.2 (i))

$$\pi_{k+1,n}(z) = \frac{z(1-z)}{n}\pi'_{k,n}(z) + z\pi_{k,n}(z),$$

$z \in \mathbb{C}$, $k = 0, 1, 2, \dots,$, $n \in \mathbb{N}$.

We will use the mathematical induction. For $k = 0$ the inequality in statement is obvious. Suppose that it is true for k. By the above recurrence, by the Bernstein's inequality (since $\pi_{k,n}(z)$ is a polynomial of degree $\leq k$) and since $\frac{k(1+r)}{n} \leq kr$ we obtain

$$\|\pi_{k+1,n}\|_r \leq \frac{r(1+r)}{n} \cdot \frac{k}{r} \cdot \|\pi_{k,n}\|_r + r\|\pi_{k,n}\|_r$$

$$\leq k!(1+r)r^k \left[k \cdot \frac{1+r}{n} + r\right]$$

$$\leq k!(1+r)r^k[(k+1)r] = (k+1)!(1+r)r^{k+1}.$$

(ii) Applying the Bernstein's inequality to the above recurrence formula , for all $|z| \leq r$ we get

$$|\pi_{k+1,n}(z)| \leq \frac{r(1+r)}{n}\frac{k}{r} \cdot \|\pi_{k,n}\|_r + r|\pi_{k,n}(z)| \leq r|\pi_{k,n}(z)|$$

$$+\frac{(1+r)}{n} \cdot k\|\pi_{k,n}\|_r \leq r|\pi_{k,n}(z)| + \frac{(1+r)k}{n},$$

since by the proof of Theorem 1.1.2 (i) we have $\|\pi_{k,n}\|_1 \leq 1$, for all $k, n \in \mathbb{N}$.

Now, taking $k = 1, 2, 3, ...$ in the above inequality, by recurrence we easily obtain for all $|z| \le r$

$$|\pi_{k,n}(z)| \le r^k + \frac{1+r}{n}[1 + 2 + ... + (k-1)] = r^k + \frac{(1+r)k(k-1)}{2n},$$

which proves (ii) and the lemma. $\qquad\square$

Also we need the following.

Lemma 1.10.2. *Let $n_0 \ge 2$, $0 < \beta \le 2/3$ and $\frac{1}{2} < r < \frac{n_0^{1-\beta}}{2}$. If we denote $r_{k,n}(z) = R_n(e_k)(z)$ then for all $n \ge \max\{n_0, \frac{1}{r^{1/\beta}}\}$, $|z| \le r$ and $k = 0, 1, 2, ...,$ we have*

$$|r_{k,n}(z)| \le (k!) \cdot (2r)^k.$$

Proof. By Remark 2 from the beginning of this section it follows

$$R_n(e_k)(z) = n^{k(1-\beta)}B_n(e_k)\left(\frac{a_n z}{1 + a_n z}\right).$$

But for all $n \ge n_0 \ge 2$ and $|z| \le r < \frac{n_0^{1-\beta}}{2}$ it is easy to see that

$$\left|\frac{a_n z}{1 + a_n z}\right| \le \frac{a_n r}{1 - a_n r} < 1.$$

Therefore, applying Lemma 1.10.1 (i) with $\frac{a_n r}{1 - a_n r}$ instead of r, for all $|z| \le \frac{a_n r}{1 - a_n r}$, $k \in \mathbb{N}\bigcup\{0\}$ and $n \ge 1 + \frac{1 - a_n r}{a_n r} = \frac{1}{a_n r}$ we get

$$|r_{k,n}(z)| = n^{k(1-\beta)}\left|B_n(e_k)\left(\frac{a_n z}{1 + a_n z}\right)\right|$$

$$\le n^{k(1-\beta)}k!\left(1 + \frac{a_n r}{1 - a_n r}\right)\left(\frac{a_n r}{1 - a_n r}\right)^k.$$

But it is easy to see that the condition $n \ge \frac{1}{a_n r}$ is equivalent with $n \ge \frac{1}{r^{1/\beta}}$. Also, the conditions $n \ge \{n_0, \frac{1}{r^{1/\beta}}\}$, $|z| \le r < \frac{n_0^{1-\beta}}{2}$ where $\frac{1}{2} < r$ implies that $\frac{n_0^{1-\beta}}{2} \le \frac{a_n r}{1 - a_n r}$.

Indeed, simple calculation shows that $\frac{n_0^{1-\beta}}{2} \le \frac{a_n r}{1 - a_n r}$ is equivalent to $n_0^{1-\beta} \le rn^{\beta-1}(2 + n_0^{1-\beta})$. But $\frac{1}{2 + n_0^{1-\beta}} < \frac{1}{2} \le r$, which implies $\left(\frac{n_0}{n}\right)^{1-\beta} \le 1 \le r(2 + n_0^{1-\beta})$, that is exactly $n_0^{1-\beta} \le rn^{\beta-1}(2 + n_0^{1-\beta})$.

In conclusion, for all $n \ge \max\{n_0, \frac{1}{r^{1/\beta}}\}$, $|z| \le r < \frac{n_0^{1-\beta}}{2}$ and $k = 0, 1, 2, ...,$ we obtain

$$|r_{k,n}(z)| \le n^{k(1-\beta)}k!\frac{n^{k(\beta-1)}r^k}{(1 - a_n r)^{k+1}} \le 2(k!)(2r)^k,$$

if we can prove that $\frac{1}{1 - a_n r} < 2$ for $n \ge n_0 \ge 2$ and $r < \frac{n_0^{1-\beta}}{2}$. But $r < \frac{n_0^{1-\beta}}{2} < \frac{n^{1-\beta}}{2}$ implies $r < \frac{1}{2a_n}$ for all $n \ge n_0$, which is equivalent to $\frac{1}{1 - a_n r} < 2$. The lemma is proved. $\qquad\square$

Now we are in position to prove the following convergence result.

Theorem 1.10.3. *Let $n_0 \geq 2$, $0 < \beta \leq 2/3$ and $\frac{1}{2} < r < R \leq \frac{n_0^{1-\beta}}{2}$. Suppose that $f : \mathbb{D}_R \bigcup [R, +\infty) \to \mathbb{C}$ is uniformly continuous and bounded on $[0, +\infty)$, is analytic in $\mathbb{D}_R$, that is $f(z) = \sum_{j=0}^{\infty} c_k z^k$ for all $z \in \mathbb{D}_R$ and there exist $M > 0$, $0 < A < \frac{1}{2r}$ with $|c_k| \leq M \frac{A^k}{k!}$ for all $k = 0, 1, 2, ...$, (which implies $|f(z)| \leq M e^{A|z|}$, for all $z \in \mathbb{D}_R$). Then the sequence $(R_n(f)(z))_{n \geq n_0}$ is uniformly convergent to f in $\mathbb{D}_r$.*

Proof.   First we prove that $R_n(f)(z) = \sum_{j=0}^{\infty} R_n(e_j)(z)$ for all $z \in \mathbb{D}_r$, where $e_j(z) = z^j$. In this sense, for any $m \in \mathbb{N}$ define

$$f_m(z) = \sum_{j=0}^{m} c_j z^j \text{ if } |z| \leq r \text{ and } f_m(x) = f(x) \text{ if } x \in (r, +\infty).$$

From the hypothesis on f it is clear that each f_m is bounded on $[0, +\infty)$, which implies that

$$|R_n(f_m)(z)| \leq \frac{1}{|1 + a_n z|^n} \sum_{j=0}^{n} (a_n |z|)^j \binom{n}{j} M(f_m) < \infty,$$

that is all $R_n(f_m)(z)$ with $n \geq n_0$, $r < \frac{n_0^{1-\beta}}{2}$, $m \in \mathbb{N}$ are well-defined for $z \in \mathbb{D}_r$.
    Denoting

$$f_{m,k}(z) = c_k e_k(z) \text{ if } |z| \leq r \text{ and } f_{m,k}(x) = \frac{f(x)}{m+1} \text{ if } x \in (r, \infty),$$

it is clear that each $f_{m,k}$ is bounded on $[0, \infty)$ and that $f_m(z) = \sum_{k=0}^{m} f_{m,k}(z)$. Since from the linearity of S_n we have

$$R_n(f_m)(z) = \sum_{k=0}^{m} c_k R_n(e_k)(z), \text{ for all } |z| \leq r,$$

it suffices to prove that $\lim_{m \to \infty} R_n(f_m)(z) = R_n(f)(z)$ for any fixed $n \in \mathbb{N}$, $n \geq n_0$ and $|z| \leq r$. But this is immediate from $\lim_{m \to \infty} \|f_m - f\|_r = 0$, from $\|f_m - f\|_{B[0,+\infty)} \leq \|f_m - f\|_r$ and from the inequality

$$|R_n(f_m)(z) - R_n(f)(z)| \leq M_{r,n} \|f_m - f\|_{B[0,\infty)} \leq M_{r,n} \|f_m - f\|_r,$$

valid for all $|z| \leq r$. Here $\| \cdot \|_{B[0,+\infty)}$ denotes the uniform norm on $C[0, +\infty)$-the space of all real-valued bounded functions on $[0, +\infty)$.
    Therefore for all $|z| \leq r$, $n \geq n_0$ we obtain

$$|R_n(f)(z)| \leq \sum_{k=0}^{\infty} |c_k| \cdot |R_n(e_k)(z)| \leq 2 \sum_{k=0}^{\infty} |c_k| \cdot k! (2r)^k \leq 2M \sum_{k=0}^{\infty} (2rA)^k < \infty.$$

Now, since we have $\lim_{n \to \infty} R_n(f)(x) = f(x)$ for all $x \in [0, r)$ (by Theorem 1 in Balázs-Szabados [34]), by the classical Vitali's theorem it follows that $R_n(f)(z)$ uniformly converges to $f(z)$ in $\mathbb{D}_r$. $\square$

As in the case of complex Bernstein polynomials, for an upper estimate in approximation by $R_n(f)(z)$ we would need a recurrence formula. For this purpose, formally differentiating $R_n(f)(z)$ we obtain

$$R'_n(f)(z) = \frac{1}{(1+a_n z)^k} \sum_{j=1}^{n} f(j/b_n) \binom{n}{j} j a_n (a_n z)^{j-1} - \frac{na_n}{1+a_n z} R_n(f)(z).$$

Taking here $f(z) = e_k(z)$ by simple calculation we obtain $R'_n(e_k)(z) = \frac{b_n}{z} R_n(e_{k+1})(z) - \frac{na_n}{1+a_n z} R_n(e_k)(z)$, or denoting $r_{k,n}(z) = R_n(e_k)(z)$ finally we easily arrive to the recurrence formula

$$r_{k+1,n}(z) = \frac{z}{b_n} r'_{k,n}(z) + \frac{z}{1+a_n z} r_{k,n}(z),$$

valid for all $k = 0, 1, 2, ...$, $|z| \leq r < \frac{n_0^{1-\beta}}{2}$ and $n \geq n_0$.

Note that from this recurrence formula we easily get that $r_{k,n}(z)$ is a rational function of the form $r_{k,n}(z) = \frac{P_{k,n}(z)}{(1+a_n z)^k}$ with $P_{k,n}(z)$ a polynomial of degree $\leq k$ in z.

Also, simple calculation leads us to the recurrence

$$r_{k+1,n}(z) - z^{k+1} = \frac{z}{b_n}[r_{k,n}(z) - z^k]' + \frac{z}{1+a_n z}[r_{k,n}(z) - z^k]$$
$$+ z^k \left[\frac{k}{b_n} - \frac{a_n z^2}{1+a_n z} \right].$$

In order to use the kinds of reasonings in the case of complex Bernstein polynomials in Section 1.1, we need a Bernstein type inequality for rational functions. The following direct consequence of the Bernstein's inequality in closed unit disk for rational functions in Borwein-Erdélyi [47], Corollary 6, will be useful.

Corollary 1.10.4. *Let $f(z) = \frac{p_k(z)}{\Pi_{j=1}^k (z-a_j)}$, where $p_k(z)$ is a polynomial of degree $\leq k$ and we suppose that $|a_j| \geq R > 1$, for all $j = 1, ..., k$. If $1 \leq r < R$ then for all $|z| \leq r$ we have*

$$|f'(z)| \leq \frac{R+r}{R-r} \cdot \frac{k}{r} \|f\|_r.$$

Proof. Denote $g(u) = f(ru)$, $|u| \leq 1$. We get

$$g(u) = \frac{p_k(ru)}{\Pi_{j=1}^k (ru - a_j)} = \frac{p_k(ru)/r^k}{\Pi_{j=1}^k (u - a_j/r)}, |u| \leq 1.$$

Since $\frac{|a_j|}{r} \geq \frac{R}{r} > 1$, we can apply Corollary 6 in Borwein-Erdélyi [47] so that we obtain

$$|g'(u)| \leq \frac{R/r+1}{R/r-1} \cdot k\|g\|_1 = \frac{R+r}{R-r} \cdot k\|g\|_1 = \frac{R+r}{R-r} \cdot k\|f\|_r.$$

But $g'(u) = rf'(ru)$, which proves the corollary. $\square$

Now we are in position to prove the following upper estimate in approximation by $R_n(f)(z)$.

Theorem 1.10.5. *Let $n_0 \geq 2$, $0 < \beta \leq 2/3$ and $\frac{1}{2} < r < R \leq \frac{n_0^{1-\beta}}{2}$. Suppose that $f : \mathbb{D}_R \bigcup [R, +\infty) \to \mathbb{C}$ is uniformly continuous and bounded on $[0, +\infty)$, is analytic in $\mathbb{D}_R$, that is $f(z) = \sum_{j=0}^{\infty} c_k z^k$ for all $z \in \mathbb{D}_R$ and there exist $M > 0$, $0 < A < \frac{1}{2r}$ with $|c_k| \leq M\frac{A^k}{k!}$ for all $k = 0, 1, 2, ...$, (which implies $|f(z)| \leq Me^{A|z|}$, for all $z \in \mathbb{D}_R$). Then for all $n \geq \max\{n_0, \frac{1}{r^{1/\beta}}\}$ and $|z| \leq r$ we have the upper estimate*

$$|R_n(f)(z) - f(z)| \leq \frac{1}{n^{1-\beta}} \cdot 2Mre^{2rA} + \frac{1}{n^\beta} \cdot \frac{8M}{r} \sum_{k=0}^{\infty} k(2rA)^k.$$

Proof.   Let $|z| \leq r < R \leq \frac{n_0^{1-\beta}}{2}$, $n \geq n_0$ and $k \in \mathbb{N} \bigcup \{0\}$. From the last proved recurrence, Corollary 1.10.4 and Lemma 1.10.2 we obtain

$$|r_{k+1,n}(z) - z^{k+1}|$$

$$\leq \frac{r}{b_n}[|r'_{k,n}(z)| + kr^{k-1}] + \frac{r}{1-a_nr}|r_{k,n}(z) - z^k| + r^k\left[\frac{k}{b_n} + \frac{a_nr^2}{1-a_nr}\right]$$

$$\leq \frac{r}{b_n}\left[\frac{k}{r} \cdot \frac{n_0^{1-\beta}+r}{n_0^{1-\beta}-r} \cdot \|r_{k,n}\|_r\right] + \frac{kr^k}{b_n} + \frac{r}{1-a_nr}|r_{k,n}(z) - z^k| + \frac{kr^k}{b_n}$$

$$+ \frac{a_nr^{k+2}}{1-a_nr}$$

$$\leq \frac{r}{1-a_nr}|r_{k,n}(z) - z^k| + \frac{2k(k!)}{b_n} \cdot (2r)^k \cdot \frac{n_0^{1-\beta}+r}{n_0^{1-\beta}-r} + \frac{2kr^k}{b_n} + \frac{a_nr^{k+2}}{1-a_nr}.$$

But from the end of the proof of Lemma 1.10.2 we have $\frac{1}{1-a_nr} < 2$ while the condition $r < \frac{n_0^{1-\beta}}{2}$ is equivalent to $\frac{n_0^{1-\beta}+r}{n_0^{1-\beta}-r} < 3$, which immediately implies

$$|r_{k+1,n}(z) - z^{k+1}| \leq 2r|r_{k,n}(z) - z^k| + \frac{6k(k!)}{b_n}(2r)^k + \frac{2kr^k}{b_n} + 2r^{k+2}a_n,$$

that is

$$|r_{k+1,n}(z) - z^{k+1}| \leq 2r|r_{k,n}(z) - z^k| + \frac{8k(k!)}{b_n}(2r)^k + 2r^{k+2}a_n,$$

for all $k = 0, 1, 2, ...$ Taking step by step $k = 0, 1, 2, ...$ we easily obtain

$$|r_{k,n}(z) - z^k| \leq r^{k+1}a_n\sum_{j=1}^{k} 2^j + \frac{8(2r)^{k-1}}{b_n} \cdot \sum_{j=1}^{k-1} j(j!) \leq 2r^{k+1}a_n(2^{k-1}-1)$$

$$+ \frac{8(2r)^{k-1}}{b_n}k(k!) \leq (2ra_n)(2r)^k + \frac{8(2r)^{k-1}}{b_n}k(k!),$$

taking into account that $\sum_{j=1}^{k-1} j(j!) \leq k(k!)$.

Taking into account the hypothesis in the coefficients of f we therefore obtain

$$|R_n(f)(z) - f(z)| \leq \sum_{k=0} |c_k| \cdot |r_{k,n}(z) - z^k|$$

$$\leq 2Mra_n \sum_{k=0}^{\infty} \frac{(2rA)^k}{k!} + \frac{1}{b_n} \cdot \frac{8M}{r} \sum_{k=0}^{\infty} k(2rA)^k$$

$$= \frac{2Mr}{n^{1-\beta}} \cdot e^{2rA} + \frac{1}{b_n} \cdot \frac{8M}{r} \sum_{k=0}^{\infty} k(2rA)^k,$$

where $\sum_{k=0}^{\infty} k(2rA)^k < \infty$ for $2rA < 1$. The theorem is proved. $\qquad\square$

Remark. The upper estimate in Theorem 1.10.5 can obviously be written in the form

$$|R_n(f)(z) - f(z)| = O\left(a_n + \frac{1}{b_n}\right).$$

The Voronovskaja-type result for $R_n(f)(x)$, $x \in [0, \infty)$ was obtained by Balázs [33]. In the case of complex variable, the Voronovskaja-type result for $R_n(f)(z)$ can be stated as follows.

Theorem 1.10.6. *Let $n_0 \geq 2$, $0 < \beta \leq 2/3$ and $\frac{1}{2} < r < R \leq \frac{n_0^{1-\beta}}{2}$. Suppose that $f : \mathbb{D}_R \bigcup [R, +\infty) \to \mathbb{C}$ is uniformly continuous and bounded on $[0, +\infty)$, is analytic in $\mathbb{D}_R$, that is $f(z) = \sum_{j=0}^{\infty} c_k z^k$ for all $z \in \mathbb{D}_R$ and there exist $M > 0$, $0 < A < \frac{1}{2r}$ with $|c_k| \leq M \frac{A^k}{k!}$ for all $k = 0, 1, 2, \ldots$, (which implies $|f(z)| \leq M e^{A|z|}$, for all $z \in \mathbb{D}_R$). Then for all $n \geq \max\{n_0, \frac{1}{r^{1/\beta}}\}$ and $|z| \leq r$ we have*

$$\left| R_n(f)(z) - f(z) + \frac{a_n z^2}{1 + a_n z} f'(z) - \frac{a_n^2 b_n z^4 + z}{2b_n(1 + a_n z)^2} f''(z) \right|$$

$$\leq C_r(f) \left(a_n + \frac{1}{b_n} \right)^2.$$

Proof. We clearly have

$$\left| R_n(f)(z) - f(z) + \frac{a_n z^2}{1 + a_n z} f'(z) - \frac{a_n^2 b_n z^4 + z}{2b_n(1 + a_n z)^2} f''(z) \right| \leq \sum_{k=0}^{\infty} |c_k| \cdot |E_{k,n}(z)|,$$

where

$$E_{k,n}(z) = r_{k,n}(z) - e_k(z) + \frac{a_n k z^{k+1}}{1 + a_n z} - \frac{a_n^2 b_n z^4 + z}{2b_n(1 + a_n z)^2} k(k-1) z^{k-2}.$$

Here we easily obtain $E_{0,n}(z) = E_{1,n}(z) = 0$ for all z.

Now, from the recurrence formula for $r_{k,n}(z)$ obtained after the proof of Theorem 1.10.3, by long but simple calculation we obtain the recurrence for $E_{k,n}(z)$ given by

$$E_{k+1,n}(z) = \frac{z}{b_n} E'_{k,n}(z) + \frac{z}{1 + a_n z} E_{k,n}(z) + A_{k,n}(z),$$

where

$$A_{k,n}(z) = \frac{ka_n^2 z^{k+3}}{(1+a_n z)^2} + kz^{k+1} \cdot \frac{a_n}{b_n} \cdot \frac{6a_n z + 2a_n^2 z^2 + 5 - k}{2(1+a_n z)^3}.$$

Under the hypothesis in the statement of Theorem 1.10.6, by the proof of Lemma 1.10.2 we have

$$\left| \frac{1}{1+a_n z} \right| \le \frac{1}{1 - a_n r} < 2,$$

which immediately implies

$$|A_{k,n}(z)| \le C_r k(k+1)(k+2)(\max\{r^{k+1}, r^{k+3}\}) \left(a_n^2 + \frac{a_n}{b_n} \right)$$

$$\le C_r k(k+1)(k+2)(2r)^k \left(a_n + \frac{1}{b_n} \right)^2.$$

Therefore, from the recurrence we obtain

$$|E_{k,n}(z)| \le \frac{r}{b_n} |E_{k-1,n}'(z)| + 2r|E_{k-1,n}(z)| + |A_{k,n}(z)|.$$

First we will estimate the quantity $|E_{k-1,n}'(z)|$. Let $r < r_1 < R < \frac{n_0^{1-\beta}}{2}$ be with $r_1 < \frac{1}{2A}$ and denote by Γ the circle of center 0 and radius r_1. By the Cauchy's theorem and taking into account the estimate for $\|r_{k-1,n} - e_{k-1}\|_{r_1}$ in the proof of Theorem 1.10.5, for all $|z| \le r$ we obtain

$$|E_{k-1,n}'(z)| \le \frac{1}{2\pi} \left| \int_{\Gamma} \frac{E_{k-1,n}(u) du}{u-z} \right| \le \frac{r}{r_1 - r} \|E_{k-1,n}\|_{r_1}$$

$$\le \frac{r}{r_1 - r} \left[\|r_{k-1,n} - e_{k-1}\|_{r_1} + C_{r_1} k(k-1)(2r_1)^k \left(a_n + \frac{1}{b_n} \right) \right]$$

$$\le C_{r,r_1} k(k-1)k!(2r_1)^k \left(a_n + \frac{1}{b_n} \right).$$

By the inequality $\frac{1}{b_n} \left(a_n + \frac{1}{b_n} \right) \le \left(a_n + \frac{1}{b_n} \right)^2$, this immediately implies

$$|E_{k,n}(z)| \le 2r|E_{k-1,n}(z)| + \frac{r}{b_n} C_{r,r_1} k(k-1)k!(2r_1)^k \left(a_n + \frac{1}{b_n} \right) + |A_{k,n}(z)|$$

$$\le 2r|E_{k-1,n}(z)| + C_{r,r_1}(k-1)k(k+1)k!(2r_1)^k \left(a_n + \frac{1}{b_n} \right)^2.$$

Now taking in this inequality $k = 1, 2, ...$, by mathematical induction we easily arrive at

$$|E_{k,n}(z)| \le C_{r,r_1} \left(a_n + \frac{1}{b_n} \right)^2 \left[\sum_{j=2}^{k} (j-1)j(j+1)j!(2r_1)^j \right]$$

$$\le C_{r,r_1} \left(a_n + \frac{1}{b_n} \right)^2 (k-1)k(k+1)(k+2)k!(2r_1)^k.$$

Here the constants C_{r,r_1} can be different at each occurrence. Finally, taking into account the hypothesis too we obtain

$$
\left| R_n(f)(z) - f(z) + \frac{a_n z^2}{1 + a_n z} f'(z) - \frac{a_n^2 b_n z^4 + z}{2 b_n (1 + a_n z)^2} f''(z) \right|
$$

$$
\leq \sum_{k=0}^{\infty} |c_k| \cdot |E_{k,n}(z)|
$$

$$
\leq C_{r,r_1} \left(a_n + \frac{1}{b_n} \right)^2 \cdot \sum_{k=2}^{\infty} |c_k| k! (k-1) k (k+1)(k+2)(2r_1)^k
$$

$$
\leq C_{r,r_1} \left(a_n + \frac{1}{b_n} \right)^2 \sum_{k=2}^{\infty} (k-1) k (k+1)(k+2)(2r_1 A)^k,
$$

which proves the theorem since $2r_1 A < 1$. $\qquad\square$

Now we are in position to obtain the exact degree of approximation for $R_n(f)(z)$.

Theorem 1.10.7. *Let* $n_0 \geq 2$, $0 < \beta \leq 2/3$, $\beta \neq 1/2$ *and* $\frac{1}{2} < r < R \leq \frac{n_0^{1-\beta}}{2}$. *Suppose that* $f : \mathbb{D}_R \bigcup [R, +\infty) \to \mathbb{C}$ *is uniformly continuous and bounded on* $[0, +\infty)$, *is analytic in* $\mathbb{D}_R$, *that is* $f(z) = \sum_{j=0}^{\infty} c_k z^k$ *for all* $z \in \mathbb{D}_R$ *and there exist* $M > 0$, $0 < A < \frac{1}{2r}$ *with* $|c_k| \leq M \frac{A^k}{k!}$ *for all* $k = 0, 1, 2, ...$, *(which implies* $|f(z)| \leq M e^{A|z|}$, *for all* $z \in \mathbb{D}_R$). *If* f *is not a polynomial of degree* ≤ 1 *then for all* $n \geq \max\{n_0, \frac{1}{r^{1/\beta}}\}$ *we have*

$$
\| R_n(f) - f \|_r \sim \left(a_n + \frac{1}{b_n} \right),
$$

where the constants in the equivalence are independent of n.

Proof. We can write

$$
R_n(f)(z) - f(z)
$$
$$
= \left(a_n + \frac{1}{b_n} \right) \left\{ \frac{1}{a_n + 1/b_n} \cdot \frac{-a_n z^2 f'(z)}{1 + a_n z} + \frac{1}{a_n + 1/b_n} \cdot \frac{(a_n^2 b_n z^4 + z) f''(z)}{2 b_n (1 + a_n z)^2} \right.
$$
$$
+ \left(a_n + \frac{1}{b_n} \right)
$$
$$
\left. \cdot \left[\frac{1}{(a_n + 1/b_n)^2} \left(R_n(f)(z) - f(z) + \frac{a_n z^2 f'(z)}{1 + a_n z} - \frac{(a_n^2 b_n z^4 + z) f''(z)}{2 b_n (1 + a_n z)^2} \right) \right] \right\}.
$$

Since $a_n + \frac{1}{b_n} \to 0$ as $n \to \infty$, taking into account the estimate in Theorem 1.10.6 and the reasonings in the cases of the previous complex Bernstein-type operators, it remains to show that for sufficiently large n and for all $|z| \leq r$ we have $|T(z)| > \rho > 0$, where ρ is independent of n and

$$
T(z) := \frac{1}{a_n + 1/b_n} \cdot \frac{-a_n z^2 f'(z)}{1 + a_n z} + \frac{1}{a_n + 1/b_n} \cdot \frac{(a_n^2 b_n z^4 + z) f''(z)}{2 b_n (1 + a_n z)^2}.
$$

But since $a_n = n^{\beta-1} \to 0$ as $n \to \infty$ and $b_n = n^{\beta}$, we obtain

$$\left| \frac{1}{a_n + 1/b_n} \cdot \frac{-a_n z^2 f'(z)}{1 + a_n z} \right| \geq \frac{n^{2\beta-1}}{1 + n^{2\beta-1}} \cdot \frac{1}{1 + a_1 r} |z^2 f'(z)|$$

$$= \frac{n^{2\beta-1}}{1 + n^{2\beta-1}} \cdot \frac{|z^2 f'(z)|}{1 + a_1 r}.$$

If $2\beta - 1 < 0$ then $\frac{n^{2\beta-1}}{1+n^{2\beta-1}} \to 0$ and therefore the term $\frac{1}{a_n+1/b_n} \cdot \frac{-a_n z^2 f'(z)}{1+a_n z} \to 0$ as $n \to \infty$, uniformly with respect to $|z| \leq r$. In this case the term does not count for our estimate.

If $2\beta - 1 > 0$ then $\frac{n^{2\beta-1}}{1+n^{2\beta-1}} \to 1$ and therefore for sufficiently large n we have

$$\left| \frac{1}{a_n + 1/b_n} \cdot \frac{-a_n z^2 f'(z)}{1 + a_n z} \right| \geq \frac{1}{2} \cdot \frac{|z^2 f'(z)|}{1 + a_1 r}$$

For the other term it follows

$$\frac{1}{a_n + 1/b_n} \cdot \frac{(a_n^2 b_n z^4 + z) f''(z)}{2 b_n (1 + a_n z)^2} = \frac{n^{3\beta-2}}{1 + n^{2\beta-1}} \cdot \frac{z^4 f''(z)}{2(1 + a_n z)^2}$$

$$+ \frac{z f''(z)}{2(1 + n^{2\beta-1})(1 + a_n z)^2}.$$

Here it is clear that if $2\beta - 1 > 0$ then this term converges to 0 (uniformly with respect to $|z| \leq r$) and therefore does not count for our estimate, while if $2\beta - 1 < 0$, then $n^{2\beta-1} \to 0$ (as $n \to \infty$) and here only the term $\frac{z f''(z)}{2(1+n^{2\beta-1})(1+a_n z)^2}$ counts for the estimate.

Therefore, concluding all the above reasonings, there exists an index n_1 depending on β, such that if $2\beta - 1 > 0$ then for all $n \geq n_1$ and all $|z| \leq r$ we have $|T(z)| \geq \frac{1}{2} \cdot \frac{|z^2 f'(z)|}{1+r}$, while if $2\beta - 1 < 0$ then for all $n \geq n_1$ and all $|z| \leq r$ it follows

$$|T(z)| \geq \frac{1}{2} \cdot \left| \frac{z f''(z)}{2(1 + a_n z)^2} \right| \geq \frac{|z f''(z)|}{4(1 + r)^2}.$$

Since f is not a polynomial of degree ≤ 1, it easily follows that $\|e_2 f'\|_r > 0$ and $\|e_1 f''\|_r > 0$, which implies that in both cases $2\beta - 1 > 0$ and $2\beta - 1 < 0$ we have $\|T\|_r > \rho > 0$, with ρ independent of n.

In the case of $2\beta - 1 = 0$, that is $\beta = 1/2$, we obtain

$$\frac{n^{3\beta-2}}{1 + n^{2\beta-1}} \cdot \frac{z^4 f''(z)}{2(1 + a_n z)^2} = \frac{n^{1/2} z^4 f''(z)}{4(1 + a_n z)^2} \to \infty,$$

as $n \to \infty$, so that the case $\beta = 1/2$ remains unsettled.

In conclusion,

$$\|R_n(f) - f\|_r \geq \left(a_n + \frac{1}{b_n} \right) [\|T\|_r - \|G_n\|_r] \geq \left(a_n + \frac{1}{b_n} \right) \frac{1}{2} \cdot \|T\|_r,$$

for all $n \geq n_1$. (Here $(G_n(z))_{n \in \mathbb{N}}$ is a sequence of analytic functions uniformly convergent to zero with respect to $|z| \leq r$).

In the case when $n = 1, 2, ..., n_1 - 1$ the lower estimate is trivial.

Finally, taking into account the upper estimate in Theorem 1.10.5 (in fact see the Remark after the proof of Theorem 1.10.5), our theorem is proved. $\square$

1.11 Bibliographical Notes and Open Problems

Theorems 1.1.8, 1.1.9, 1.4.2, Lemmas 1.4.3, 1.4.4, Theorems 1.4.5, 1.5.5, 1.6.14, 1.6.15, 1.7.6, Lemma 1.8.9, Corollary 1.8.10, Lemma 1.10.1, 1.10.2, Theorem 1.10.3, Corollary 1.10.4, Theorems 1.10.5, 1.10.6, 1.10.7 are new and appear for the first time here.

Note 1.11.1. From a very long list, some references concerning the Bernstein-type operators of one real variable, different from those considered in this Chapter 1, for which would be possible to develop similar results are, for example : Altomare [8], Altomare-Campiti [9], Altomare-Mangino [10], Altomare-Raşa [11], Bleimann-Butzer-Hahn [46], Cimoca-Lupaş [55], Derriennic [56; 57; 59], Durrmeyer [67], Lupaş [129; 130; 131], Lupaş-Müller [132], Meyer-König and Zeller [136], Moldovan G. [140], Raşa [159], Soardi [168], Stancu [176].

Open Problem 1.11.2. A Voronovskaja-type formula with a quantitative estimate and the exact orders of approximation in Theorem 1.6.9 for the complex polynomials $S_n^{<\gamma>}(f)(z)$ remain as open questions.

Open Problem 1.11.3. It is well-known that if $f : [a, b] \to \mathbb{R}$, where $a, b \in \mathbb{R}$, $a < b$, the Bernstein polynomials attached to f are given by the formula

$$B_n(f; [a, b])(x) = \frac{1}{(b - a)^n} \sum_{k=0}^{n} f(a_k) \binom{n}{k} (x - a)^k (b - x)^{n-k},$$

$a_k = a + k\frac{b-a}{n}$. Now, if we consider the complexified form $B_n(f; [a, b])(z)$ for z belonging to a disk containing the real interval $[a, b]$ and we suppose that f is analytic in that disk, then similar results with those for $B_n(f; [0, 1])(z)$ in Sections 1.1, 1.2, 1.3 and 1.4 can be obtained. But more interesting seems to be the case when $[a, b]$ is a "complex" interval in $\mathbb{C}$, that is $a, b \in \mathbb{C}$ and $[a, b] = \{z = \lambda a + (1 - \lambda)b; \lambda \in [0, 1]\}$. In this case, for the complexified form $B_n(f; [a, b])(z)$ as above, it would be of interest to study its approximation properties in a disk containing the "complex" interval $[a, b]$, when f is supposed to be analytic in that disk.

Open Problem 1.11.4. Let $L_n : C[a, b] \to C[a, b]$, $n \in \mathbb{N}$, $a, b \in \mathbb{R}$, $a < b$, be a sequence of positive and linear operators (of one real variable) attached to $f \in C[a, b]$, satisfying for example the conditions in the classical Korovkin theorem, that is $\lim_{n \to \infty} L_n(e_k)(x) = e_k(x)$, uniformly for $x \in [a, b]$, for $k = 0, 1, 2$ (here we recall that $e_k(x) = x^k$). Taking into account the results in this chapter, it is natural to ask if the convergence properties of the sequence $(L_n(f))_{n \in \mathbb{N}}$ remain valid if we complexify $L_n(f)(x)$ (that is if we replace $x \in [a, b]$ by $z \in \mathbb{D}_R$), supposing in addition that f can be extended to an analytic function in the disk $\mathbb{D}_R$ containing the real segment $[a, b]$.

In general this does not happen. For example, if one considers as $L_n(f)(x)$ the Hermite-Fejér interpolation polynomials on an infinite interpolatory matrix in $[-1, 1]$ consisting of the roots of orthogonal polynomials, or more general on an arbitrary infinite triangular matrix in $[-1, 1]$, then for $f(z) = z$ the polynomials $L_{2n-1}(f)(z)$ diverges for all $z \in \mathbb{C} \setminus [-1, 1]$, see Brutman-Gopengauz [49] and Brutman-Gopengauz-Vértesi [50].

However, for other sequences of positive and linear operators (possibly of noninterpolatory kind) the question remains open.

More exactly, it would be interesting to see if the Korovkin theory for the complexified sequence of a sequence of positive and linear operators (possibly under some additional hypothesis) still holds, that is if under some possible additional hypothesis on $L_n(e_0)$, $L_n(e_1)$ and $L_n(e_2)$, the conditions

$$\lim_{n \to \infty} L_n(f)(z) = e_i(z), i = 0, 1, 2, \text{ uniformly in } \mathbb{D}_r,$$

would imply $L_n(f)(z) \to f(z)$ (as $n \to \infty$), uniformly in $\mathbb{D}_r$, for any analytic function in $\mathbb{D}_R$ and $r < R$.

Remark. From Vitali's theorem it is immediate that if $(L_n)_{n \in \mathbb{N}}$ is a sequence of positive and linear operators satisfying the Korovkin theorem and if in addition the complexified sequence satisfies

$$|L_n(e_k)(z)| \leq M_r r^k, \text{ for all } |z| \leq r, \ k \in \mathbb{N} \cup \{0\}, n \in \mathbb{N},$$

then for any analytic function f we have $\lim_{n \to \infty} L_n(f)(z) = f(z)$, uniformly in $\mathbb{D}_r$. The problem is how to reduce the additional conditions on the set $\{L_n(e_0), L_n(e_1), L_n(e_2)\}$ only. In the case of Bernstein-type operators this seems to be possible because of a recurrence formula with respect to k satisfied by $L_n(e_k)$.

Open Problem 1.11.5. It is left to the reader to prove that the upper estimates in the Voronovskaja-type results for q-Bernstein polynomials in Theorem 1.5.2 (ii) in fact hold with equivalence of order $\frac{1}{[n]_q^2}$.

More general, let us consider the exponential-type operators introduced in May [134] by the general formula $M_n(f)(x) = \int_a^b W_n(x, t) f(t) dt$, where the kernel $W_n : I(a, b) \times I(a, b) \to \mathbb{R}$ has the following properties : 1) $W_n(x, t) \geq 0$ for all $(x, t) \in I(a, b) \times I(a, b)$; 2) $\int_a^b W_n(x, t) dt = 1$, for all $x \in I(a, b)$; 3) $\frac{\partial}{\partial x} W_n(x, t) = \frac{n(t-x)}{p(x)} W_n(x, t)$, where $p(x)$ is a strictly positive polynomial for $x \in I(a, b)$. Here $I(a, b)$ can be of the form $[a, b]$, $(-\infty, b]$, $[a, +\infty)$ or $(-\infty, +\infty)$. Also, denote

$$A_{n,m}(x) = n^m \int_a^b W_n(x, t)(t - x)^m dt.$$

Note that for $a = 0$, $b = 1$ and $p(x) = x(1 - x)$ we obtain the Bernstein polynomials, for $a = 0$, $b = \infty$ and $p(x) = x$ we obtain the Favard-Szász-Mirakjan operators, for $a = 0$, $b = \infty$ and $p(x) = x(1 + x)$ we obtain the Baskakov operators, for $a = 0$, $b = \infty$ and $p(x) = x^2$ we obtain the Post-Widder operators, for $a = -\infty$, $b = \infty$ and $p(x) = 1$ we obtain the Gauss-Weierstrass operators.

In Pop [156] the following generalized Voronovskaja's theorem was proved : if f is $2p$ continuous differentiable in $I(a, b)$ then for all $x \in I(a, b)$ we have

$$\left| M_n(f)(x) - \sum_{i=0}^{2p} \frac{1}{n^i i!} A_{n,i}(x) f^{(i)}(x) \right| = o(1/n^p).$$

Our conjecture is that for the complex operators

$$M_n(f)(z) = \int_a^b W_n(z, t) f(t) dt,$$

where f is supposed to be analytic in a suitable disk $\overline{\mathbb{D}}_r$ depending on the operator (see the theorems mentioned at the beginning of this Open Problem), we have the equivalence

$$\left\| M_n(f) - \sum_{i=0}^{2p} \frac{1}{n^i i!} A_{n,i} f^{(i)} \right\|_r \sim \frac{1}{n^{p+1}}.$$

For general $p \in \mathbb{N}$ and general operators $M_n(f)$, it could be useful the ideas in the proof for the complex Bernstein polynomials (that is the proof of Corollary 1.3.4), see also the proof of Theorem 1.8.2 (ii) for the case of complex Favard-Szász-Mirakjan operators.

Open Problem 1.11.6. The complex Bernstein-Stancu polynomials depending on the parameter $\gamma \geq 0$ defined for disks of center in origin are given by the formula

$$S_n^{<\gamma>}(f)(z) = \sum_{p=0}^n \binom{n}{p} \frac{z(z+\gamma)...(z+(p-1)\gamma)}{(1+\gamma)...(1+(p-1)\gamma)} \Delta_{1/n}^p f(0), |z| \leq r.$$

Writing now

$$\frac{z(z+\gamma)...(z+(p-1)\gamma)}{(1+\gamma)...(1+(p-1)\gamma)} = \sum_{j=0}^p A_{p,j} z^j,$$

where by identification of the coefficients we evidently can explicitly find each $A_{p,j} \in \mathbb{R}$, it follows

$$S_n^{<\gamma>}(f)(z) = \sum_{p=0}^n \binom{n}{p} \left[\sum_{j=0}^p A_{p,j} z^j \right] \Delta_{1/n}^p f(0), |z| \leq r.$$

Then, for $G \subset \mathbb{C}$ a compact set such that $\tilde{\mathbb{C}} \setminus G$ is connected and by using the Faber polynomials $F_p(z)$ attached to G (see Definition 1.0.10), for $f \in A(\overline{G})$ we can introduce the Bernstein-Stancu-Faber polynomials depending on the parameter $\gamma \geq 0$, given by the formula

$$\mathcal{S}_n^{<\gamma>}(f; \overline{G})(z) = \sum_{p=0}^n \binom{n}{p} \left[\sum_{j=0}^p A_{p,j} F_j(z) \right] \Delta_{1/n}^p F(0), z \in G, \ n \in \mathbb{N},$$

where Ψ is the unique conformal mapping Ψ of $\tilde{\mathbb{C}} \setminus \overline{\mathbb{D}}_1$ onto $\tilde{\mathbb{C}} \setminus G$ satisfying $\Psi(\infty) = \infty$ and $\Psi'(\infty) > 0$ and $F(w) = \frac{1}{2\pi i} \int_{|u|=1} \frac{f(\Psi(u))}{u-w} du$, $w \in \mathbb{D}_1$. Here, since $F(1)$ is involved in $\Delta_{1/n}^n F(0)$ and therefore in the definition of $\mathcal{S}_n^{<\gamma>}(f; G)(z)$ too, in addition we will suppose that F can be extended by continuity on the boundary $\partial \mathbb{D}_1$.

It is an open problem to extend the results in Section 1.6 (corresponding there to $S_n^{<\gamma>}(f)(z)$, $|z| \leq r$) to the Bernstein-Stancu-Faber polynomials $\mathcal{S}_n^{<\gamma>}(f; \overline{G})(z)$, $z \in G$ (exactly as we did for the complex Bernstein polynomials and Bernstein-Stancu polynomials depending on two parameters $0 \leq \alpha \leq \beta$).

Open Problem 1.11.7. It is known that the Kantorovich variant of the complex Bernstein polynomial for compact disks is given by

$$K_n(f)(z) = B'_{n+1}(g)(z), \quad g(z) = \int_0^z f(u) du, \quad |z| \leq r,$$

where $B_{n+1}(g)(z) = \sum_{k=0}^{n+1} \binom{n+1}{k} \Delta_{1/(n+1)}^k g(0) z^k$ denotes the complex classical Bernstein polynomial of degree $n + 1$. This immediately implies the representation

$$K_n(f)(z) = \sum_{j=0}^{n} \binom{n+1}{j+1} (j+1) \Delta_{1/(n+1)}^{j+1} g(0) z^j, \quad |z| \leq r,$$

and suggests the following expression for the Kantorovich-Faber polynomials attached to a set $G \subset \mathbb{C}$

$$\mathcal{K}_n(f; \overline{G})(z) = \sum_{j=0}^{n} \binom{n+1}{j+1} (j+1) \Delta_{1/(n+1)}^{j+1} F(0) F_j(z), \quad z \in \overline{G}, n \in \mathbb{N},$$

with F defined as in the above Open Problem 1.11.6 (with g instead of f).

Following the same procedure and taking into account that the complex Stancu-Kantorovich polynomials depending on two parameters $0 \leq \alpha \leq \beta$ are given by

$$K_n^{(\alpha,\beta)}(f)(z) = \frac{n+1+\beta}{n+1} \sum_{k=0}^{n} \binom{n+1}{k+1} \Delta_{1/(n+\beta)}^{k+1} g[\alpha/(n+\beta)] z^j, |z| \leq r,$$

the expression of Stancu-Kantorovich-Faber polynomials attached to a set $G \subset \mathbb{C}$, can be given by

$$\mathcal{K}_n^{(\alpha,\beta)}(f; \overline{G})(z) = \frac{n+1+\beta}{n+1} \sum_{j=0}^{n} \binom{n+1}{j+1} (j+1) \Delta_{1/(n+\beta)}^{j+1} F[\alpha/(n+\beta)] F_j(z),$$

$z \in \overline{G}$, where F is defined as above.

Remain as open questions the approximation properties of the polynomials $\mathcal{K}_n(f; \overline{G})(z)$ and $\mathcal{K}_n^{(\alpha,\beta)}(f; \overline{G})(z)$.

Open Problem 1.11.8. In a similar manner, taking into account that the complex Favard-Szász-Mirakjan operators for compact disks can be written in the form

$$S_n(f)(z) = \sum_{j=0}^{\infty} [0, 1/n, ..., j/n; f] z^j, |z| \leq r,$$

for a set $G \subset \mathbb{C}$ we can formally define the Favard-Szász-Mirakjan-Faber operators by

$$S_n(f; \overline{G})(z) = \sum_{j=0}^{\infty} [0, 1/n, ..., j/n; f] F_j(z), z \in \overline{G},$$

where $F(w) = \frac{1}{2\pi i} \int_{|u|=1} \frac{f(\Psi(u))}{u-w} du$, $w \in \mathbb{D}_1$ and we suppose that F can be extended by continuity on the boundary $\partial \mathbb{D}_1$. Let us observe that for any a with $|a| > 1$, $F(a)$ is well-defined.

It is an open question to find the approximation properties of the Favard-Szász-Mirakjan-Faber operators $S_n(f; \overline{G})(z)$.

Open Problem 1.11.9. Prove Theorem 1.10.7 for the case $\beta = 1/2$.

Open Problem 1.11.10. As in the case of Favard-Szász-Mirakjan operators, it is clear that in the case of Baskakov operators too, the domain of definition $[R, +\infty) \bigcup \overline{\mathbb{D}}_R$ is rather unusual. Taking into account the form of complex Baskakov operators, more natural seems to consider and approximate the analytic function f on an annulus $A_{r,\infty} = \{z \in \mathbb{C}; r \leq |z + 1| < \infty\}$, with $r > 0$, where f can be represented as a Laurent series. The study of this problem is left as an open question.

Open Problem 1.11.11. It is well known that in the case of Bernstein-type operators of real variables, starting with the paper of Altomare [7] a theory of strongly continuous contraction semigroups on Banach spaces obtained as a limit of the iterations of these operators is much developed. Then it would be interesting to consider the Altomare's idea in the case of Bernstein-type operators of complex variables. Note that taking into account, for example, the considerations from the beginning of Section 1.2, in the complex case would correspond a Trotter's theorem and a theory of limit semigroups on a Fréchet space (i.e. a metrizable complete locally convex space) with respect to the topology of uniform convergence on compacts subsets of the open disk $\mathbb{D}_R$.

Chapter 2

Bernstein-Type Operators of Several Complex Variables

The results for the complex univariate operators in Chapter 1 can be extended to the case of several complex variables. We illustrate here the theory on three important operators : Bernstein, Favard-Szász-Mirakjan and Baskakov.

For simplicity, the results are presented for bivariate case, but from the proofs it is easy to see that they remain valid for several complex variables.

2.1   Introduction

In this section we present some requisites in the theory of holomorphic (analytic) functions of several complex variables.

Definition 2.1.1. (see e.g. Andreian Cazacu [23] or Kohr [117]) Let $\mathbb{C}^p$ denote the space of p-complex variables $z = (z_1, ..., z_p), z_j \in \mathbb{C}, j = 1, ..., p$.

(i) The open polydisk of center $a = (a_1, ..., a_p) \in \mathbb{C}^p$ and radius $R = (R_1, ..., R_p) \in \mathbb{R}_+^p$ is defined by

$$P(a; R) = \{z = (z_1, ..., z_n) \in \mathbb{C}; |z_j - a_j| < R_j, \forall j = 1, ..., p\}.$$

(ii) Let Ω be a domain in $\mathbb{C}^p$ and $f : \Omega \to \mathbb{C}$. We say that f is holomorphic (or analytic) on Ω if f is continuous in Ω and holomorphic (analytic) in each variable separately (when the others are fixed). Equivalently, f is holomorphic (analytic) in Ω, if for each point $a = (a_1, ..., a_p) \in \Omega$, there exists a neighborhood of a such that we have

$$f(z) = \sum_{j_1,...,j_p=0}^{\infty} c_{j_1,...,j_p}(z - a_1)^{j_1}...(z - a_p)^{j_p}, \forall z \in \Omega,$$

where the series converges absolutely and uniform on each compact subset of Ω.

By using the multi-index notations $|j| = j_1 + j_2 + ... + j_p$, $z^j = z_1^{j_1} z_2^{j_2}...z_p^{j_p}$, $j! = j_1! j_2!...j_p!$, for any $j = (j_1, ..., j_p), j_k \in \{0, 1, ..., \}$, it is well known that for e.g. $a = 0$, we can write $c_{j_1,...,j_p} = \frac{\partial^{|j|} f(0)}{\partial z_1^{j_1}...\partial z_p^{j_p}} \frac{1}{j!} := D^j f(0)/j!$ and therefore we get the Taylor form $f(z) = \sum_{j_1,...,j_p=0}^{\infty} \frac{D^j f(0)}{j!} z^j$.

Also, the following result called the Cauchy's formula on polydisk is classical.

Theorem 2.1.2. (see e.g. Kohr [117]) *Let* $i^2 = -1$, $a = (a_1, ..., a_p) \in \mathbb{C}^p$, $R = (R_1, ..., R_p)$, $R_j > 0$, $j = 1, ..., p$ *and* $f : \overline{P(a; R)} \to \mathbb{C}$ *be a holomorphic function in* $P(a; R)$ *and continuous in* $\overline{P(a; R)}$. *Then for all* $z = (z_1, ..., z_p) \in P(a; R)$ *and all* $k_j \in \mathbb{N} \bigcup \{0\}$, $j = 1, ..., p$, *we have :*

(i)

$$f(z_1, ..., z_p) = \frac{1}{(2\pi i)^p} \int_{|u_p - z_p| = R_p} \cdots \int_{|u_1 - z_1| = R_1} \frac{f(u_1, ..., u_p) du_1 ... du_p}{(u_1 - z_1)...(u_p - z_p)}.$$

(ii)

$$\frac{\partial^{k_1 + ... + k_p} f}{\partial z_1^{k_1} ... \partial z_p^{k_p}}(z_1, ..., z_p)$$

$$= \frac{(k_1!)...(k_p!)}{(2\pi i)^p} \int_{|u_p - z_p| = R_p} \cdots \int_{|u_1 - z_1| = R_1} \frac{f(u_1, ..., u_p) du_1 ... du_p}{(u_1 - z_1)^{k_1 + 1}...(u_p - z_p)^{k_p + 1}}.$$

Remarks. 1) The Vitali's theorem in univariate case (Theorem 1.0.1) holds for several complex variables (see e.g. Kohr [117], p. 29, Theorem 1.3.7.).

2) The Weierstrass's theorem in univariate case (Theorem 1.0.3) holds for several complex variables, by replacing the derivatives with partial derivatives of any order (see e.g. Kohr [117], p. 26, Theorem 1.3.2).

3) The Maximum Modulus Theorem in univariate case (Theorem 1.0.5) holds for several complex variables on polydisks (see e.g. Kohr [117], p. 23, Corollary 1.2.5).

4) The identity's theorem in univariate case (Theorem 1.0.7) holds in several complex variables in the following slightly modified form : if $\Omega \subset \mathbb{C}^n$ is a domain and the analytic functions $f, g : \Omega \to \mathbb{C}$ coincide on a nonempty open set $G \subset \Omega$, then they coincide on Ω (see e.g. Kohr [117], p. 20, Theorem 1.2.1).

This chapter can be described as follows :

- in Section 2.2 we study the complex Bernstein polynomials, both tensor product and non-tensor product types ;

- in Section 2.3 the complex Favard-Szász-Mirakjan operator of tensor product kind, without exponential growth conditions on f is studied ;

- Section 2.3 deals with the complex Baskakov operator of tensor product kind.

2.2 Bernstein Polynomials

For $f(z_1, z_2)$ analytic in the polydisc $P(0; R) = \mathbb{D}_{R_1} \times \mathbb{D}_{R_2}$ where $R = (R_1, R_2)$ and $|z_1| \leq r_1$, $|z_2| \leq r_2$, $1 \leq r_1 < R_1$ with $1 \leq r_2 < R_2$, we can define two kinds of bivariate complex Bernstein polynomials :

1) the tensor product kind defined by

$$B_{n,m}(f)(z_1, z_2) = \sum_{k=0}^{n} \sum_{j=0}^{m} p_{n,k}(z_1) p_{m,j}(z_2) f(k/n, j/m),$$

where $p_{n,k}(z) = \binom{n}{k} z^k (1-z)^{n-k}$, and

2) the non-tensor product (or simplex) kind given by

$$B_n(f)(z_1, z_2) = \sum_{k=0}^{n} \sum_{j=0}^{n-k} p_{n,k,j}(z_1, z_2) f(k/n, j/n),$$

where

$$p_{n,k,j}(z_1, z_2) = \binom{n}{k}\binom{n-k}{j} z_1^k z_2^j (1 - z_1 - z_2)^{n-k-j}$$

$$= \frac{n!}{k!j!(n-k-j)!} z_1^k z_2^j (1 - z_1 - z_2)^{n-k-j}.$$

Note that in the case of several real variables, the tensor product Bernstein polynomial was first introduced and studied in Hildebrandt-Schoenberg [108] and Butzer [52], while the simplex kind Bernstein polynomial of several real variables was first introduced and studied in Dinghas [61], Lorentz [125], p. 51 and Stancu [175].

In this section, approximation properties of the above bivariate complex polynomials will be proved. We begin with the properties of the tensor-product kind. In this sense first we present :

Theorem 2.2.1. *Suppose that* $f : P(0; R) \to \mathbb{C}$ *is analytic in* $P(0; R) = \mathbb{D}_{R_1} \times \mathbb{D}_{R_2}$, *that is* $f(z_1, z_2) = \sum_{k=0}^{\infty} \sum_{j=0}^{\infty} c_{k,j} z_1^k z_2^j$, *for all* $(z_1, z_2) \in P(0; R)$, $R = (R_1, R_2)$.

(i) For all $|z_1| \le r_1$, $|z_2| \le r_2$ *with* $1 \le r_1 < R_1$, $1 \le r_2 < R_2$ *and* $n, m \in \mathbb{N}$ *we have*

$$|B_{n,m}(f)(z_1, z_2) - f(z_1, z_2)| \le C_{r_1, r_2, n, m}(f),$$

where

$$C_{r_1, r_2, n, m}(f) = \frac{3r_2(1+r_2)}{2m} \cdot \sum_{k=0}^{\infty} \sum_{j=0}^{\infty} |c_{k,j}| \cdot j(j-1) r_2^{j-2} r_1^k$$

$$+ \frac{3r_1(1+r_1)}{2n} \cdot \sum_{k=0}^{\infty} \sum_{j=0}^{\infty} |c_{k,j}| \cdot k(k-1) r_1^{k-2} r_2^j.$$

(ii) Let $k_1, k_2 \in \mathbb{N}$ *be with* $k_1 + k_2 \ge 1$ *and* $1 \le r_1 < r_1^* < R_1$, $1 \le r_2 < r_2^* < R_2$. *Then for all* $|z_1| \le r_1$, $|z_2| \le r_2$ *and* $n, m \in \mathbb{N}$ *we have*

$$\left| \frac{\partial^{k_1+k_2} B_{n,m}(f)}{\partial z_1^{k_1} \partial z_2^{k_2}}(z_1, z_2) - \frac{\partial^{k_1+k_2} f}{\partial z_1^{k_1} \partial z_2^{k_2}}(z_1, z_2) \right|$$

$$\le C_{r_1^*, r_2^*, n, m}(f) \cdot \frac{(k_1)!}{(r_1^* - r_1)^{k_1+1}} \cdot \frac{(k_2)!}{(r_2^* - r_2)^{k_2+1}}.$$

Proof.   (i) Denote $e_{k,j}(z_1, z_2) = e_k(z_1) \cdot e_j(z_2)$, where $e_k(u) = u^k$. Clearly we get

$$|B_{n,m}(f)(z_1, z_2) - f(z_1, z_2)| \leq \sum_{k=0}^{\infty} \sum_{j=0}^{\infty} |c_{k,j}| \cdot |B_{n,m}(e_{k,j})(z_1, z_2) - e_{k,j}(z_1, z_2)|.$$

But taking into account the estimates in the proof of Theorem 1.1.2, (i), for all $|z_1| \leq r_1$ and $|z_2| \leq r_2$ we obtain

$$
\begin{aligned}
|B_{n,m}(e_{k,j})(z_1, z_2) - e_{k,j}(z_1, z_2)| &= |B_n(e_k)(z_1) \cdot B_m(e_j)(z_2) - z_1^k \cdot z_2^j| \\
&\leq |B_n(e_k)(z_1) \cdot B_m(e_j)(z_2) - B_n(e_k)(z_1)z_2^j| + |B_n(e_k)(z_1)z_2^j - z_1^k z_2^j| \\
&\leq |B_n(e_k)(z_1)| \cdot |B_m(e_j)(z_2) - z_2^j| + |z_2^j| \cdot |B_n(e_k)(z_1) - z_1^k| \\
&\leq r_1^k \cdot \frac{3r_2(1 + r_2)}{2m} j(j-1)r_2^{j-2} + r_2^j \frac{3r_1(1 + r_1)}{2n} k(k-1)r_1^{k-2},
\end{aligned}
$$

which immediately implies the estimate in (i).

(ii) Let $1 \leq r_1 < r_1^* < R_1$, $1 \leq r_2 < r_2^* < R_2$. By the Cauchy's formula in Theorem 2.1.2, (ii), we get

$$\frac{\partial^{k_1+k_2} B_{n,m}(f)}{\partial z_1^{k_1} \partial z_2^{k_2}}(z_1, z_2) - \frac{\partial^{k_1+k_2} f}{\partial z_1^{k_1} \partial z_2^{k_2}}(z_1, z_2)$$

$$= \frac{(k_1!)(k_2!)}{(2\pi i)^2} \int_{|u_2 - z_2| = r_2^*} \int_{|u_1 - z_1| = r_1^*} \frac{[B_{n,m}(u_1, u_2) - f(u_1, u_2)]du_1 du_2}{(u_1 - z_1)^{k_1+1}(u_2 - z_2)^{k_2+1}}.$$

Passing to absolute value with $|z_1| \leq r_1$, $|z_2| \leq r_2$ and taking into account that $|u_1 - z_1| \geq r_1^* - r_1$, $|u_2 - z_2| \geq r_2^* - r_2$, by applying the estimate in (i) we easily obtain

$$\left| \frac{\partial^{k_1+k_2} B_{n,m}(f)}{\partial z_1^{k_1} \partial z_2^{k_2}}(z_1, z_2) - \frac{\partial^{k_1+k_2} f}{\partial z_1^{k_1} \partial z_2^{k_2}}(z_1, z_2) \right|$$

$$\leq C_{r_1^*, r_2^*, n, m}(f) \cdot \frac{(k_1)!}{(r_1^* - r_1)^{k_1+1}} \cdot \frac{(k_2)!}{(r_2^* - r_2)^{k_2+1}},$$

which proves the theorem. $\square$

In what follows a Voronovskaja's result for $B_{n,m}(f)$ is presented. It will be the product of the parametric extensions generated by the Voronovskaja's formula in univariate case in Theorem 1.1.3. Indeed, for $f(z_1, z_2)$ defining the parametric extensions of the Voronovskaja's formula by

$$_{z_1}L_n(f)(z_1, z_2) := B_n(f(\cdot, z_2))(z_1) - f(z_1, z_2) - \frac{z_1(1 - z_1)}{2n} \cdot \frac{\partial^2 f}{\partial z_1^2}(z_1, z_2),$$

$$_{z_2}L_m(f)(z_1, z_2) := B_m(f(z_1, \cdot))(z_2) - f(z_1, z_2) - \frac{z_2(1 - z_2)}{2m} \cdot \frac{\partial^2 f}{\partial z_2^2}(z_1, z_2),$$

their product (composition) gives

$$
_{z_2}L_m(f)(z_1, z_2) \circ_{\cdot z_1} L_n(f)(z_1, z_2)
$$
$$
= B_m\left[B_n(f(\cdot, \cdot))(z_1) - f(z_1, \cdot) - \frac{z_1(1-z_1)}{2n} \cdot \frac{\partial^2 f}{\partial z_1^2}(z_1, \cdot) \right](z_2)
$$
$$
- \left[B_n(f(\cdot, z_2))(z_1) - f(z_1, z_2) - \frac{z_1(1-z_1)}{2n} \cdot \frac{\partial^2 f}{\partial z_1^2}(z_1, z_2) \right] - \frac{z_2(1-z_2)}{2m}
$$
$$
\cdot \left[B_n(\frac{\partial^2 f}{\partial z_2^2}(\cdot, z_2))(z_1) - \frac{\partial^2 f}{\partial z_2^2}(z_1, z_2) - \frac{z_1(1-z_1)}{2n} \cdot \frac{\partial^2}{\partial z_1^2}\left[\frac{\partial^2 f}{\partial z_2^2}\right](z_1, z_2) \right]
$$
$$
:= E_1 - E_2 - E_3.
$$

After simple calculation evidently that we can write

$$
_{z_2}L_m(f)(z_1, z_2) \circ_{z_1} L_n(f)(z_1, z_2)
$$
$$
= B_{n,m}(f)(z_1, z_2) - B_m(f(z_1, \cdot))(z_2) - \frac{z_1(1-z_1)}{2n} \cdot B_m\left(\frac{\partial^2 f}{\partial z_1^2}(z_1, \cdot) \right)(z_2)
$$
$$
- B_n(f(\cdot, z_2))(z_1) + f(z_1, z_2) + \frac{z_1(1-z_1)}{2n}\frac{\partial^2 f}{\partial z_1^2}(z_1, z_2)
$$
$$
- \frac{z_2(1-z_2)}{2m} \cdot B_n\left(\frac{\partial^2 f}{\partial z_2^2}(\cdot, z_2) \right)(z_1) + \frac{z_2(1-z_2)}{2m} \cdot \frac{\partial^2 f}{\partial z_2^2}(z_1, z_2)
$$
$$
+ \frac{z_1(1-z_1)}{2n} \cdot \frac{z_2(1-z_2)}{2m} \cdot \frac{\partial^4 f}{\partial z_1^2 \partial z_2^2}(z_1, z_2),
$$

from which immediately can be derived the commutativity property

$$
_{z_2}L_m(f)(z_1, z_2) \circ_{z_1} L_n(f)(z_1, z_2) = {}_{z_1}L_n(f)(z_1, z_2) \circ_{z_2} L_m(f)(z_1, z_2).
$$

The Voronovskaja-type formula can be stated as follows.

Theorem 2.2.2. *Suppose that* $f : P(0; R) \to \mathbb{C}$ *is analytic in* $P(0; R) = \mathbb{D}_{R_1} \times \mathbb{D}_{R_2}$, *that is* $f(z_1, z_2) = \sum_{k=0}^{\infty} \sum_{j=0}^{\infty} c_{k,j} z_1^k z_2^j$, *for all* $(z_1, z_2) \in P(0; R)$, $R = (R_1, R_2)$.
For all $|z_1| \le r_1$, $|z_2| \le r_2$ *with* $1 \le r_1 < R_1$, $1 \le r_2 < R_2$ *and* $n, m \in \mathbb{N}$ *we have*

$$
|_{z_2}L_m(f)(z_1, z_2) \circ_{z_1} L_n(f)(z_1, z_2)| \le C_{r_1, r_2}(f)\left[\frac{1}{n^2} + \frac{1}{m^2} \right],
$$

where

$$
C_{r_1, r_2}(f) = \max\left\{ \sum_{k=2}^{\infty} \sum_{j=0}^{\infty} |c_{k,j}| r_2^j \left[\frac{5(1 + r_1^2)k(k-1)(k-2)^2 r_1^{k-2}}{2} \right], \right.
$$
$$
\left. \frac{r_2(1 + r_2)}{2} \sum_{k=2}^{\infty} \sum_{j=2}^{\infty} |c_{k,j}| j(j-1) r_2^{j-2} \cdot \left[\frac{5(1 + r_1^2)k(k-1)(k-2)^2 r_1^{k-2}}{2} \right] \right\}.
$$

Proof. First by hypothesis we can write $f(z_1, z_2) = \sum_{k=0}^{\infty} f_k(z_2) z_1^k$, where $f_k(z_2) = \sum_{j=0}^{\infty} c_{k,j} z_2^j$. It follows $\frac{\partial^2 f}{\partial z_1^2}(z_1, z_2) = \sum_{k=2}^{\infty} f_k(z_2)k(k-1)z_1^{k-2}$ and

$\frac{\partial^2 f}{\partial z_2^2}(z_1, z_2) = \sum_{k=0}^{\infty} \frac{\partial^2 f_k}{\partial z_2^2}(z_2) z_1^k$, where $\frac{\partial^2 f_k}{\partial z_2^2}(z_2) = \sum_{j=2}^{\infty} c_{k,j} j(j-1) z_2^{j-2}$. This implies $B_n(f(\cdot, z_2))(z_1) = \sum_{k=0}^{\infty} f_k(z_2) B_n(e_1^k)(z_1)$ and

$$B_n(f(\cdot, z_2))(z_1) - f(z_1, z_2) - \frac{z_1(1-z_1)}{2n} \cdot \frac{\partial^2 f}{\partial z_1^2}(z_1, z_2)$$

$$= \sum_{k=2}^{\infty} f_k(z_2) \left[B_n(e_1^k)(z_1) - e_1^k(z_1) - \frac{z_1^{k-1}(1-z_1)k(k-1)}{2n} \right].$$

Applying now B_m to the last expression with respect to z_2, we obtain

$$E_1 = \sum_{k=2}^{\infty} B_m(f_k)(z_2) \left[B_n(e_1^k)(z_1) - e_1^k(z_1) - \frac{z_1^{k-1}(1-z_1)k(k-1)}{2n} \right]$$

$$= \sum_{k=2}^{\infty} \left(\sum_{j=0}^{\infty} c_{k,j} B_m(e_1^j)(z_2) \right) \left[B_n(e_1^k)(z_1) - e_1^k(z_1) - \frac{z_1^{k-1}(1-z_1)k(k-1)}{2n} \right].$$

Passing now to absolute value with $|z_1| \leq r_1$ and $|z_2| \leq r_2$ and taking into account the estimates in the proofs of Theorem 1.1.2, (i) and Theorem 1.1.3, (ii), it follows

$$|E_1| \leq \sum_{k=2}^{\infty} \sum_{j=0}^{\infty} |c_{k,j}| r_2^j \left[\frac{5(1+r_1^2)k(k-1)(k-2)^2 r_1^{k-2}}{2n^2} \right].$$

Similarly,

$$|E_2| \leq \sum_{k=2}^{\infty} |f_k(z_2)| \cdot \left[\frac{5(1+r_1^2)k(k-1)(k-2)^2 r_1^{k-2}}{2n^2} \right]$$

$$\leq \sum_{k=2}^{\infty} \sum_{j=0}^{\infty} |c_{k,j}| r_2^j \left[\frac{5(1+r_1^2)k(k-1)(k-2)^2 r_1^{k-2}}{2n^2} \right].$$

Then

$$B_n \left(\frac{\partial^2 f}{\partial z_2^2}(\cdot, z_2) \right)(z_1)$$

$$= \sum_{k=0}^{\infty} \frac{\partial^2 f_k}{\partial z_2^2}(z_2) B_n(e_1^k)(z_1) = \sum_{k=0}^{\infty} \sum_{j=2}^{\infty} c_{k,j} j(j-1) z_2^{j-2} B_n(e_1^k)(z_1),$$

and

$$\left[B_n(\frac{\partial^2 f}{\partial z_2^2}(\cdot, z_2))(z_1) - \frac{\partial^2 f}{\partial z_2^2}(z_1, z_2) - \frac{z_1(1-z_1)}{2n} \cdot \frac{\partial^2}{\partial z_1^2} \left[\frac{\partial^2 f}{\partial z_2^2} \right](z_1, z_2) \right]$$

$$= \sum_{k=2}^{\infty} \sum_{j=2}^{\infty} c_{k,j} j(j-1) z_2^{j-2} \left[B_n(e_1^k)(z_1) - e_1^k(z_1) - \frac{z_1^{k-1}(1-z_1)k(k-1)}{2n} \right],$$

which again by Theorem 1.1.3, (ii) implies

$$|E_3| \leq \frac{r_2(1+r_2)}{2m} \sum_{k=2}^{\infty} \sum_{j=2}^{\infty} |c_{k,j}| j(j-1) r_2^{j-2} \cdot \left[\frac{5(1+r_1^2)k(k-1)(k-2)^2 r_1^{k-2}}{2n^2} \right].$$

Note that if we estimate now $|_{z_1} L_n(f)(z_1, z_2) \circ _{z_2} L_m(f)(z_1, z_2)|$, then by reasons of symmetry we get a similar order of approximation, simply interchanging above the places of n and m.

In conclusion,

$$|_{z_2} L_m(f)(z_1, z_2) \circ _{z_1} L_n(f)(z_1, z_2)| \leq |E_1| + |E_2| + |E_3| \leq C_{r_1, r_2}(f) \left[\frac{1}{n^2} + \frac{1}{m^2} \right],$$

with $C_{r_1, r_2}(f)$ given by the statement. $\qquad\square$

Theorem 2.2.2 will be used to find the exact order in approximation by $B_{n,n}(f)$. In this sense we have the following.

Theorem 2.2.3. *Suppose that $f : P(0; R) \to \mathbb{C}$ is analytic in $P(0; R) = \mathbb{D}_{R_1} \times \mathbb{D}_{R_2}$, that is $f(z_1, z_2) = \sum_{k=0}^{\infty} \sum_{j=0}^{\infty} c_{k,j} z_1^k z_2^j$, for all $(z_1, z_2) \in P(0; R)$, $R = (R_1, R_2)$. Denoting $\|f\|_{r_1, r_2} = \sup\{|f(z_1, z_2)|; |z_1| \leq r_1; |z_2| \leq r_2\}$, if f is not a solution of the complex partial differential equation*

$$z_1(1-z_1) \cdot \frac{\partial^2 f}{\partial z_1^2}(z_1, z_2) + z_2(1-z_2) \cdot \frac{\partial^2 f}{\partial z_2^2}(z_1, z_2) = 0, (z_1, z_2) \in P(0, R),$$

then we have

$$\|B_{n,n}(f) - f\|_{r_1, r_2} \geq \frac{K_{r_1, r_2, f}}{n}, \text{ for all } n \in \mathbb{N},$$

where $K_{r_1, r_2, f}$ is independent on n.

Proof. We can write

$$B_{n,n}(f)(z_1, z_2) - f(z_1, z_2)$$

$$= \frac{2}{n} \left\{ \frac{z_1(1-z_1)}{4} \cdot \frac{\partial^2 f}{\partial z_1^2}(z_1, z_2) + \frac{z_2(1-z_2)}{4} \cdot \frac{\partial^2 f}{\partial z_2^2}(z_1, z_2) \right.$$

$$\left. + \frac{2}{n} \left[\frac{n^2}{4} \left(_{z_2} L_n(f) \circ _{z_1} L_n(f) \right) (z_1, z_2) \right] + R_n(f)(z_1, z_2) \right\},$$

where

$$R_n(f)(z_1, z_2)$$

$$= \frac{n}{2} \left(B_n(f(z_1, \cdot))(z_2) - f(z_1, z_2) - \frac{z_2(1-z_2)}{2n} \cdot \frac{\partial^2 f}{\partial z_2^2}(z_1, z_2) \right)$$

$$+ \frac{n}{2} \left(B_n(f(\cdot, z_2))(z_1) - f(z_1, z_2) - \frac{z_1(1-z_1)}{2n} \cdot \frac{\partial^2 f}{\partial z_1^2}(z_1, z_2) \right)$$

$$+ \frac{z_1(1-z_1)}{4} \left(B_n \left(\frac{\partial^2 f}{\partial z_1^2}(z_1, \cdot) \right)(z_2) - \frac{\partial^2 f}{\partial z_1^2}(z_1, z_2) \right)$$

$$+ \frac{z_2(1-z_2)}{4} \left(B_n \left(\frac{\partial^2 f}{\partial z_2^2}(\cdot, z_2) \right)(z_1) - \frac{\partial^2 f}{\partial z_2^2}(z_1, z_2) \right)$$

$$- \frac{z_1(1-z_1)z_2(1-z_2)}{8n} \cdot \frac{\partial^4 f}{\partial z_1^2 \partial z_2^2}(z_1, z_2).$$

By using Theorem 1.1.3 and the reasonings in the above Theorem 2.2.2, it is immediate that $\|R_n(f)\|_{r_1,r_2} \to 0$ as $n \to \infty$. Also, by Theorem 2.2.2 we obtain

$$\frac{n^2}{4}\|(z_2 L_m(f) \circ z_1 L_n(f))\|_{r_1,r_2} \le \frac{C_{r_1,r_2}(f)}{2},$$

which implies

$$\left\|\frac{2}{n}\left[\frac{n^2}{4}(z_2 L_n(f) \circ z_1 L_n(f))\right] + R_n(f)\right\|_{r_1,r_2} \to 0, \text{ as } n \to \infty.$$

Denoting $H(z_1,z_2) = \frac{z_1(1-z_1)}{4} \cdot \frac{\partial^2 f}{\partial z_1^2}(z_1,z_2) + \frac{z_2(1-z_2)}{4} \cdot \frac{\partial^2 f}{\partial z_2^2}(z_1,z_2)$ and taking into account the inequalities

$$\|F+G\|_{r_1,r_2} \ge |\, \|F\|_{r_1,r_2} - \|G\|_{r_1,r_2} \,| \ge \|F\|_{r_1,r_2} - \|G\|_{r_1,r_2},$$

it follows

$$\|B_{n,n}(f) - f\|_{r_1,r_2}$$

$$\ge \frac{2}{n}\left\{\|H\|_{r_1,r_2} - \left\|\frac{2}{n}\left[\frac{n^2}{4}(z_2 L_n(f) \circ z_1 L_n(f))\right] + R_n(f)\right\|_{r_1,r_2}\right\}$$

$$\ge \frac{2}{n} \cdot \frac{1}{2}\|H\|_{r_1,r_2} = \frac{1}{n} \cdot \|H\|_{r_1,r_2},$$

for all $n \ge n_0$, with n_0 depending only on f, r_1 and r_2. We used here that by hypothesis we have $\|H\|_{r_1,r_2} > 0$.

For $n \in \{1, 2, ..., n_0-1\}$ we reason exactly as for the complex univariate Bernstein polynomials in the proof of Theorem 1.1.4. $\qquad\square$

Combining Theorem 2.2.2 with Theorem 2.2.3 we immediately obtain the following.

Corollary 2.2.4. *Suppose that $f : P(0;R) \to \mathbb{C}$ is analytic in $P(0;R) = \mathbb{D}_{R_1} \times \mathbb{D}_{R_2}$, that is $f(z_1,z_2) = \sum_{k=0}^{\infty} \sum_{j=0}^{\infty} c_{k,j} z_1^k z_2^j$, for all $(z_1, z_2) \in P(0;R)$, $R = (R_1, R_2)$. If the Taylor series of f contains at least one term of the form $c_{k,0} z_1^k$ with $c_{k,0} \neq 0$ and $k \ge 2$ or of the form $c_{0,j} z_2^j$ with $c_{0,j} \neq 0$ and $j \ge 2$, then we have*

$$\|B_{n,n}(f) - f\|_{r_1,r_2} \sim \frac{1}{n}, \text{ for all } n \in \mathbb{N}.$$

Proof. It suffices to prove that under the hypothesis on f, it cannot be a solution of the complex partial differential equation

$$z_1(1-z_1) \cdot \frac{\partial^2 f}{\partial z_1^2}(z_1,z_2) + z_2(1-z_2) \cdot \frac{\partial^2 f}{\partial z_2^2}(z_1,z_2) = 0, (z_1,z_2) \in P(0,R).$$

Indeed, suppose the contrary. Since simple calculation give

$$z_1(1-z_1) \cdot \frac{\partial^2 f}{\partial z_1^2} = \sum_{j=0}^{\infty} 2c_{2,j} z_1 z_2^j + \sum_{k=2}^{\infty} \sum_{j=0}^{\infty} z_1^k z_2^j [c_{k+1,j} k(k+1) - c_{k,j} k(k-1)]$$

and

$$z_2(1-z_2)\cdot\frac{\partial^2 f}{\partial z_2^2}=\sum_{k=0}^{\infty}2c_{k,2}z_1^k z_2+\sum_{k=0}^{\infty}\sum_{j=2}^{\infty}z_1^k z_2^j[c_{k,j+1}j(j+1)-c_{k,j}j(j-1)]$$

by summing and equaling with zero, we obtain

$$2c_{0,2}z_2+2c_{1,2}z_1z_2+\sum_{k=2}^{\infty}z_1^k z_2[c_{k+1,j}k(k+1)-c_{k,1}k(k-1)+2c_{k,2}]$$

$$+\sum_{j=2}^{\infty}z_2^j[c_{0,j+1}j(j+1)-c_{0,j}j(j-1)]$$

$$+\sum_{j=2}^{\infty}z_1z_2^j[c_{1,j+1}j(j+1)-c_{1,j}j(j-1)+2c_{2,j}]$$

$$+2c_{2,0}z_1+2c_{2,1}z_1z_2+\sum_{k=2}^{\infty}z_1^k[c_{k+1,0}k(k+1)-c_{k,0}k(k-1)]$$

$$+\sum_{k=2}^{\infty}\sum_{j=2}^{\infty}z_1^k z_2^j\,[c_{k+1,j}k(k+1)-c_{k,j}k(k-1)$$

$$+c_{k,j+1}j(j+1)-c_{k,j}j(j-1)]=0.$$

By the identification of coefficients, among others we immediately get $c_{0,2}=c_{2,0}=0$ and from the terms under the second and fourth sign Σ, it follows

$$c_{0,j+1}j(j+1)-c_{0,j}j(j-1)=0,\ c_{k+1,0}k(k+1)-c_{k,0}k(k-1)=0,$$

for all $j=2,3,...,\ k=2,3,...,$. From here by recurrence it is easy to deduce that $c_{0,k}=c_{j,0}=0$ for all $j=2,3,...,\ k=2,3,...,$ which contradicts the hypothesis on f. Therefore the hypothesis and the lower estimate in Theorem 2.2.3 holds, which ends the proof. $\qquad\Box$

In what follows we study the approximation properties of the non-tensor product kind bivariate complex Bernstein polynomials $B_n(f)(z_1,z_2)$. Note that obviously we have $B_n(f)(z_1,z_2)\neq B_{n,n}(f)(z_1,z_2)$.

For $f(z_1,z_2)=\sum_{k=0}^{\infty}\sum_{j=0}^{\infty}c_{k,j}z_1^k z_2^j$, as in the case of tensor product Bernstein polynomials, denoting $e_{k,j}(z_1,z_2)=e_k(z_1)\cdot e_j(z_2)=z_1^k\cdot z_2^j$ we can write

$$B_n(f)(z_1,z_2)-f(z_1,z_2)=\sum_{k=0}^{\infty}\sum_{j=0}^{\infty}c_{k,j}[B_n(e_{k,j})(z_1,z_2)-e_{k,j}(z_1,z_2)].$$

Therefore, in order to obtain approximation results a representation formula for $B_n(e_{k,j})(z_1,z_2)$ will be important. In this sense we present

Theorem 2.2.5. *(i)* $B_n(e_{p,q})(z_1,z_2)=e_{p,q}(z_1,z_2)$, *for all* p,q *with* $0\le p+q\le 1$ *and* $n\in\mathbb{N}$. *Also,*

$$B_n(e_{p,0})(z_1,z_2)=B_n(e_p)(z_1)\ and\ B_n(e_{0,q})(z_1,z_2)=B_n(e_q)(z_2),$$

where B_n in the above right-hand sides denotes the Bernstein polynomial of one variable.

(ii) Denoting $m = \min\{n, p\}$ and $s = \min\{n, q\}$, for all $n, p, q \in \mathbb{N}$ we can write

$$B_n(e_{p,q})(z_1, z_2) = \sum_{i=1}^{m} \sum_{j=1}^{s} C_{n,i,j,p,q} z_1^i z_2^j,$$

where

$$C_{n,i,j,p,q} = \binom{n}{i}\binom{n}{j} \Delta_{1/n}^i e_p(0) \cdot \Delta_{1/n}^j e_q(0) \cdot \frac{n^{[i+j]}}{n^{[i]} n^{[j]}} \geq 0,$$

and $n^{[i]} = n(n-1)...(n-i+1)$;

(iii) Denoting $m = \min\{n, p\}$ and $s = \min\{n, q\}$, for all $n, p, q \in \mathbb{N}$ we have

$$\sum_{i=1}^{m} \sum_{j=1}^{s} C_{n,i,j,p,q} = \sum_{k=0}^{n} \binom{n}{k} \Delta_{1/n}^k e_q(0) \left(\frac{n-k}{n}\right)^p \leq 1.$$

Proof. (i) We have the possibilities $p = q = 0$, or $p = 0, q = 1$ or $p = 1, q = 0$, for which by simple calculation we obtain $B_n(e_{p,q})(z_1, z_2) = e_{p,q}(z_1, z_2)$. Also, the next two equalities follow by simple calculation.

(ii) Denoting by $S(p, i)$ and $S(q, j)$ the Stirling numbers of second kind and taking into account that $S(p, i) = S(q, j) = 0$ for all $i > p$ and $j > q$, by using Proposition 1 in Farcaş [71], we have

$$B_n(e_{p,q})(z_1, z_2) = \frac{1}{n^{p+q}} \sum_{i=1}^{m} \sum_{j=1}^{s} n^{[i+j]} S(p, i) S(q, j) z_1^i z_2^j$$

$$= \sum_{i=1}^{m} \frac{n^{[i]} S(p, i)}{n^p} z_1^i \sum_{j=1}^{s} \frac{n^{[j]} S(p, j)}{n^q} z_2^j \frac{n^{[i+j]}}{n^{[i]} n^{[j]}}$$

$$= \sum_{i=1}^{m} \sum_{j=1}^{s} C_{n,i,j,p,q} z_1^i z_2^j,$$

where $C_{n,i,j,p,q}$ is given by the formula in statement. Note that we used here the formula

$$\frac{n^{[i]} S(p, i)}{n^p} = \binom{n}{i} \Delta_{1/n}^i e_p(0) = \binom{n}{i} [0, 1/n, ..., i/n; e_p] i! / n^i.$$

Since e_p is convex of any order, from the last formula it is clear that all $C_{n,i,j,p,q} \geq 0$.

(iii) From the definition formula for $B_n(f)(z_1, z_2)$, we obtain

$$B_n(e_{p,q})(1, 1)$$

$$= \sum_{k=0}^{n} \sum_{j=0}^{n-k} \binom{n}{k}\binom{n-k}{j} (-1)^{n-k-j} \frac{k^p}{n^p} \cdot \frac{j^q}{n^q} = \sum_{k=0}^{n} \binom{n}{k} \frac{k^p}{n^p} \Delta_{1/n}^{n-k} e_q(0)$$

$$= \sum_{k=0}^{n} \binom{n}{k} \Delta_{1/n}^k e_q(0) \left(\frac{n-k}{n}\right)^p \leq \sum_{k=0}^{n} \binom{n}{k} \Delta_{1/n}^k e_q(0) = 1.$$

On the other hand, replacing $x = y = 1$ in the formula from the above point (ii), it follows exactly the required inequality. $\qquad \square$

The approximation results are expressed by the following.

Theorem 2.2.6. *Suppose that $f : P(0; R) \to \mathbb{C}$ is analytic in $P(0; R) = \mathbb{D}_{R_1} \times \mathbb{D}_{R_2}$, that is $f(z_1, z_2) = \sum_{k=0}^{\infty} \sum_{j=0}^{\infty} c_{k,j} z_1^k z_2^j$, for all $(z_1, z_2) \in P(0; R)$, $R = (R_1, R_2)$.*
(i) For all $|z_1| \leq r_1$, $|z_2| \leq r_2$ with $1 \leq r_1 < R_1$, $1 \leq r_2 < R_2$ and $n \in \mathbb{N}$ we have

$$|B_n(f)(z_1, z_2) - f(z_1, z_2)| \leq \frac{C_{r_1, r_2}(f)}{n},$$

where

$$C_{r_1, r_2}(f) = 2 \sum_{p=0}^{\infty} \sum_{q=0}^{\infty} |c_{p,q}|(p+q)(p+q-1)r_1^p r_2^q.$$

(ii) Let $k_1, k_2 \in \mathbb{N}$ be with $k_1 + k_2 \geq 1$ and $1 \leq r_1 < r_1^* < R_1$, $1 \leq r_2 < r_2^* < R_2$. Then for all $|z_1| \leq r_1$, $|z_2| \leq r_2$ and $n \in \mathbb{N}$ we have

$$\left| \frac{\partial^{k_1+k_2} B_{n,m}(f)}{\partial z_1^{k_1} \partial z_2^{k_2}}(z_1, z_2) - \frac{\partial^{k_1+k_2} f}{\partial z_1^{k_1} \partial z_2^{k_2}}(z_1, z_2) \right|$$

$$\leq \frac{C_{r_1^*, r_2^*}(f)}{n} \cdot \frac{(k_1)!}{(r_1^* - r_1)^{k_1+1}} \cdot \frac{(k_2)!}{(r_2^* - r_2)^{k_2+1}}.$$

Proof. (i) It is immediate the inequality

$$|B_n(f)(z_1, z_2) - f(z_1, z_2)| \leq \sum_{p=0}^{\infty} \sum_{q=0}^{\infty} |c_{p,q}| \cdot |B_n(e_{p,q})(z_1, z_2) - e_{p,q}(z_1, z_2)|.$$

To estimate $|B_n(e_{p,q})(z_1, z_2) - e_{p,q}(z_1, z_2)|$ we have four possibilities :
1) $0 \leq p \leq n$ and $0 \leq q \leq n$; 2) $0 \leq p \leq n$ and $q > n$; 3)$p > n$ and $0 \leq q \leq n$; 4) $p > n$ and $q > n$.

Case 1). It follows $m = p$, $s = q$. By Theorem 2.2.5, (i) we may consider $p \geq 1$ and $q \geq 1$. Also, by Theorem 2.2.5, (ii), (iii), for all $1 \leq |z_1| \leq r_1$ and $1 \leq |z_2| \leq r_2$ we obtain

$$|B_n(e_{p,q})(z_1, z_2) - e_{p,q}(z_1, z_2)| = |[C_{n,p,q,p,q} - 1]e_{p,q}(z_1, z_2)]$$

$$+ \sum_{i=1}^{p-1} \sum_{j=1}^{q} C_{n,i,j,p,q} x^i y^j + \sum_{i=1}^{p} \sum_{j=1}^{q-1} C_{n,i,j,p,q} x^i y^j| \leq [1 - C_{n,p,q,p,q}]r_1^p r_2^q$$

$$+ [B_n(e_{p,q})(1, 1) - \sum_{j=1}^{q} C_{n,p,j,p,q}]r_1^p r_2^q + [B_n(e_{p,q})(1, 1) - \sum_{i=1}^{p} C_{n,i,q,p,q}]r_1^p r_2^q$$

$$\leq [1 - C_{n,p,q,p,q}]r_1^p r_2^q + [1 - \sum_{j=1}^{q} C_{n,p,j,p,q}]r_1^p r_2^q + [1 - \sum_{i=1}^{p} C_{n,i,q,p,q}]r_1^p r_2^q$$

$$\leq 3[1 - C_{n,p,q,p,q}]r_1^p r_2^q.$$

But

$$C_{n,p,q,p,q} = \binom{n}{p}\binom{n}{q}\frac{p!}{n^p}\cdot\frac{q!}{n^q}\cdot\frac{n^{[p+q]}}{n^{[p]}n^{[q]}} = \frac{n^{[p+q]}}{n^{p+q}} = \Pi_{j=1}^{p+q}\frac{n-j+1}{n},$$

which implies

$$1 - C_{n,p,q,p,q} \le \frac{(p+q)(p+q-1)}{2n}.$$

We used here the inequality

$$1 - \prod_{j=1}^{p+q} y_j \le \sum_{j=1}^{p+q}(1-y_j), 0 \le y_j \le 1, j = 1, 2, ..., p+q,$$

applied for $y_j = \frac{n-j+1}{n}$. This implies

$$|B_n(e_{p,q})(z_1, z_2) - e_{p,q}(z_1, z_2)| \le \frac{3(p+q)(p+q-1)}{2n}r_1^p r_2^q.$$

Case 2). It follows $m = p$ and $s = n$ and again by Theorem 2.2.5, (i) we may consider $p \ge 1$. By Theorem 2.2.5, (ii), (iii), for all $1 \le |z_1| \le r_1$ and $1 \le |z_2| \le r_2$ we obtain

$$|B_n(e_{p,q})(z_1, z_2) - e_{p,q}(z_1, z_2)| \le |B_n(e_{p,q})(z_1, z_2)| + |e_{p,q}(z_1, z_2)|$$

$$\le \sum_{i=1}^{p}\sum_{j=1}^{n} C_{n,i,j,p,n}r_1^p r_2^n + r_1^p r_2^q \le 2r_1^p r_2^q \le 2nr_1^p r_2^q \le 2(q-1)r_1^p r_2^q$$

$$\le 2\frac{(q-1)q}{n}r_1^p r_2^q \le 2\frac{(p+q)(p+q-1)}{n}r_1^p r_2^q.$$

Case 3). By reason of symmetry, the Case 3) is identical with the Case 2.

Case 4). Identical with the Cases 2 and 3.

(ii) By using the Cauchy's formula in Theorem 2.1.2, (ii), we reason exactly as in the proof of Theorem 2.2.1, (ii). $\square$

2.3   Favard-Szász-Mirakjan Operators

In this section we consider the bivariate complex Favard-Szász-Mirakjan operators of tensor product kind given by

$$S_{n,m}(f)(z_1, z_2) = \sum_{k=0}^{\infty}\sum_{j=0}^{\infty} s_{n,k}(z_1)s_{m,j}(z_2)f(k/n, j/m),$$

where $s_{n,k}(z) = e^{-nz}\frac{(nz)^k}{k!}$, $f : ([R_1, +\infty) \cup \overline{\mathbb{D}}_{R_1}) \times ([R_2, +\infty) \cup \overline{\mathbb{D}}_{R_2}) \to \mathbb{C}$ is supposed to be bounded on $[0, +\infty) \times [0, \infty)$ and analytic in $\mathbb{D}_{R_1} \times \mathbb{D}_{R_2}$.

Note that in the case of two real variables, these operators were first considered by Leonte-Virtopeanu [121].

We will use similar methods with those for complex bivariate Bernstein polynomials of tensor product kind in Section 2.2. The results in univariate case obtained for complex Favard-Szász-Mirakjan in Chapter 1 will be useful.

First we present the following result.

Theorem 2.3.1. *Suppose* $2 < R_1 < \infty$, $2 < R_2 < \infty$ *and that* $f : ([R_1, +\infty) \cup \overline{\mathbb{D}}_{R_1}) \times ([R_2, +\infty) \cup \overline{\mathbb{D}}_{R_2}) \to \mathbb{C}$ *is bounded on* $[R_1, \infty) \times [R_2, \infty)$ *and analytic in* $\mathbb{D}_{R_1} \times \mathbb{D}_{R_2}$, *i.e* $f(z_1, z_2) = \sum_{k=0}^{\infty} \sum_{j=0}^{\infty} c_{k,j} z_1^k z_2^j$, *for all* $|z_1| < R_1$, $|z_2| < R_2$.

(i) Let $1 \le r_1 < \frac{R_1}{2}$ *and* $1 \le r_2 < \frac{R_2}{2}$. *For all* $|z_1| \le r_1$, $|z_2| \le r_2$ *and* $n, m \in \mathbb{N}$ *we have*

$$|S_{n,m}(f)(z_1, z_2) - f(z_1, z_2)| \le C_{r_1, r_2, n, m}(f),$$

where

$$C_{r_1, r_2, n, m}(f) = \frac{6}{m} \cdot \sum_{k=0}^{\infty} \sum_{j=0}^{\infty} |c_{k,j}| \cdot (j-1)(2r_1)^k (2r_2^{j-1})$$

$$+ \frac{6}{n} \cdot \sum_{k=0}^{\infty} \sum_{j=0}^{\infty} |c_{k,j}| \cdot (k-1)(2r_1)^{k-1}(2r_2)^j.$$

(ii) Let $k_1, k_2 \in \mathbb{N}$ *be with* $k_1 + k_2 \ge 1$ *and* $1 \le r_1 < r_1^* < \frac{R_1}{2}$, $1 \le r_2 < r_2^* < \frac{R_2}{2}$. *Then for all* $|z_1| \le r_1$, $|z_2| \le r_2$ *and* $n, m \in \mathbb{N}$ *we have*

$$\left| \frac{\partial^{k_1+k_2} S_{n,m}(f)}{\partial z_1^{k_1} \partial z_2^{k_2}}(z_1, z_2) - \frac{\partial^{k_1+k_2} f}{\partial z_1^{k_1} \partial z_2^{k_2}}(z_1, z_2) \right|$$

$$\le C_{r_1^*, r_2^*, n, m}(f) \cdot \frac{(k_1)!}{(r_1^* - r_1)^{k_1+1}} \cdot \frac{(k_2)!}{(r_2^* - r_2)^{k_2+1}},$$

where $C_{r_1^*, r_2^*, n, m}(f)$ *is given at the above point (i).*

Proof. (i) Denote $e_{k,j}(z_1, z_2) = e_k(z_1) \cdot e_j(z_2)$, where $e_k(u) = u^k$. Clearly we get

$$|S_{n,m}(f)(z_1, z_2) - f(z_1, z_2)| \le \sum_{k=0}^{\infty} \sum_{j=0}^{\infty} |c_{k,j}| \cdot |S_{n,m}(e_{k,j})(z_1, z_2) - e_{k,j}(z_1, z_2)|.$$

But taking into account the estimates in the proof of Theorem 1.8.4, (i), for all $|z_1| \le r_1$ and $|z_2| \le r_2$ we obtain

$$|S_{n,m}(e_{k,j})(z_1, z_2) - e_{k,j}(z_1, z_2)| = |S_n(e_k)(z_1) \cdot S_m(e_j)(z_2) - z_1^k \cdot z_2^j|$$

$$\le |S_n(e_k)(z_1) \cdot S_m(e_j)(z_2) - S_n(e_k)(z_1) z_2^j| + |S_n(e_k)(z_1) z_2^j - z_1^k z_2^j|$$

$$\le |S_n(e_k)(z_1)| \cdot |S_m(e_j)(z_2) - z_2^j| + |z_2^j| \cdot |S_n(e_k)(z_1) - z_1^k|$$

$$\le (2r_1)^k \cdot \frac{6(j-1)}{m}(2r_2)^{j-1} + r_2^j \cdot \frac{6(k-1)}{n}(2r_1)^{k-1},$$

which immediately implies the estimate in (i).

(ii) Let $1 \le r_1 < r_1^* < \frac{R_1}{2}$, $1 \le r_2 < r_2^* < \frac{R_2}{2}$. By the Cauchy's formula in Theorem 2.1.2, (ii), we obtain

$$\frac{\partial^{k_1+k_2} S_{n,m}(f)}{\partial z_1^{k_1} \partial z_2^{k_2}}(z_1, z_2) - \frac{\partial^{k_1+k_2} f}{\partial z_1^{k_1} \partial z_2^{k_2}}(z_1, z_2)$$

$$= \frac{(k_1!)(k_2!)}{(2\pi i)^2} \int_{|u_2 - z_2| = r_2^*} \int_{|u_1 - z_1| = r_1^*} \frac{[S_{n,m}(u_1, u_2) - f(u_1, u_2)]du_1 du_2}{(u_1 - z_1)^{k_1+1}(u_2 - z_2)^{k_2+1}}.$$

Passing to absolute value with $|z_1| \leq r_1$, $|z_2| \leq r_2$ and taking into account that $|u_1 - z_1| \geq r_1^* - r_1$, $|u_2 - z_2| \geq r_2^* - r_2$, by applying the estimate in (i) we easily obtain

$$\left| \frac{\partial^{k_1+k_2} S_{n,m}(f)}{\partial z_1^{k_1} \partial z_2^{k_2}}(z_1, z_2) - \frac{\partial^{k_1+k_2} f}{\partial z_1^{k_1} \partial z_2^{k_2}}(z_1, z_2) \right|$$

$$\leq C_{r_1^*, r_2^*, n, m}(f) \cdot \frac{(k_1)!}{(r_1^* - r_1)^{k_1+1}} \cdot \frac{(k_2)!}{(r_2^* - r_2)^{k_2+1}},$$

which proves the theorem. $\qquad\qquad\qquad\qquad\qquad\qquad\qquad\qquad\qquad\qquad\square$

In what follows a Voronovskaja-type formula for $S_{n,m}(f)$ is presented. It will be the product of the parametric extensions generated by the Voronovskaja's formula in univariate case in Theorem 1.8.5. Indeed, for $f(z_1, z_2)$ defining the parametric extensions of the Voronovskaja's formula by

$$_{z_1}L_n(f)(z_1, z_2) := S_n(f(\cdot, z_2))(z_1) - f(z_1, z_2) - \frac{z_1}{2n} \cdot \frac{\partial^2 f}{\partial z_1^2}(z_1, z_2),$$

$$_{z_2}L_m(f)(z_1, z_2) := S_m(f(z_1, \cdot))(z_2) - f(z_1, z_2) - \frac{z_2}{2m} \cdot \frac{\partial^2 f}{\partial z_2^2}(z_1, z_2),$$

their product (composition) give

$$_{z_2}L_m(f)(z_1, z_2) \circ {}_{z_1}L_n(f)(z_1, z_2)$$

$$= S_m \left[S_n(f(\cdot, \cdot))(z_1) - f(z_1, \cdot) - \frac{z_1}{2n} \cdot \frac{\partial^2 f}{\partial z_1^2}(z_1, \cdot) \right] (z_2)$$

$$- \left[S_n(f(\cdot, z_2))(z_1) - f(z_1, z_2) - \frac{z_1}{2n} \cdot \frac{\partial^2 f}{\partial z_1^2}(z_1, z_2) \right] - \frac{z_2}{2m}$$

$$\cdot \left[S_n(\frac{\partial^2 f}{\partial z_2^2}(\cdot, z_2))(z_1) - \frac{\partial^2 f}{\partial z_2^2}(z_1, z_2) - \frac{z_1}{2n} \cdot \frac{\partial^2}{\partial z_1^2} \left[\frac{\partial^2 f}{\partial z_2^2} \right](z_1, z_2) \right]$$

$$:= E_1 - E_2 - E_3.$$

After simple calculation evidently that we can write

$$_{z_2}L_m(f)(z_1, z_2) \circ {}_{z_1}L_n(f)(z_1, z_2)$$

$$= S_{n,m}(f)(z_1, z_2) - S_m(f(z_1, \cdot))(z_2) - \frac{z_1}{2n} \cdot S_m \left(\frac{\partial^2 f}{\partial z_1^2}(z_1, \cdot) \right)(z_2)$$

$$- S_n(f(\cdot, z_2))(z_1) + f(z_1, z_2) + \frac{z_1}{2n} \frac{\partial^2 f}{\partial z_1^2}(z_1, z_2)$$

$$- \frac{z_2}{2m} \cdot S_n \left(\frac{\partial^2 f}{\partial z_2^2}(\cdot, z_2) \right)(z_1) + \frac{z_2}{2m} \cdot \frac{\partial^2 f}{\partial z_2^2}(z_1, z_2)$$

$$+ \frac{z_1}{2n} \cdot \frac{z_2}{2m} \cdot \frac{\partial^4 f}{\partial z_1^2 \partial z_2^2}(z_1, z_2),$$

from which immediately can be derived the commutativity property

$$_{z_2}L_m(f)(z_1, z_2) \circ {}_{z_1}L_n(f)(z_1, z_2) = {}_{z_1}L_n(f)(z_1, z_2) \circ {}_{z_2}L_m(f)(z_1, z_2).$$

The Voronovskaja's result can be stated as follows.

Theorem 2.3.2. *Suppose* $2 < R_1 < \infty$, $2 < R_2 < \infty$ *and that* $f : ([R_1, +\infty) \cup \overline{\mathbb{D}}_{R_1}) \times ([R_2, +\infty) \cup \overline{\mathbb{D}}_{R_2}) \to \mathbb{C}$ *is bounded on* $[R_1, \infty) \times [R_2, \infty)$ *and analytic in* $\mathbb{D}_{R_1} \times \mathbb{D}_{R_2}$, *i.e* $f(z_1, z_2) = \sum_{k=0}^{\infty} \sum_{j=0}^{\infty} c_{k,j} z_1^k z_2^j$, *for all* $|z_1| < R_1$, $|z_2| < R_2$.

For all $|z_1| \le r_1$, $|z_2| \le r_2$ *with* $1 \le r_1 < \frac{R_1}{2}$, $1 \le r_2 < \frac{R_2}{2}$ *and* $n, m \in \mathbb{N}$ *we have*

$$|_{z_2} L_m(f)(z_1, z_2) \circ {}_{z_1} L_n(f)(z_1, z_2)| \le C_{r_1, r_2}(f) \left[\frac{1}{n^2} + \frac{1}{m^2} \right],$$

where

$$C_{r_1, r_2}(f) = \max \left\{ \sum_{k=2}^{\infty} \sum_{j=0}^{\infty} |c_{k,j}| (2r_2)^j \left[26 r_1 (2r_1)^{k-3} (k-1)^2 (k-2) \right], \right.$$

$$\left. \sum_{k=2}^{\infty} \sum_{j=2}^{\infty} |c_{k,j}| j(j-1) r_2^{j-1} \cdot \left[13 r_1 (2r_1)^{k-3} (k-1)^2 (k-2) \right] \right\}.$$

Proof. First by hypothesis we can write $f(z_1, z_2) = \sum_{k=0}^{\infty} f_k(z_2) z_1^k$, where $f_k(z_2) = \sum_{j=0}^{\infty} c_{k,j} z_2^j$. It follows $\frac{\partial^2 f}{\partial z_1^2}(z_1, z_2) = \sum_{k=2}^{\infty} f_k(z_2) k(k-1) z_1^{k-2}$ and $\frac{\partial^2 f}{\partial z_2^2}(z_1, z_2) = \sum_{k=0}^{\infty} \frac{\partial^2 f_k}{\partial z_2^2}(z_2) z_1^k$, where $\frac{\partial^2 f_k}{\partial z_2^2}(z_2) = \sum_{j=2}^{\infty} c_{k,j} j(j-1) z_2^{j-2}$. This implies $S_n(f(\cdot, z_2))(z_1) = \sum_{k=0}^{\infty} f_k(z_2) S_n(e_1^k)(z_1)$ and

$$S_n(f(\cdot, z_2))(z_1) - f(z_1, z_2) - \frac{z_1}{2n} \cdot \frac{\partial^2 f}{\partial z_1^2}(z_1, z_2)$$

$$= \sum_{k=2}^{\infty} f_k(z_2) \left[S_n(e_1^k)(z_1) - e_1^k(z_1) - \frac{z_1^{k-1} k(k-1)}{2n} \right].$$

Applying now S_m to the last expression with respect to z_2, we obtain

$$E_1 = \sum_{k=2}^{\infty} S_m(f_k)(z_2) \left[S_n(e_1^k)(z_1) - e_1^k(z_1) - \frac{z_1^{k-1} k(k-1)}{2n} \right]$$

$$= \sum_{k=2}^{\infty} \left(\sum_{j=0}^{\infty} c_{k,j} S_m(e_1^j)(z_2) \right) \left[S_n(e_1^k)(z_1) - e_1^k(z_1) - \frac{z_1^{k-1} k(k-1)}{2n} \right].$$

Passing now to absolute value with $|z_1| \le r_1$ and $|z_2| \le r_2$ and taking into account the estimates in the proofs of Theorem 1.8.4, (i) and Theorem 1.8.5, it follows

$$|E_1| \le \sum_{k=2}^{\infty} \sum_{j=0}^{\infty} |c_{k,j}| (2r_2)^j \left[\frac{26 r_1 (2r_1)^{k-3} (k-1)^2 (k-2)}{n^2} \right].$$

Similarly,

$$|E_2| \leq \sum_{k=2}^{\infty} |f_k(z_2)| \cdot \left[\frac{26 r_1 (2r_1)^{k-3}(k-1)^2(k-2)}{n^2} \right]$$

$$\leq \sum_{k=2}^{\infty} \sum_{j=0}^{\infty} |c_{k,j}| r_2^j \left[\frac{26 r_1 (2r_1)^{k-3}(k-1)^2(k-2)}{n^2} \right].$$

Then

$$S_n \left(\frac{\partial^2 f}{\partial z_2^2}(\cdot, z_2) \right)(z_1)$$

$$= \sum_{k=0}^{\infty} \frac{\partial^2 f_k}{\partial z_2^2}(z_2) S_n(e_1^k)(z_1) = \sum_{k=0}^{\infty} \sum_{j=2}^{\infty} c_{k,j} j(j-1) z_2^{j-2} S_n(e_1^k)(z_1),$$

and

$$\left[S_n(\frac{\partial^2 f}{\partial z_2^2}(\cdot, z_2))(z_1) - \frac{\partial^2 f}{\partial z_2^2}(z_1, z_2) - \frac{z_1}{2n} \cdot \frac{\partial^2}{\partial z_1^2} \left[\frac{\partial^2 f}{\partial z_2^2} \right](z_1, z_2) \right]$$

$$= \sum_{k=2}^{\infty} \sum_{j=2}^{\infty} c_{k,j} j(j-1) z_2^{j-2} \left[S_n(e_1^k)(z_1) - e_1^k(z_1) - \frac{z_1^{k-1} k(k-1)}{2n} \right],$$

which again by Theorem 1.8.5 implies

$$|E_3| \leq \frac{r_2}{2m} \sum_{k=2}^{\infty} \sum_{j=2}^{\infty} |c_{k,j}| j(j-1) r_2^{j-2} \cdot \left[\frac{26 r_1 (2r_1)^{k-3}(k-1)^2(k-2)}{n^2} \right].$$

Note that if we estimate now $|_{z_1} L_n(f)(z_1, z_2) \circ _{z_2} L_m(f)(z_1, z_2)|$, then by reasons of symmetry we get a similar order of approximation, simply interchanging above the places of n and m.

In conclusion,

$$|_{z_2} L_m(f)(z_1, z_2) \circ _{z_1} L_n(f)(z_1, z_2)| \leq |E_1| + |E_2| + |E_3| \leq C_{r_1, r_2}(f) \left[\frac{1}{n^2} + \frac{1}{m^2} \right],$$

with $C_{r_1, r_2}(f)$ given by the statement. $\square$

Theorem 2.3.2 will be used to find the exact order in approximation by $B_{n,n}(f)$. In this sense we have the following.

Theorem 2.3.3. *Suppose $2 < R_1 < \infty$, $2 < R_2 < \infty$ and that $f : ([R_1, +\infty) \cup \overline{\mathbb{D}}_{R_1}) \times ([R_2, +\infty) \cup \overline{\mathbb{D}}_{R_2}) \to \mathbb{C}$ is bounded on $[R_1, \infty) \times [R_2, \infty)$ and analytic in $\mathbb{D}_{R_1} \times \mathbb{D}_{R_2}$, i.e $f(z_1, z_2) = \sum_{k=0}^{\infty} \sum_{j=0}^{\infty} c_{k,j} z_1^k z_2^j$, for all $|z_1| < R_1$, $|z_2| < R_2$. Denoting $\|f\|_{r_1, r_2} = \sup\{|f(z_1, z_2)|; |z_1| \leq r_1; |z_2| \leq r_2\}$, if f is not a solution of the complex partial differential equation*

$$z_1 \cdot \frac{\partial^2 f}{\partial z_1^2}(z_1, z_2) + z_2 \cdot \frac{\partial^2 f}{\partial z_2^2}(z_1, z_2) = 0, |z_1| < R_1, |z_2| < R_2,$$

then we have

$$\|S_{n,n}(f) - f\|_{r_1, r_2} \geq \frac{K_{r_1, r_2, f}}{n}, \text{ for all } n \in \mathbb{N},$$

where $K_{r_1,r_2,f}$ is independent on n.

Proof. We can write

$$S_{n,n}(f)(z_1,z_2) - f(z_1,z_2)$$
$$= \frac{2}{n} \left\{ \frac{z_1}{4} \cdot \frac{\partial^2 f}{\partial z_1^2}(z_1,z_2) + \frac{z_2}{4} \cdot \frac{\partial^2 f}{\partial z_2^2}(z_1,z_2) \right.$$
$$\left. + \frac{2}{n} \left[\frac{n^2}{4} \left({}_{z_2}L_n(f) \circ {}_{z_1}L_n(f) \right)(z_1,z_2) \right] + R_n(f)(z_1,z_2) \right\},$$

where

$$R_n(f)(z_1,z_2)$$
$$= \frac{n}{2} \left(S_n(f(z_1,\cdot))(z_2) - f(z_1,z_2) - \frac{z_2}{2n} \cdot \frac{\partial^2 f}{\partial z_2^2}(z_1,z_2) \right)$$
$$+ \frac{n}{2} \left(S_n(f(\cdot,z_2))(z_1) - f(z_1,z_2) - \frac{z_1}{2n} \cdot \frac{\partial^2 f}{\partial z_1^2}(z_1,z_2) \right)$$
$$+ \frac{z_1}{4} \left(S_n \left(\frac{\partial^2 f}{\partial z_1^2}(z_1,\cdot) \right)(z_2) - \frac{\partial^2 f}{\partial z_1^2}(z_1,z_2) \right)$$
$$+ \frac{z_2}{4} \left(S_n \left(\frac{\partial^2 f}{\partial z_2^2}(\cdot,z_2) \right)(z_1) - \frac{\partial^2 f}{\partial z_2^2}(z_1,z_2) \right)$$
$$- \frac{z_1 z_2}{8n} \cdot \frac{\partial^4 f}{\partial z_1^2 \partial z_2^2}(z_1,z_2).$$

By using Theorem 1.8.5 and the reasonings in the above Theorem 2.3.2, it is immediate that $\|R_n(f)\|_{r_1,r_2} \to 0$ as $n \to \infty$. Also, by Theorem 2.3.2 we obtain

$$\frac{n^2}{4} \|({}_{z_2}L_m(f) \circ {}_{z_1}L_n(f))\|_{r_1,r_2} \leq \frac{C_{r_1,r_2}(f)}{2},$$

which implies

$$\left\| \frac{2}{n} \left[\frac{n^2}{4} \left({}_{z_2}L_n(f) \circ {}_{z_1}L_n(f) \right) \right] + R_n(f) \right\|_{r_1,r_2} \to 0, \text{ as } n \to \infty.$$

Denoting $H(z_1,z_2) = \frac{z_1}{4} \cdot \frac{\partial^2 f}{\partial z_1^2}(z_1,z_2) + \frac{z_2}{4} \cdot \frac{\partial^2 f}{\partial z_2^2}(z_1,z_2)$ and taking into account the inequalities

$$\|F + G\|_{r_1,r_2} \geq |\|F\|_{r_1,r_2} - \|G\|_{r_1,r_2}| \geq \|F\|_{r_1,r_2} - \|G\|_{r_1,r_2},$$

it follows

$$\|S_{n,n}(f) - f\|_{r_1,r_2}$$
$$\geq \frac{2}{n} \left\{ \|H\|_{r_1,r_2} - \left\| \frac{2}{n} \left[\frac{n^2}{4} \left({}_{z_2}L_n(f) \circ {}_{z_1}L_n(f) \right) \right] + R_n(f) \right\|_{r_1,r_2} \right\}$$
$$\geq \frac{2}{n} \cdot \frac{1}{2} \|H\|_{r_1,r_2} = \frac{1}{n} \cdot \|H\|_{r_1,r_2},$$

for all $n \geq n_0$, with n_0 depending only on f, r_1 and r_2. We used here that by hypothesis we have $\|H\|_{r_1,r_2} > 0$.

For $n \in \{1, 2, ..., n_0-1\}$ we reason exactly as for the complex univariate Bernstein polynomials in the proof of Theorem 1.1.4. $\square$

Combining now Theorem 2.3.2 with Theorem 2.3.3 we immediately obtain the following.

Corollary 2.3.4. *Suppose $2 < R_1 < \infty$, $2 < R_2 < \infty$ and that $f : ([R_1, +\infty) \cup \overline{\mathbb{D}}_{R_1}) \times ([R_2, +\infty) \cup \overline{\mathbb{D}}_{R_2}) \to \mathbb{C}$ is bounded on $[R_1, \infty) \times [R_2, \infty)$ and analytic in $\mathbb{D}_{R_1} \times \mathbb{D}_{R_2}$, i.e $f(z_1, z_2) = \sum_{k=0}^{\infty} \sum_{j=0}^{\infty} c_{k,j} z_1^k z_2^j$, for all $|z_1| < R_1$, $|z_2| < R_2$.*

If the Taylor series of f contains at least one term of the form $c_{k,0} z_1^k$ with $c_{k,0} \neq 0$ and $k \geq 2$ or of the form $c_{0,j} z_2^j$ with $c_{0,j} \neq 0$ and $j \geq 2$, then we have

$$\|S_{n,n}(f) - f\|_{r_1, r_2} \sim \frac{1}{n}, \text{ for all } n \in \mathbb{N}.$$

Proof. It suffices to prove that under the hypothesis on f, it cannot be a solution of the complex partial differential equation

$$z_1 \cdot \frac{\partial^2 f}{\partial z_1^2}(z_1, z_2) + z_2 \cdot \frac{\partial^2 f}{\partial z_2^2}(z_1, z_2) = 0, |z_1| < R_1, |z_2| < R_2.$$

Indeed, suppose the contrary. Since simple calculation give

$$z_1 \cdot \frac{\partial^2 f}{\partial z_1^2} = \sum_{k=0}^{\infty} \sum_{j=0}^{\infty} c_{k,j} k(k-1) z_1^{k-1} z_2^j = \sum_{k=1}^{\infty} \sum_{j=0}^{\infty} c_{k+1,j} k(k+1) z_1^k z_2^j$$

$$= \sum_{k=1}^{\infty} c_{k+1,0} k(k+1) z_1^k + \sum_{k=1}^{\infty} \sum_{j=1}^{\infty} c_{k+1,j} k(k+1) z_1^k z_2^j,$$

and

$$z_2 \cdot \frac{\partial^2 f}{\partial z_2^2} = \sum_{j=1}^{\infty} c_{0,j+1} j(j+1) z_2^j + \sum_{k=1}^{\infty} \sum_{j=1}^{\infty} c_{k+1,j} j(j+1) z_1^k z_2^j,$$

by summing and equaling with zero, by the identification of coefficients, we immediately obtain that $c_{0,k} = c_{j,0} = 0$ for all $j = 2, 3, ..., k = 2, 3, ...$, which contradicts the hypothesis on f.

Therefore the hypothesis and the lower estimate in Theorem 2.3.3 holds, which ends the proof. □

2.4 Baskakov Operators

In this section we consider the bivariate complex Baskakov operators of tensor product kind given by

$$V_{n,m}(f)(z_1, z_2) = \sum_{k=0}^{\infty} \sum_{j=0}^{\infty} \frac{n(n+1)...(n+k-1)}{n^k} \cdot \frac{m(m+1)...(m+j-1)}{m^j}$$

$$\cdot \, [0, 1/n, ..., k/n; [0, 1/n, ..., j/m; f(\cdot, \cdot)]_{z_2}]_{z_1} z_1^k z_2^j,$$

where $f : ([R_1, +\infty) \cup \overline{\mathbb{D}}_{R_1}) \times ([R_2, +\infty) \cup \overline{\mathbb{D}}_{R_2}) \to \mathbb{C}$ has all the partial derivatives bounded in $[0, \infty) \times [0, \infty)$, by the same constant, is supposed to be analytic in $\mathbb{D}_{R_1} \times \mathbb{D}_{R_2}$ and satisfies some exponential growth conditions.

Note that in the case of two real variables, these operators were first considered by Stancu [184].

We will use similar methods with those for complex bivariate Bernstein polynomials of tensor product kind in Section 2.2. The results in univariate case obtained for complex Baskakov operators in Chapter 1 will be useful.

Theorem 2.4.1. *Let $n_0, m_0 \in \mathbb{N}$ and $3 \leq n_0 < 2R_1 < \infty$, $3 \leq m_0 < 2R_2 < \infty$. Suppose that $f : ([R_1, +\infty) \cup \overline{\mathbb{D}}_{R_1}) \times ([R_2, +\infty) \cup \overline{\mathbb{D}}_{R_2}) \to \mathbb{C}$ has all the partial derivatives bounded in $[0, \infty) \times [0, \infty)$ by the same constant, f is supposed to be analytic in $\mathbb{D}_{R_1} \times \mathbb{D}_{R_2}$, i.e. $f(z_1, z_2) = \sum_{k=0}^{\infty} \sum_{j=0}^{\infty} c_{k,j} z_1^k z_2^j$, for all $|z_1| < R_1$, $|z_2| < R_2$ and there exist the constants $M > 0$, $A_i \in \left(\frac{1}{R_i}, 1 \right)$, $i = 1, 2$, with the property $|c_{k,j}| \leq M \frac{A_1^k A_2^j}{k! j!}$, for all $k, j = 0, 1, 2, ...$, (which implies $|f(z_1, z_2)| \leq M e^{A_1 |z_1| + A_2 |z_2|}$, for all $|z_1| < R_1$, $|z_2| < R_2$).*

(i) Let $1 \leq r_1 < \min\{\frac{n_0}{2}, \frac{1}{A_1}\}$, $1 \leq r_2 < \min\{\frac{m_0}{2}, \frac{1}{A_2}\}$ be arbitrary fixed. For all $|z_1| \leq r_1$, $|z_2| \leq r_2$, $n > n_0$ and $m > m_0$ we have

$$|V_{n,m}(f)(z_1, z_2) - f(z_1, z_2)| \leq C_{r_1, r_2, n, m}(f),$$

where

$$C_{r_1, r_2, n, m}(f) = \frac{6M}{n} \sum_{k=0}^{\infty} \sum_{j=0}^{\infty} (r_1 A_1)^k \frac{(r_2 A_2)^j}{j!} (k+1)(k-1)$$

$$+ \frac{6M}{m} \sum_{k=0}^{\infty} \sum_{j=0}^{\infty} (r_1 A_1)^k (r_2 A_2)^j (j+1)(j-1).$$

(ii) Let $k_1, k_2 \in \mathbb{N}$ be with $k_1 + k_2 \geq 1$ and $1 \leq r_1 < r_1^* < \min\{\frac{n_0}{2}, \frac{1}{A_1}\}$, $1 \leq r_2 < r_2^* < \min\{\frac{m_0}{2}, \frac{1}{A_2}\}$ be arbitrary fixed.
Then for all $|z_1| \leq r_1$, $|z_2| \leq r_2$, $n > n_0$ and $m > m_0$ we have

$$\left| \frac{\partial^{k_1 + k_2} V_{n,m}(f)}{\partial z_1^{k_1} \partial z_2^{k_2}}(z_1, z_2) - \frac{\partial^{k_1 + k_2} f}{\partial z_1^{k_1} \partial z_2^{k_2}}(z_1, z_2) \right|$$

$$\leq C_{r_1^*, r_2^*, n, m}(f) \cdot \frac{(k_1)!}{(r_1^* - r_1)^{k_1 + 1}} \cdot \frac{(k_2)!}{(r_2^* - r_2)^{k_2 + 1}},$$

where $C_{r_1^*, r_2^*, n, m}(f)$ is given as at the above point (i).

Proof.   (i) Denote $e_{k,j}(z_1, z_2) = e_k(z_1) \cdot e_j(z_2)$, where $e_k(u) = u^k$. Clearly we get

$$|V_{n,m}(f)(z_1, z_2) - f(z_1, z_2)| \leq \sum_{k=0}^{\infty} \sum_{j=0}^{\infty} |c_{k,j}| \cdot |V_{n,m}(e_{k,j})(z_1, z_2) - e_{k,j}(z_1, z_2)|.$$

But taking into account the estimates in the proof of Theorem 1.9.1, (i), for all $|z_1| \leq r_1$ and $|z_2| \leq r_2$ we obtain

$$|V_{n,m}(e_{k,j})(z_1, z_2) - e_{k,j}(z_1, z_2)| = |V_n(e_k)(z_1) \cdot V_m(e_j)(z_2) - z_1^k \cdot z_2^j|$$

$$\leq |V_n(e_k)(z_1) \cdot V_m(e_j)(z_2) - V_n(e_k)(z_1) z_2^j| + |V_n(e_k)(z_1) z_2^j - z_1^k z_2^j|$$

$$\leq |V_n(e_k)(z_1)| \cdot |V_m(e_j)(z_2) - z_2^j| + |z_2^j| \cdot |V_n(e_k)(z_1) - z_1^k|$$

$$\leq r_1^k (k+1)! \frac{6 r_2^j (j+1)!(j-1)}{m} + r_2^j \frac{6 r_1^k (k+1)!(k-1)}{n},$$

which from the conditions on the coefficients $c_{k,j}$ implies

$$|V_{n,m}(f)(z_1, z_2) - f(z_1, z_2)| \leq \sum_{k=0}^{\infty} \sum_{j=0}^{\infty} |c_{k,j}| \cdot |V_{n,m}(e_{k,j})(z_1, z_2) - e_{k,j}(z_1, z_2)|$$

$$\leq \frac{6M}{n} \sum_{k=0}^{\infty} \sum_{j=0}^{\infty} (r_1 A_1)^k \frac{(r_2 A_2)^j}{j!} (k+1)(k-1)$$

$$+ \frac{6M}{m} \sum_{k=0}^{\infty} \sum_{j=0}^{\infty} (r_1 A_1)^k (r_2 A_2)^j (j+1)(j-1),$$

which proves (i).

(ii) Let $1 \leq r_1 < r_1^* < R_1$, $1 \leq r_2 < r_2^* < R_2$. By the Cauchy's formula in Theorem 2.1.2, (ii), we get

$$\frac{\partial^{k_1+k_2} V_{n,m}(f)}{\partial z_1^{k_1} \partial z_2^{k_2}}(z_1, z_2) - \frac{\partial^{k_1+k_2} f}{\partial z_1^{k_1} \partial z_2^{k_2}}(z_1, z_2)$$

$$= \frac{(k_1!)(k_2!)}{(2\pi i)^2} \int_{|u_2 - z_2| = r_2^*} \int_{|u_1 - z_1| = r_1^*} \frac{[V_{n,m}(u_1, u_2) - f(u_1, u_2)]du_1 du_2}{(u_1 - z_1)^{k_1+1}(u_2 - z_2)^{k_2+1}}.$$

Passing to absolute value with $|z_1| \leq r_1$, $|z_2| \leq r_2$ and taking into account that $|u_1 - z_1| \geq r_1^* - r_1$, $|u_2 - z_2| \geq r_2^* - r_2$, by applying the estimate in (i) we easily obtain

$$\left| \frac{\partial^{k_1+k_2} V_{n,m}(f)}{\partial z_1^{k_1} \partial z_2^{k_2}}(z_1, z_2) - \frac{\partial^{k_1+k_2} f}{\partial z_1^{k_1} \partial z_2^{k_2}}(z_1, z_2) \right|$$

$$\leq C_{r_1^*, r_2^*, n, m}(f) \cdot \frac{(k_1)!}{(r_1^* - r_1)^{k_1+1}} \cdot \frac{(k_2)!}{(r_2^* - r_2)^{k_2+1}},$$

which proves the theorem. $\qquad \square$

In what follows a Voronovskaja-type formula for $V_{n,m}(f)$ is presented. It will be the product of the parametric extensions generated by the Voronovskaja's formula in univariate case in Theorem 1.9.3. Indeed, for $f(z_1, z_2)$ defining the parametric extensions of the Voronovskaja's formula by

$$_{z_1}L_n(f)(z_1, z_2) := V_n(f(\cdot, z_2))(z_1) - f(z_1, z_2) - \frac{z_1(1+z_1)}{2n} \cdot \frac{\partial^2 f}{\partial z_1^2}(z_1, z_2),$$

$$_{z_2}L_m(f)(z_1, z_2) := V_m(f(z_1, \cdot))(z_2) - f(z_1, z_2) - \frac{z_2(1+z_2)}{2m} \cdot \frac{\partial^2 f}{\partial z_2^2}(z_1, z_2),$$

their product (composition) give

$$_{z_2}L_m(f)(z_1, z_2) \circ _{z_1}L_n(f)(z_1, z_2)$$

$$= V_m \left[V_n(f(\cdot, \cdot))(z_1) - f(z_1, \cdot) - \frac{z_1(1+z_1)}{2n} \cdot \frac{\partial^2 f}{\partial z_1^2}(z_1, \cdot) \right](z_2)$$

$$- \left[V_n(f(\cdot, z_2))(z_1) - f(z_1, z_2) - \frac{z_1(1+z_1)}{2n} \cdot \frac{\partial^2 f}{\partial z_1^2}(z_1, z_2) \right] - \frac{z_2(1+z_2)}{2m}$$

$$\cdot \left[V_n(\frac{\partial^2 f}{\partial z_2^2}(\cdot, z_2))(z_1) - \frac{\partial^2 f}{\partial z_2^2}(z_1, z_2) - \frac{z_1(1+z_1)}{2n} \cdot \frac{\partial^2}{\partial z_1^2} \left[\frac{\partial^2 f}{\partial z_2^2} \right](z_1, z_2) \right]$$

$$:= E_1 - E_2 - E_3.$$

After simple calculation evidently that we can write

$$_{z_2}L_m(f)(z_1, z_2) \circ {}_{z_1}L_n(f)(z_1, z_2)$$

$$= V_{n,m}(f)(z_1, z_2) - V_m(f(z_1, \cdot))(z_2) - \frac{z_1(1 + z_1)}{2n} \cdot V_m\left(\frac{\partial^2 f}{\partial z_1^2}(z_1, \cdot)\right)(z_2)$$

$$- V_n(f(\cdot, z_2))(z_1) + f(z_1, z_2) + \frac{z_1(1 + z_1)}{2n} \frac{\partial^2 f}{\partial z_1^2}(z_1, z_2)$$

$$- \frac{z_2(1 + z_2)}{2m} \cdot V_n\left(\frac{\partial^2 f}{\partial z_2^2}(\cdot, z_2)\right)(z_1) + \frac{z_2(1 + z_2)}{2m} \cdot \frac{\partial^2 f}{\partial z_2^2}(z_1, z_2)$$

$$+ \frac{z_1(1 + z_1)}{2n} \cdot \frac{z_2(1 + z_2)}{2m} \cdot \frac{\partial^4 f}{\partial z_1^2 \partial z_2^2}(z_1, z_2),$$

from which immediately can be derived the commutativity property

$$_{z_2}L_m(f)(z_1, z_2) \circ {}_{z_1}L_n(f)(z_1, z_2) = {}_{z_1}L_n(f)(z_1, z_2) \circ {}_{z_2}L_m(f)(z_1, z_2).$$

The Voronovskaja's result can be stated as follows.

Theorem 2.4.2. *Suppose that the hypothesis on the function f and the constants n_0, m_0, R_1, R_2, M, A_1, A_2 in the statement of Theorem 2.4.1 hold. Let $1 \le r_1 < \min\{\frac{n_0}{2}, \frac{1}{A_1}\}$, $1 \le r_2 < \min\{\frac{m_0}{2}, \frac{1}{A_2}\}$ be arbitrary fixed. Then for all $|z_1| \le r_1$, $|z_2| \le r_2$, $n > n_0$ and $m > m_0$ we have*

$$|_{z_2}L_m(f)(z_1, z_2) \circ {}_{z_1}L_n(f)(z_1, z_2)| \le C_{r_1, r_2}(f)\left[\frac{1}{n^2} + \frac{1}{m^2}\right],$$

where

$$C_{r_1, r_2}(f) = 16M \sum_{k=2}^{\infty} \sum_{j=0}^{\infty} (r_1 A_1)^k (r_2 A_2)^j (j + 1)(k - 1)(k - 2)^2.$$

Proof. First by hypothesis we can write $f(z_1, z_2) = \sum_{k=0}^{\infty} f_k(z_2) z_1^k$, where $f_k(z_2) = \sum_{j=0}^{\infty} c_{k,j} z_2^j$. It follows $\frac{\partial^2 f}{\partial z_1^2}(z_1, z_2) = \sum_{k=2}^{\infty} f_k(z_2) k(k - 1) z_1^{k-2}$ and $\frac{\partial^2 f}{\partial z_2^2}(z_1, z_2) = \sum_{k=0}^{\infty} \frac{\partial^2 f_k}{\partial z_2^2}(z_2) z_1^k$, where $\frac{\partial^2 f_k}{\partial z_2^2}(z_2) = \sum_{j=2}^{\infty} c_{k,j} j(j - 1) z_2^{j-2}$. This implies $V_n(f(\cdot, z_2))(z_1) = \sum_{k=0}^{\infty} f_k(z_2) V_n(e_1^k)(z_1)$ and

$$V_n(f(\cdot, z_2))(z_1) - f(z_1, z_2) - \frac{z_1(1 + z_1)}{2n} \cdot \frac{\partial^2 f}{\partial z_1^2}(z_1, z_2)$$

$$= \sum_{k=2}^{\infty} f_k(z_2)\left[V_n(e_1^k)(z_1) - e_1^k(z_1) - \frac{z_1^{k-1}(1 + z_1)k(k - 1)}{2n}\right].$$

Applying now V_m to the last expression with respect to z_2, we obtain

$$E_1 = \sum_{k=2}^{\infty} V_m(f_k)(z_2)\left[V_n(e_1^k)(z_1) - e_1^k(z_1) - \frac{z_1^{k-1}(1 + z_1)k(k - 1)}{2n}\right]$$

$$= \sum_{k=2}^{\infty}\left(\sum_{j=0}^{\infty} c_{k,j} V_m(e_1^j)(z_2)\right)\left[V_n(e_1^k)(z_1) - e_1^k(z_1) - \frac{z_1^{k-1}(1 + z_1)k(k - 1)}{2n}\right].$$

Passing now to absolute value with $|z_1| \leq r_1$ and $|z_2| \leq r_2$ and taking into account the estimates in the proofs of Theorem 1.9.1, (i) and Theorem 1.9.3, it follows

$$|E_1| \leq \sum_{k=2}^{\infty} \sum_{j=0}^{\infty} |c_{k,j}| r_2^j (j+1)! \left[\frac{16 r_1^k k! (k-1)(k-2)^2}{n^2}\right]$$

$$\leq \frac{16M}{n^2} \sum_{k=2}^{\infty} \sum_{j=0}^{\infty} [(r_1 A_1)^k (r_2 A_2)^j (j+1)(k-1)(k-2)^2].$$

Similarly,

$$|E_2| \leq \sum_{k=2}^{\infty} |f_k(z_2)| \cdot \left[\frac{16 r_1^k k! (k-1)(k-2)^2}{n^2}\right]$$

$$\leq \frac{16}{n^2} \sum_{k=2}^{\infty} \sum_{j=0}^{\infty} |c_{k,j}| r_2^j \left[r_1^k k! (k-1)(k-2)^2\right]$$

$$\leq \frac{16M}{n^2} \sum_{k=2}^{\infty} \sum_{j=0}^{\infty} (r_1 A_1)^k \frac{(r_2 A_2)^j}{j!} (k-1)(k-2)^2.$$

Then

$$V_n \left(\frac{\partial^2 f}{\partial z_2^2}(\cdot, z_2)\right)(z_1)$$

$$= \sum_{k=0}^{\infty} \frac{\partial^2 f_k}{\partial z_2^2}(z_2) V_n(e_1^k)(z_1) = \sum_{k=0}^{\infty} \sum_{j=2}^{\infty} c_{k,j} j(j-1) z_2^{j-2} V_n(e_1^k)(z_1),$$

and

$$\left[V_n(\frac{\partial^2 f}{\partial z_2^2}(\cdot, z_2))(z_1) - \frac{\partial^2 f}{\partial z_2^2}(z_1, z_2) - \frac{z_1(1+z_1)}{2n} \cdot \frac{\partial^2}{\partial z_1^2}\left[\frac{\partial^2 f}{\partial z_2^2}\right](z_1, z_2)\right]$$

$$= \sum_{k=2}^{\infty} \sum_{j=2}^{\infty} c_{k,j} j(j-1) z_2^{j-2} \left[V_n(e_1^k)(z_1) - e_1^k(z_1) - \frac{z_1^{k-1}(1+z_1)k(k-1)}{2n}\right],$$

which again by Theorem 1.9.3, implies

$$|E_3| \leq \frac{r_2(1+r_2)}{2m} \sum_{k=2}^{\infty} \sum_{j=2}^{\infty} |c_{k,j}| j(j-1) r_2^{j-2} \cdot \left[\frac{16 r_1^k k! (k-1)(k-2)^2}{n^2}\right]$$

$$\leq \frac{8(1+r_2)M}{r_2 n^2 m} \sum_{k=2}^{\infty} \sum_{j=2}^{\infty} (r_1 A_1)^k \frac{(r_2 A_2)^j}{(j-2)!} (k-1)(k-2)^2.$$

Note that if we estimate now $|_{z_1} L_n(f)(z_1, z_2) \circ _{z_2} L_m(f)(z_1, z_2)|$, then by reasons of symmetry we get a similar order of approximation, simply interchanging above the places of n and m.

In conclusion,

$$|_{z_2} L_m(f)(z_1, z_2) \circ _{z_1} L_n(f)(z_1, z_2)| \leq |E_1| + |E_2| + |E_3| \leq C_{r_1, r_2}(f) \left[\frac{1}{n^2} + \frac{1}{m^2}\right],$$

with $C_{r_1, r_2}(f)$ given by the statement. $\qquad\square$

Theorem 2.4.2 will be used to find the exact order in approximation by $V_{n,n}(f)$. In this sense we have the following.

Theorem 2.4.3. *Suppose that $n_0 = m_0$ and that the hypothesis on the function f and the constants n_0, m_0, R_1, R_2, M, A_1, A_2 in the statement of Theorem 2.4.1 hold. Let $1 \le r_1 < \min\{\frac{n_0}{2}, \frac{1}{A_1}\}$, $1 \le r_2 < \min\{\frac{n_0}{2}, \frac{1}{A_2}\}$ be arbitrary fixed. Denoting $\|f\|_{r_1,r_2} = \sup\{|f(z_1, z_2)|; |z_1| \le r_1; |z_2| \le r_2\}$, if f is not a solution of the complex partial differential equation*

$$z_1(1+z_1) \cdot \frac{\partial^2 f}{\partial z_1^2}(z_1, z_2) + z_2(1+z_2) \cdot \frac{\partial^2 f}{\partial z_2^2}(z_1, z_2) = 0, |z_1| < R_1, |z_2| < |R_2|,$$

then for all $n > n_0$ we have

$$\|V_{n,n}(f) - f\|_{r_1,r_2} \ge \frac{K_{r_1,r_2,f}}{n},$$

where $K_{r_1,r_2,f}$ is independent on n and m.

Proof. We can write

$$V_{n,n}(f)(z_1, z_2) - f(z_1, z_2)$$
$$= \frac{2}{n} \left\{ \frac{z_1(1+z_1)}{4} \cdot \frac{\partial^2 f}{\partial z_1^2}(z_1, z_2) + \frac{z_2(1+z_2)}{4} \cdot \frac{\partial^2 f}{\partial z_2^2}(z_1, z_2) \right.$$
$$\left. + \frac{2}{n} \left[\frac{n^2}{4} \left(_{z_2}L_n(f) \circ \,_{z_1}L_n(f)\right)(z_1, z_2) \right] + R_n(f)(z_1, z_2) \right\},$$

where

$$R_n(f)(z_1, z_2)$$
$$= \frac{n}{2} \left(V_n(f(z_1, \cdot))(z_2) - f(z_1, z_2) - \frac{z_2(1+z_2)}{2n} \cdot \frac{\partial^2 f}{\partial z_2^2}(z_1, z_2) \right)$$
$$+ \frac{n}{2} \left(V_n(f(\cdot, z_2))(z_1) - f(z_1, z_2) - \frac{z_1(1+z_1)}{2n} \cdot \frac{\partial^2 f}{\partial z_1^2}(z_1, z_2) \right)$$
$$+ \frac{z_1(1+z_1)}{4} \left(V_n \left(\frac{\partial^2 f}{\partial z_1^2}(z_1, \cdot) \right)(z_2) - \frac{\partial^2 f}{\partial z_1^2}(z_1, z_2) \right)$$
$$+ \frac{z_2(1+z_2)}{4} \left(V_n \left(\frac{\partial^2 f}{\partial z_2^2}(\cdot, z_2) \right)(z_1) - \frac{\partial^2 f}{\partial z_2^2}(z_1, z_2) \right)$$
$$- \frac{z_1(1+z_1)z_2(1+z_2)}{8n} \cdot \frac{\partial^4 f}{\partial z_1^2 \partial z_2^2}(z_1, z_2).$$

By using Theorem 1.9.3 and the reasonings in the above Theorem 2.4.2, it is immediate that $\|R_n(f)\|_{r_1,r_2} \to 0$ as $n \to \infty$. Also, by Theorem 2.4.2 we obtain

$$\frac{n^2}{4} \|(_{z_2}L_m(f) \circ \,_{z_1}L_n(f))\|_{r_1,r_2} \le \frac{C_{r_1,r_2}(f)}{2},$$

which implies

$$\left\| \frac{2}{n} \left[\frac{n^2}{4} \left(_{z_2}L_n(f) \circ \,_{z_1}L_n(f)\right) \right] + R_n(f) \right\|_{r_1,r_2} \to 0, \text{ as } n \to \infty.$$

Denoting $H(z_1, z_2) = \frac{z_1(1+z_1)}{4} \cdot \frac{\partial^2 f}{\partial z_1^2}(z_1, z_2) + \frac{z_2(1+z_2)}{4} \cdot \frac{\partial^2 f}{\partial z_2^2}(z_1, z_2)$ and taking into account the inequalities

$$\|F + G\|_{r_1, r_2} \geq |\ \|F\|_{r_1, r_2} - \|G\|_{r_1, r_2}\ | \geq \|F\|_{r_1, r_2} - \|G\|_{r_1, r_2},$$

it follows

$$\|V_{n,n}(f) - f\|_{r_1, r_2}$$

$$\geq \frac{2}{n} \left\{ \|H\|_{r_1, r_2} - \left\| \frac{2}{n} \left[\frac{n^2}{4} \left({}_{z_2}L_n(f) \circ {}_{z_1}L_n(f) \right) \right] + R_n(f) \right\|_{r_1, r_2} \right\}$$

$$\geq \frac{2}{n} \cdot \frac{1}{2} \|H\|_{r_1, r_2} = \frac{1}{n} \cdot \|H\|_{r_1, r_2},$$

for all $n \geq n_0$, with n_0 depending only on f, r_1 and r_2. We used here that by hypothesis we have $\|H\|_{r_1, r_2} > 0$.

For $n \in \{1, 2, ..., n_0 - 1\}$ we reason exactly as for the complex univariate Bernstein polynomials in the proof of Theorem 1.1.4. $\qquad \square$

Combining Theorem 2.4.2 with Theorem 2.4.3 we immediately obtain the following.

Corollary 2.4.4. *Suppose that the hypothesis in the statement of Theorem 2.4.3 hold. If the Taylor series of f contains at least one term of the form $c_{k,0} z_1^k$ with $c_{k,0} \neq 0$ and $k \geq 2$ or of the form $c_{0,j} z_2^j$ with $c_{0,j} \neq 0$ and $j \geq 2$, then we have*

$$\|V_{n,n}(f) - f\|_{r_1, r_2} \sim \frac{1}{n}, \text{ for all } n > n_0.$$

Proof. It suffices to prove that under the hypothesis on f, it cannot be a solution of the complex partial differential equation

$$z_1(1 + z_1) \cdot \frac{\partial^2 f}{\partial z_1^2}(z_1, z_2) + z_2(1 + z_2) \cdot \frac{\partial^2 f}{\partial z_2^2}(z_1, z_2) = 0, |z_1| < R_1, |z_2| < R_2.$$

Indeed, suppose the contrary. Since simple calculation give

$$z_1(1 + z_1) \cdot \frac{\partial^2 f}{\partial z_1^2} = \sum_{j=0}^{\infty} 2c_{2,j} z_1 z_2^j + \sum_{k=2}^{\infty} \sum_{j=0}^{\infty} z_1^k z_2^j [c_{k+1,j} k(k+1) + c_{k,j} k(k-1)]$$

and

$$z_2(1 + z_2) \cdot \frac{\partial^2 f}{\partial z_2^2} = \sum_{k=0}^{\infty} 2c_{k,2} z_1^k z_2 + \sum_{k=0}^{\infty} \sum_{j=2}^{\infty} z_1^k z_2^j [c_{k,j+1} j(j+1) + c_{k,j} j(j-1)]$$

by summing and equaling with zero, we obtain

$$2c_{0,2} z_2 + 2c_{1,2} z_1 z_2 + \sum_{k=2}^{\infty} z_1^k z_2 [c_{k+1,j} k(k+1) + c_{k,1} k(k-1) + 2c_{k,2}]$$

$$+ \sum_{j=2}^{\infty} z_2^j [c_{0,j+1} j(j+1) + c_{0,j} j(j-1)] + \sum_{j=2}^{\infty} z_1 z_2^j [c_{1,j+1} j(j+1) + c_{1,j} j(j-1) + 2c_{2,j}]$$

$$+ 2c_{2,0}z_1 + 2c_{2,1}z_1z_2 + \sum_{k=2}^{\infty} z_1^k[c_{k+1,0}k(k+1) + c_{k,0}k(k-1)]$$

$$+ \sum_{k=2}^{\infty}\sum_{j=2}^{\infty} z_1^k z_2^j[c_{k+1,j}k(k+1) + c_{k,j}k(k-1) + c_{k,j+1}j(j+1) + c_{k,j}j(j-1)] = 0.$$

By the identification of coefficients, among others we immediately get $c_{0,2} = c_{2,0} = 0$ and from the terms under the second and fourth sign Σ, it follows

$$c_{0,j+1}j(j+1) + c_{0,j}j(j-1) = 0, \ c_{k+1,0}k(k+1) + c_{k,0}k(k-1) = 0,$$

for all $j = 2, 3, ..., k = 2, 3, ...,$. From here by recurrence it is easy to deduce that $c_{0,k} = c_{j,0} = 0$ for all $j = 2, 3, ..., k = 2, 3, ...$, which contradicts the hypothesis on f. Therefore the hypothesis and the lower estimate in Theorem 2.4.3 holds, which ends the proof. $\qquad\square$

2.5 Bibliographical Notes and Open Problems

All the results in Sections 2.2, 2.3 and 2.4 are new and appear for the first time here.

Note 2.5.1. From a long list, some references concerning Bernstein-type operators of two or several real variables, different from those considered in this Chapter 2, for which would be possible to develop similar results, are : Stancu [177; 178; 179; 180; 181; 182], Stancu-Vernescu [183], Vlaic [196].

Open Problem 2.5.2. Suppose that $f : P(0;R) \to \mathbb{C}$ is analytic in $P(0;R) = \mathbb{D}_{R_1} \times \mathbb{D}_{R_2}$, that is $f(z_1, z_2) = \sum_{k=0}^{\infty}\sum_{j=0}^{\infty} c_{k,j}z_1^k z_2^j$, for all $(z_1, z_2) \in P(0;R)$, $R = (R_1, R_2)$. Prove the following Voronovskaja-type formula for

$$B_n(f)(z_1, z_2) = \sum_{k=0}^{n}\sum_{j=0}^{n-k} \binom{n}{k}\binom{n-k}{j} z_1^k z_2^j (1 - z_1 - z_2)^{n-k-j} f(k/n, j/n) :$$

for all $|z_1| \le r_1, |z_2| \le r_2$ with $1 \le r_1 < R_1, 1 \le r_2 < R_2$ and $n \in \mathbb{N}$ we have

$$\left| B_n(f)(z_1, z_2) - f(z_1, z_2) - \frac{z_1(1 - z_1)}{2n} \cdot \frac{\partial^2 f}{\partial z_1^2}(z_1, z_2) + \frac{z_1 z_2}{n} \cdot \frac{\partial^2 f}{\partial z_1 \partial z_2}(z_1, z_2) \right.$$
$$\left. - \frac{z_2(1 - z_2)}{2n} \cdot \frac{\partial^2 f}{\partial z_2^2}(z_1, z_2) \right| \le \frac{K_{r_1, r_2}(f)}{n^2}.$$

Chapter 3

Complex Convolutions

This chapter deals with approximation of complex analytic functions by linear and non-linear complex convolutions, of polynomial or of non-polynomial kind.

3.1 Linear Polynomial Convolutions

It is known that upper estimates in terms of the moduli of smoothness of various orders in approximation in the unit disk by the complex convolution polynomials of de la Vallée Poussin, Fejér, Riesz-Zygmund, Jackson and Beatson were obtained in Gaier [75] (see also Gaier [76], p. 53) and Gal [93]. Since most of these details can be found in the recent book Gal [77], we will not reproduce here the proofs of the above mentioned results and the geometric properties of these complex convolutions.

Also, the saturation order of approximation for the complex Fejér, Riesz-Zygmund and Rogosinski convolutions even in more general Jordain domains in $\mathbb{C}$ were obtained in e.g. Bruj-Schmieder [48].

In this section first we present upper estimates (of the same order as those given by the moduli of smoothness) in approximation in compact disks with explicit constants depending on the coefficients of Taylor's series of the approximated function f for the complex convolution polynomials of de la Vallée Poussin, Fejér, Riesz-Zygmund, Jackson and Beatson. Then, in addition, as new results we prove Voronovskaja-type formulas with quantitative estimates which allow us to derive the exact orders of approximation not only for the convolutions but also for their derivatives of any order in compact disks.

Therefore, as in the case of complex Bernstein-type operators in Chapter 1, the exact degrees of approximation for the above mentioned complex convolution polynomials will be obtained by three steps : 1) upper estimates ; 2) quantitative Voronovskaja's theorems ; 3) lower estimates by using step 2. An exception is the case of complex convolutions of Beatson-type when the exact order of approximation is obtained by a different method.

The classical de la Vallée Poussin convolution polynomials of real variable attached to a 2π-periodic function g are defined by (see e.g. Butzer-Nessel [53], pp. 299-300)

$$P_n^*(g)(x) = \frac{1}{2\pi} \int_{-\pi}^{\pi} g(t) K_n(x-t) dt, x \in \mathbb{R},$$

where the kernel $K_n(u)$ is given by

$$K_n(u) = \frac{(n!)^2}{(2n)!} \left(2 \cos \frac{u}{2}\right)^{2n} = \sum_{j=-n}^{n} \frac{(n!)^2}{(n-j)!(n+j)!} e^{iju}, i^2 = -1, u \in \mathbb{R}.$$

Their complex form attached to an analytic function f in a disk $\mathbb{D}_R = \{z \in \mathbb{C},$ $|z| < R\}$ and given by

$$P_n(f)(z) = \frac{1}{2\pi} \int_{-\pi}^{\pi} f(ze^{iu}) K_n(u) du = \frac{1}{\binom{2n}{n}} \sum_{j=0}^{n} c_j \binom{2n}{n+j} z^j$$

$$= \sum_{j=0}^{n} c_j \frac{(n!)^2}{(n-j)!(n+j)!} z^j,$$

have nice shape-preserving properties, that is preserve the starlikeness and convexity of f in the unit disk, for all $n \in \mathbb{N}$ (see Pólya-Schoenberg [150]). For this reason the de la Vallée Poussin convolutions are considered as the "trigonometric" analogues of the classical Bernstein polynomials. Also, in Gal [93] as an approximation order of f by $P_n(f)$ was found $\omega_1(f; 1/\sqrt{n})_{\partial \mathbb{D}_1}$. But probably it would be possible to obtain the approximation order in terms of $\omega_2(f; 1/\sqrt{n})_{\partial \mathbb{D}_1}$.

Firstly, below upper estimates in approximation by $P_n(f)(z)$ with explicit constants depending on the coefficients of Taylor's series of the approximated function $f(z)$ are presented.

Theorem 3.1.1. *Let $R > 1$, $\mathbb{D}_R = \{z \in \mathbb{C}; |z| < R\}$ and let us suppose that $f : \mathbb{D}_R \to \mathbb{C}$ is analytic in $\mathbb{D}_R$, that is we can write $f(z) = \sum_{k=0}^{\infty} c_k z^k$, for all $z \in \mathbb{D}_R$.*

(i) For any $r \in [1, R)$ we have

$$\|P_n(f) - f\|_r \leq \frac{M_r(f)}{n}, n \in \mathbb{N},$$

where $\|f\|_r = \sup\{|f(z)|; |z| \leq r\}$ and $M_r(f) = \sum_{k=1}^{\infty} |c_k| k^2 r^k < \infty$.

(ii) If $1 \leq r < r_1 < R$ and $p \in \mathbb{N}$ then we have

$$\|P_n^{(p)}(f) - f^{(p)}\|_r \leq \frac{r_1 p! M_{r_1}(f)}{(r_1 - r)^{p+1} n}, n \in \mathbb{N}.$$

Proof. (i) Denote $e_k(z) = z^k$. Since $P_n(e_0)(z) = 1$ and obviously we have $P_n(f)(z) = \sum_{k=0}^{\infty} c_k P_n(e_k)(z)$, it follows that $\|P_n(f)(z) - f(z)\|_r \leq \sum_{k=1}^{\infty} |c_k| \cdot \|P_n(e_k)(z) - e_k(z)\|_r$.

But by the formula of $P_n(f)(z)$ in Introduction, since we can write $e_k(z) = \sum_{j=0}^{\infty} c_j z^j$ with $c_k = 1$ and $c_j = 0$ for all $j \neq k$, it is immediate that $P_n(e_k)(z) = 0$ if $k > n$ and

$$P_n(e_k)(z) = \frac{(n!)^2}{(n-k)!(n+k)!} e_k(z) = e_k(z) \prod_{j=1}^{k} \left(1 - \frac{k}{n+j}\right), \text{ if } k \leq n.$$

This immediately implies that $\|P_n(e_k) - e_k\|_r \leq r^k \leq \frac{k}{n}r^k \leq \frac{k^2}{n}r^k$ if $k > n$, while for $k \leq n$ we get

$$\|P_n(e_k)(z) - e_k\|_r \leq \left| 1 - \prod_{j=1}^{k} \left(1 - \frac{k}{n+j} \right) \right| r^k$$

$$\leq \sum_{j=1}^{k} \left[1 - \left(1 - \frac{k}{n+j} \right) \right] r^k$$

$$\leq k r^k \sum_{j=1}^{k} \frac{1}{n+j} \leq \frac{k^2 r^k}{n+1} \leq \frac{k^2 r^k}{n}.$$

Here we have applied the inequality $1 - \Pi_{i=1}^{k} x_i \leq \sum_{i=1}^{k}(1 - x_i)$, valid for $0 \leq x_i \leq 1$, $i = 1, ..., k$.

In conclusion we have

$$\|P_n(e_k) - e_k\|_r \leq \frac{k^2}{n}r^k, \text{ for all } k, n \in \mathbb{N},$$

which implies the estimate in (i).

(ii) Denoting by γ the circle of radius $r_1 > r$ and center 0, since for any $|z| \leq r$ and $v \in \gamma$, we have $|v - z| \geq r_1 - r$, by the Cauchy's formulas it follows that for all $|z| \leq r$ and $n \in \mathbb{N}$, we have

$$|P_n^{(p)}(f)(z) - f^{(p)}(z)| = \frac{p!}{2\pi} \left| \int_{\gamma} \frac{P_n(f)(v) - f(v)}{(v-z)^{p+1}} dv \right| \leq \frac{M_{r_1}(f)}{n} \frac{p!}{2\pi} \frac{2\pi r_1}{(r_1 - r)^{p+1}},$$

which proves (ii) and the theorem. $\qquad\square$

The Voronovskaja-type formula for the real de la Vallée Poussin convolutions $P_n^*(g)(x)$ attached to an integrable 2π-periodic function g admitting a derivative of second order at x is due to Natanson (see e.g. Natanson [144], Chapter 10, Section 3, Satz 3) and is given by

$$P_n^*(g)(x) - g(x) - \frac{g''(x)}{n} = \frac{\alpha_n}{n},$$

where $\alpha_n \to 0$ as $n \to \infty$.

In what follows we present its analogue for the complex convolutions $P_n(f)(z)$, which will be used in the proof of the exactness degree of approximation.

Theorem 3.1.2. *Let $R > 1$, $\mathbb{D}_R = \{z \in \mathbb{C}; |z| < R\}$ and let us suppose that $f : \mathbb{D}_R \to \mathbb{C}$ is analytic in $\mathbb{D}_R$, that is we can write $f(z) = \sum_{k=0}^{\infty} c_k z^k$, for all $z \in \mathbb{D}_R$. For any $r \in [1, R)$ we have*

$$\left\| P_n(f) - f + \frac{e_2 f''}{n} + \frac{e_1 f'}{n} \right\|_r \leq \frac{A_r(f)}{n^2}, n \in \mathbb{N}$$

where $A_r(f) = \sum_{k=1}^{\infty} |c_k| k^4 r^k < \infty$.

Proof. Denote $e_k(z) = z^k$ and

$$|E_{k,n}(z)| = \left| P_n(e_k)(z) - e_k(z) + \frac{z^k k(k-1)}{n} + \frac{z^k k}{n} \right|.$$

For all $|z| \leq r$ we get

$$
\left| P_n(f)(z) - f(z) + \frac{e_2(z) f''(z)(z)}{n} + \frac{e_1 f'(z)}{n} \right|
$$

$$
\leq \sum_{k=0}^{\infty} |c_k| \cdot |E_{k,n}(z)| = \sum_{k=1}^{n} |c_k| \cdot |E_{k,n}(z)| + \sum_{k=n+1}^{\infty} |c_k| \cdot |E_{k,n}(z)|
$$

$$
= \sum_{k=1}^{n} |c_k| \cdot |E_{k,n}(z)| + \sum_{k=n+1}^{\infty} |c_k| \left| -z^k + \frac{z^k k(k-1)}{n} + \frac{z^k k}{n} \right|.
$$

But for $|z| \leq r$ we have

$$
\sum_{k=n+1}^{\infty} |c_k| \left| -z^k + \frac{z^k k(k-1)}{n} + \frac{z^k k}{n} \right| = \sum_{k=n+1}^{\infty} |c_k| r^k \left| -1 + \frac{k(k-1)}{n} + \frac{k}{n} \right| \leq
$$

$$
\sum_{k=n+1}^{\infty} |c_k| r^k \frac{k^2}{n} \leq \sum_{k=n+1}^{\infty} |c_k| \frac{k}{n} r^k \frac{k^2}{n} = \frac{1}{n^2} \sum_{k=n+1}^{\infty} |c_k| k^3 r^k \leq \frac{1}{n^2} \sum_{k=n+1}^{\infty} |c_k| k^4 r^k.
$$

Therefore, it remains to estimate $|E_{k,n}(z)|$ for $|z| \leq r$ and $0 \leq k \leq n$. Since it is immediate that $E_{0,n}(z) = 0$, it suffices to consider $1 \leq k \leq n$. We obtain

$$
|E_{k,n}(z)| = \left| \frac{(n!)^2}{(n-k)!(n+k)!} z^k - z^k + \frac{z^k k(k-1)}{n} + \frac{kz^k}{n} \right|
$$

$$
= |z|^k \left| \frac{(n!)^2}{(n-k)!(n+k)!} - 1 + \frac{k^2}{n} \right|.
$$

In what follows we prove by mathematical induction that for all $k = 1, 2, ..., n$ and $n \in \mathbb{N}$ we have

$$
0 \leq \frac{(n!)^2}{(n-k)!(n+k)!} - 1 + \frac{k^2}{n} \leq \frac{k^4}{n^2}. \tag{3.1}
$$

Since (3.1) is immediate for $k = n$, we may suppose that $1 \leq k \leq n - 1$. First we can write

$$
\frac{(n!)^2}{(n-k-1)!(n+k+1)!} - 1 + \frac{(k+1)^2}{n}
$$

$$
= \left[\frac{(n!)^2}{(n-k)!(n+k)!} - 1 + \frac{k^2}{n} \right] \cdot \frac{n-k}{n+k+1} + \frac{n-k}{n+k+1} \left[1 - \frac{k^2}{n} \right]
$$

$$
- \left(1 - \frac{k^2}{n} \right) + \frac{2k+1}{n} = \left[\frac{(n!)^2}{(n-k)!(n+k)!} - 1 + \frac{k^2}{n} \right] \cdot \frac{n-k}{n+k+1}
$$

$$
+ \left(1 - \frac{k^2}{n} \right) \left(-\frac{(2k+1)}{n+k+1} \right) + \frac{2k+1}{n}
$$

$$
= \left[\frac{(n!)^2}{(n-k)!(n+k)!} - 1 + \frac{k^2}{n} \right] \cdot \frac{n-k}{n+k+1} + \frac{(2k+1)(k^2+k+1)}{n(n+k+1)}.
$$

Now suppose that both inequalities in (3.1) are valid for k. From the above relationship first it follows that the left-hand side in (3.1) is valid for $k+1$. Also, passing above to the upper estimate, we obtain

$$\frac{(n!)^2}{(n-k-1)!(n+k+1)!} - 1 + \frac{(k+1)^2}{n} \leq \frac{k^4}{n^2} \cdot \frac{n-k}{n+k+1} + \frac{(2k+1)(k^2+k+1)}{n(n+k+1)}$$

$$\leq \frac{k^4}{n^2} + \frac{(2k+1)(k^2+k+1)}{n^2} = \frac{k^4+2k^3+3k^2+3k+1}{n^2} \leq \frac{(k+1)^4}{n^2},$$

which proves that (3.1) is valid for $k+1$ too and that $\|E_{k,n}\|_r \leq r^k \frac{k^4}{n^2}$, for all $1 \leq k \leq n$ and $n \in \mathbb{N}$.

Replacing this in the first inequality of the proof we obtain the theorem. $\qquad\square$

By using Theorems 3.1.1 and 3.1.2 now we are in position to obtain the exact degree of approximation by $P_n(f)(z)$ and its derivatives. The first main result is a lower estimate in Theorem 3.1.1, (i).

Theorem 3.1.3. *Let $R > 1$, $\mathbb{D}_R = \{z \in \mathbb{C}; |z| < R\}$ and let us suppose that $f : \mathbb{D}_R \to \mathbb{C}$ is analytic in $\mathbb{D}_R$, that is we can write $f(z) = \sum_{k=0}^{\infty} c_k z^k$, for all $z \in \mathbb{D}_R$. If f is not a polynomial of degree 0, then for any $r \in [1, R)$ we have*

$$\|P_n(f) - f\|_r \geq \frac{C_r(f)}{n}, n \in \mathbb{N},$$

where the constant $C_r(f)$ depends only on f and r.

Proof. For all $z \in \mathbb{D}_R$ and $n \in \mathbb{N}$ we have

$$P_n(f)(z) - f(z)$$

$$= \frac{1}{n}\left\{ -[z^2 f''(z) + z f'(z)] + \frac{1}{n}\left[n^2\left(P_n(f)(z) - f(z) + \frac{z^2}{n}f''(z) + \frac{z f'(z)}{n} \right)\right]\right\}.$$

In what follows we will apply to this identity the following obvious property :

$$\|F + G\|_r \geq |\,\|F\|_r - \|G\|_r\,| \geq \|F\|_r - \|G\|_r.$$

It follows

$$\|P_n(f) - f\|_r$$

$$\geq \frac{1}{n}\left\{ \|e_2 f'' + e_1 f'\|_r - \frac{1}{n}\left[n^2 \left\| P_n(f) - f + \frac{e_2 f''}{n} + \frac{e_1 f'}{n} \right\|_r\right]\right\}.$$

Taking into account that by hypothesis f is not a polynomial of degree 0 in $\mathbb{D}_R$, we get $\|e_2 f'' + e_1 f'\|_r > 0$. Indeed, supposing the contrary it follows that $z^2 f''(z) + z f'(z) = 0$ for all $z \in \overline{\mathbb{D}}_r$. But it is easy to see (by using the form of f as a power series and by identifying the coefficients) that the only analytic solution of this differential equation is $f'(z) = 0$, for all $z \in \overline{\mathbb{D}}_r$, which contradicts the hypothesis.

But by Theorem 3.1.2 we have

$$n^2 \left\| P_n(f) - f + \frac{e_2}{n}f'' + \frac{e_1}{n}f' \right\|_r \leq A_r(f).$$

Therefore, there exists an index n_0 depending only on f and r, such that for all $n \geq n_0$ we have

$$\|e_2 f'' + e_1 f'\|_r - \frac{1}{n} \left[n^2 \left\| P_n(f) - f + \frac{e_2}{n} f'' + \frac{e_1}{n} f' \right\|_r \right]$$

$$\geq \frac{1}{2} \|e_2 f'' + e_1 f'\|_r \,,$$

which immediately implies

$$\|P_n(f) - f\|_r \geq \frac{1}{n} \cdot \frac{1}{2} \|e_2 f'' + e_1 f'\|_r \,, \forall n \geq n_0.$$

For $n \in \{1, ..., n_0 - 1\}$ we obviously have $\|P_n(f) - f\|_r \geq \frac{M_{r,n}(f)}{n}$ with $M_{r,n}(f) = n \cdot \|P_n(f) - f\|_r > 0$, which finally implies $\|P_n(f) - f\|_r \geq \frac{C_r(f)}{n}$ for all n, where $C_r(f) = \min\{M_{r,1}(f), ..., M_{r,n_0-1}(f), \frac{1}{2} \|e_2 f'' + e_1 f'\|_r\}$. This completes the proof. $\square$

Combining now Theorem 3.1.3 with Theorem 3.1.1, (i) we immediately get the following.

Corollary 3.1.4. *Let $R > 1$, $\mathbb{D}_R = \{z \in \mathbb{C}; |z| < R\}$ and let us suppose that $f : \mathbb{D}_R \to \mathbb{C}$ is analytic in $\mathbb{D}_R$. If f is not a polynomial of degree 0, then for any $r \in [1, R)$ we have*

$$\|P_n(f) - f\|_r \sim \frac{1}{n}, n \in \mathbb{N},$$

where the constants in the equivalence depend on f and r.

In the case of approximation by the derivatives of $P_n(f)(z)$ the following result holds.

Theorem 3.1.5. *Let $\mathbb{D}_R = \{z \in \mathbb{C}; |z| < R\}$ be with $R > 1$ and let us suppose that $f : \mathbb{D}_R \to \mathbb{C}$ is analytic in $\mathbb{D}_R$, i.e. $f(z) = \sum_{k=0}^{\infty} c_k z^k$, for all $z \in \mathbb{D}_R$. Also, let $1 \leq r < r_1 < R$ and $p \in \mathbb{N}$ be fixed. If f is not a polynomial of degree $\leq p - 1$, then we have*

$$\|P_n^{(p)}(f) - f^{(p)}\|_r \sim \frac{1}{n},$$

where the constants in the equivalence depend on f, r, r_1 and p.

Proof. Denoting by Γ the circle of radius r_1 and center 0 (where $r_1 > r \geq 1$), by the Cauchy's formulas it follows that for all $|z| \leq r$ and $n \in \mathbb{N}$ we have

$$P_n^{(p)}(f)(z) - f^{(p)}(z) = \frac{p!}{2\pi i} \int_\Gamma \frac{P_n(f)(v) - f(v)}{(v - z)^{p+1}} dv,$$

where the inequality $|v - z| \geq r_1 - r$ is valid for all $|z| \leq r$ and $v \in \Gamma$.

Taking into account Theorem 3.1.1, (ii), it remains to prove the lower estimate for $\|P_n^{(p)}(f) - f^{(p)}\|_r$. For this purpose, as in the proof of Theorem 3.1.3, for all $v \in \Gamma$ and $n \in \mathbb{N}$ we have

$$P_n(f)(v) - f(v)$$

$$= \frac{1}{n} \left\{ -[v^2 f''(v) + v f'(v)] + \frac{1}{n} \left[n^2 \left(P_n(f)(v) - f(v) + \frac{v^2}{n} f''(v) + \frac{v f'(v)}{n} \right) \right] \right\},$$

which replaced in the above Cauchy's formula implies

$$P_n^{(p)}(f)(z) - f^{(p)}(z) = \frac{1}{n} \left\{ \frac{p!}{2\pi i} \int_\Gamma \frac{-[v^2 f''(v) + v f'(v)]}{(v-z)^{p+1}} dv \right.$$

$$+ \frac{1}{n} \cdot \frac{p!}{2\pi i} \int_\Gamma \frac{n^2 \left(P_n(f)(v) - f(v) + \frac{v^2}{n} f''(v) + \frac{v}{n} f'(v) \right)}{(v-z)^{p+1}} dv \right\}$$

$$= \frac{1}{n} \left\{ \left[-z^2 f''(z) - z f'(z) \right]^{(p)} \right.$$

$$+ \frac{1}{n} \cdot \frac{p!}{2\pi i} \int_\Gamma \frac{n^2 \left(P_n(f)(v) - f(v) + \frac{v^2}{n} f''(v) + \frac{v}{n} f'(v) \right)}{(v-z)^{p+1}} dv \right\}.$$

Passing now to $\| \cdot \|_r$ it follows

$$\|P_n^{(p)}(f) - f^{(p)}\|_r \geq \frac{1}{n} \left\{ \left\| [e_2 f'' + e_1 f']^{(p)} \right\|_r \right.$$

$$- \frac{1}{n} \left\| \frac{p!}{2\pi} \int_\Gamma \frac{n^2 \left(P_n(f)(v) - f(v) + \frac{v^2}{n} f''(v) + \frac{v}{n} f' \right)}{(v-z)^{p+1}} dv \right\|_r \right\},$$

where by using Theorem 3.1.2 we obtain

$$\left\| \frac{p!}{2\pi} \int_\Gamma \frac{n^2 \left(P_n(f)(v) - f(v) + \frac{v^2}{n} f''(v) + \frac{v}{n} f' \right)}{(v-z)^{p+1}} dv \right\|_r$$

$$\leq \frac{p!}{2\pi} \cdot \frac{2\pi r_1 n^2}{(r_1 - r)^{p+1}} \left\| P_n(f) - f + \frac{e_2}{n} f'' + \frac{e_1}{n} f' \right\|_{r_1}$$

$$\leq \frac{A_{r_1}(f) p! r_1}{(r_1 - r)^{p+1}}.$$

But by hypothesis on f we have $\left\| [e_2 f'' + e_1 f']^{(p)} \right\|_r > 0$.

Indeed, supposing the contrary it follows that $z^2 f''(z) + z f'(z) = Q_{p-1}(z)$, for all $z \in \overline{\mathbb{D}}_r$, where $Q_{p-1}(z) = \sum_{j=1}^{p-1} A_j z^j$ necessarily is a polynomial of degree $\leq p - 1$, vanishing at $z = 0$. Denoting $f'(z) = g(z)$ the differential equation becomes $z^2 g'(z) + z g(z) = Q_{p-1}(z)$, for all $z \in \overline{\mathbb{D}}_r$. Seeking now the analytic solution in the form $g(z) = \sum_{j=0}^\infty \alpha_j z^j$, replacing in the differential equation, by the identification of coefficients we easily obtain that $g(z)$ necessarily is a polynomial of degree $\leq p - 2$, which will imply that $f(z)$ necessarily is a polynomial of degree $\leq p - 1$, in contradiction with the hypothesis.

Finally, reasoning exactly as in the proof of Theorem 3.1.3, we immediately get the desired conclusion. $\square$

An important trigonometric mean of real variable in approximation theory is the Fejér mean $F_n^*(g)(x)$, defined as the arithmetic mean of the sequence of partial sums of the Fourier series of g and given by

$$F_n^*(g)(x) = \frac{1}{2\pi} \int_{-\pi}^{\pi} g(t) K_n(x-t) dt, x \in \mathbb{R},$$

where the kernel $K_n(u)$ is given by $K_n(u) = \frac{1}{n} \left(\frac{\sin(nu/2)}{\sin(u/2)} \right)^2$. Concerning this mean Fejér [73] proved that if $f : \mathbb{R} \to \mathbb{C}$ is a continuous function with period 2π, then the sequence $(F_n(g)(x))_n$ converges uniformly to g on $[-\pi, \pi]$.

Its complex form attached to an analytic function $f(z) = \sum_{k=0}^{\infty} c_k z^k$ in a disk $\mathbb{D}_R = \{z \in \mathbb{C}, |z| < R\}$ and given by

$$F_n(f)(z) = \frac{1}{2\pi} \int_{-\pi}^{\pi} f(ze^{iu}) K_n(u) du$$

$$= \frac{1}{2\pi n} \int_{-\pi}^{\pi} f(ze^{iu}) \left(\frac{\sin(nu/2)}{\sin(u/2)} \right)^2 du = \sum_{k=0}^{n-1} c_k \frac{n-k}{n} z^k,$$

(for the last formula see e.g. Gal [86], Theorem 3.1) also has nice approximation properties, satisfying the estimate (see Gaier [75], Theorem 1)

$$\|F_n(f) - f\|_r \le M\omega_1 \left(f; \frac{1}{n} \right)_{\overline{\mathbb{D}}_r} \le \frac{M\|f'\|_r}{n} := \frac{C_r(f)}{n}, \text{for all } n \in \mathbb{N},$$

where $M > 0$, $0 < r < R$, $\|f'\|_r = \sup\{|f(z)|; |z| \le r\}$ and $\omega_1 \left(f; \frac{1}{n} \right)_{\overline{\mathbb{D}}_r}$ is the classical modulus of continuity of f in $\overline{\mathbb{D}}_r$.

In fact, from the saturation result in Bruj-Schmieder [48], pp. 161-162, it follows that if f is not a constant then the approximation order by $F_n(f)$ is exactly $\frac{1}{n}$.

The Fejér means, both trigonometric and complex cases were generalized by Riesz-Zygmund, their complex form being defined for any $s \in \mathbb{N}$ by

$$R_{n,s}(f)(z) = \sum_{k=0}^{n-1} c_k \left[1 - \left(\frac{k}{n} \right)^s \right] z^k, n \in \mathbb{N}.$$

For $s = 1$ one recapture the Fejér means and for $s = 2$ one get the means introduced by Riesz.

From the saturation result in Bruj-Schmieder [48], pp. 161-162, it follows that if $f(z)$ is not a constant then the approximation order by $R_{n,s}(f)(z)$ is exactly $\frac{1}{n^s}$.

In what follows we complete this result by proving that the approximation order by the derivatives of $R_{n,s}(f)$ also is exactly $\frac{1}{n^s}$. Useful in the proof will be the Voronovskaja's formula for $R_{n,s}(f)(z)$ with a quantitative estimate. It is worth noting that our method is different from that in Bruj-Schmieder [1]. Also, in addition we obtain here a Voronovskaja result with a quantitative upper estimate and the exact orders in simultaneous approximation by derivatives.

First we obtain an upper estimate in approximation by the derivatives of $R_{n,s}(f)(z)$.

Theorem 3.1.6. (Gal [85]) *Let $R > 1$, $\mathbb{D}_R = \{z \in \mathbb{C}; |z| < R\}$ and let us suppose that $f : \mathbb{D}_R \to \mathbb{C}$ is analytic in $\mathbb{D}_R$. If $1 \leq r < r_1 < R$ and $s, p \in \mathbb{N}$ then we have*

$$\|R_{n,s}^{(p)}(f) - f^{(p)}\|_r \leq \frac{r_1 p! C_{r_1,s}(f)}{(r_1 - r)^{p+1} n^s}, n \in \mathbb{N},$$

where $C_{r_1,s}(f) = \sum_{k=0}^{\infty} |c_k| k^s r^k < \infty$.

Proof. An estimate of the form

$$\|R_{n,s}(f) - f\|_r \leq \frac{C_{r,s}(f)}{n^s}, n \in \mathbb{N},$$

essentially follows from Bruj-Schmieder [48]. Below we reprove this estimate in a different and simple way with an explicit constant $C_{r,s}(f)$. Thus, denoting $e_k(z) = z^k$ and writing $f(z) = \sum_{k=0}^{\infty} c_k e_k(z)$, for all $|z| \leq r$ we obtain

$$
\begin{aligned}
|R_{n,s}(f)(z) - f(z)| &\leq \sum_{k=0}^{\infty} |c_k| \cdot |R_{n,s}(e_k)(z) - e_k(z)| \\
&= \sum_{k=0}^{n-1} |c_k| \cdot |R_{n,s}(e_k)(z) - e_k(z)| \\
&\quad + \sum_{k=n}^{\infty} |c_k| \cdot |R_{n,s}(e_k)(z) - e_k(z)| \\
&\leq \frac{1}{n^s} \sum_{k=0}^{n-1} |c_k| k^s r^k + \sum_{k=n}^{\infty} |c_k| r^k.
\end{aligned}
$$

Here we used that $e_k(z) = \sum_{j=0}^{\infty} c_j z^j$, with $c_k = 1$ and $c_j = 0$ for all $j \neq k$.

Taking into account that for $k \geq n$ we have $1 \leq \frac{k^s}{n^s}$, from the previous inequality we get

$$|R_{n,s}(f)(z) - f(z)| \leq \frac{1}{n^s} \sum_{k=0}^{n-1} |c_k| k^s r^k + \frac{1}{n^s} \sum_{k=n}^{\infty} |c_k| k^s r^k = \frac{1}{n^s} \sum_{k=0}^{\infty} |c_k| k^s r^k,$$

for all $|z| \leq r$, therefore we can take $C_{r,s}(f) = \sum_{k=0}^{\infty} |c_k| k^s r^k < \infty$.

Now, denoting by γ the circle of radius $r_1 > r$ and center 0, since for any $|z| \leq r$ and $v \in \gamma$, we have $|v - z| \geq r_1 - r$, by the Cauchy's formulas it follows that for all $|z| \leq r$ and $n \in \mathbb{N}$, we have

$$
\begin{aligned}
|R_{n,s}^{(p)}(f)(z) - f^{(p)}(z)| &= \frac{p!}{2\pi} \left| \int_{\gamma} \frac{R_{n,s}(f)(v) - f(v)}{(v - z)^{p+1}} dv \right| \\
&\leq \frac{C_{r_1,s}(f)}{n^s} \cdot \frac{p!}{2\pi} \cdot \frac{2\pi r_1}{(r_1 - r)^{p+1}},
\end{aligned}
$$

which proves the theorem. $\qquad\square$

Useful in the proof of the exact degree of approximation will be the Voronovskaja's formula for $R_{n,s}(f)(z)$. For this purpose, we need the following simple lemma.

Lemma 3.1.7. (Gal [85]) *Let $k, s \in \mathbb{N}$. If we denote $k^s = \sum_{j=1}^{s} \alpha_{j,s} k(k-1)...(k-(j-1))$, then the coefficients $\alpha_{j,s}$ can be chosen independent of k and to satisfy $\alpha_{1,s} = \alpha_{s,s} = 1$ for all $s \geq 1$ and $\alpha_{j,s+1} = \alpha_{j-1,s} + j\alpha_{j,s}$, $j = 2, ..., s$, $s \geq 2$.*

Proof. We have

$$k^{s+1}$$

$$= \sum_{j=1}^{s+1} \alpha_{j,s+1} k(k-1)...(k-(j-1)) = k \sum_{j=1}^{s} \alpha_{j,s} k(k-1)...(k-(j-1))$$

$$= \sum_{j=1}^{s} \alpha_{j,s} (k-j+j) k(k-1)...(k-(j-1)) = \sum_{j=1}^{s} \alpha_{j,s} k(k-1)...(k-j)$$

$$+ \sum_{j=1}^{s} j\alpha_{j,s} k(k-1)...(k-(j-1)) = \sum_{j=2}^{s+1} \alpha_{j-1,s} k(k-1)...(k-(j-1))$$

$$+ \sum_{j=1}^{s} j\alpha_{j,s} k(k-1)...(k-(j-1)) = \alpha_{s,s} + \sum_{j=2}^{s} \alpha_{j-1,s} k(k-1)...(k-(j-1))$$

$$+ \sum_{j=2}^{s} j\alpha_{j,s} k(k-1)...(k-(j-1)) + \alpha_{1,s},$$

which implies

$$\sum_{j=1}^{s+1} \alpha_{j,s+1} k(k-1)...(k-(j-1)) = \alpha_{1,s+1}$$

$$+ \sum_{j=2}^{s} \alpha_{j,s+1} k(k-1)...(k-(j-1))$$

$$+ \alpha_{s+1,s+1} = \alpha_{s,s} + \sum_{j=2}^{s} \alpha_{j-1,s} k(k-1)...(k-(j-1))$$

$$+ \sum_{j=2}^{s} j\alpha_{j,s} k(k-1)...(k-(j-1)) + \alpha_{1,s},$$

and proves the lemma. $\square$

Now, the Voronovskaja-type formula for $R_{n,s}(f)(z)$ one states as follows.

Theorem 3.1.8. (Gal [85]) *Let $R > 1$, $\mathbb{D}_R = \{z \in \mathbb{C}; |z| < R\}$ and let us suppose that $f : \mathbb{D}_R \to \mathbb{C}$ is analytic in $\mathbb{D}_R$, that is we can write $f(z) = \sum_{k=0}^{\infty} c_k z^k$, for all $z \in \mathbb{D}_R$. For any $r \in [1, R)$ we have*

$$\left\| R_{n,s}(f) - f + \frac{1}{n^s} \sum_{j=1}^{s} \alpha_{j,s} e_j f^{(j)} \right\|_r \leq \frac{A_{r,s}(f)}{n^{s+1}}, n \in \mathbb{N}$$

where $A_{r,s}(f) = \sum_{k=1}^{\infty} |c_k| k^{s+1} r^k < \infty$, $e_j(z) = z^j$ and $\alpha_{j,s}$ are defined by Lemma 3.1.7.

Proof. Denoting

$$E_{k,n,s}(z) = R_{n,s}(e_k)(z) - e_k(z) + \frac{1}{n^s} \sum_{j=1}^{s} \alpha_{j,s} k(k-1)...(k-(j-1)) z^k,$$

we obtain

$$\left| R_{n,s}(f)(z) - f(z) + \frac{1}{n^s} \sum_{j=1}^{s} \alpha_{j,s} z^j f^j(z) \right| \leq \sum_{k=0}^{\infty} |c_k| \cdot |E_{k,n,s}(z)|$$

$$= \sum_{k=1}^{n-1} |c_k| \cdot |E_{k,n,s}(z)| + \sum_{k=n}^{\infty} |c_k| \cdot |E_{k,n,s}(z)|$$

$$= \sum_{k=1}^{n-1} |c_k| \cdot \left| \left(1 - \left(\frac{k}{n} \right)^s \right) - 1 + \frac{1}{n^s} \sum_{j=1}^{s} \alpha_{j,s} k(k-1)...(k-(j-1)) \right| \cdot |z|^k$$

$$+ \sum_{k=n}^{\infty} |c_k| \cdot \left| -1 + \frac{1}{n^s} \sum_{j=1}^{s} \alpha_{j,s} k(k-1)...(k-(j-1)) \right| \cdot |z|^k$$

$$= 0 + \sum_{k=n}^{\infty} |c_k| \cdot \left| -1 + \frac{1}{n^s} \sum_{j=1}^{s} \alpha_{j,s} k(k-1)...(k-(j-1)) \right| \cdot |z|^k,$$

where for the first sum we used Lemma 3.1.7.

Taking into account that by Lemma 3.1.7 we have $\sum_{j=1}^{s} \alpha_{j,s} n(n-1)...(n-(j-1)) = n^s$, for all $|z| \leq r$ it follows

$$\left| R_{n,s}(f)(z) - f(z) + \frac{1}{n^s} \sum_{j=1}^{s} \alpha_{j,s} z^j f^j(z) \right|$$

$$\leq \left(\sum_{k=n}^{\infty} |c_k| r^k \right) \cdot \left(\sum_{j=1}^{s} \alpha_{j,s} \frac{k(k-1)...(k-(j-1))}{n^s} \right)$$

$$= \frac{1}{n^s} \sum_{k=n}^{\infty} |c_k| k^s r^k \leq \frac{1}{n^{s+1}} \sum_{k=n}^{\infty} |c_k| k^{s+1} r^k \leq \frac{1}{n^{s+1}} \sum_{k=1}^{\infty} |c_k| k^{s+1} r^k,$$

which proves the theorem. $\qquad \square$

By using Theorems 3.1.6. and 3.1.8 we are in position to obtain the exact degree of approximation by the derivatives of $R_{n,s}(f)(z)$.

Theorem 3.1.9. (Gal [85]) *Let $\mathbb{D}_R = \{z \in \mathbb{C}; |z| < R\}$ be with $R > 1$ and let us suppose that $f : \mathbb{D}_R \to \mathbb{C}$ is analytic in $\mathbb{D}_R$, i.e. $f(z) = \sum_{k=0}^{\infty} c_k z^k$, for all $z \in \mathbb{D}_R$.*

Also, let $1 \le r < r_1 < R$ and $p, s \in \mathbb{N}$ be fixed. If f is not a polynomial of degree $\le p - 1$ then we have

$$\|R_{n,s}^{(p)}(f) - f^{(p)}\|_r \sim \frac{1}{n^s},$$

where the constants in the equivalence depend on f, r, r_1, s and p.

Proof.   Denoting by Γ the circle of radius $r_1 >$ and center 0 (where $r_1 > r \ge 1$), by the Cauchy's formulas it follows that for all $|z| \le r$ and $n \in \mathbb{N}$ we have

$$R_{n,s}^{(p)}(f)(z) - f^{(p)}(z) = \frac{p!}{2\pi i} \int_\Gamma \frac{R_{n,s}(f)(v) - f(v)}{(v - z)^{p+1}} dv,$$

where the inequality $|v - z| \ge r_1 - r$ is valid for all $|z| \le r$ and $v \in \Gamma$.

Taking into account Theorem 3.1.6, it remains to prove the lower estimate for $\|R_{n,s}^{(p)}(f) - f^{(p)}\|_r$. For this purpose, for all $v \in \Gamma$ and $n \in \mathbb{N}$ we have

$$R_{n,s}(f)(v) - f(v) = \frac{1}{n^s} \left\{ -\sum_{j=1}^{s} \alpha_{j,s} v^j f^{(j)}(v) \right.$$
$$\left. + \frac{1}{n} \left[n^{s+1} \left(R_{n,s}(f)(v) - f(v) + \frac{1}{n^s} \sum_{j=1}^{s} \alpha_{j,s} v^j f^{(j)}(v) \right) \right] \right\},$$

which replaced in the above Cauchy's formula implies

$$R_{n,s}^{(p)}(f)(z) - f^{(p)}(z) = \frac{1}{n^s} \left\{ \frac{p!}{2\pi i} \int_\Gamma \frac{-\sum_{j=1}^{s} \alpha_{j,s} v^j f^{(j)}(v)}{(v - z)^{p+1}} dv \right.$$
$$\left. + \frac{1}{n} \cdot \frac{p!}{2\pi i} \int_\Gamma \frac{n^{s+1} \left(R_{n,s}(f)(v) - f(v) + \frac{1}{n^s} \sum_{j=1}^{s} \alpha_{j,s} v^j f^{(j)}(v) \right)}{(v - z)^{p+1}} dv \right\}$$

$$= \frac{1}{n^s} \left\{ \left[-\sum_{j=1}^{s} \alpha_{j,s} v^j f^{(j)}(v) \right]^{(p)} \right.$$
$$\left. + \frac{1}{n} \cdot \frac{p!}{2\pi i} \int_\Gamma \frac{n^{s+1} \left(R_{n,s}(f)(v) - f(v) + \frac{1}{n^s} \sum_{j=1}^{s} \alpha_{j,s} v^j f^{(j)}(v) \right)}{(v - z)^{p+1}} dv \right\}.$$

Passing now to $\|\cdot\|_r$ it follows

$$\|R_{n,s}^{(p)}(f) - f^{(p)}\|_r \ge \frac{1}{n^s} \left\{ \left\| \left[\sum_{j=1}^{s} \alpha_{j,s} e_j f^{(j)} \right]^{(p)} \right\|_r \right.$$
$$\left. - \frac{1}{n} \left\| \frac{p!}{2\pi} \int_\Gamma \frac{n^{s+1} \left(R_{n,s}(f)(v) - f(v) + \frac{1}{n^s} \sum_{j=1}^{s} \alpha_{j,s} v^j f^{(j)}(v) \right)}{(v - z)^{p+1}} dv \right\|_r \right\},$$

where by using Theorem 3.1.8 we obtain

$$\left\| \frac{p!}{2\pi} \int_\Gamma \frac{n^{s+1} \left(R_{n,s}(f)(v) - f(v) + \frac{1}{n^s} \sum_{j=1}^s \alpha_{j,s} v^j f^{(j)}(v) \right)}{(v-z)^{p+1}} dv \right\|_r$$

$$\leq \frac{p!}{2\pi} \cdot \frac{2\pi r_1 n^{s+1}}{(r_1 - r)^{p+1}} \left\| R_{n,s}(f) - f + \frac{1}{n^s} \sum_{j=1}^s \alpha_{j,s} e_j f^{(j)} \right\|_{r_1} \leq \frac{A_{r_1,s}(f) p! r_1}{(r_1 - r)^{p+1}}.$$

But by hypothesis on f we have $\left\| \left[\sum_{j=1}^s \alpha_{j,s} e_j f^{(j)} \right]^{(p)} \right\|_r > 0.$

Indeed, supposing the contrary it follows that $\sum_{j=1}^s \alpha_{j,s} z^j f^{(j)}(z) = Q_{p-1}(z)$, for all $z \in \overline{\mathbb{D}}_r$, where $Q_{p-1}(z) = \sum_{j=1}^{p-1} A_j z^j$ necessarily is a polynomial of degree $\leq p - 1$.

Seeking now the analytic solution in the form $f(z) = \sum_{k=0}^\infty \beta_k z^k$, replacing in the differential equation and taking into account again Lemma 3.1.7, by identification of the coefficients β_k we easily obtain $\beta_k = 0$, for all $k \geq p$, that is $f(z)$ necessarily is a polynomial of degree $\leq p - 1$, in contradiction with the hypothesis.

Therefore, there exists an index n_0 depending only on f, s, p, r and r_1, such that for all $n \geq n_0$ we have

$$\left\| \left[\sum_{j=1}^s \alpha_{j,s} e_j f^{(j)} \right]^{(p)} \right\|_r$$

$$- \frac{1}{n} \left\| \frac{p!}{2\pi} \int_\Gamma \frac{n^{s+1} \left(R_{n,s}(f)(v) - f(v) + \frac{1}{n^s} \sum_{j=1}^s \alpha_{j,s} v^j f^{(j)}(v) \right)}{(v-z)^{p+1}} dv \right\|_r$$

$$\geq \frac{1}{2} \left\| \left[\sum_{j=1}^s \alpha_{j,s} e_j f^{(j)} \right]^{(p)} \right\|_r,$$

which immediately implies

$$\| R_{n,s}^{(p)}(f) - f^{(p)} \|_r \geq \frac{1}{n^s} \cdot \frac{1}{2} \left\| \left[\sum_{j=1}^s \alpha_{j,s} e_j f^{(j)} \right]^{(p)} \right\|_r,$$

for all $n \geq n_0$.

For $n \in \{1, ..., n_0 - 1\}$ we obviously have $\| R_{n,s}^{(p)}(f) - f^{(p)} \|_r \geq \frac{M_{r,s,p,n}(f)}{n}$ with $M_{r,s,,p,n}(f) = n \cdot \| R_{n,s}^{(p)}(f) - f^{(p)} \|_r > 0$, which finally implies $\| R_{n,s}^{(p)}(f) - f^{(p)} \|_r \geq \frac{C_{r,s,p}(f)}{n}$ for all n, where

$$C_{r,s,p}(f) = \min \left\{ M_{r,s,p,1}(f), ..., M_{r,s,p,n_0-1}(f), \frac{1}{2} \left\| \left[\sum_{j=1}^s \alpha_{j,s} e_j f^{(j)} \right]^{(p)} \right\|_r \right\}.$$

This completes the proof. $\square$

In what follows we will make similar considerations on the complex convolutions of Jackson and of Beatson type given by

$$J_n(f)(z) = \frac{1}{\pi} \int_{-\pi}^{\pi} f(ze^{it}) K_{n,2}(t) dt,$$

and

$$P_{n,2,p}(f)(z) = \frac{1}{\pi} \int_{-\pi}^{\pi} f(ze^{iu}) B_{n,2,p}(u) du,$$

respectively, where $p \in \mathbb{N}$,

$$K_{n,2}(t) = \frac{3}{2n(2n^2+1)} \left(\frac{sin(nt/2)}{sin(t/2)} \right)^4, \quad B_{n,2,1}(u) = \frac{n}{2\pi} \int_{u-\pi/n}^{u+\pi/n} K_{n,2}(t) dt,$$

$$B_{n,2,p}(u) = \frac{n}{2\pi} \int_{u-\pi/n}^{u+\pi/n} B_{n,2,p-1}(t) dt, \quad p = 2, 3, \dots.$$

First we present without proofs the following results.

Theorem 3.1.10. (Gal [93], Theorem 3.1.) *We have :*
(i) $K_{n,2}(t) = \frac{1}{2} + \sum_{k=1}^{2n-2} \lambda_{k,n} cos(kt)$, *where*

$$\lambda_{k,n} = \frac{4n^3 - 6k^2 n + 3k^3 - 3k + 2n}{2n(2n^2+1)}, \quad \text{if } 1 \le k \le n,$$

$$\lambda_{k,n} = \frac{(k-2n) - (k-2n)^3}{2n(2n^2+1)}, \quad \text{if } n \le k \le 2n-2.$$

(ii) $B_{n,2,p}(t) = \frac{1}{2} + \sum_{k=1}^{2n-2} \left[\frac{n}{k\pi} sin(k\pi/n) \right]^p \lambda_{k,n} cos(kt)$, $p = 1, 2, \dots$
(iii) If $f(z) = \sum_{k=0}^{\infty} c_k z^k$ *is analytic in* $|z| < R$ *with* $R > 1$ *then*

$$J_n(f)(z) = c_0 + \sum_{k=1}^{2n-2} c_k \lambda_{k,n} z^k,$$

$$P_{n,2,p}(f)(z) = c_0 + \sum_{k=1}^{2n-2} c_k \left[\frac{n}{k\pi} sin(k\pi/n) \right]^p \lambda_{k,n} z^k.$$

(iv) (Gal [86], p. 423) If f is analytic in $\mathbb{D}_1$ and continuous in $\overline{\mathbb{D}}_1$ then for all $|z| \le 1$ *and* $n \in \mathbb{N}$ *we have*

$$|J_n(f)(z) - f(z)| \le C\omega_2(f; 1/n)_{\partial \mathbb{D}_1},$$

where $C > 0$ is an absolute constant and

$$\omega_2(f; \delta)_{\partial \mathbb{D}_1} = \sup\{|\Delta_u^2 f(e^{it})|; |t| \le \pi, |u| \le \delta\},$$

$\Delta_u^2 g(t) = g(x) - 2g(x+u) + g(x+2u)$.
(v) (Gal [93], Corollary 2.4.) If f is analytic in $\mathbb{D}_1$ and continuous in $\overline{\mathbb{D}}_1$ then for all $|z| \le 1$ *and* $n \in \mathbb{N}$ *we have*

$$|P_{n,2,p}(f)(z) - f(z)| \le C_p \omega_2(f; 1/n)_{\partial \mathbb{D}_1},$$

where $C_p > 0$ is an absolute constant independent of n, f and z.

Remarks. 1) The proofs of all the results in Theorem 3.1.10 can also be found in the recent book Gal [77].

2) Analysing the proofs of Theorem 3.1.10, (iv), (v), it easily follows that for $|z| \leq r$ with $1 \leq r < R$, the approximation errors can be expressed in terms of

$$\omega_2(f; 1/n)_{\partial \mathbb{D}_r} = \sup\{|\Delta_u^2 f(re^{it})|; |t| \leq \pi, |u| \leq 1/n\},$$

with constants in front of ω_2 depending on r.

3) We have $\lambda_{k,n} \geq 0$ for all $1 \leq k \leq 2n - 2$. Indeed, first let us suppose that $1 \leq k < j \leq n$. We get

$$\begin{aligned}
\lambda_{j,n} - \lambda_{k,n} &= \frac{-6j^2 n + 3j^3 - 3j + 6k^2 n - 3k^3 + 3k}{4n^3 + 2n} \\
&= \frac{-6n(j^2 - k^2) + 3(j^3 - k^3) - 3(j - k)}{4n^3 + 2n} \\
&= \frac{(j - k)}{4n^3 + 2n}[-6n(j + k) + 3j^2 + 3jk + 3k^2 - 3] \\
&\leq \frac{(j - k)}{4n^3 + 2n}[-6j(j + k) + 3j^2 + 3jk + 3k^2 - 3] \\
&= \frac{(j - k)}{4n^3 + 2n}[-3j^2 - 3jk + 3k^2 - 3] \leq 0,
\end{aligned}$$

which means that for $1 \leq k < j \leq n$, the sequence $\lambda_{k,n}$ is decreasing, that is

$$0 \leq \frac{n^3 - n}{4n^3 + 2n} = \lambda_{n,n} \leq \lambda_{k,n} \leq \lambda_{1,n} = \frac{4(n^3 - n)}{4n^3 + 2n} < 1, \text{ for all } 1 \leq k \leq n.$$

Now, let us suppose that $2 \leq n \leq k < j \leq 2n - 2$. We get

$$\begin{aligned}
\lambda_{j,n} - \lambda_{k,n} &= \frac{(j - 2n) - (j - 2n)^3 - (k - 2n) + (k - 2n)^3}{4n^3 + 2n} \\
&= \frac{(j - k)}{4n^3 + 2n}[1 - (j - 2n)^2 - (j - 2n)(k - 2n) - (k - 2n)^2] \leq 0,
\end{aligned}$$

which for all $2 \leq n \leq k \leq 2n - 2$ implies

$$0 \leq \frac{6}{4n^3 + 2n} = \lambda_{2n-2,n} \leq \lambda_{k,n} \leq \lambda_{n,n} = \frac{n^3 - n}{4n^3 + 2n} < 1.$$

An immediate consequence of Theorem 3.1.10, (iv), (v), of the above Remark 2 and of the Cauchy's formula is the following.

Corollary 3.1.11. *Let $f(z) = \sum_{k=0}^{\infty} c_k z^k$ be analytic in $|z| < R$ with $R > 1$ and $1 \leq r < R$.*

(i) For all $|z| \leq r$ and $n \in \mathbb{N}$ we have

$$|J_n(f)(z) - f(z)| \leq \frac{C_r}{n^2} \sum_{k=2}^{\infty} |c_k| k(k - 1) r^{k-2},$$

$$|P_{n,2,p}(f)(z) - f(z)| \leq \frac{C_{p,r}}{n^2} \sum_{k=2}^{\infty} |c_k| k(k-1) r^{k-2}.$$

(ii) If $1 \leq r < r_1 < R$ then for all $|z| \leq r$ and $n, p, q \in \mathbb{N}$ we have

$$|J_n^{(q)}(f)(z) - f^{(q)}(z)| \leq \frac{r_1 q! M_{r_1}(f)}{(r_1 - r)^{q+1} n^2},$$

$$|P_{n,2,p}^{(q)}(f)(z) - f^{(q)}(z)| \leq \frac{r_1 q! M_{r_1,p}(f)}{(r_1 - r)^{q+1} n^2},$$

where

$$M_{r_1}(f) = C_{r_1} \sum_{k=2}^{\infty} |c_k| k(k-1) r_1^{k-2} \text{ and } M_{r_1,p}(f) = C_{p,r_1} \sum_{k=2}^{\infty} |c_k| k(k-1) r_1^{k-2}.$$

Proof. (i) Since $\Delta_u^2 g(t) = 2u^2[t, t+u, t+2u; g]$, by the mean value theorem for divided differences in Complex Analysis (see e.g. Stancu [172], p. 258, Exercise 4.20) we get

$$|\Delta_u^2 f(re^{it})| \leq u^2 \|f''\|_r,$$

where

$$\|f''\|_r = \sup\{|f''(z)|; |z| \leq r\} \leq \sum_{k=2}^{\infty} |c_k| k(k-1) r^{k-2},$$

which immediately proves (i).

(ii) Denoting by γ the circle of radius $r_1 > r$ and center 0, since for any $|z| \leq r$ and $v \in \gamma$, we have $|v - z| \geq r_1 - r$, by the Cauchy's formulas it follows that for all $|z| \leq r$ and $n \in \mathbb{N}$, we have

$$|J_n^{(q)}(f)(z) - f^{(q)}(z)| = \frac{q!}{2\pi} \left| \int_{\gamma} \frac{J_n(f)(v) - f(v)}{(v-z)^{q+1}} dv \right| \leq \frac{M_{r_1}(f)}{n^2} \frac{q!}{2\pi} \frac{2\pi r_1}{(r_1 - r)^{q+1}}.$$

The reasoning in the case of $P_{n,2,p}^{(q)}(f)(z)$ is similar, which proves (ii) and the corollary. $\square$

For the complex Jackson polynomials $J_n(f)$, the following Voronovskaja's result hold.

Theorem 3.1.12. *If $f(z) = \sum_{k=0}^{\infty} c_k z^k$ is analytic in $|z| < R$ with $R > 1$ then for all $|z| \leq r$, $1 \leq r < R$, we have*

$$\left| J_n(f)(z) - f(z) + \frac{3z^2 f''(z)}{2n^2} + \frac{3z f'(z)}{2n^2} \right| \leq \frac{A_r(f)}{n^3},$$

where $A_r(f) = \frac{15}{2} \sum_{k=0}^{\infty} |c_k| k^4 r^k$.

Proof. Since we can write $J_n(f)(z) = \sum_{k=0}^{\infty} c_k J_n(e_k)(z)$, it follows that

$$\left| J_n(f)(z) - f(z) + \frac{3z^2 f''(z)}{2n^2} + \frac{3z f'(z)}{2n^2} \right| \leq \sum_{k=0}^{\infty} |c_k| \cdot |E_{k,n}(z)|,$$

where $E_{k,n}(z) = J_n(e_k)(z) - z^k + \frac{3k(k-1)z^k}{2n^2} + \frac{3kz^k}{2n^2}$.

Since

$$\sum_{k=0}^{\infty} |c_k| \cdot |E_{k,n}(z)| = \sum_{k=0}^{n} |c_k| \cdot |E_{k,n}(z)| + \sum_{k=n+1}^{2n-2} |c_k| \cdot |E_{k,n}(z)|$$

$$+ \sum_{k=2n-1}^{\infty} |c_k| \cdot |E_{k,n}(z)| := A + B + C,$$

by Theorem 3.1.10, (i), (iii), we obtain

$$A =$$

$$= \sum_{k=0}^{n} |c_k| \cdot \left| \frac{4n^3 - 6k^2n + 3k^3 - 3k + 2n}{2n(2n^2+1)} z^k - z^k + \frac{3k(k-1)z^k}{2n^2} + \frac{3kz^k}{2n^2} \right|$$

$$\leq \sum_{k=0}^{n} |c_k| r^k \cdot \left| \frac{4n^3 - 6k^2n + 3k^3 - 3k + 2n}{2n(2n^2+1)} - 1 + \frac{3k^2}{2n^2} \right|$$

$$= \sum_{k=0}^{n} |c_k| r^k \cdot \left| \frac{3n(k^3-k) + 3k^2}{2n(2n^2+1)} \right| \leq \sum_{k=0}^{n} |c_k| r^k \frac{k^3 + k^2}{n^3} \leq \sum_{k=0}^{n} |c_k| r^k \frac{2k^4}{n^3},$$

$$B = \sum_{k=n+1}^{2n-2} |c_k| \cdot \left| \frac{(k-2n) - (k-2n)^3}{2n(2n^2+1)} z^k - z^k + \frac{3k^2 z^k}{2n^2} \right|$$

$$\leq \sum_{k=n+1}^{2n-2} |c_k| r^k \left| \frac{(k-2n) - (k-2n)^3}{2n(2n^2+1)} - 1 + \frac{3k^2}{2n^2} \right|$$

$$= \sum_{k=n+1}^{2n-2} |c_k| r^k \left| \frac{4n^4 - 12kn^3 + n^2(12k^2 - 4) + n(k - k^3) + 3k^2}{2n^2(2n^2+1)} \right|$$

$$\leq \sum_{k=n+1}^{2n-2} |c_k| r^k \frac{4k^4 + 12k^4 + k^2(12k^2 - 4) + k(k + k^3) + 3k^2}{2n^2(2n^2+1)}$$

$$\leq \sum_{k=n+1}^{2n-2} |c_k| r^k \frac{30k^4}{4n^4} \leq \frac{15}{2n^2} \sum_{k=n+1}^{2n-2} |c_k| r^k k^4.$$

In the proof of inequality satisfied by B we applied the fact that $n < k$.

Finally,

$$C = \sum_{k=2n-1}^{\infty} |c_k| \cdot \left| -z^k + \frac{3z^k k(k-1)}{2n^2} + \frac{3z^k k}{2n^2} \right|$$

$$\leq \sum_{k=2n-1}^{\infty} |c_k| r^k \left| \frac{3k^2}{2n^2} - 1 \right| \leq \frac{3}{2n^2} \sum_{k=2n-1}^{\infty} |c_k| r^k k^2$$

$$\leq \frac{3}{2n^3} \sum_{k=2n-1}^{\infty} |c_k| r^k k^3 \leq \frac{3}{2n^3} \sum_{k=2n-1}^{\infty} |c_k| r^k k^4.$$

In conclusion,

$$\sum_{k=0}^{\infty} |c_k| \cdot |E_{k,n}(z)| = A + B + C \le \frac{15}{2n^3} \sum_{k=0}^{\infty} |c_k| r^k k^4,$$

which proves the theorem. $\square$

By using Corollary 3.1.11 and Theorem 3.1.12 now we are in position to obtain the exact degree of approximation by $J_n(f)(z)$ and its derivatives. The first main result is a lower estimate in Theorem 3.1.11, (i) for the complex Jackson polynomials.

Theorem 3.1.13. *Let $R > 1$, $\mathbb{D}_R = \{z \in \mathbb{C}; |z| < R\}$ and let us suppose that $f : \mathbb{D}_R \to \mathbb{C}$ is analytic in $\mathbb{D}_R$, that is we can write $f(z) = \sum_{k=0}^{\infty} c_k z^k$, for all $z \in \mathbb{D}_R$. If f is not a polynomial of degree 0, then for any $r \in [1, R)$ we have*

$$\|J_n(f) - f\|_r \ge \frac{C_r(f)}{n^2}, n \in \mathbb{N},$$

where the constant $C_r(f)$ depends only on f and r.

Proof.   For all $z \in \mathbb{D}_R$ and $n \in \mathbb{N}$ we have

$$J_n(f)(z) - f(z) = \frac{1}{n^2} \left\{ -\left[\frac{3z^2 f''(z)}{2} + \frac{3z f'(z)}{2} \right] \right.$$
$$\left. + \frac{1}{n} \left[n^3 \left(J_n(f)(z) - f(z) + \frac{3z^2}{2n^2} f''(z) + \frac{3z f'(z)}{2n^2} \right) \right] \right\}.$$

In what follows we will apply to this identity the following obvious property :

$$\|F + G\|_r \ge |\, \|F\|_r - \|G\|_r \,| \ge \|F\|_r - \|G\|_r.$$

It follows

$$\|J_n(f) - f\|_r$$
$$\ge \frac{1}{n^2} \left\{ \left\| \frac{3e_2 f''}{2} + \frac{3e_1 f'}{2} \right\|_r - \frac{1}{n} \left[n^3 \left\| J_n(f) - f + \frac{3e_2 f''}{2n^2} + \frac{3e_1 f'}{2n^2} \right\|_r \right] \right\}.$$

Taking into account that by hypothesis f is not a polynomial of degree 0 in $\mathbb{D}_R$, by the proof of Theorem 3.1.3, we get $\left\| \frac{3e_2 f''}{2} + \frac{3e_1 f'}{2} \right\|_r > 0$.

But by Theorem 3.1.12 we have

$$n^3 \left\| J_n(f) - f + \frac{3e_2}{2n^2} f'' + \frac{3e_1}{2n^2} f' \right\|_r \le A_r(f).$$

Therefore, there exists an index n_0 depending only on f and r, such that for all $n \ge n_0$ we have

$$\left\| \frac{3e_2 f''}{2} + \frac{3e_1 f'}{2} \right\|_r - \frac{1}{n} \left[n^3 \left\| J_n(f) - f + \frac{3e_2}{2n^2} f'' + \frac{3e_1}{2n^2} f' \right\|_r \right]$$
$$\ge \frac{1}{2} \left\| \frac{3e_2 f''}{2} + \frac{3e_1 f'}{2} \right\|_r,$$

which immediately implies

$$\|J_n(f) - f\|_r \geq \frac{1}{n^2} \cdot \frac{1}{2} \left\| \frac{3e_2 f''}{2} + \frac{3e_1 f'}{2} \right\|_r, \forall n \geq n_0.$$

For $n \in \{1, ..., n_0 - 1\}$ we obviously have $\|J_n(f) - f\|_r \geq \frac{M_{r,n}(f)}{n^2}$ with $M_{r,n}(f) = n^2 \cdot \|J_n(f) - f\|_r > 0$, which finally implies $\|J_n(f) - f\|_r \geq \frac{C_r(f)}{n^2}$ for all n, where $C_r(f) = \min\{M_{r,1}(f), ..., M_{r,n_0-1}(f), \frac{1}{2} \left\| \frac{3e_2 f''}{2} + \frac{3e_1 f'}{2} \right\|_r\}$. This completes the proof. $\square$

Combining now Theorem 3.1.13 with Corollary 3.1.11, (i) we immediately get the following.

Corollary 3.1.14. *Let $R > 1$, $\mathbb{D}_R = \{z \in \mathbb{C}; |z| < R\}$ and let us suppose that $f : \mathbb{D}_R \to \mathbb{C}$ is analytic in $\mathbb{D}_R$. If f is not a polynomial of degree 0, then for any $r \in [1, R)$ we have*

$$\|J_n(f) - f\|_r \sim \frac{1}{n^2}, n \in \mathbb{N},$$

where the constants in the equivalence depend on f and r.

In the case of approximation by the derivatives of $J_n(f)(z)$ the following result holds.

Theorem 3.1.15. *Let $\mathbb{D}_R = \{z \in \mathbb{C}; |z| < R\}$ be with $R > 1$ and let us suppose that $f : \mathbb{D}_R \to \mathbb{C}$ is analytic in $\mathbb{D}_R$, i.e. $f(z) = \sum_{k=0}^{\infty} c_k z^k$, for all $z \in \mathbb{D}_R$. Also, let $1 \leq r < r_1 < R$ and $q \in \mathbb{N}$ be fixed. If f is not a polynomial of degree $\leq q - 1$, then we have*

$$\|J_n^{(q)}(f) - f^{(q)}\|_r \sim \frac{1}{n^2},$$

where the constants in the equivalence depend on f, r, r_1 and q.

Proof. Denoting by Γ the circle of radius r_1 and center 0 (where $r_1 > r \geq 1$), by the Cauchy's formulas it follows that for all $|z| \leq r$ and $n \in \mathbb{N}$ we have

$$J_n^{(q)}(f)(z) - f^{(q)}(z) = \frac{q!}{2\pi i} \int_{\Gamma} \frac{J_n(f)(v) - f(v)}{(v - z)^{q+1}} dv,$$

where the inequality $|v - z| \geq r_1 - r$ is valid for all $|z| \leq r$ and $v \in \Gamma$.

Taking into account Corollary 3.1.11, (ii), it remains to prove the lower estimate for $\|J_n^{(q)}(f) - f^{(q)}\|_r$.

For this purpose, as in the proof of Theorem 3.1.13, for all $v \in \Gamma$ and $n \in \mathbb{N}$ we have

$$J_n(f)(v) - f(v) = \frac{1}{n^2} \left\{ -\left[\frac{3v^2 f''(v)}{2} + \frac{3v f'(v)}{2} \right] \right.$$
$$\left. + \frac{1}{n} \left[n^3 \left(J_n(f)(v) - f(v) + \frac{3v^2}{2n^2} f''(v) + \frac{3v f'(v)}{2n^2} \right) \right] \right\},$$

which replaced in the above Cauchy's formula implies

$$J_n^{(q)}(f)(z) - f^{(q)}(z) = \frac{1}{n^2}\left\{ \frac{q!}{2\pi i}\int_\Gamma \frac{-\left[\frac{3v^2 f''(v)}{2} + \frac{3v f'(v)}{2}\right]}{(v-z)^{q+1}}dv \right.$$

$$\left. + \frac{1}{n}\cdot\frac{q!}{2\pi i}\int_\Gamma \frac{n^3\left(J_n(f)(v) - f(v) + \frac{3v^2}{2n^2}f''(v) + \frac{3v}{2n^2}f'(v)\right)}{(v-z)^{q+1}}dv \right\}$$

$$= \frac{1}{n^2}\left\{ \left[-\frac{3z^2 f''(z)}{2} - \frac{3z f'(z)}{2}\right]^{(q)} \right.$$

$$\left. + \frac{1}{n}\cdot\frac{q!}{2\pi i}\int_\Gamma \frac{n^3\left(J_n(f)(v) - f(v) + \frac{3v^2}{2n^2}f''(v) + \frac{3v}{2n^2}f'(v)\right)}{(v-z)^{q+1}}dv \right\}.$$

Passing now to $\|\cdot\|_r$ it follows

$$\|J_n^{(q)}(f) - f^{(q)}\|_r \geq \frac{1}{n^2}\left\{ \left\|\left[\frac{3e_2}{2}f'' + \frac{3e_1}{2}f'\right]^{(q)}\right\|_r \right.$$

$$\left. - \frac{1}{n}\left\|\frac{q!}{2\pi}\int_\Gamma \frac{n^3\left(J_n(f)(v) - f(v) + \frac{3v^2}{2n^2}f''(v) + \frac{3v}{2n^2}f'\right)}{(v-z)^{q+1}}dv\right\|_r \right\},$$

where by using Theorem 3.1.12 we obtain

$$\left\|\frac{q!}{2\pi}\int_\Gamma \frac{n^3\left(J_n(f)(v) - f(v) + \frac{3v^2}{2n^2}f''(v) + \frac{3v}{2n^2}f'\right)}{(v-z)^{q+1}}dv\right\|_r$$

$$\leq \frac{q!}{2\pi}\cdot\frac{2\pi r_1 n^3}{(r_1-r)^{q+1}}\left\|J_n(f) - f + \frac{3e_2}{2n^2}f'' + \frac{3e_1}{2n^2}f'\right\|_{r_1} \leq \frac{A_{r_1}(f)q!r_1}{(r_1-r)^{q+1}}.$$

But by the hypothesis on f and by the proof of Theorem 3.1.5 we have

$$\left\|\left[\frac{3e_2}{2}f'' + \frac{3e_1}{2}f'\right]^{(q)}\right\|_r > 0.$$

Finally, reasoning exactly as in the proof of Theorem 3.1.13, we immediately get the desired conclusion. □

In the case of Beatson-type convolutions $P_{n,2,p}(f)(z)$, $p \geq 2$, the following result concerning the exact degree of approximation holds.

Theorem 3.1.16. *Let $\mathbb{D}_R = \{z \in \mathbb{C}; |z| < R\}$ be with $R > 1$ and let us suppose that $f : \mathbb{D}_R \to \mathbb{C}$ is analytic in $\mathbb{D}_R$, i.e. $f(z) = \sum_{k=0}^\infty c_k z^k$, for all $z \in \mathbb{D}_R$. Also, let $1 \leq r < r_1 < R$, $p \in \mathbb{N}$, $p \geq 2$ even number and $q \in \mathbb{N}\bigcup\{0\}$ be fixed. If f is not a constant for $q = 0$ and is not a polynomial of degree $\leq q - 1$ for $q \in \mathbb{N}$, then for all $n \in \mathbb{N}$ we have*

$$\|P_{n,2,p}^{(q)}(f) - f^{(q)}\|_r \sim \frac{1}{n^2},$$

where the constants in the equivalence depend on f, r, r_1, p and q.

Proof. Denoting by Γ the circle of radius r_1 and center 0 (where $r_1 > r \geq 1$), by the Cauchy's formulas it follows that for all $|z| \leq r$ and $n, q \in \mathbb{N}$ we have

$$P_{n,2,p}^{(q)}(f)(z) - f^{(q)}(z) = \frac{q!}{2\pi i} \int_\Gamma \frac{P_{n,2,p}(f)(v) - f(v)}{(v-z)^{q+1}} dv,$$

where the inequality $|v - z| \geq r_1 - r$ is valid for all $|z| \leq r$ and $v \in \Gamma$.

Taking into account Corollary 3.1.11, (ii), for $q \in \mathbb{N}$ and Corollary 3.1.11, (i) for $q = 0$, in both cases it remains to prove the lower estimate for $\|P_{n,2,p}^{(q)}(f) - f^{(q)}\|_r$.

We will use here a different method of proof, for which we don't need a Voronovskaja-type formula.

By Theorem 3.1.10, (iii), $P_{n,2,p}(f)(z)$ is a polynomial of degree $\leq 2n - 2$. Since $f^{(q)}(z) = \sum_{k=q}^\infty k(k-1)...(k-q+1)c_k z^{k-q}$ and by Theorem 3.1.10, (iii), we have

$$P_{n,2,p}^{(q)}(f)(z) = \sum_{k=q}^{2n-2} c_k k(k-1)...(k-q+1) \left[\frac{n}{k\pi} \sin(k\pi/n)\right]^p \lambda_{k,n} z^{k-q},$$

with $\lambda_{k,n}$ given by Theorem 3.1.10, (i), for $z = re^{i\varphi}$ we can write (here $i^2 = -1$)

$$[P_{n,2,p}^{(q)}(f)(z) - f^{(q)}(z)]e^{-is\varphi}$$

$$= \sum_{k=q}^\infty c_k k(k-1)...(k-q+1)r^{k-q} \left\{\left[\frac{n}{k\pi}\sin(k\pi/n)\right]^p \lambda_{k,n} - 1\right\} e^{i\varphi(k-q-s)}$$

$$=: E_{n,p,q,s}(z).$$

Integrating from $-\pi$ to π we immediately obtain

$$\frac{1}{2\pi} \int_{-\pi}^\pi E_{n,p,q,s}(z) d\varphi$$

$$= c_{q+s}(q+s)(q+s-1)...(s+1)r^s \left\{\left[\frac{n}{(q+s)\pi}\sin((q+s)\pi/n)\right]^p \lambda_{q+s,n} - 1\right\}.$$

Taking into account that by the Remark 3 after Theorem 3.1.10 we have $0 \leq \lambda_{k,n} \leq 1$ for all k, n, we immediately obtain

$$\|P_{n,2,p}^{(q)}(f) - f^{(q)}\|_r$$

$$\geq |c_{q+s}| \frac{(q+s)!}{s!} r^s \left\{1 - \left[\frac{n}{(q+s)\pi}\sin((q+s)\pi/n)\right]^p \lambda_{q+s,n}\right\}$$

$$\geq |c_{q+s}| \frac{(q+s)!}{s!} r^s [1 - \lambda_{q+s,n}],$$

since $0 \leq \left[\frac{n}{(q+s)\pi}\sin((q+s)\pi/n)\right]^p \leq 1$ by the known inequality $|\sin(x)| \leq x$ for all $x \geq 0$.

First consider $q = 0$ and denote $V_n = \inf_{1 \leq s}(1 - \lambda_{s,n})$. By Remark 3 after Theorem 3.1.10 we get $V_n = 1 - \frac{4n^3 - 4n}{4n^3 + 2n} = \frac{3}{2n^2 + 1} \geq \frac{1}{n^2}$. By the above lower estimate for $\|P_{n,2,p}(f) - f\|_r$, for all $s \geq 1$ and $n \in \mathbb{N}$ it follows

$$n^2 \|P_{n,2,p}(f) - f\|_r \geq \frac{\|P_{n,2,p}(f) - f\|_r}{V_n} \geq \frac{\|P_{n,2,p}(f) - f\|_r}{1 - \lambda_{s,n}} \geq |c_s| r^s.$$

This implies that if there exists a subsequence $(n_k)_k$ with $\lim_{k\to\infty} n_k = +\infty$ and such that $\lim_{k\to\infty} n_k^2 \|P_{n_k,2,p}(f) - f\|_r = 0$ then $c_s = 0$ for all $s \geq 1$, that is f is constant on $\overline{\mathbb{D}}_r$.

Therefore, if f is not a constant then $\liminf n^2 \|P_{n,2,p}(f) - f\|_r > 0$, which implies that there exists a constant $C_{r,p}(f) > 0$ such that $n^2 \|P_{n,2,p}(f) - f\|_r \geq C_{r,p}(f)$, for all $n \in \mathbb{N}$, that is

$$\|P_{n,2,p}(f) - f\|_r \geq \frac{C_{r,p}(f)}{n^2}, n \in \mathbb{N}.$$

Now, consider $q \geq 1$ and denote $V_{q,n} = \inf_{s\geq 0}(1 - \lambda_{q+s,n})$. Evidently that we have $V_{q,n} \geq \inf_{s\geq 1}(1 - \lambda_{s,n}) \geq \frac{1}{n^2}$. Reasoning as in the case of $q = 0$ we obtain

$$n^2 \|P_{n,2,p}^{(q)}(f) - f^{(q)}\|_r \geq \frac{\|P_{n,2,p}^{(q)}(f) - f^{(q)}\|_r}{V_{q,n}} \geq \frac{\|P_{n,2,p}^{(q)}(f) - f^{(q)}\|_r}{1 - \lambda_{q+s,n}}$$

$$\geq |c_{q+s}| \frac{(q+s)!}{s!} r^s,$$

for all $s \geq 0$ and $n \in \mathbb{N}$. Now, if there exists a subsequence $(n_k)_k$ with $\lim_{k\to\infty} n_k = +\infty$ and such that $\lim_{k\to\infty} n_k^2 \|P_{n_k,2,p}^{(q)}(f) - f^{(q)}\|_r = 0$ then $c_{q+s} = 0$ for all $s \geq 0$, that is f is a polynomial of degree $\leq q - 1$ on $\overline{\mathbb{D}}_r$.

Therefore $\liminf n^2 \|P_{n,2,p}^{(q)}(f) - f^{(q)}\|_r > 0$ when f is not a polynomial of degree $\leq q - 1$, which implies that there exists a constant $C_{r,p,q}(f) > 0$ such that $n^2 \|P_{n,2,p}^{(q)}(f) - f^{(q)}\|_r \geq C_{r,p,q}(f)$, for all $n \in \mathbb{N}$. That is

$$\|P_{n,2,p}^{(q)}(f) - f^{(q)}\|_r \geq \frac{C_{r,p,q}(f)}{n^2}, n \in \mathbb{N},$$

which proves the theorem. $\square$

Remarks. 1) The case when p is an odd number in Theorem 3.1.16 remains unsettled.

2) By using the considerations in Section 1.0, the above complex convolutions can be generalized to approximation in a Jordan domain G whose boundary is rectifiable and of bounded rotation. Indeed, keeping the notations there, let $f \in A(\overline{G})$ and $F \in A(\overline{\mathbb{D}}_1)$ such that $f = T[F]$. As in e.g. Gal [93], p. 318, Theorem 3.1, (i), attach to $F(w) = \sum_{k=0}^{\infty} a_k w^k, w \in \overline{\mathbb{D}}_1$, the complex convolutions through a certain trigonometric kernel $O(t) = \frac{1}{2} + \sum_{k=1}^{m_n} \rho_{k,n} \cos(kt)$,

$$Q_n(F)(w) = a_0 + \sum_{k=1}^{m_n} a_k \rho_{k,n} w^k.$$

Then, suggested by the Remark after Theorem 1.0.12, attach to $f = T[F]$ the complex polynomial $L_n(f) = T[Q_n(F)]$ given by

$$L_n(f)(z) = \sum_{k=0}^{m_n} a_k \rho_{k,n} F_k(z), \ z \in \overline{G},$$

where $F_k(z)$ denotes the Faber polynomial of degree k attached to G.

But according to Theorem 1.0.11, we have $a_k = a_k(f)$-the Faber coefficients of f on G and according to Definition 1.0.10, (ii), we can write

$$a_k(f) = \frac{1}{2\pi i} \int_{|u|=1} \frac{f[\Psi(u)]}{u^{k+1}} du = \frac{1}{2\pi i} \int_{-\pi}^{\pi} f[\Psi(e^{it})] e^{-ikt} dt, k \in \mathbb{N} \cup \{0\}.$$

Therefore, the complex convolutions attached to f in $\overline{G}$ can be written as

$$L_n(f)(z) = \sum_{k=0}^{m_n} a_k(f) \rho_{k,n} F_k(z), \; z \in \overline{G}.$$

The exact orders of approximation in the cases when the trigonometric kernels are those of Fejér, Riesz-Zygmund and Rogosinski were obtained in Bruj-Schmieder [48], for details see Theorem 7 and its Remarks in Section 6 there.

Below we present with proof a saturation result for the complex convolutions $L_n(f)(z), n \in \mathbb{N}$, where for simplicity we take $m_n = n$.

Theorem 3.1.17. (Bruj-Schmieder [48], Theorem 4) *Let $G \subset \mathbb{C}$ be a Jordan domain with rectifiable boundary and $f \in A(\overline{G})$ a non-constant function. If $\rho_{k,n} \neq 1$ for all $n \in \mathbb{N}$, $k = 1, 2, ..., n$, then*

$$\max_{z \in \overline{G}} |f(z) - L_n(f)(z)| \neq o(\min_{1 \leq k \leq n} |1 - \rho_{k,n}|).$$

Recall here that by definition $a_n = o(b_n)$ if $\lim_{n \to \infty} \frac{a_n}{b_n} = 0$.

Proof. By Definition 1.0.10 (ii) we have $a_m(f) = \frac{1}{2\pi i} \int_{|u|=1} \frac{f(\Psi(u))}{u^{m+1}} du$. Also, taking into account that by Kövari-Pommerenke [120], p. 198, Lemma 2, the integral $\frac{1}{2\pi i} \int_{|u|=1} \frac{F_n(\Psi(u))}{u^{m+1}} du$ is equal to zero for $m \neq n$ and equal to 1 for $m = n$, we immediately obtain

$$\frac{1}{2\pi i} \int_{|u|=1} \frac{f(\Psi(u)) - L_n(f)(\Psi(u))}{u^{m+1}} du = (1 - \rho_{m,n}) a_m(f), n \geq m \geq 0.$$

This implies

$$|1 - \rho_{m,n}| \cdot |a_m(f)| \leq \max_{z \in \overline{G}} |f(z) - L_n(f)(z)|.$$

Now, let us assume that the conclusion of Theorem 3.1.17 is false, that is

$$\max_{z \in \overline{G}} |f(z) - L_n(f)(z)| = o(\min_{1 \leq k \leq n} |1 - \rho_{k,n}|).$$

It implies

$$|1 - \rho_{m,n}| \cdot |a_m(f)| = o(\min_{1 \leq k \leq n} |1 - \rho_{k,n}|)$$

and therefore

$$\lim_{n \to \infty} \frac{|1 - \rho_{m,n}| \cdot |a_m(f)|}{\min_{1 \leq k \leq n} |1 - \rho_{k,n}|} = 0,$$

which means $a_m(f) = 0$, for all $m \geq 1$. Then, by Theorem 1, p. 44 in Gaier [76] we get $f(z) = a_0(f)$ for all $z \in \overline{G}$, a contradiction.

In conclusion, the statement of Theorem 3.1.17 is valid. $\square$

Remarks. 1) (Example A in Bruj-Schmieder [48], p. 161) If we take $\rho_{k,n} :=$ $\max\{0, 1 - k/(n+1)\}$, that is the Fejér-type convolution, then $L_n(f)(z)$ approximate f on $\overline{G}$ not better than the order $\min_{1 \leq m \leq n} |1 - \rho_{m,n}| = 1/(n+1)$.

2) (Example B in Bruj-Schmieder [48], p. 161) If we take $\rho_{k,n} := \max\{0, 1 - [k/(n+1)]^s\}$, where $s \geq 1$ is fixed, that is the Riesz-Zygmund-type convolution, then $L_n(f)(z)$ approximate f on $\overline{G}$ not better than the order $\min_{1 \leq m \leq n} |1 - \rho_{m,n}| = [1/(n+1)]^s$.

3) For the so-called Rogosinki-type convolution, in Bruj-Schmieder [48], p. 162, similarly it is proved that approximate f on $\overline{G}$ not better than the order $1/n^2$.

3.2   Linear Non-Polynomial Convolutions

In several recent papers, the geometric properties of complex linear integral transforms generated by the Hadamard product (convolution) of normalized analytic functions f in the unit disk with various kinds of kernels, when the kernels are special functions such as hypergeometric functions or Hurwitz kind functions were intensively studied, see e.g. Choi-Kim-Saigo [54], Dziok-Srivastava [68], Kanas-Srivastava [111], Murugusundaramoorthy-Magesh [142], Ponnusamy-Ronning [152], Prajapat-Raina-Srivastava [157], Răducanu-Srivastava [160], Srivastava-Attiya [169], Srivastava et al [170]. A common characteristic of these complex convolutions is that although they preserve the geometric properties of f like starlikeness or convexity, they do not have good approximation properties.

In this section we study some complex linear integral convolutions of complex analytic f with various real non-trigonometric kernels, which besides the preservations of some geometric properties of f have also good approximation properties. More exactly, we will refer to the complex linear integral convolutions of Picard kind, Gauss-Weierstrass kind, Poisson-Cauchy kind, Post-Widder kind, rotational-invariant kind, Sikkema kind and spline kind. Most of the results were obtained in Anastassiou-Gal [18], Anastassiou-Gal [16], Anastassiou-Gal [17], Anastassiou-Gal [19], Anastassiou-Gal [20], but also new results are presented. Applications to heat and Laplace equation with complex spatial variables are presented, see Gal-Gal-Goldstein [97].

The above mentioned results hold in compact disks of the complex plane, but extensions to Jordan domains whose boundaries are rectifiable and of bounded rotation also are presented.

For $\mathbb{D}_1 = \{z \in \mathbb{C}; |z| < 1\}$ let us define

$$A(\overline{\mathbb{D}}_1) = \{f : \overline{\mathbb{D}}_1 \to \mathbb{C}; f \text{ is analytic on } \mathbb{D}_1, \text{ and continuous on } \overline{\mathbb{D}}_1\},$$

endowed with the uniform norm $\|f\|_1 = \sup\{|f(z)|; z \in \overline{\mathbb{D}}_1\}$. Is well-known that $(A(\overline{\mathbb{D}}_1), \|\cdot\|_1)$ is a Banach space. Therefore if $f \in A(\overline{\mathbb{D}}_1)$ then we have $f(z) = \sum_{k=0}^{\infty} a_k z^k$, for all $z \in \mathbb{D}_1$.

Also, define the subclass of $A(\overline{\mathbb{D}}_1)$ of normalized functions by

$$A^*(\overline{\mathbb{D}}_1) = \{f \in A(\overline{\mathbb{D}}_1); f(0) = 0, f'(0) = 1\}.$$

For the geometric properties, the following classes of analytic functions will be of interest :

$$S^*(\mathbb{D}_1) = \{f \in A^*(\overline{\mathbb{D}}_1); Re\left[\frac{zf'(z)}{f(z)}\right] > 0, \text{ for all } z \in \mathbb{D}_1\},$$

$$S_\gamma^*(\mathbb{D}_1) = \{f \in A^*(\overline{\mathbb{D}}_1); Re\left[e^{i\gamma}\frac{zf'(z)}{f(z)}\right] > 0, \text{ for all } z \in \mathbb{D}_1\}, |\gamma| < \pi/2,$$

$$K(\mathbb{D}_1) = \{f \in A^*(\overline{\mathbb{D}}_1); Re\left[\frac{zf''(z)}{f'(z)} + 1\right] > 0, \text{ for all } z \in \mathbb{D}_1\},$$

$$S_1 = \left\{f \in A^*(\overline{\mathbb{D}}_1); \; f(z) = z + \sum_{k=2}^\infty a_k z^k, \; \sum_{k=2}^\infty k|a_k| \le 1\right\},$$

$$S_2 = \left\{f \text{ is analytic in } \mathbb{D}_1, \; f(z) = \sum_{k=1}^\infty a_k z^k, \; z \in \mathbb{D}_1, \quad |a_1| \ge \sum_{k=2}^\infty |a_k|\right\},$$

$$S_3 = \{f \in A^*(\overline{\mathbb{D}}_1); |f''(z)| \le 1, \; \forall z \in \mathbb{D}_1\},$$

$$\mathcal{P} = \{f \colon \overline{\mathbb{D}}_1 \to \mathbb{C}; f \text{ is analytic on } \mathbb{D}_1, f(0) = 1, \mathrm{Re}[f(z)] > 0, \; \forall z \in \mathbb{D}_1\},$$

$$\mathcal{R} = \{f \in A^*(\overline{\mathbb{D}}_1); \; \mathrm{Re}[f'(z)] > 0, \; \forall z \in \mathbb{D}_1\},$$

$$S_M = \{f \in A^*(\overline{\mathbb{D}}_1); |f'(z)| < M, \; \forall z \in \mathbb{D}_1\}, \quad M > 1.$$

According to e.g. Mocanu-Bulboacă-Sălăgean [138], p. 97, Exercise 4.9.1, if $f \in S_1$ then $\left|\frac{zf''(z)}{f(z)} - 1\right| < 1, \forall z \in \mathbb{D}_1$ and therefore f is starlike (and univalent) on $\mathbb{D}_1$.

According to Alexander [5], p. 22, if $f \in S_2$ then f is starlike (and univalent) on $\mathbb{D}_1$.

By Obradović [145], if $f \in S_3$ then f is starlike (and univalent) on $\mathbb{D}_1$. Also, it is well known that $\mathcal{R}$ is the class of functions with bounded turn (i.e. $|\arg f'(z)| < \frac{\pi}{2}$, $\forall z \in \mathbb{D}_1$) and that $f \in \mathcal{R}$ implies the univalence of f on $\mathbb{D}_1$.

According to e.g. Mocanu-Bulboacă-Sălăgean [138], p. 111, Exercise 5.4.1, $f \in S_M$ implies that f is univalent in $\left\{z \in \mathbb{C}; |z| < \frac{1}{M}\right\}$.

3.2.1 *Picard, Poisson-Cauchy and Gauss-Weierstrass Complex Convolutions*

In this subsection for $f \in A(\overline{\mathbb{D}}_1)$, $z \in \overline{\mathbb{D}}_1$ and $\xi \in \mathbb{R}$, $\xi > 0$, we will study the approximation and geometric properties of the following complex linear convolutions

$$P_\xi(f)(z) = \frac{1}{2\xi}\int_{-\infty}^{+\infty} f(ze^{iu})e^{-|u|/\xi}\,du,$$

$$Q_\xi(f)(z) = \frac{\xi}{\pi}\int_{-\pi}^{\pi} \frac{f(ze^{iu})}{u^2 + \xi^2}\,du, \quad Q_\xi^*(f)(z) = \frac{\xi}{\pi}\int_{-\infty}^{+\infty} \frac{f(ze^{-iu})}{u^2 + \xi^2}\,du,$$

$$R_\xi(f)(z) = \frac{2\xi^3}{\pi} \int_{-\infty}^{+\infty} \frac{f(ze^{iu})}{(u^2 + \xi^2)^2} \, du,$$

$$W_\xi(f)(z) = \frac{1}{\sqrt{\pi\xi}} \int_{-\pi}^{\pi} f(ze^{iu})e^{-u^2/\xi} \, du,$$

$$W_\xi^*(f)(z) = \frac{1}{\sqrt{\pi\xi}} \int_{-\infty}^{+\infty} f(ze^{-iu})e^{-u^2/\xi} \, du,$$

and their Jackson-type generalizations

$$P_{n,\xi}(f)(z) = -\frac{1}{2\xi} \int_{-\infty}^{+\infty} \left[e^{-|u|/\xi} \sum_{k=1}^{n+1} (-1)^k \binom{n+1}{k} f(ze^{iku}) \right] du,$$

$$Q_{n,\xi}(f)(z) = -\frac{1}{\frac{2}{\xi}\operatorname{arctg}\left(\frac{\pi}{\xi}\right)} \cdot \sum_{k=1}^{n+1} (-1)^k \binom{n+1}{k} \int_{-\pi}^{\pi} \frac{f(ze^{iku})}{u^2 + \xi^2} \, du,$$

$$W_{n,\xi}(f)(z) = -\frac{1}{2C(\xi)} \cdot \sum_{k=1}^{n+1} (-1)^k \binom{n+1}{k} \int_{-\pi}^{\pi} f(ze^{iku})e^{-u^2/\xi^2} \, du,$$

$$W_{n,\xi}^*(f)(z) = -\frac{1}{2C^*(\xi)} \cdot \sum_{k=1}^{n+1} (-1)^k \binom{n+1}{k} \int_{-\infty}^{+\infty} f(ze^{iku})e^{-u^2/\xi^2} \, du,$$

where $n \in \mathbb{N}$, $C(\xi) = \int_0^\pi e^{-u^2/\xi^2} \, du$ and $C^*(\xi) = \int_0^\infty e^{-u^2/\xi^2} \, du$.

Here $P_\xi(f)$ is called of Picard–type, $Q_\xi(f)$, $Q_\xi^*(f)$ and $R_\xi(f)$ are called of Poisson–Cauchy–type, $W_\xi(f)$, $W_\xi^*(f)$ are called of Gauss–Weierstrass–type, $P_{n,\xi}(f)(z)$ is called of generalized Picard–type, $Q_{n,\xi}(f)(z)$ is called of generalized Poisson–Cauchy–type and $W_{n,\xi}(f)(z)$, $W_{n,\xi}^*(f)(z)$ are called of generalized Gauss–Weierstrass–type.

The approximation properties of the Picard–type convolution $P_\xi(f)$ are expressed by the following result. Note that while below Theorem 3.2.1, (i), (ii) and (iii) were obtained in Anastassiou–Gal [18], Theorem 3.2.1, (iv) is new.

Theorem 3.2.1. *Let $\xi \in \mathbb{R}$, $\xi > 0$.*

(i) If $f \in A(\overline{\mathbb{D}}_1)$ then $P_\xi(f) \in A(\overline{\mathbb{D}}_1)$, that is it is continuous on $\overline{\mathbb{D}}_1$ and analytic on $\mathbb{D}_1$. Moreover, if $f(z) = \sum_{k=0}^\infty a_k z^k$ for all $z \in \mathbb{D}_R$, $R \geq 1$, then

$$P_\xi(f)(z) = \sum_{k=0}^\infty \frac{a_k}{1 + \xi^2 k^2} z^k,$$

for all $z \in \mathbb{D}_R$, if $f(0) = 0$ then $P_\xi(f)(0) = 0$ and if $f'(0) = 1$ then $P_\xi'(f)(0) = \frac{1}{1+\xi^2} \neq 1$, for all $\xi > 0$.

(ii) If f is continuous on $\overline{\mathbb{D}}_1$ then for all $\delta \geq 0$ and $\xi > 0$

$$\omega_1(P_\xi(f); \delta)_{\overline{\mathbb{D}}_1} \leq \omega_1(f; \delta)_{\overline{\mathbb{D}}_1},$$

where $\omega_1(f; \delta)_{\overline{\mathbb{D}}_1} = \sup\{|f(z_1) - f(z_2)|; z_1, z_2 \in \overline{\mathbb{D}}_1, |z_1 - z_2| \leq \delta\}$;

(iii) If $f \in A(\overline{\mathbb{D}}_1)$ then for all $z \in \overline{\mathbb{D}}_1$ and $\xi > 0$,

$$|P_\xi(f)(z) - f(z)| \leq C\omega_2(f; \xi)_{\partial \mathbb{D}_1},$$

where

$$\omega_2(f; \xi)_{\partial \mathbb{D}_1} = \sup\{|f(e^{i(x+u)}) - 2f(e^{iu}) + f(e^{i(x-u)})|; \ x \in \mathbb{R}, \ |u| \leq \xi\}.$$

(iv) Let us suppose that $f(z) = \sum_{k=0}^\infty a_k z^k$ for all $z \in \mathbb{D}_R$, $R > 1$. If f is not constant for $q = 0$ and not a polynomial of degree $\leq q - 1$ for $q \in \mathbb{N}$, then for all $1 \leq r < r_1 < R$, $\xi \in (0, 1]$ and $q \in \mathbb{N} \bigcup \{0\}$ we have

$$\|P_\xi^{(q)}(f) - f^{(q)}\|_r \sim \xi^2,$$

where the constants in the equivalence depend only on f, q, r, r_1.

Proof. (i) Let $z_0, z_n \in \overline{\mathbb{D}}_1$ be with $\lim_{n \to \infty} z_n = z_0$. We get

$$|P_\xi(f)(z_n) - P_\xi(f)(z_0)| \leq \frac{1}{2\xi} \int_{-\infty}^{+\infty} |f(z_n e^{iu}) - f(z_0 e^{iu})| e^{-|u|/\xi} \, du$$

$$\leq \frac{1}{2\xi} \int_{-\infty}^{+\infty} \omega_1(f; |z_n e^{iu} - z_0 e^{iu}|)_{\overline{\mathbb{D}}_1} e^{-|u|/\xi} \, du$$

$$= \frac{1}{2\xi} \int_{-\infty}^{+\infty} \omega_1(f; |z_n - z_0|)_{\overline{\mathbb{D}}_1} e^{-|u|/\xi} \, du$$

$$= \omega_1(f; |z_n - z_0|)_{\overline{\mathbb{D}}_1}.$$

Passing to limit with $n \to \infty$, it follows that $P_\xi(f)(z)$ is continuous at $z_0 \in \overline{\mathbb{D}}_1$, since f is continuous on $\overline{\mathbb{D}}_1$. It remains to prove that $P_\xi(f)(z)$ is analytic on $\mathbb{D}$. For $f \in A(\overline{\mathbb{D}}_1)$, we can write $f(z) = \sum_{k=0}^\infty a_k z^k$, $z \in \mathbb{D}_1$. For fixed $z \in \mathbb{D}_1$, we get $f(ze^{iu}) = \sum_{k=0}^\infty a_k e^{iku} z^k$ and since $|a_k e^{iku}| = |a_k|$, for all $u \in \mathbb{R}$ and the series $\sum_{k=0}^\infty a_k z^k$ is absolutely convergent, it follows that the series $\sum_{k=0}^\infty a_k e^{iku} z^k$ is uniformly convergent with respect to $u \in \mathbb{R}$. This immediately implies that the series can be integrated term by term, i.e.

$$P_\xi(f)(z) = \frac{1}{2\xi} \sum_{k=0}^\infty a_k z^k \left(\int_{-\infty}^\infty e^{iku} e^{-|u|/\xi} \, du \right).$$

Now let $f(z) = \sum_{k=0}^\infty a_k z^k$ for all $z \in \mathbb{D}_R$, $R \geq 1$.

From the above reasonings we can write

$$P_\xi(f)(z) = \sum_{k=0}^\infty a_k z^k \left[\frac{1}{2\xi} \int_{-\infty}^{+\infty} e^{iku} e^{-|u|/\xi} \, du \right], \quad \forall z \in \mathbb{D}_1.$$

But

$$\frac{1}{2\xi} \int_{-\infty}^{+\infty} e^{iku} e^{-|u|/\xi} \, du$$

$$= \frac{1}{2\xi} \int_{-\infty}^{+\infty} \cos(ku) \cdot e^{-|u|/\xi} \, du = \frac{1}{\xi} \int_{0}^{+\infty} \cos(ku) e^{-u/\xi} \, du$$

$$= \frac{1}{\xi} \cdot \frac{e^{-u/\xi} \left[-\frac{1}{\xi} \cos(ku) + k \sin(ku) \right]}{\frac{1}{\xi^2} + k^2} \Bigg|_{0}^{\infty} = \frac{1}{1 + k^2 \xi^2}.$$

Also, if $a_0 = 0$ then $P_\xi(f)(0) = 0$ and if $a_1 = 0$ then $P_\xi(f)(0) = \frac{1}{1+\xi^2}$, which proves (i).

(ii) Let $z_1, z_2 \in \overline{\mathbb{D}}_1$, $|z_1 - z_2| \leq \delta$. We get

$$|P_\xi(f)(z_1) - P_\xi(f)(z_2)| \leq \frac{1}{2\xi} \int_{-\infty}^{+\infty} |f(z_1 e^{iu}) - f(z_2 e^{iu})| e^{-|u|/\xi} \, du$$

$$\leq \omega_1(f; |z_1 - z_2|)_{\overline{\mathbb{D}}_1} \leq \omega_1(f; \delta)_{\overline{\mathbb{D}}_1}.$$

Passing to sup with $|z_1 - z_2| < \delta$, it follows the desired inequality.

(iii) We have

$$P_\xi(f)(z) - f(z) = \frac{1}{2\xi} \int_{-\infty}^{+\infty} [f(ze^{iu}) - f(z)] e^{-|u|/\xi} \, du$$

$$= \frac{1}{2\xi} \int_{0}^{\infty} [f(ze^{iu}) - 2f(z) + f(ze^{-iu})] e^{-u/\xi} \, du,$$

which implies

$$|P_\xi(f)(z) - f(z)| \leq \frac{1}{2\xi} \int_{0}^{\infty} |f(ze^{iu}) - 2f(z) + f(ze^{-iu})| e^{-u/\xi} \, du,$$

for all $z \in \overline{\mathbb{D}}_1$.

By the maximum modulus principle we can take $|z| = 1$, case when

$$|f(ze^{iu}) - 2f(z) + f(ze^{-iu})| \leq \omega_2(f; u)_{\partial \mathbb{D}_1},$$

which implies that for all $z \in \overline{\mathbb{D}}_1$ we have

$$|P_\xi(f)(z) - f(z)| \leq \frac{1}{2\xi} \int_{0}^{+\infty} \omega_2(f; u)_{\partial \mathbb{D}_1} e^{-u/\xi} du$$

$$= \frac{1}{2\xi} \int_{0}^{+\infty} \omega_2\left(f; \frac{u}{\xi} \cdot \xi\right)_{\partial \mathbb{D}_1} e^{-u/\xi} \, du$$

$$\leq \left(\frac{1}{2\xi} \int_{0}^{+\infty} \left[1 + \frac{u}{\xi} \right]^2 e^{-u/\xi} du \right) \omega_2(f; \xi)_{\partial \mathbb{D}_1}$$

$$\leq C \omega_2(f; \xi)_{\partial \mathbb{D}_1},$$

(for the last inequalities see e.g. Gal [94], p. 252, proof of Theorem 2.1, (i)).

(iv) Analysing the proofs of the above points (i)-(iii), it easily follows that they hold by replacing everywhere $\mathbb{D}_1$ with $\mathbb{D}_r$, where $1 \leq r < R$. Therefore the approximation error in (iii) can be expressed in terms of

$$\omega_2(f; \xi)_{\partial \mathbb{D}_r} = \sup\{|\Delta_u^2 f(re^{it})|; |t| \leq \pi, \ |u| \leq \xi\},$$

with constants in front of ω_2 depending on $r \geq 1$.

Since $\Delta_u^2 g(t) = 2u^2[t, t+u, t+2u; g]$, by the mean value theorem for divided differences in Complex Analysis (see e.g. Stancu [172], p. 258, Exercise 4.20) we get

$$|\Delta_u^2 f(re^{it})| \leq u^2 \|f''\|_r,$$

where

$$\|f''\|_r = \sup\{|f''(z)|; |z| \leq r\} \leq \sum_{k=2}^{\infty} |a_k| k(k-1) r^{k-2}.$$

For all $\xi \in (0, 1]$ from (iii) it follows

$$\|P_\xi(f) - f\|_r \leq C_r(f)\xi^2.$$

Now denoting by γ the circle of radius $r_1 > 1$ and center 0, since for any $|z| \leq r$ and $v \in \gamma$, we have $|v - z| \geq r_1 - r$, by the Cauchy's formulas it follows that for all $|z| \leq r$ and $\xi \in (0, 1]$, we have

$$|P_\xi^{(q)}(f)(z) - f^{(q)}(z)| = \frac{q!}{2\pi} \left| \int_\gamma \frac{P_\xi(f)(v) - f(v)}{(v - z)^{q+1}} dv \right|$$

$$\leq C_{r_1}(f)\xi^2 \frac{q!}{2\pi} \frac{2\pi r_1}{(r_1 - r)^{q+1}}$$

$$= C_{r_1}(f)\xi^2 \frac{q! r_1}{(r_1 - r)^{q+1}},$$

which proves the upper estimate in approximation by $P_\xi^{(q)}(f)(z)$.

It remains to prove the lower estimate for $\|P_\xi^{(q)}(f) - f^{(q)}\|_r$. For this purpose, take $z = re^{i\varphi}$ and $p \in \mathbb{N} \bigcup \{0\}$. We have

$$\frac{1}{2\pi}[f^{(q)}(z) - P_\xi^{(q)}(f)(z)]e^{-ip\varphi}$$

$$= \frac{1}{2\pi} \sum_{k=q}^{\infty} a_k k(k-1)...(k-q+1) r^{k-q} e^{i\varphi(k-q-p)} \left[1 - \frac{1}{1 + \xi^2 k^2}\right].$$

Integrating from $-\pi$ to π we obtain

$$\frac{1}{2\pi} \int_{-\pi}^{\pi} [f^{(q)}(z) - P_\xi^{(q)}(f)(z)]e^{-ip\varphi} d\varphi$$

$$= a_{q+p}(q+p)(q+p-1)...(p+1) r^p \frac{\xi^2 (q+p)^2}{1 + \xi^2 (q+p)^2}.$$

Passing now to absolute value we easily obtain

$$|a_{q+p}|(q+p)(q+p-1)...(p+1) r^p \frac{\xi^2 (q+p)^2}{1 + \xi^2 (q+p)^2} \leq \|f^{(q)} - P_\xi^{(q)}(f)\|_r.$$

First consider $q = 0$ and denote $V_\xi = \inf_{1 \leq p}(\frac{\xi^2 p^2}{1 + \xi^2 p^2})$. We get $V_\xi = \frac{\xi^2}{1 + \xi^2}$ and for all $\xi \in (0, 1]$ it follows $V_\xi \geq \xi^2/2$.

By the above lower estimate for $\|P_\xi(f) - f\|_r$, for all $p \geq 1$ and $\xi \in (0,1]$ it follows

$$\frac{2\|P_\xi(f) - f\|_r}{\xi^2} \geq \frac{\|P_\xi(f) - f\|_r}{V_\xi} \geq \frac{\|P_\xi(f) - f\|_r}{\frac{\xi^2 p^2}{1+\xi^2 p^2}} \geq |a_p| r^p.$$

This implies that if there exists a subsequence $(\xi_k)_k$ in $(0,1]$ with $\lim_{k\to\infty} \xi_k = 0$ and such that $\lim_{k\to\infty} \frac{\|P_{\xi_k}(f) - f\|_r}{\xi_k^2} = 0$ then $a_p = 0$ for all $p \geq 1$, that is f is constant on $\overline{\mathbb{D}}_r$.

Therefore, if f is not a constant then $\inf_{\xi \in (0,1]} \frac{\|P_\xi(f) - f\|_r}{\xi^2} > 0$, which implies that there exists a constant $C_r(f) > 0$ such that $\frac{\|P_\xi(f) - f\|_r}{\xi^2} \geq C_r(f)$, for all $\xi \in (0,1]$, that is

$$\|P_\xi(f) - f\|_r \geq C_r(f)\xi^2, \text{ for all } \xi \in (0,1].$$

Now, consider $q \geq 1$ and denote $V_{q,\xi} = \inf_{p \geq 0}(\frac{\xi^2(q+p)^2}{1+\xi^2(q+p)^2})$. Evidently that we have $V_{q,\xi} \geq \inf_{p \geq 1}(\frac{\xi^2 p^2}{1+\xi^2 p^2}) \geq \xi^2/2$.

Reasoning as in the case of $q = 0$ we obtain

$$\frac{2\|P_\xi^{(q)}(f) - f^{(q)}\|_r}{\xi^2} \geq \frac{\|P_\xi^{(q)}(f) - f^{(q)}\|_r}{V_{q,\xi}} \geq \frac{\|P_\xi^{(q)}(f) - f^{(q)}\|_r}{\frac{\xi^2(q+p)^2}{1+\xi^2(q+p)^2}}$$

$$\geq |a_{q+p}| \frac{(q+p)!}{p!} r^p,$$

for all $p \geq 0$ and $\xi \in (0,1]$.

This implies that if there exists a subsequence $(\xi_k)_k$ in $(0,1]$ with $\lim_{k\to\infty} \xi_k = 0$ and such that $\lim_{k\to\infty} \frac{\|P_{\xi_k}^{(q)}(f) - f^{(q)}\|_r}{\xi_k^2} = 0$ then $a_{q+p} = 0$ for all $p \geq 0$, that is f is a polynomial of degree $\leq q - 1$ on $\overline{\mathbb{D}}_r$.

Therefore, $\inf_{\xi \in (0,1]} \frac{\|P_\xi^{(q)}(f) - f^{(q)}\|_r}{\xi^2} > 0$ when f is not a polynomial of degree $\leq q - 1$, which implies that there exists a constant $C_{r,q}(f) > 0$ such that $\frac{\|P_\xi^{(q)}(f) - f^{(q)}\|_r}{\xi^2} \geq C_{r,q}(f)$, for all $\xi \in (0,1]$, that is

$$\|P_\xi^{(q)}(f) - f^{(q)}\|_r \geq C_{r,q}(f)\xi^2, \text{ for all } \xi \in (0,1],$$

which proves the theorem. $\qquad \square$

In what follows, we present some geometric properties of $P_\xi(f)(z)$.

Theorem 3.2.2. (Anastassiou-Gal [18]) *For all $\xi > 0$ we have*

$$P_\xi(S_2) \subset S_2 \quad and \quad P_\xi(\mathcal{P}) \subset \mathcal{P}.$$

Proof. By Theorem 3.2.1, for $f(z) = \sum_{k=1}^{\infty} a_k z^k \in S_2$, we get

$$\sum_{k=2}^{\infty} \left| \frac{a_k}{1+\xi^2 k^2} \right| = \sum_{k=2}^{\infty} \frac{|a_k|}{1+\xi^2} \cdot \frac{1+\xi^2}{1+\xi^2 k^2} \leq \frac{1}{1+\xi^2} \sum_{k=2}^{\infty} |a_k| \leq \frac{|a_1|}{1+\xi^2}$$

and since $P_\xi(f)(z) = \sum_{k=0}^{\infty} \frac{a_k}{1+\xi^2 k^2} z^k$, it follows $P_\xi(f) \in S_2$.

Now, let $f(z) = \sum_{k=0}^{\infty} a_k z^k \in \mathcal{P}$, that is $a_0 = 1$ and if $f(z) = U(x,y) + iV(x,y)$, $z = x + iy \in \mathbb{D}_1$, then $U(x,y) > 0$, $\forall z = x + iy \in \mathbb{D}_1$.

We get $P_\xi(f)(0) = a_0 = 1$ and

$$P_\xi(f)(z) = \frac{1}{2\xi} \int_{-\infty}^{+\infty} U(r\cos(u+t), r\sin(u+t)) e^{-|u|/\xi}\, du$$

$$+ i \cdot \frac{1}{2\xi} \int_{-\infty}^{+\infty} V(r\cos(u+t), r\sin(u+t)) e^{-|u|/\xi}\, du,$$

for all $z = re^{it} \in \mathbb{D}_1$, which immediately implies

$$\mathrm{Re}[P_\xi(f)(z)] = \frac{1}{2\xi} \int_{-\infty}^{+\infty} U(r\cos(u+t), r\sin(u+t)) e^{-|u|/\xi}\, du > 0,$$

that is $P_\xi(f) \in \mathcal{P}$. $\qquad\square$

Theorem 3.2.3. (Anastassiou-Gal [18]) *For all $\xi > 0$ we have $(1+\xi^2)P_\xi(S_1) \subset S_1$, $(1+\xi^2)P_\xi(S_M) \subset S_{M(1+\xi^2)}$ and $(1+\xi^2)P_\xi(S_{3,\xi}) \subset S_3$, where*

$$S_{3,\xi} = \left\{ f \in S_3;\ |f''(z)| \le \frac{1}{1+\xi^2},\ \forall z \in D \right\} \subset S_3.$$

Proof. Let $f \in S_1$. By Theorem 3.2.1, (i) we obtain

$$(1+\xi^2)P_\xi(f)(z) = \sum_{k=1}^{\infty} a_k \frac{1+\xi^2}{1+\xi^2 k^2} z^k,$$

if $f(z) = \sum_{k=1}^{\infty} a_k z^k \in S_1$. It follows $(1+\xi^2)P_\xi'(f)(0) = a_1 = 1$, that is

$$(1+\xi^2)P_\xi(f)(z) = z + \sum_{k=2}^{\infty} a_k \cdot \frac{1+\xi^2}{1+\xi^2 k^2} z^k$$

and

$$\sum_{k=2}^{\infty} k|a_k| \frac{1+\xi^2}{1+\xi^2 k^2} \le \sum_{k=2}^{\infty} k|a_k| \le 1,$$

that is $(1+\xi^2)P_\xi(f) \in S_1$.

Let $f \in S_M$. We get

$$|(1+\xi^2)P_\xi'(f)(z)| = (1+\xi^2) \cdot \left| \frac{1}{2\xi} \int_{-\infty}^{+\infty} f'(ze^{iu}) e^{iu} e^{-|u|/\xi}\, du \right|$$

$$\le (1+\xi^2) \frac{1}{2\xi} \int_{-\infty}^{+\infty} |f'(ze^{iu})| e^{-|u|/\xi}\, du$$

$$< M(1+\xi^2), \quad z \in \mathbb{D}_1.$$

Also, $P_\xi(f)(0) = 0$ and $(1 + \xi^2)P'_\xi(f)(0) = 1$, which implies that $(1 + \xi^2)P_\xi(f) \in S_{M(1+\xi^2)}$.

Now, let $f \in S_{3,\xi}$. We have

$$(1 + \xi^2)P''_\xi(f)(z) = (1 + \xi^2) \cdot \frac{1}{2\xi} \int_{-\infty}^{+\infty} f''(ze^{iu})e^{2iu}e^{-|u|/\xi}\, du,$$

which implies

$$|(1 + \xi^2)P''_\xi(f)(z)| \le (1 + \xi^2)\frac{1}{2\xi} \cdot \int_{-\infty}^{+\infty} |f''(ze^{iu})|e^{-|u|/\xi}\, du \le 1,$$

that is $(1 + \xi^2)P_\xi(f) \in S_3$. $\qquad\qquad\qquad\qquad\qquad\qquad\qquad$ $\square$

Remarks. 1) Since the constant $(1+\xi^2)$ does not influence the geometric properties of $P_\xi(f)$, it follows that for all $\xi > 0$ we have:

if $f \in S_1$ then $P_\xi(f)$ is starlike (and univalent) in $\mathbb{D}_1$;

if $f \in S_M$ then $P_\xi(f)$ is univalent in $\{z \in \mathbb{C};\, |z| < \frac{1}{M(1+\xi^2)}\}$;

if $f \in S_{3,\xi} \subset S_3$ then $P_\xi(f)$ is starlike and univalent in $\mathbb{D}_1$.

2) Since

$$P'_\xi(f)(z) = \frac{1}{2\xi} \int_{-\infty}^{+\infty} f'(ze^{iu})e^{iu}e^{-|u|/\xi}\, du,$$

it is obvious that the condition $\mathrm{Re}[f'(z)] > 0$, $\forall z \in \mathbb{D}_1$, does not imply $\mathrm{Re}[P'_\xi(f)(z)] > 0$ on $\mathbb{D}_1$.

In this case, we may follow the idea in e.g. Gal [93] Theorem 3.4, to construct another singular integral, as follows: for $f \in A^*(\overline{\mathbb{D}}_1)$, we define $S_\xi(f)(z) = \int_0^z Q_n(u)\, du$ with

$$Q_n(z) = \frac{1}{2\xi} \int_{-\infty}^{+\infty} f'(ze^{it})e^{-|t|/\xi}\, dt.$$

Then, it is an easy task to show that $(1 + \xi^2)S_\xi(\mathcal{R}) \subset \mathcal{R}$, for all $\xi > 0$ and the following estimate holds

$$|S_\xi(f)(z) - f(z)| \le C\omega_2(f';\xi)_{\partial\mathbb{D}_1}, \quad \forall z \in \mathbb{D}_1,\ \xi > 0.$$

Since $\inf\{\frac{1}{1+\xi^2}; \xi \in [0,1]\} = \frac{1}{2}$, by Theorem 3.2.3 it is immediate the following result.

Corollary 3.2.4. (Anastassiou-Gal [18]) $P_\xi(S_{3,\frac{1}{2}}) \subset S_3$ and $f \in S_M$ implies that $P_\xi(f)$ is univalent in $\{z \in \mathbb{C};\, |z| < \frac{1}{2M}\}$, for all $\xi \in [0,1]$.

Remark. Of course that if we consider, for example, $\xi \in [0, \frac{1}{2}]$, then $\inf\{\frac{1}{1+\xi^2}; x \in [0, \frac{1}{2}]\} = \frac{4}{5}$ and by Theorem 3.2.3 get $P_\xi(S_{3,\frac{4}{5}}) \subset S_3$ and $f \in S_M$ implies $P_\xi(f)$ is univalent in $\{z \in \mathbb{C};\, |z| < \frac{4}{5M}\}$, $\forall \xi \in [0, \frac{1}{2}]$.

Obviously $S_{3,\frac{1}{2}} \subset S_{3,\frac{5}{4}}$ and $\{z \in \mathbb{C};\, |z| < \frac{1}{2M}\} \subset \{z \in \mathbb{C};\, |z| < \frac{4}{5M}\}$.

In what follows the properties of $Q_\xi(f)$, $Q^*_\xi(f)$ and $R_\xi(f)$ will be studied.

First we present their approximation properties. Note that Theorem 3.2.5, (i)-(iii) were proved in Anastassiou-Gal [18], while Theorem 3.2.5, (iv) is new.

Theorem 3.2.5. *(i) If $f(z) = \sum_{k=0}^{\infty} a_k z^k$ is analytic in $\mathbb{D}_1$, then for all $\xi > 0$, $Q_\xi(f)(z)$, $Q_\xi^*(f)(z)$, $R_\xi(f)(z)$ are analytic in $\mathbb{D}_1$ and for all $z \in \mathbb{D}_1$ we have,*

$$Q_\xi(f)(z) = \sum_{k=0}^{\infty} a_k b_k(\xi) z^k, \text{ with } b_k(\xi) = \frac{2\xi}{\pi} \int_0^\pi \frac{\cos ku}{u^2 + \xi^2} \, du,$$

$$Q_\xi^*(f)(z) = \sum_{k=0}^{\infty} a_k b_k^*(\xi) z^k, \text{ with } b_k^*(\xi) = \frac{2\xi}{\pi} \int_0^{+\infty} \frac{\cos ku}{u^2 + \xi^2} \, du,$$

$$R_\xi(f)(z) = \sum_{k=0}^{\infty} a_k c_k(\xi) z^k, \text{ with } c_k(\xi) = \frac{4\xi^3}{\pi} \int_0^\infty \frac{\cos ku}{(u^2 + \xi^2)^2} \, du.$$

Also, if f is continuous on $\overline{\mathbb{D}}_1$ then $Q_\xi(f)$, $Q_\xi^*(f)$ and $R_\xi(f)$ are also continuous on $\overline{\mathbb{D}}_1$;

Here $b_1(\xi) > 0$, $\forall \xi > 0$, $b_k^*(\xi) = e^{-k\xi}$, $c_k(\xi) = (1 + k\xi)e^{-k\xi}$, for all $\xi > 0$ and $k \in \mathbb{N} \bigcup \{0\}$.

(ii) If f is continuous on $\overline{\mathbb{D}}_1$ then for all $\delta > 0$ and $\xi > 0$ it follows

$$\omega_1(Q_\xi^*(f); \delta)_{\overline{\mathbb{D}}_1} \le \omega_1(f; \delta)_{\overline{\mathbb{D}}_1},$$

$$\omega_1(Q_\xi(f); \delta)_{\overline{\mathbb{D}}_1} \le \omega_1(f; \delta)_{\overline{\mathbb{D}}_1},$$

$$\omega_1(R_\xi(f); \delta)_{\overline{\mathbb{D}}_1} \le \omega_1(f; \delta)_{\overline{\mathbb{D}}_1}.$$

(iii) Let $f \in A(\overline{\mathbb{D}}_1)$. For all $z \in \overline{\mathbb{D}}_1$ and $\xi > 0$ it follows

$$|Q_\xi(f)(z) - f(z)| \le C \frac{\omega_2(f; \xi)_{\partial \mathbb{D}_1}}{\xi} + C\xi \|f\|_{\overline{\mathbb{D}}_1}$$

$$|R_\xi(f)(z) - f(z)| \le C\omega_2(f; \xi)_{\partial \mathbb{D}_1}.$$

If $f \in Lip_\alpha(M; \overline{\mathbb{D}}_1)$ with $0 < \alpha < 1$, that is $|f(u) - f(v)| \le M|u - v|^\alpha$ for all $u, v \in \overline{\mathbb{D}}_1$, then

$$|Q_\xi^*(f)(z) - f(z)| \le C\xi^\alpha, \forall |z| \le 1, \xi > 0,$$

where $C > 0$ is independent of ξ and z but depends on f.

Moreover, if there exists an $R > 1$ such that f is analytic in $\mathbb{D}_R$ then for any $1 \le r < R$ we have

$$|Q_\xi^*(f)(z) - f(z)| \le C\xi, \forall |z| \le r, \xi > 0,$$

where $C > 0$ is independent of ξ and z but depends on f and r.

(iv) In addition, let us suppose that $f(z) = \sum_{k=0}^{\infty} a_k z^k$ for all $z \in \mathbb{D}_R$, $R > 1$. If f is not constant for $q = 0$ and not a polynomial of degree $\le q - 1$ for $q \in \mathbb{N}$, then for all $1 \le r < r_1 < R$, $\xi \in (0, 1]$ and $q \in \mathbb{N} \bigcup \{0\}$ we have

$$\|[Q_\xi^*]^{(q)}(f) - f^{(q)}\|_r \sim \xi,$$

$$\|R_\xi^{(q)}(f) - f^{(q)}\|_r \sim \xi^2,$$

where the constants in the equivalences depend only on f, q, r, r_1.

Proof.   (i) Let $f(z) = \sum_{k=0}^\infty a_k z^k$, $z \in \mathbb{D}_1$. Reasoning as for the case of Picard-type integral in Theorem 3.2.1, (i), we obtain:

$$Q_\xi(f)(z) = \sum_{k=0}^\infty a_k z^k \left[\frac{\xi}{\pi} \int_{-\pi}^\pi e^{iku} \cdot \frac{1}{u^2 + \xi^2} \, du \right],$$

where

$$\frac{\xi}{\pi} \int_{-\pi}^\pi e^{iku} \cdot \frac{1}{u^2 + \xi^2} \, du = \frac{\xi}{\pi} \int_{-\pi}^\pi \frac{\cos ku}{u^2 + \xi^2} \, du + i\frac{\xi}{\pi} \int_{-\pi}^\pi \frac{\sin ku}{u^2 + \xi^2} \, du$$

$$= \frac{2\xi}{\pi} \int_0^\pi \frac{\cos ku}{u^2 + \xi^2} \, du = b_k(\xi).$$

Since obviously we can write $Q_\xi^*(f)(z) = \frac{\xi}{\pi} \int_{-\infty}^\infty \frac{f(ze^{iu})}{u^2+\xi^2} \, du$, it follows

$$Q_\xi^*(f)(z) = \sum_{k=0}^\infty a_k z^k \left[\frac{\xi}{\pi} \int_{-\infty}^{+\infty} e^{iku} \cdot \frac{1}{u^2 + \xi^2} \, du \right],$$

where

$$\frac{\xi}{\pi} \int_{-\infty}^{+\infty} e^{iku} \cdot \frac{1}{u^2 + \xi^2} \, du = \frac{2\xi}{\pi} \int_0^\infty \frac{\cos ku}{u^2 + \xi^2} \, du = b_k^*(\xi);$$

$$R_\xi(f)(z) = \sum_{k=0}^\infty a_k z^k \left[\frac{2\xi^3}{\pi} \int_{-\infty}^{+\infty} \frac{e^{iku}}{(u^2 + \xi^2)^2} \, du \right],$$

and where

$$\frac{2\xi^3}{\pi} \int_{-\infty}^{+\infty} e^{iku} \cdot \frac{1}{(u^2 + \xi^2)^2} \, du = \frac{4\xi^3}{\pi} \int_0^\infty \frac{\cos ku}{(u^2 + \xi^2)^2} \, du.$$

The continuity of f on $\overline{\mathbb{D}}_1$ implies the continuity of $Q_\xi(f)$, $Q_\xi^*(f)$ and $R_\xi(f)$ as in the proof of Theorem 3.2.1, (i) for $P_\xi(f)$.

It remains to show that $b_1(\xi) > 0$, $b_k^*(\xi) = e^{-k\xi}$ and $c_k(\xi) = (1 + k\xi)e^{-k\xi}$, for all $\xi > 0$ and $k \geq 0$. Indeed, firstly we have

$$b_1(\xi) = \frac{2\xi}{\pi} \int_0^\pi \frac{\cos u}{u^2 + \xi^2} du = \frac{2\xi}{\pi} \left[\int_0^{\pi/2} \frac{\cos u}{u^2 + \xi^2} du + \int_{\pi/2}^\pi \frac{\cos u}{u^2 + \xi^2} du \right]$$

$$= \frac{2\xi}{\pi} \left[\int_0^{\pi/2} \frac{\cos u}{u^2 + \xi^2} du - \int_0^{\pi/2} \frac{\sin u}{\left(u + \frac{\pi}{2}\right)^2 + \xi^2} du \right]$$

$$> \frac{2\xi}{\pi} \int_0^{\pi/2} \frac{\cos u - \sin u}{u^2 + \xi^2} du$$

$$= \frac{2\xi}{\pi} \left[\int_0^{\pi/4} \frac{\cos u - \sin u}{u^2 + \xi^2} du + \int_{\pi/4}^{\pi/2} \frac{\cos u - \sin u}{u^2 + \xi^2} du \right]$$

$$:= \frac{2\xi}{\pi} [I_1 + I_2].$$

Here

$$0 < I_1 = \int_0^{\pi/4} \frac{\cos u - \sin u}{u^2 + \xi^2} du > \int_0^{\pi/4} \frac{\cos u - \sin u}{(\pi^2/16) + \xi^2} du$$

$$= \frac{16}{\pi^2 + 16\xi^2} [\sin u + \cos u]_0^{\pi/4} = \frac{16(\sqrt{2} - 1)}{\pi^2 + 16\xi^2}.$$

Also, $I_2 < 0$ and

$$|I_2| = -I_2 = \int_{\pi/4}^{\pi/2} \frac{\sin u - \cos u}{u^2 + \xi^2} du \leq \frac{1}{(\pi^2/16) + \xi^2} \cdot \int_{\pi/4}^{\pi/2} [\sin u - \cos u] du$$

$$= \frac{16}{\pi^2 + 16\xi^2} [-\cos u - \sin u]_{\pi/4}^{\pi/2} = \frac{16(\sqrt{2} - 1)}{\pi^2 + 16\xi^2},$$

which implies $I_1 + I_2 \geq 0$. Therefore, it follows $b_1(\xi) > \frac{2\xi}{\pi}[I_1 + I_2] \geq 0$, for all $\xi > 0$.
For $b_k^*(\xi)$ we obtain

$$b_k^*(\xi) = \frac{2\xi}{\pi} \int_0^{+\infty} \frac{1}{u^2 + \xi^2} \cos ku\, du = \frac{2\xi}{\pi} \int_0^{+\infty} \frac{1}{v^2/k^2 + \xi^2} \cos v\, \frac{dv}{k}$$

$$= \frac{2k\xi}{\pi} \int_0^{+\infty} \frac{1}{v^2 + k^2\xi^2} \cos v\, dv = e^{-k\xi}.$$

Applying now the classical residue's theorem to $f(z) = \frac{e^{iz}}{z^2+1}$, it is immediate that $\int_0^\infty \frac{\cos(u\xi)}{u^2+1} du = \frac{\pi}{2} e^{-\xi}$, which implies $b_1^*(\xi) = \frac{2}{\pi} \cdot \frac{\pi}{2} e^{-\xi} = e^{-\xi}$, for all $\xi > 0$. For

$$c_k(\xi) = \frac{4\xi^3}{\pi} \cdot \int_0^\infty \frac{\cos ku}{(u^2 + \xi^2)^2} du = \frac{4k^3\xi^3}{\pi} \int_0^\infty \frac{\cos v}{(v^2 + k^2\xi^2)^2} dv,$$

applying the residue's theorem to $f(z) = \frac{e^{iz}}{(z^2+\eta^2)^2}$, we immediately get

$$\int_0^\infty \frac{\cos u}{(u^2 + \eta^2)^2} du = \frac{\pi}{4\eta^3}(1 + \eta)e^{-\eta},$$

that is replacing $\eta = k\xi$ it follows $c_k(\xi) = (1 + k\xi)e^{-k\xi}$, $\forall \xi > 0$.

(ii) Let $z_1, z_2 \in \overline{\mathbb{D}}_1$ be with $|z_1 - z_2| \leq \delta$. We get

$$|Q_\xi^*(f)(z_1) - Q_\xi^*(f)(z_2)| \leq \frac{\xi}{\pi} \int_{-\infty}^{+\infty} \frac{|f(z_1 e^{-iu}) - f(z_2 e^{-iu})|}{u^2 + \xi^2} du$$

$$\leq \omega_1(f; |z_1 - z_2|)_{\overline{\mathbb{D}}_1} \frac{\xi}{\pi} \int_{-\infty}^{+\infty} \frac{du}{u^2 + \xi^2} \leq \omega_1(f; \delta)_{\overline{\mathbb{D}}_1},$$

where from passing to supremum after z_1, z_2 it follows $\omega_1(Q_\xi^*(f); \delta)_{\overline{\mathbb{D}}_1} \leq \omega_1(f; \delta)_{\overline{\mathbb{D}}_1}$.
Also

$$|Q_\xi(f)(z_1) - Q_\xi(f)(z_2)| \leq \frac{\xi}{\pi} \int_{-\pi}^{\pi} \frac{|f(z_1 e^{iu}) - f(z_2 e^{iu})|}{u^2 + \xi^2} du$$

$$\leq \omega_1(f; |z_1 - z_2|)_{\overline{\mathbb{D}}_1} \cdot \frac{\xi}{\pi} \int_{-\pi}^{\pi} \frac{du}{u^2 + \xi^2}$$

$$\leq \omega_1(f; \delta)_{\overline{\mathbb{D}}_1} \cdot \frac{\xi}{\pi} \int_{-\infty}^{+\infty} \frac{du}{u^2 + \xi^2} = \omega_1(f; \delta)_{\overline{\mathbb{D}}_1}.$$

The reasonings for $R_\xi(f)$ are similar, which proves (ii).

(iii) We can write

$$Q_\xi(f)(z) - f(z) = \frac{\xi}{\pi} \int_0^\pi \frac{f(ze^{iu}) - 2f(z) + f(ze^{-iu})}{u^2 + \xi^2} \, du - f(z)E(\xi),$$

where

$$|E(\xi)| = E(\xi) = 1 - \frac{2\xi}{\pi} \int_0^\pi \frac{du}{u^2 + \xi^2} = 1 - \frac{2}{\pi} \arctan \frac{\pi}{\xi} \leq \frac{2}{\pi^2}\xi$$

(for the last estimate $|E(\xi)| \leq \frac{2}{\pi^2}\xi$ see e.g. Gal [94], p. 257).

Passing to modulus, it follows

$$|Q_\xi(f)(z) - f(z)| \leq \frac{\xi}{\pi} \int_0^\pi \frac{|f(ze^{iu}) - 2f(z) + f(ze^{-iu})|}{u^2 + \xi^2} \, du + \|f\|_{\overline{\mathbb{D}}_1}|E(\xi)|$$

$$\leq \frac{\xi}{\pi} \int_0^\pi \frac{\omega_2(f; u)_{\partial \mathbb{D}_1}}{u^2 + \xi^2} \, du + \|f\|_{\overline{D}} \cdot |E(\xi)|$$

$$\leq C\frac{\xi}{\pi} \cdot \omega_2(f; \xi)_{\partial \mathbb{D}_1} \cdot \int_0^\pi \left[1 + \frac{u}{\xi}\right]^2 \frac{1}{u^2 + \xi^2} \, du + C\xi\|f\|_{\overline{\mathbb{D}}_1}.$$

Reasoning as in the proof of Theorem 3.1, pp. 257–258 in Gal [94], we arrive at the desired estimate.

For $Q_\xi^*(f)(z)$ we have

$$Q_\xi^*(f)(z) - f(z) = \frac{\xi}{\pi} \int_{-\infty}^\infty \frac{[f(ze^{-iu}) - f(z)]}{u^2 + \xi^2} \, du.$$

By $f \in Lip_\alpha(M; \overline{\mathbb{D}}_1)$ and since by the maximum modulus theorem we can take $|z| = 1$, we obtain

$$|Q_\xi^*(f)(z) - f(z)| \leq \frac{\xi}{\pi} \int_{-\infty}^\infty \frac{|f(ze^{-iu}) - f(z)|}{u^2 + \xi^2} \, du$$

$$\leq M\frac{\xi}{\pi} \int_{-\infty}^\infty \frac{|e^{-iu} - 1|^\alpha}{u^2 + \xi^2} \, du = M\frac{\xi}{\pi} \int_{-\infty}^\infty \frac{(2|\sin(u/2)|)^\alpha}{u^2 + \xi^2} \, du$$

$$\leq 2M\frac{\xi}{\pi} \int_0^\infty \frac{u^\alpha}{u^2 + \xi^2} \, du$$

$$= \frac{2M}{\pi}\xi^\alpha \int_0^\infty \frac{v^\alpha}{v^2 + 1} \, dv,$$

where it is easy to show that $\int_0^\infty \frac{v^\alpha}{v^2+1} dv < \infty$.

Now, if f is supposed to be analytic in $\mathbb{D}_R$ with $R > 1$, we can write $f(z) = \sum_{k=0}^\infty a_k z^k$, for all $|z| \leq r$ with $1 \leq r < R$, which immediately implies that the series is uniformly and absolutely convergent, that is $\sum_{k=0}^\infty |a_k| \cdot |z|^k \leq \sum_{k=0}^\infty |a_k| \cdot r^k < \infty$.

Taking into account the above point (i), since by the mean value theorem we have $|e^{-k\xi} - 1| \leq \xi$ for all $k = 0, 1, ...,$, for any $|z| \leq r$ we get

$$|Q_\xi^*(f)(z) - f(z)| = \left|\sum_{k=0}^\infty a_k[e^{-k\xi} - 1]z^k\right| \leq \xi \sum_{k=0}^\infty |a_k|r^k,$$

therefore by the above considerations we can take $C = \sum_{k=0}^{\infty} |a_k| r^k < \infty$.

For $R_\xi(f)(z)$ we obtain

$$|R_\xi(f)(z) - f(z)| \leq \frac{2\xi^3}{\pi} \int_0^{+\infty} \frac{|f(ze^{iu}) - 2f(z) + f(ze^{-iu})|}{(u^2 + \xi^2)^2} \, du$$

$$\leq C \frac{2\xi^3}{\pi} \int_0^{+\infty} \frac{\omega_2(f; u)_{\partial \mathbb{D}_1}}{(u^2 + \xi^2)^2} \, du$$

$$= C \frac{2\xi^3}{\pi} \int_0^{+\infty} \frac{\omega_2(f; \xi(u/\xi))_{\partial \mathbb{D}_1}}{(u^2 + \xi^2)^2} \, du$$

$$\leq C\omega_2(f; \xi)_{\partial \mathbb{D}_1} \cdot \frac{2\xi^3}{\pi} \int_0^{\infty} \left[1 + \frac{u}{\xi}\right]^2 \cdot \frac{1}{(u^2 + \xi^2)^2} \, du$$

$$\leq C\omega_2(f; \xi)_{\partial \mathbb{D}_1},$$

since by easy calculation we get that

$$\frac{2\xi^3}{\pi} \int_0^{\infty} \left[1 + \frac{u}{\xi}\right]^2 \cdot \frac{1}{(u^2 + \xi^2)^2} \, du \leq C,$$

where $C > 0$ is independent of ξ.

(iv) Analysing the proofs of the above points (i)-(iii), it easily follows that they hold by replacing everywhere $\mathbb{D}_1$ with $\mathbb{D}_r$, where $1 \leq r < R$. Therefore the approximation errors in (iii) can be expressed in terms of

$$\omega_2(f; \xi)_{\partial \mathbb{D}_r} = \sup\{|\Delta_u^2 f(re^{it})|; |t| \leq \pi, \ |u| \leq \xi\},$$

with constants in front of ω_2 depending on $r \geq 1$.

Since $\Delta_u^2 g(t) = 2u^2[t, t+u, t+2u; g]$, by the mean value theorem for divided differences in Complex Analysis (see e.g. Stancu [172], p. 258, Exercise 4.20) we get

$$|\Delta_u^2 f(re^{it})| \leq u^2 \|f''\|_r,$$

where

$$\|f''\|_r = \sup\{|f''(z)|; |z| \leq r\} \leq \sum_{k=2}^{\infty} |c_k| k(k-1) r^{k-2}.$$

For all $\xi \in (0, 1]$ from (iii) it follows

$$\|Q_\xi^*(f) - f\|_r \leq C_r(f)\xi.$$

Also, from (iii) we immediately obtain

$$\|R_\xi(f) - f\|_r \leq C_r(f)\xi^2, \text{ for all } \xi \in (0, 1].$$

Now denoting by γ the circle of radius $r_1 > 1$ and center 0, since for any $|z| \leq r$ and $v \in \gamma$, we have $|v - z| \geq r_1 - r$, by the Cauchy's formulas it follows that for all $|z| \leq r$ and $\xi \in (0, 1]$, we have

$$|[Q_\xi^*]^{(q)}(f)(z) - f^{(q)}(z)| = \frac{q!}{2\pi} \left| \int_\gamma \frac{Q_\xi^*(f)(v) - f(v)}{(v-z)^{q+1}} dv \right|$$

$$\leq C_{r_1}(f)\xi \frac{q!}{2\pi} \frac{2\pi r_1}{(r_1 - r)^{q+1}}$$

$$= C_{r_1}(f)\xi \frac{q! r_1}{(r_1 - r)^{q+1}},$$

which proves the upper estimate for $\|[Q_\xi^*]^{(q)}(f) - f^{(q)}\|_r$.

In a similar manner it follows an upper estimate of the form

$$\|R_\xi^{(q)}(f)(z) - f^{(q)}(z)\|_r \leq C_{q,r,r_1}(f)\xi^2, \text{ for all } \xi \in (0,1].$$

It remains to prove the lower estimates for $\|[Q_\xi^*]^{(q)}(f) - f^{(q)}\|_r$ and $\|R_\xi^{(q)}(f) - f^{(q)}\|_r$.

First we deal with the Q_ξ^* operator. For this purpose, take $z = re^{i\varphi}$ and $p \in \mathbb{N}\bigcup\{0\}$. We have

$$\frac{1}{2\pi}[f^{(q)}(z) - (Q_\xi^*)^{(q)}(f)(z)]e^{-ip\varphi}$$

$$= \frac{1}{2\pi}\sum_{k=q}^{\infty} a_k k(k-1)...(k-q+1)r^{k-q}e^{i\varphi(k-q-p)}[1 - b_k^*(\xi)]$$

$$= \frac{1}{2\pi}\sum_{k=q}^{\infty} a_k k(k-1)...(k-q+1)r^{k-q}e^{i\varphi(k-q-p)}[1 - e^{-k\xi}].$$

Integrating from $-\pi$ to π we obtain

$$\frac{1}{2\pi}\int_{-\pi}^{\pi}[f^{(q)}(z) - (Q_\xi^*)^{(q)}(f)(z)]e^{-ip\varphi}d\varphi$$

$$= a_{q+p}(q+p)(q+p-1)...(p+1)r^p[1 - e^{-(q+p)\xi}].$$

Passing now to absolute value we easily obtain

$$|a_{q+p}|(q+p)(q+p-1)...(p+1)r^p[1 - e^{-(q+p)\xi}] \leq \|f^{(q)} - (Q_\xi^*)^{(q)}(f)\|_r.$$

First consider $q = 0$ and denote $V_\xi = \inf_{1 \leq p}(1 - e^{-p\xi})$. We get $V_\xi = 1 - e^{-\xi}$ and by the mean value theorem applied to $h(x) = e^{-x}$ on $[0, \xi]$ there exists $\eta \in (0, \xi)$ such that for all $\xi \in (0, 1]$ we have

$$V_\xi = h(0) - h(\xi) = (-\xi)h'(\eta) = (\xi)e^{-\eta} \geq (\xi)e^{-\xi} \geq e^{-1}\xi \geq \xi/3.$$

By the above lower estimate for $\|Q_\xi^*(f) - f\|_r$, for all $p \geq 1$ and $\xi \in (0, 1]$ it follows

$$\frac{3\|Q_\xi^*(f) - f\|_r}{\xi} \geq \frac{\|Q_\xi^*(f) - f\|_r}{V_\xi} \geq \frac{\|Q_\xi^*(f) - f\|_r}{1 - e^{-p\xi}} \geq |a_p|r^p.$$

This implies that if there exists a subsequence $(\xi_k)_k$ in $(0, 1]$ with $\lim_{k\to\infty}\xi_k = 0$ and such that $\lim_{k\to\infty}\frac{\|Q_{\xi_k}^*(f)-f\|_r}{\xi_k} = 0$ then $a_p = 0$ for all $p \geq 1$, that is f is constant on $\overline{\mathbb{D}}_r$.

Therefore, if f is not a constant then $\inf_{\xi \in (0,1]}\frac{\|Q_\xi^*(f)-f\|_r}{\xi} > 0$, which implies that there exists a constant $C_r(f) > 0$ such that $\frac{\|Q_\xi^*(f)-f\|_r}{\xi} \geq C_r(f)$, for all $\xi \in (0, 1]$, that is

$$\|Q_\xi^*(f) - f\|_r \geq C_r(f)\xi, \text{ for all } \xi \in (0, 1].$$

Now, consider $q \geq 1$ and denote $V_{q,\xi} = \inf_{p \geq 0}(1 - e^{-(p+q)\xi})$. Evidently that we have $V_{q,\xi} \geq \inf_{p \geq 1}(1 - e^{-p\xi}) \geq \xi/3$.

Reasoning as in the case of $q = 0$ we obtain

$$\frac{3\|[Q_\xi^*]^{(q)}(f) - f^{(q)}\|_r}{\xi} \geq \frac{\|[Q_\xi^*]^{(q)}(f) - f^{(q)}\|_r}{V_{q,\xi}} \geq \frac{\|[Q_\xi^*]^{(q)}(f) - f^{(q)}\|_r}{1 - e^{-(p+q)\xi}}$$

$$\geq |a_{q+p}|\frac{(q+p)!}{p!}r^p,$$

for all $p \geq 0$ and $\xi \in (0, 1]$.

This implies that if there exists a subsequence $(\xi_k)_k$ in $(0, 1]$ with $\lim_{k\to\infty} \xi_k = 0$ and such that $\lim_{k\to\infty} \frac{\|[Q_{\xi_k}^*]^{(q)}(f) - f^{(q)}\|_r}{\xi_k} = 0$ then $a_{q+p} = 0$ for all $p \geq 0$, that is f is a polynomial of degree $\leq q - 1$ on $\overline{\mathbb{D}}_r$.

Therefore, $\inf_{\xi \in (0,1]} \frac{\|[Q_\xi^*]^{(q)}(f) - f^{(q)}\|_r}{\xi} > 0$ when f is not a polynomial of degree $\leq q - 1$, which implies that there exists a constant $C_{r,q}(f) > 0$ such that $\frac{\|[Q_\xi^*]^{(q)}(f) - f^{(q)}\|_r}{\xi} \geq C_{r,q}(f)$, for all $\xi \in (0, 1]$, that is

$$\|[Q_\xi^*]^{(q)}(f) - f^{(q)}\|_r \geq C_{r,q}(f)\xi, \text{ for all } \xi \in (0, 1].$$

Now, let us consider the lower estimate in the case of R_ξ operator. For this purpose, take $z = re^{i\varphi}$ and $p \in \mathbb{N}\bigcup\{0\}$. We have

$$\frac{1}{2\pi}[f^{(q)}(z) - R_\xi^{(q)}(f)(z)]e^{-ip\varphi}$$

$$= \frac{1}{2\pi}\sum_{k=q}^{\infty} a_k k(k-1)...(k-q+1)r^{k-q}e^{i\varphi(k-q-p)}[1 - c_k(\xi)]$$

$$= \frac{1}{2\pi}\sum_{k=q}^{\infty} a_k k(k-1)...(k-q+1)r^{k-q}e^{i\varphi(k-q-p)}[1 - (1 + k\xi)e^{-k\xi}].$$

Integrating from $-\pi$ to π we obtain

$$\frac{1}{2\pi}\int_{-\pi}^{\pi}[f^{(q)}(z) - R_\xi^{(q)}(f)(z)]e^{-ip\varphi}d\varphi$$

$$= a_{q+p}(q+p)(q+p-1)...(p+1)r^p[1 - (1 + (q+p)\xi)e^{-(q+p)\xi}].$$

Passing now to absolute value we easily obtain

$$|a_{q+p}|(q+p)(q+p-1)...(p+1)r^p[1 - (1 + (q+p)\xi)e^{-(q+p)\xi}] \leq \|f^{(q)} - R_\xi^{(q)}(f)\|_r.$$

First consider $q = 0$ and denote $V_\xi = \inf_{1 \leq p}(1 - (1 + p\xi)e^{-p\xi})$. We easily get $V_\xi = 1 - (1 + \xi)e^{-\xi}$ and we have

$$V_\xi \geq \frac{\xi^2}{4} \text{ for all } \xi \in (0, 1].$$

Indeed, denoting $F(\xi) = 1 - (1+\xi)e^{-\xi} - \frac{\xi^2}{4}$ we get $F'(\xi) = 1 - \frac{\xi}{2} + \xi e^{-\xi} > 0$, for all $\xi \in (0, 1]$. That is F is increasing and since $F(0) = 0$ it follows the above inequality for V_ξ.

By the above lower estimate for $\|R_\xi(f) - f\|_r$, for all $p \geq 1$ and $\xi \in (0, 1]$ it follows

$$\frac{4\|R_\xi(f) - f\|_r}{\xi^2} \geq \frac{\|R_\xi(f) - f\|_r}{V_\xi} \geq \frac{\|R_\xi(f) - f\|_r}{1 - (1 + p\xi)e^{-p\xi}} \geq |a_p|r^p.$$

This implies that if there exists a subsequence $(\xi_k)_k$ in $(0, 1]$ with $\lim_{k\to\infty} \xi_k = 0$ and such that $\lim_{k\to\infty} \frac{\|R_{\xi_k}(f) - f\|_r}{\xi_k^2} = 0$ then $a_p = 0$ for all $p \geq 1$, that is f is constant on $\overline{\mathbb{D}}_r$.

Therefore, if f is not a constant then $\inf_{\xi\in(0,1]} \frac{\|R_\xi(f) - f\|_r}{\xi^2} > 0$, which implies that there exists a constant $C_{r,p}(f) > 0$ such that $\frac{\|R_\xi(f) - f\|_r}{\xi^2} \geq C_{r,p}(f)$, for all $\xi \in (0, 1]$, that is

$$\|R_\xi(f) - f\|_r \geq C_{r,p}\xi^2, \text{ for all } \xi \in (0, 1].$$

Now, consider $q \geq 1$ and denote $V_{q,\xi} = \inf_{p\geq 0}(1 - (1 + (q+p)\xi)e^{-(q+p)\xi})$. Evidently that we have $V_{q,\xi} \geq \inf_{p\geq 1}(1 - (1 + p\xi)e^{-p\xi}) \geq \xi^2/4$.

Reasoning as in the case of $q = 0$ we obtain

$$\frac{4\|R_\xi^{(q)}(f) - f^{(q)}\|_r}{\xi^2} \geq \frac{\|R_\xi^{(q)}(f) - f^{(q)}\|_r}{V_{q,\xi}} \geq \frac{\|R_\xi^{(q)}(f) - f^{(q)}\|_r}{1 - (1 + (q+p)\xi)e^{-(q+p)\xi}}$$

$$\geq |a_{q+p}|\frac{(q+p)!}{p!}r^p,$$

for all $p \geq 0$ and $\xi \in (0, 1]$.

This implies that if there exists a subsequence $(\xi_k)_k$ in $(0, 1]$ with $\lim_{k\to\infty} \xi_k = 0$ and such that $\lim_{k\to\infty} \frac{\|R_{\xi_k}^{(q)}(f) - f^{(q)}\|_r}{\xi_k^2} = 0$ then $a_{q+p} = 0$ for all $p \geq 0$, that is f is a polynomial of degree $\leq q - 1$ on $\overline{\mathbb{D}}_r$.

Therefore, $\inf_{\xi\in(0,1]} \frac{\|R_\xi^{(q)}(f) - f^{(q)}\|_r}{\xi^2} > 0$ when f is not a polynomial of degree $\leq q - 1$, which implies that there exists a constant $C_{r,q}(f) > 0$ such that $\frac{\|R_\xi^{(q)}(f) - f^{(q)}\|_r}{\xi^2} \geq C_{r,q}(f)$, for all $\xi \in (0, 1]$, that is

$$\|R_\xi^{(q)}(f) - f^{(q)}\|_r \geq C_{r,q}(f)\xi^2, \text{ for all } \xi \in (0, 1],$$

which proves the theorem. $\qquad\qquad\qquad\qquad\qquad\qquad\qquad\qquad\qquad\qquad \square$

Remark. The estimates in approximation by $Q_\xi^*(f)(z)$ contained in Theorem 3.2.5 (iii), correct the wrong estimate $|Q_\xi^*(f)(z) - f(z)| \leq C\frac{\omega_2(f;\xi)_{\partial\mathbb{D}_1}}{\xi}$ obtained in the paper Anastassiou-Gal [18], Theorem 3.1, (ii) (and also stated without proof in the book Gal [77], p. 318, Theorem 5.4.2, 2)). Also, note that Theorem 3.2 (ii) in Anastassiou-Gal [18] still remains valid since we can use there the second estimate for $|Q_\xi^*(f)(z) - f(z)|$ in Theorem 3.2.5 (iii), which implies the uniform convergence of $Q_\xi^*(f)$ to f on $\overline{\mathbb{D}}_1$.

In what follows we present some geometric properties of the complex Poisson–Cauchy convolutions.

Theorem 3.2.6. (Anastassiou-Gal [18]) *(i) If* $f(z) = \sum_{k=0}^{\infty} a_k z^k$, $z \in \mathbb{D}_1$ *and* $T_\xi(f)(z) = \sum_{k=0}^{\infty} A_k z^k$, *is any from* $Q_\xi(f)$, $Q_\xi^*(f)$ *and* $R_\xi(f)$, *then*

$$|A_k| \le |a_k|, \qquad \forall k = 0, 1, \dots, ;$$

(ii) If $f(z) = \sum_{k=1}^{\infty} a_k z^k$, $z \in \mathbb{D}_1$ *is univalent in* $\mathbb{D}_1$ *and* $f(\mathbb{D}_1)$ *is convex, then for any* $\xi > 0$, $Q_\xi(f)(z)$ *is close-to-convex on* $\mathbb{D}_1$ *;*

(iii) For all $\xi > 0$ *we have :* $Q_\xi^*(\mathcal{P}) \subset \mathcal{P}$, $R_\xi(\mathcal{P}) \subset \mathcal{P}$;

$$\frac{1}{b_1(\xi)} \cdot Q_\xi(S_{3,b_1(\xi)}) \subset S_3, \quad \frac{1}{b_1^*(\xi)} \cdot Q_\xi^*(S_{3,b_1^*(\xi)}) \subset S_3,$$

$$\frac{1}{c_1(\xi)} \cdot R_\xi(S_{3,c_1(\xi)}) \subset S_3, \quad \frac{1}{b_1(\xi)} Q_\xi(S_M) \subset S_{M/|b_1(\xi)|},$$

$$\frac{1}{b_1^*(\xi)} Q_\xi^*(S_M) \subset S_{M/|b_1^*(\xi)|}, \quad \frac{1}{c_1(\xi)} R_\xi(S_M) \subset S_{M/|c_1(\xi)|},$$

where $S_{3,a} = \{f \in S_3; |f''(z)| \le |a|\}$ *and* $S_B = \{f \in A^*(\overline{\mathbb{D}}_1); |f'(z)| < B, z \in \mathbb{D}_1\}$.

Proof. (i) With the notations in the statement of Theorem 3.2.5, (i), for all $k = 0, 1, 2, \dots$, we obtain

$$|b_k(\xi)| \le \frac{2\xi}{\pi} \int_0^\pi \frac{|\cos ku|}{u^2 + \xi^2} \, du \le \frac{2\xi}{\pi} \int_0^\pi \frac{du}{u^2 + \xi^2}$$

$$= \frac{2\xi}{\pi} \cdot \frac{1}{\xi} \operatorname{arctg} \frac{u}{\xi} \Big|_0^\pi = \frac{2}{\pi} \operatorname{arctg} \frac{\pi}{\xi} \le 1,$$

$$|b_k^*(\xi)| \le \frac{2\xi}{\pi} \cdot \frac{1}{\xi} \operatorname{arctg} \frac{u}{\xi} \Big|_0^\infty = 1,$$

$$|c_k(\xi)| \le \frac{4\xi^3}{\pi} \int_0^\infty \frac{du}{(u^2 + \xi^2)^2} = 1,$$

which immediately implies (i).

(ii) First, it is immediate that we can write

$$Q_\xi(f)(z) = \frac{\xi}{\pi} \int_{-\pi}^\pi \frac{f(ze^{-iu})}{u^2 + \xi^2} \, du.$$

Since $h(u) = \frac{1}{u^2 + \xi^2}$ satisfies $h(\pi) = h(-\pi)$, we may extend it by 2π-periodicity on the whole $\mathbb{R}$, such that this extension is continuous on $\mathbb{R}$.

By $h'(u) = \frac{-2u}{(u^2 + \xi^2)^2}$, it follows that h is non-decreasing on $[-\pi, 0]$ and non-increasing on $[0, \pi]$. Then by Suffridge [187], p. 799, Theorem 3, it follows that $Q_\xi(f)(z)$ is close-to-convex on $\mathbb{D}_1$.

(iii) Let $f \in \mathcal{P}$, $f = U + iV$, $U > 0$. Then by definitions, it easily follows that $Q_\xi(f)$, $Q_\xi^*(f)$, $R_\xi(f) \in \mathcal{P}$. We take here into account that by Theorem 3.2.5, (i), the condition $a_0 = f(0) = 1$, implies

$$Q_\xi^*(f)(0) = a_0 b_0^*(\xi) = b_0^*(\xi) = \frac{\xi}{\pi} \int_{-\infty}^{+\infty} \frac{du}{u^2 + \xi^2} = 1$$

and

$$R_\xi(f)(0) = a_0 c_0(\xi) = \frac{2\xi^3}{\pi} \int_{-\infty}^{+\infty} \frac{du}{(u^2 + \xi^2)^2} = 1.$$

Now, let $f(z) = \sum_{k=0}^{\infty} a_k z^k$, with $a_0 = 0$, $a_1 = 1$. First, by Theorem 3.2.5, (i), we get

$$\frac{1}{b_1(\xi)} Q_\xi(f)(0) = 0, \quad \frac{1}{b_1(\xi)} Q'_\xi(f)(0) = 1, \quad \frac{1}{b_1^*(\xi)} Q_\xi^*(f)(0) = 0,$$

$$\frac{1}{b_1^*(\xi)} \cdot [Q_\xi^*(f)]'(0) = 1, \quad \frac{1}{c_1(\xi)} R_\xi(f)(0) = 0, \quad \frac{1}{c_1(\xi)} \cdot R'_\xi(f)(0) = 1.$$

Then

$$Q''_\xi(f)(z) = \frac{\xi}{\pi} \int_{-\pi}^{\pi} f''(ze^{iu}) e^{2iu} \cdot \frac{1}{u^2 + \xi^2} \, du,$$

$$[Q_\xi^*(f)]''(z) = \frac{\xi}{\pi} \int_{-\infty}^{+\infty} f''(ze^{-iu}) e^{-2iu} \cdot \frac{1}{u^2 + \xi^2} \, du,$$

$$[R_\xi(f)]''(z) = \frac{2\xi^3}{\pi} \int_{-\infty}^{+\infty} f''(ze^{iu}) e^{2iu} \cdot \frac{1}{(u^2 + \xi^2)^2} \, du.$$

Let $f \in S_{3,b_1(\xi)}$. We get

$$\left| \frac{1}{b_1(\xi)} \cdot Q''_\xi(f)(z) \right| \leq \frac{1}{|b_1(\xi)|} \cdot \frac{\xi}{\pi} \int_{-\pi}^{\pi} |f''(ze^{iu})| \cdot \frac{1}{u^2 + \xi^2} \, du$$

$$\leq \frac{\xi}{\pi} \int_{-\pi}^{\pi} \frac{du}{u^2 + \xi^2} = \frac{2}{\pi} \operatorname{arctg} \frac{\pi}{\xi} \leq 1,$$

that is $\frac{1}{b_1(\xi)} \cdot Q_\xi(f) \in S_3$.

Let $f \in S_{3,b_1^*(\xi)}$. We get

$$\left| \frac{1}{b_1^*(\xi)} \cdot [Q^*(f)]''(z) \right| \leq \frac{1}{|b_1^*(\xi)|} \cdot \frac{\xi}{\pi} \int_{-\infty}^{+\infty} |f''(ze^{iu})| \cdot \frac{1}{u^2 + \xi^2} \, du$$

$$\leq \frac{\xi}{\pi} \int_{-\infty}^{+\infty} \frac{du}{u^2 + \xi^2} = 1,$$

that is $\frac{1}{b_1^*(\xi)} Q_\xi^*(f) \in S_3$. The proof in the case of $\frac{1}{c_1(\xi)} \cdot R_\xi(f)$ is similar.

Now, let $f \in S_M$. It follows

$$\left| \frac{1}{b_1(\xi)} Q'_\xi(f)(z) \right| \leq \frac{1}{|b_1(\xi)|} \frac{\xi}{\pi} \int_{-\pi}^{\pi} |f'(ze^{iu})| \cdot \frac{1}{u^2 + \xi^2} \, du$$

$$< \frac{M}{|b_1(\xi)|} \cdot \frac{2}{\pi} \operatorname{arctg} \frac{\pi}{\xi} \leq \frac{M}{|b_1(\xi)|}.$$

The proofs in the case of $\frac{1}{b_1^*(\xi)} \cdot Q_\xi^*(f)$ and $\frac{1}{c_1(\xi)} \cdot R_\xi(f)$ are similar, which proves the theorem. $\qquad\square$

Remarks. 1) Theorem 3.2.6 (iii) says that if $f \in S_{3,b_1(\xi)}$ then $Q_\xi(f)$ is starlike and univalent on $\mathbb{D}_1$ and if $f \in S_{M/|b_1(\xi)|}$ then $Q_\xi(f)$ is univalent in the disk

$$\left\{ z \in \mathbb{C}; \ |z| < \frac{|b_1(\xi)|}{M} \right\} \subset \left\{ z \in \mathbb{C}; \ |z| < \frac{1}{M} \right\}.$$

Similar properties hold for $Q_\xi^*(f)$, $b_1^*(\xi)$, and $R_\xi(f)$, $c_1(\xi)$.

2) Let us denote $B = \inf\{|b_1(\xi)|; \ \xi \in (0,1]\}$. If $B > 0$, then by Theorem 3.2.6 (iii), the following properties hold: $f \in S_{3,B}$ implies $Q_\xi(f) \in S_3$, for all $\xi \in (0,1]$, $f \in S_M$, $(M > 1)$ implies $Q_\xi(f)$ is univalent in $\left\{|z| < \frac{B}{M}\right\}$, $\forall \xi \in (0,1]$. Therefore it remains to calculate B, to check if $B > 0$, problems which are left to the reader as open questions.

Now, since $\inf\{|b_1^*(\xi)|; \ \xi \in (0,1]\} = \inf\{e^{-\xi}; \ \xi \in (0,1]\} = \frac{1}{e}$ and $\inf\{|c_1(\xi)|;$ $\xi \in (0,1]\} = \inf\{(1+\xi)e^{-\xi}; \ \xi \in (0,1]\} = \frac{2}{e}$ (since $h(\xi) = (1+\xi)e^{-\xi}$ is decreasing on $[0,1]$), from Theorem 3.2.5 (i) and Theorem 3.2.6 (iii) we immediately get the following.

Corollary 3.2.7. (Anastassiou-Gal [18]) *(i) If $f \in S_{3,\frac{1}{e}}$ then $Q_\xi^*(f) \in S_3$, for all $\xi \in (0,1]$ and if $f \in S_M$, $(M > 1)$, then $Q_\xi^*(f)$ is univalent in $\left\{z \in \mathbb{C}; \ |z| < \frac{1}{eM}\right\}$, for all $\xi \in (0,1]$.*

(ii) If $f \in S_{3,\frac{2}{e}}$ then $R_\xi(f) \in S_3$ for all $\xi \in (0,1]$ and if $f \in S_M$ then $R_\xi(f)$ is univalent in $\left\{|z| < \frac{2}{eM}\right\}$, for all $\xi \in (0,1]$.

The complex convolutions $W_\xi(f)(z)$ and $W_\xi^*(f)(z)$ are studied in what follows.

Concerning the Theorem 3.2.8 below, note here that (i), (ii) and the first estimate in (iii) were obtained in Anastassiou-Gal [18], while the second estimate (that for $W_\xi^*(f)(z)$) in (iii) and both estimates in (iv) are new.

Theorem 3.2.8. *(i) If $f(z) = \sum\limits_{k=0}^{\infty} a_k z^k$ is analytic in $\mathbb{D}_1$, then for all $\xi > 0$, $W_\xi(f)(z)$ and $W_\xi^*(f)(z)$ are analytic in $\mathbb{D}_1$ and for all $z \in \mathbb{D}_1$ we have*

$$W_\xi(f)(z) = \sum_{k=0}^{\infty} a_k d_k(\xi) z^k,$$

with

$$d_k(\xi) = \frac{1}{\sqrt{\pi \xi}} \cdot \int_{-\pi}^{\pi} e^{-u^2/\xi} \cos ku \, du$$

$$W_\xi^*(f)(z) = \sum_{k=0}^{\infty} a_k d_k^*(\xi) z^k,$$

with

$$d_k^*(\xi) = \frac{1}{\sqrt{\pi \xi}} \int_{-\infty}^{+\infty} e^{-u^2/\xi} \cos ku \, du.$$

Also, if f is continuous on $\overline{\mathbb{D}}_1$ then $W_\xi(f)$ and $W_\xi^*(f)$ are continuous on $\overline{\mathbb{D}}_1$. Here $d_1(\xi) > 0$ and $d_k^*(\xi) = e^{-k^2\xi/4} > 0$ for all $\xi > 0$ and $k \in \mathbb{N} \bigcup \{0\}$.

(ii) Let f be continuous on $\overline{\mathbb{D}}_1$. For all $\delta > 0$ and $\xi > 0$ we have

$$\omega_1(W_\xi^*(f); \delta)_{\overline{\mathbb{D}}_1} \leq \omega_1(f; \delta)_{\overline{\mathbb{D}}_1},$$
$$\omega_1(W_\xi(f); \delta)_{\overline{\mathbb{D}}_1} \leq \omega_1(f; \delta)_{\overline{\mathbb{D}}_1}.$$

(iii) Let $f \in A(\overline{\mathbb{D}}_1)$. For all $\xi \in (0, 1]$ and $z \in \overline{\mathbb{D}}_1$ it follows

$$|W_\xi(f)(z) - f(z)| \leq C \frac{\omega_2(f; \xi)_{\partial \mathbb{D}_1}}{\xi} + C\sqrt{\xi}\|f\|_{\overline{\mathbb{D}}_1},$$

$$|W_\xi^*(f)(z) - f(z)| \leq C\omega_2(f; \sqrt{\xi})_{\partial \mathbb{D}_1}.$$

(iv) In addition, let us suppose that $f(z) = \sum_{k=0}^\infty a_k z^k$ for all $z \in \mathbb{D}_R$, $R > 1$. If f is not constant for $q = 0$ and not a polynomial of degree $\leq q - 1$ for $q \in \mathbb{N}$, then for all $1 \leq r < r_1 < R$, $\xi \in (0, 1]$ and $q \in \mathbb{N} \bigcup\{0\}$ we have

$$\|[W_\xi^*]^{(q)}(f) - f^{(q)}\|_r \sim \xi,$$

where the constants in the equivalences depend only on f, q, r, r_1.

Proof. (i) Reasoning as for the $P_\xi(f)$ operator, we can write

$$W_\xi^*(f)(z) = \frac{1}{\sqrt{\pi\xi}} \int_{-\infty}^{+\infty} \sum_{k=0}^\infty a_k z^k e^{iuk} e^{-u^2/\xi}\, du$$

$$= \sum_{k=0}^\infty a_k z^k \cdot \frac{1}{\sqrt{\pi\xi}} \int_{-\infty}^{+\infty} [\cos(ku) + i\sin(ku)] e^{-u^2/\xi}\, du$$

$$= \sum_{k=0}^\infty a_k d_k^*(\xi) z^k,$$

where

$$d_k^*(\xi) = \frac{1}{\sqrt{\pi\xi}} \int_{-\infty}^{+\infty} \cos(ku) e^{-u^2/\xi}\, du.$$

The reasonings in the case of $W_\xi(f)(z)$ are similar. The proof of continuity on $\overline{\mathbb{D}}_1$ of $W_\xi(f)$ and $W_\xi^*(f)$ is similar to that for $P_\xi(f)$ in the proof of Theorem 3.2.1, (i).

It remains to prove that $d_1(\xi) > 0$, $\forall \xi > 0$ and that $d_k^*(\xi) = e^{-k^2\xi/4} > 0$, for all $\xi > 0$ and $k \geq 0$.

Indeed, firstly we have

$$d_1(\xi) = \frac{1}{\sqrt{\pi\xi}} \int_{-\pi}^\pi e^{-u^2/\xi} \cos u\, du = \frac{2}{\sqrt{\pi}\eta} \int_0^\pi \cos u\, e^{-(u/\eta)^2}\, du,$$

where $\eta = \sqrt{\xi} > 0$. We obtain

$$d_1(\eta) = \frac{2}{\sqrt{\pi}\eta} \cdot \left[\int_0^{\pi/2} \cos u\, e^{-(u/\eta)^2}\, du + \int_{\pi/2}^\pi \cos u\, e^{-(u/\eta)^2}\, du \right]$$

$$= \frac{2}{\sqrt{\pi}\eta} \left[\int_0^{\pi/2} \cos u\, e^{-(u/\eta)^2}\, du - \int_0^{\pi/2} \sin u\, e^{-\left(\frac{u+\pi/2}{\eta}\right)^2}\, du \right]$$

$$> \frac{2}{\sqrt{\pi}\eta} \left[\int_0^{\pi/2} (\cos u - \sin u) e^{-(u/\eta)^2}\, du \right] := \frac{2}{\sqrt{\pi}\eta}[I_1 + I_2],$$

where

$$I_1 = \int_0^{\pi/4} (\cos u - \sin u) e^{-(u/\eta)^2} \, du > 0$$

and

$$I_2 = - \int_{\pi/4}^{\pi/2} (\sin u - \cos u) e^{-(u/\eta)^2} \, du < 0.$$

It follows

$$I_1 > \int_0^{\pi/4} (\cos u - \sin u) e^{-(\pi/(4\eta))^2} \, du = (\sqrt{2} - 1) e^{-(\pi/(4\eta))^2}$$

and

$$|I_2| = -I_2 < e^{-(\pi/(4\eta))^2} \int_{\pi/4}^{\pi/2} (\sin u - \cos u) \, du = (\sqrt{2} - 1) e^{-(\pi/(4\eta))^2}.$$

Therefore,

$$d_1(\eta) > I_1 + I_2 \geq (\sqrt{2} - 1) e^{-(\pi/(4\eta))^2} - (\sqrt{2} - 1) e^{-(\pi/(4\eta))^2} = 0,$$

for any $\eta > 0$.

Now, for $d_k^*(\xi)$ we have

$$d_k^*(\xi) = \frac{1}{\sqrt{\pi\xi}} \int_{-\infty}^{+\infty} e^{-u^2/(\xi)} \cos ku \, du = e^{-k^2 \xi/4},$$

for all $\xi > 0$ and $k \geq 0$.

(ii) For $|z_1 - z_2| < \delta$, we get

$$|W_\xi^*(f)(z_1) - W_\xi^*(f)(z_2)| \leq \frac{1}{\sqrt{\pi\xi}} \cdot \int_{-\infty}^{+\infty} |f(z_1 e^{-iu}) - f(z_2 e^{-iu})| e^{-u^2/\xi} \, du$$
$$\leq \omega_1(f; |z_1 - z_2|)_{\overline{\mathbb{D}}_1} \leq \omega_1(f; \delta)_{\overline{\mathbb{D}}_1},$$

and

$$|W_\xi(f)(z_1) - W_\xi(f)(z_2)| \leq \frac{1}{\sqrt{\pi\xi}} \int_{-\pi}^{+\pi} |f(z_1 e^{iu}) - f(z_2 e^{iu})| e^{-u^2/\xi} \, du$$

$$\leq \omega_1(f; |z_1 - z_2|)_{\overline{\mathbb{D}}_1} \cdot \frac{1}{\sqrt{\pi\xi}} \int_{-\pi}^{\pi} e^{-u^2/\xi} \, du$$

$$\leq \omega_1(f; \delta)_{\overline{\mathbb{D}}_1} \cdot \frac{1}{\sqrt{\pi\xi}} \int_{-\infty}^{+\infty} e^{-u^2/\xi} \, du$$

$$= \omega_1(f; \delta)_{\overline{\mathbb{D}}_1}.$$

(iii) We can write

$$W_\xi(f)(z) - f(z) = \frac{1}{\sqrt{\pi\xi}} \int_0^\pi [f(z e^{iu}) - 2f(z) + f(z e^{-iu})] e^{-u^2/\xi} \, du$$

$$-f(z) \left[1 - \frac{1}{\sqrt{\pi\xi}} \int_{-\pi}^{\pi} e^{-u^2/\xi} \, du \right].$$

Here

$$\left| f(z) \left[1 - \frac{1}{\sqrt{\pi\xi}} \cdot \int_{-\pi}^{\pi} e^{-u^2/\xi} \, du \right] \right| = \left| f(z) \left[1 - \frac{2}{\sqrt{\pi\xi}} \int_{0}^{\pi} e^{-u^2/\xi} \, du \right] \right|$$

$$= \left| f(z) \left[\frac{2}{\sqrt{\pi\xi}} \int_{0}^{\infty} e^{-u^2/\xi} \, du - \frac{2}{\sqrt{\pi\xi}} \int_{0}^{\pi} e^{-u^2/\xi} \, du \right] \right|$$

$$= |f(z)| \cdot \left| \frac{2}{\sqrt{\pi\xi}} \int_{\pi}^{\infty} e^{-u^2/\xi} \, du \right|$$

$$\leq \|f\|_{\overline{\mathbb{D}}_1} \cdot \frac{2}{\sqrt{\pi\xi}} \int_{\pi}^{\infty} \frac{\xi}{u^2} \, du = 2\|f\|_{\overline{\mathbb{D}}_1} \sqrt{\xi} \cdot \frac{1}{\pi\sqrt{\pi}}.$$

By the maximum modulus principle we can take $|z| = 1$, which implies

$$|W_\xi(f)(z) - f(z)| \leq \frac{1}{\sqrt{\pi\xi}} \int_{0}^{\pi} \omega_2(f; u)_{\partial \mathbb{D}_1} e^{-u^2/\xi} \, du + 2\|f\|_{\overline{\mathbb{D}}_1} \sqrt{\xi} \frac{1}{\pi\sqrt{\pi}}$$

$$\leq \text{(reasoning as in Gal [94], p. 258)}$$

$$\leq \frac{C\omega_2(f; \xi)_{\partial \mathbb{D}_1}}{\xi} + 2\|f\|_{\overline{\mathbb{D}}_1} \sqrt{\xi} \cdot \frac{1}{\pi\sqrt{\pi}}.$$

Also, writing $\int_{-\infty}^{\infty} = \int_{-\infty}^{0} + \int_{0}^{\infty}$, we easily get

$$W_\xi^*(f)(z) - f(z) = \frac{1}{\sqrt{\pi\xi}} \int_{0}^{\infty} [f(ze^{iu}) - 2f(z) + f(ze^{-iu})] e^{-u^2/\xi} du,$$

which for $|z| = 1$ implies

$$|W_\xi^*(f)(z) - f(z)| \leq \frac{1}{\sqrt{\pi\xi}} \int_{0}^{+\infty} \omega_2(f; u)_{\partial \mathbb{D}_1} e^{-u^2/\xi} du$$

$$\leq \omega_2(f; \sqrt{\xi})_{\partial \mathbb{D}_1} \frac{1}{\sqrt{\pi\xi}} \int_{0}^{\infty} \left[\frac{u}{\sqrt{\xi}} + 1 \right]^2 e^{-u^2/\xi} du$$

$$\leq C\omega_2(f; \sqrt{\xi})_{\partial \mathbb{D}_1},$$

since

$$\frac{1}{\sqrt{\pi\xi}} \int_{0}^{\infty} \frac{u^2}{\xi} e^{-u^2/\xi} du = \frac{1}{\sqrt{\pi}} \int_{0}^{\infty} v^2 e^{-v^2} dv < \infty$$

and

$$\frac{2}{\sqrt{\pi\xi}} \int_{0}^{\infty} \frac{u}{\sqrt{\xi}} e^{-u^2/\xi} du = \frac{2}{\sqrt{\pi\xi}} \sqrt{\xi} \int_{0}^{\infty} v e^{-v^2} dv = \frac{2}{\sqrt{\pi}} \int_{0}^{\infty} v e^{-v^2} dv < \infty.$$

(iv) Analysing the proofs of the above points (i)-(iii), it easily follows that they hold by replacing everywhere $\mathbb{D}_1$ with $\mathbb{D}_r$, where $1 \leq r < R$. Therefore the approximation error in (iii) can be expressed in terms of

$$\omega_2(f; \sqrt{\xi})_{\partial \mathbb{D}_r} = \sup\{|\Delta_u^2 f(re^{it})|; |t| \leq \pi, |u| \leq \sqrt{\xi}\},$$

with constants in front of ω_2 depending on $r \geq 1$.

Since $\Delta_u^2 g(t) = 2u^2[t, t+u, t+2u; g]$, by the mean value theorem for divided differences in Complex Analysis (see e.g. Stancu [172], p. 258, Exercise 4.20) we get

$$|\Delta_u^2 f(re^{it})| \leq u^2 \|f''\|_r,$$

where

$$\|f''\|_r = \sup\{|f''(z)|; |z| \leq r\} \leq \sum_{k=2}^{\infty} |a_k| k(k-1) r^{k-2}.$$

For all $\xi \in (0, 1]$ from (iii) it follows

$$\|W_\xi^*(f) - f\|_r \leq C_r(f)\xi.$$

Now denoting by γ the circle of radius $r_1 > 1$ and center 0, since for any $|z| \leq r$ and $v \in \gamma$, we have $|v - z| \geq r_1 - r$, by the Cauchy's formulas it follows that for all $|z| \leq r$ and $\xi \in (0, 1]$, we have

$$|[W_\xi^*]^{(q)}(f)(z) - f^{(q)}(z)| = \frac{q!}{2\pi} \left| \int_\gamma \frac{W_\xi^*(f)(v) - f(v)}{(v - z)^{q+1}} dv \right|$$

$$\leq C_{r_1}(f)\xi \frac{q!}{2\pi} \frac{2\pi r_1}{(r_1 - r)^{q+1}}$$

$$= C_{r_1}(f)\xi \frac{q! r_1}{(r_1 - r)^{q+1}},$$

which proves the upper estimate in approximation by $[W_\xi^*]^{(q)}(f)(z)$.

It remains to prove the lower estimate for $\|[W_\xi^*]^{(q)}(f) - f^{(q)}\|_r$. For this purpose, take $z = re^{i\varphi}$ and $p \in \mathbb{N} \bigcup \{0\}$. We have

$$\frac{1}{2\pi}[f^{(q)}(z) - (W_\xi^*)^{(q)}(f)(z)]e^{-ip\varphi}$$

$$= \frac{1}{2\pi} \sum_{k=q}^{\infty} a_k k(k-1)...(k-q+1)r^{k-q}e^{i\varphi(k-q-p)}[1 - d_k^*(\xi)]$$

$$= \frac{1}{2\pi} \sum_{k=q}^{\infty} a_k k(k-1)...(k-q+1)r^{k-q}e^{i\varphi(k-q-p)}[1 - e^{-k^2\xi/4}].$$

Integrating from $-\pi$ to π we obtain

$$\frac{1}{2\pi} \int_{-\pi}^{\pi} [f^{(q)}(z) - (W_\xi^*)^{(q)}(f)(z)]e^{-ip\varphi} d\varphi$$

$$= a_{q+p}(q+p)(q+p-1)...(p+1)r^p[1 - e^{-(q+p)^2\xi/4}].$$

Passing now to absolute value we easily obtain

$$|a_{q+p}|(q+p)(q+p-1)...(p+1)r^p[1 - e^{-(q+p)^2\xi/4}] \leq \|f^{(q)} - (W_\xi^*)^{(q)}(f)\|_r.$$

First consider $q = 0$ and denote $V_\xi = \inf_{1 \leq p}(1 - e^{-p^2\xi/4})$. We get $V_\xi = 1 - e^{-\xi/4}$ and by the mean value theorem applied to $h(x) = e^{-x/4}$ on $[0, \xi]$ there exists $\eta \in (0, \xi)$ such that for all $\xi \in (0, 1]$ we have

$$V_\xi = h(0) - h(\xi) = (-\xi)h'(\eta) = (\xi/4)e^{-\eta/4} \geq (\xi/4)e^{-\xi/4} \geq \frac{e^{-1/4}}{4}\xi \geq \xi/8.$$

By the above lower estimate for $\|W_\xi^*(f) - f\|_r$, for all $p \geq 1$ and $\xi \in (0, 1]$ it follows

$$\frac{8\|W_\xi^*(f) - f\|_r}{\xi} \geq \frac{\|W_\xi^*(f) - f\|_r}{V_\xi} \geq \frac{\|W_\xi^*(f) - f\|_r}{1 - e^{-p^2\xi/4}} \geq |a_p| r^p.$$

This implies that if there exists a subsequence $(\xi_k)_k$ in $(0, 1]$ with $\lim_{k \to \infty} \xi_k = 0$ and such that $\lim_{k \to \infty} \frac{\|W_{\xi_k}^*(f) - f\|_r}{\xi_k} = 0$ then $a_p = 0$ for all $p \geq 1$, that is f is constant on $\overline{\mathbb{D}}_r$.

Therefore, if f is not a constant then $\inf_{\xi \in (0,1]} \frac{\|W_\xi^*(f) - f\|_r}{\xi} > 0$, which implies that there exists a constant $C_r(f) > 0$ such that $\frac{\|W_\xi^*(f) - f\|_r}{\xi} \geq C_r(f)$, for all $\xi \in (0, 1]$, that is

$$\|W_\xi^*(f) - f\|_r \geq C_r(f)\xi, \text{ for all } \xi \in (0, 1].$$

Now, consider $q \geq 1$ and denote $V_{q,\xi} = \inf_{p \geq 0}(1 - e^{-(p+q)^2\xi/4})$. Evidently that we have $V_{q,\xi} \geq \inf_{p \geq 1}(1 - e^{-p^2\xi/4}) \geq \xi/8$.

Reasoning as in the case of $q = 0$ we obtain

$$\frac{8\|[W_\xi^*]^{(q)}(f) - f^{(q)}\|_r}{\xi} \geq \frac{\|[W_\xi^*]^{(q)}(f) - f^{(q)}\|_r}{V_{q,\xi}} \geq \frac{\|[W_\xi^*]^{(q)}(f) - f^{(q)}\|_r}{1 - e^{-(p+q)^2\xi/4}}$$
$$\geq |a_{q+p}| \frac{(q+p)!}{p!} r^p,$$

for all $p \geq 0$ and $\xi \in (0, 1]$.

This implies that if there exists a subsequence $(\xi_k)_k$ in $(0, 1]$ with $\lim_{k \to \infty} \xi_k = 0$ and such that $\lim_{k \to \infty} \frac{\|[W_{\xi_k}^*]^{(q)}(f) - f^{(q)}\|_r}{\xi_k} = 0$ then $a_{q+p} = 0$ for all $p \geq 0$, that is f is a polynomial of degree $\leq q - 1$ on $\overline{\mathbb{D}}_r$.

Therefore, $\inf_{\xi \in (0,1]} \frac{\|[W_\xi^*]^{(q)}(f) - f^{(q)}\|_r}{\xi} > 0$ when f is not a polynomial of degree $\leq q - 1$, which implies that there exists a constant $C_{r,q}(f) > 0$ such that $\frac{\|[W_\xi^*]^{(q)}(f) - f^{(q)}\|_r}{\xi} \geq C_{r,q}(f)$, for all $\xi \in (0, 1]$, that is

$$\|[W_\xi^*]^{(q)}(f) - f^{(q)}\|_r \geq C_{r,q}(f)\xi, \text{ for all } \xi \in (0, 1],$$

which proves the theorem. $\qquad\qquad\qquad\qquad\qquad\qquad\qquad\qquad\qquad\qquad\square$

Concerning the geometric properties of the complex Gauss–Weierstrass convolutions, we present the following results.

Theorem 3.2.9. (Anastassiou-Gal [18]) *(i) If* $f(z) = \sum_{k=0}^{\infty} a_k z^k$, $z \in \mathbb{D}_1$ *and* $T_\xi(f)(z) = \sum_{k=0}^{\infty} A_k z^k$ *is any from* $W_\xi(f)(z)$ *and* $W_\xi^*(f)(z)$ *then*

$$|A_k| \leq |a_k|, \quad \forall k = 0, 1, \ldots;$$

(ii) If $f(z) = \sum_{k=1}^{\infty} a_k z^k$, $z \in \mathbb{D}_1$ *is univalent in* $\mathbb{D}_1$ *and* $f(\mathbb{D}_1)$ *is convex, then for any* $\xi > 0$, $W_\xi(f)(z)$ *is univalent in* $\mathbb{D}_1$ *and* $W_\xi(f)(\mathbb{D}_1)$ *is convex.*

Similarly, if $f(z)$ is univalent in $\mathbb{D}_1$ and $f(\mathbb{D}_1)$ is starlike with respect to the origin, then for any $\xi > 0$, $W_\xi(f)(z)$ is univalent in $\mathbb{D}_1$ and $W_\xi(f)(\mathbb{D}_1)$ is starlike with respect to the origin.

(iii) For all $\xi > 0$ we have :

$$W_\xi^*(\mathcal{P}) \subset \mathcal{P}, \quad \frac{1}{d_1(\xi)} W_\xi(S_{3,d_1(\xi)}) \subset S_3,$$

$$\frac{1}{d_1^*(\xi)} W_\xi^*(S_{3,d_1^*(\xi)}) \subset S_3, \quad \frac{1}{d_1(\xi)} W_\xi(S_M) \subset S_{M/|d_1(\xi)|},$$

$$\frac{1}{d_1^*(\xi)} W_\xi^*(S_M) \subset S_{M/|d_1^*(\xi)|}.$$

Proof.   (i) By Theorem 3.2.8 (i), we get

$$|a_k d_k(\xi)| \le |a_k| \cdot |d_k(\xi)| \le |a_k| \cdot \frac{1}{\sqrt{\pi\xi}} \int_{-\pi}^{\pi} e^{-u^2/\xi} |\cos ku| \, du$$

$$\le |a_k| \cdot \frac{1}{\sqrt{\pi\xi}} \int_{-\infty}^{+\infty} e^{-u^2/\xi} \, du = |a_k|, \quad \text{for all } k = 0, 1, 2, \dots.$$

Also, by the same theorem, we obtain

$$|a_k d_k^*(\xi)| = |a_k| \cdot |d_k^*(\xi)| \le |a_k|, \quad \text{for all } k = 0, 1, 2, \dots.$$

Also, note that $|d_0(\xi)| = d_0(\xi) \le 1$ and $|d_0^*(\xi)| = d_0(\xi) = 1$.

(ii) Let $g(u) = e^{-u^2/\xi}$, $u \in [-\pi, \pi]$. Since $g(-\pi) = g(\pi)$, we can extend $g(u)$ by 2π-periodicity on the whole $\mathbb{R}$, such that the extension, denoted by $h(u)$, is continuous on $\mathbb{R}$.

It is easy to check that $\log|h'(u)|$ is concave in each interval $[k\pi, (k+1)\pi]$, $h'(u) = 0$ if and only if $u = 2k\pi$, $k \in \mathbb{Z}$, and in $u_k = k\pi$, $k \in \mathbb{Z}$, h takes its minimum and maximum values.

Then, applying Ruscheweyh-Salinas [161], Theorem, p. 130, we get that h is PMP as in Ruscheweyh-Salinas [161], which implies that $W_\xi(f)$ preserves the convexity of f.

Also, by similar reasoning with those in Pólya-Schoenberg [150], p. 321, Lemma 5 and Corollary 5, it follows that $W_\xi(f)(z)$ preserve the starlikeness of $f(z)$ (with respect to origin) too.

(iii) The proofs are similar to the proofs in Theorem 3.2.6 (iii), which proves Theorem 3.2.9 too. $\square$

Remarks. 1) From the results presented above, it follows that $W_\xi(f)(z)$ has the best preservation property among the classes of complex singular integrals studied by the present book.

2) Let us denote $D = \inf\{|d_1(\xi)|; \, \xi \in (0,1]\}$. If $D > 0$ then by Theorem 3.2.9 (iii) we get:

if $f \in S_{3,D}$ then $W_\xi(f) \in S_3$, for all $\xi \in (0,1]$, if $f \in S_M$, $(M > 1)$, then $W_\xi(f)$ is univalent in $\{z \in \mathbb{C}; |z| < \frac{D}{M}\}$, for all $\xi \in (0,1]$. Therefore it remains to calculate D, to check if $D > 0$, problems which are left to the reader as an open question.

Since $\inf\{|d_1^*(\xi)|; \xi \in (0,1]\} = 1/(e^{1/4})$, applying now Theorems 3.2.8 (i) and 3.2.9 (iii) to $W_\xi^*(f)(z)$, we immediately get the following result.

Corollary 3.2.10. (Anastassiou-Gal [18]) *If $f \in S_{3, \frac{1}{e^{1/4}}}$ then $W_\xi^*(f) \in S_3$, for all $\xi \in (0,1]$ and if $f \in S_M$, $(M > 1)$, then $W_\xi^*(f)$ is univalent in $\{z \in \mathbb{C}; |z| < \frac{1}{Me^{1/4}}\}$, for all $\xi \in (0,1]$.*

In what follows we study the Jackson-type generalizations of Picard-type $P_{n,\xi}(f)(z)$, of Poisson–Cauchy–type $Q_{n,\xi}(f)(z)$, and of generalized Gauss–Weierstrass–type $W_{n,\xi}(f)(z)$ and $W_{n,\xi}^*(f)(z)$.

The approximation results are expressed by the following theorem. Note that Theorem 3.2.11, (i), (ii) and (iii) were obtained in Anastassiou-Gal [16] while Theorem 3.2.11 (iv) is new.

Theorem 3.2.11. *(i) Suppose that f is continuous on $\overline{\mathbb{D}}_1$. For all $\delta > 0$, $\xi > 0$ and $n \in \mathbb{N}$, we have*

$$\omega_1\big(P_{n,\xi}(f); \delta\big)_{\overline{\mathbb{D}}_1} \leq (2^{n+1} - 1)\omega_1(f; \delta)_{\overline{\mathbb{D}}_1},$$

$$\omega_1\big(W_{n,\xi}(f); \delta\big)_{\overline{\mathbb{D}}_1} \leq (2^{n+1} - 1)\omega_1(f; \delta)_{\overline{\mathbb{D}}_1},$$

$$\omega_1\big(W_{n,\xi}^*(f); \delta\big)_{\overline{\mathbb{D}}_1} \leq (2^{n+1} - 1)\omega_1(f; \delta)_{\overline{\mathbb{D}}_1},$$

$$\omega_1\big(Q_{n,\xi}(f); \delta\big)_{\overline{\mathbb{D}}_1} \leq (2^{n+1} - 1)\omega_1(f; \delta)_{\overline{\mathbb{D}}_1}.$$

(ii) Suppose that $f \in A(\overline{\mathbb{D}}_1)$. For all $z \in \overline{\mathbb{D}}_1$ and $\xi \in (0,1]$ we have

$$|P_{n,\xi}(f)(z) - f(z)| \leq \left[\sum_{k=0}^{n+1}\binom{n+1}{k}k!\right]\omega_{n+1}(f; \xi)_{\partial\mathbb{D}_1},$$

$$|W_{n,\xi}(f)(z) - f(z)| \leq C_n\omega_{n+1}(f; \xi)_{\partial\mathbb{D}_1}, \quad C_n = \frac{\int_0^\infty (1+u)^{n+1}e^{-u^2}\,du}{\int_0^\pi e^{-u^2}\,du},$$

$$|W_{n,\xi}^*(f)(z) - f(z)| \leq C_n^*\omega_{n+1}(f; \xi)_{\partial\mathbb{D}_1}, \quad C_n^* = \frac{\int_0^\infty (1+u)^{n+1}e^{-u^2}\,du}{\int_0^\infty e^{-u^2}\,du},$$

$$|Q_{n,\xi}(f)(z) - f(z)| \leq K(n,\xi)\omega_{n+1}(f; \xi)_{\partial\mathbb{D}_1}, \quad K(n,\xi) = \frac{\int_0^{\pi/\xi}\frac{(u+1)^{n+1}}{u^2+1}\,du}{\tan^{-1}\big(\frac{\pi}{\xi}\big)},$$

where

$$\omega_{n+1}(f; \xi)_{\partial\mathbb{D}_1} = \sup\{|\Delta_u^{n+1}f(e^{ix})|; |x| \leq \pi, |u| \leq \xi\}.$$

(iii) If $f(z) = \sum_{k=0}^\infty a_k z^k$ is analytic in $\mathbb{D}_1$ and continuous in $\overline{\mathbb{D}}_1$, then $P_{n,\xi}(f)(z)$, $W_{n,\xi}(f)(z)$, $W_{n,\xi}^*(f)(z)$ and $Q_{n,\xi}(f)(z)$ are analytic in $\mathbb{D}_1$ and continuous in $\overline{\mathbb{D}}_1$, for all $\xi > 0$ and $n \geq 2$.

Also, we can write

$$P_{n,\xi}(f)(z) = \sum_{p=0}^{\infty} a_p b_{p,n}(\xi) z^p,$$

with

$$b_{p,n}(\xi) = \sum_{k=1}^{n+1} (-1)^{k+1} \binom{n+1}{k} \cdot \frac{1}{\xi^2 k^2 p^2 + 1},$$

$$W_{n,\xi}(f)(z) = \sum_{p=0}^{\infty} a_p c_{p,n}(\xi) z^p,$$

with

$$c_{p,n}(\xi) = \frac{1}{C(\xi)} \sum_{k=1}^{n+1} (-1)^{k+1} \binom{n+1}{k} \cdot \int_0^{\pi} \cos(kpu) e^{-u^2/\xi^2} \, du,$$

$$W_{n,\xi}^*(f)(z) = \sum_{p=0}^{\infty} a_p c_{p,n}^*(\xi) z^p,$$

with

$$c_{p,n}^*(\xi) = \frac{1}{C^*(\xi)} \sum_{k=1}^{n+1} (-1)^{k+1} \binom{n+1}{k} \cdot \int_0^{\infty} \cos(kpu) e^{-u^2/\xi^2} \, du,$$

$$Q_{n,\xi}(f)(z) = \sum_{p=0}^{\infty} a_p d_{p,n}(\xi) z^p,$$

with

$$d_{p,n}(\xi) = \frac{\xi}{\operatorname{arctg} \frac{\pi}{\xi}} \sum_{k=1}^{n+1} (-1)^{k+1} \binom{n+1}{k} \int_0^{\pi} \frac{\cos(kpu)}{u^2 + \xi^2} \, du.$$

Here for all $\xi \in (0, \xi_n]$ we have,

$$b_{1,n}(\xi) = \sum_{k=1}^{n+1} (-1)^{k+1} \binom{n+1}{k} \frac{1}{\xi^2 k^2 + 1} > 0,$$

$$c_{1,n}(\xi) = \frac{1}{C(\xi)} \cdot \sum_{k=1}^{n+1} (-1)^{k+1} \binom{n+1}{k} \int_0^{\pi} \cos(ku) e^{-u^2/\xi^2} \, du > 0,$$

$$c_{1,n}^*(\xi) = \frac{1}{C^*(\xi)} \sum_{k=1}^{n+1} (-1)^{k+1} \binom{n+1}{k} \int_0^{\infty} \cos(ku) e^{-u^2/\xi^2} \, du$$

$$= \sum_{k=1}^{n+1} (-1)^{k+1} \binom{n+1}{k} e^{-k^2 \xi^2/4} > 0,$$

and

$$d_{1,n}(\xi) = \frac{\xi}{\arctan \frac{\pi}{\xi}} \cdot \sum_{k=1}^{n+1} (-1)^{k+1} \binom{n+1}{k} \int_0^\pi \frac{\cos ku}{u^2 + \xi^2} \, du > 0,$$

where $0 < \xi_n$ is independent of k and f (but may depend on n).

(iv) Let us suppose that $f(z) = \sum_{k=0}^\infty a_k z^k$ for all $z \in \mathbb{D}_R$, $R > 1$. For all $1 \le r < r_1 < R$, $\xi \in (0,1]$, $n \in \mathbb{N}$ and $q \in \mathbb{N} \bigcup \{0\}$ we have

$$\|P_{n,\xi}^{(q)}(f) - f^{(q)}\|_r \le C_1 \xi^{n+1}, \text{ for all } \xi \in (0,1],$$

$$\|[W_{n,\xi}^*]^{(q)}(f) - f^{(q)}\|_r \le C_2 \xi^{n+1}, \text{ for all } \xi \in (0,1],$$

where the constants C_1, C_2 depend only on f, q, r, r_1 and n.

Proof.   (i) Let $|z_1 - z_2| \le \delta$, $z_1, z_2 \in \overline{\mathbb{D}}_1$. We have

$$|P_{n,\xi}(f)(z_1) - P_{n,\xi}(f)(z_2)|$$

$$\le \omega_1\big(f; |z_1 - z_2|\big)_{\overline{\mathbb{D}}_1} \cdot \frac{1}{2\xi} \int_{-\infty}^{+\infty} \sum_{k=1}^{n+1} \binom{n+1}{k} e^{-|u|/\xi} \, du$$

$$\le \sum_{k=1}^{n+1} \binom{n+1}{k} \omega_1(f; \delta)_{\overline{\mathbb{D}}_1} = (2^{n+1} - 1)\omega_1(f; \delta)_{\overline{\mathbb{D}}_1}.$$

As above, we obtain

$$|W_{n,\xi}(f)(z_1) - W_{n,\xi}(f)(z_2)| \le \frac{1}{2C(\xi)} \cdot \int_{-\pi}^{\pi} \sum_{k=1}^{n+1} \binom{n+1}{k} e^{-u^2/\xi} \, du \cdot \omega_1(f; \delta)_{\overline{\mathbb{D}}_1}$$

$$\le \sum_{k=1}^{n+1} \binom{n+1}{k} \omega_1(f; \delta)_{\overline{\mathbb{D}}_1} = (2^{n+1} - 1)\omega_1(f; \delta)_{\overline{\mathbb{D}}_1}$$

and analogously

$$|W_{n,\xi}^*(f)(z_1) - W_{n,\xi}^*(f)(z_2)| \le (2^{n+1} - 1)\omega_1(f; \delta)_{\overline{\mathbb{D}}_1}.$$

Finally,

$$|Q_{n,\xi}(f)(z_1) - Q_{n,\xi}(f)(z_2)| \le \frac{1}{\frac{2}{\xi}\arctan \frac{\pi}{\xi}} \int_{-\pi}^{\pi} \frac{du}{u^2 + \xi^2} \cdot \sum_{k=1}^{n+1} \binom{n+1}{k} \omega_1(f; \delta)_{\overline{\mathbb{D}}_1}$$

$$= (2^{n+1} - 1)\omega_1(f; \delta)_{\overline{\mathbb{D}}_1}.$$

Passing in all the above inequalities to sup with $|z_1 - z_2| \le \delta$, we obtain the required relations in (i).

(ii) Let $|z| = 1$, $\xi > 0$ be fixed. Because of the maximum modulus principle, it suffices to estimate $|P_{n,\xi}(f)(z) - f(z)|$, for this $|z| = 1$, $z = e^{ix}$. We get

$$f(z) - P_{n,\xi}(f)(z) = f(z) \cdot \frac{1}{2\xi} \int_{-\infty}^{+\infty} e^{-|u|/\xi} \, du$$

$$+ \frac{1}{2\xi} \int_{-\infty}^{+\infty} \left[\sum_{k=1}^{n+1} (-1)^k \binom{n+1}{k} \right] f(e^{i(x+ku)}) e^{-|u|/\xi} \, du$$

$$= \frac{1}{2\xi} \int_{-\infty}^{+\infty} (-1)^{n+1} \Delta_u^{n+1} f(e^{ix}) e^{-|u|/\xi} \, du,$$

where from

$$|f(z) - P_{n,\xi}(f)(z)| \leq \frac{1}{2\xi} \int_{-\infty}^{+\infty} \omega_{n+1}(f; |u|)_{\partial \mathbb{D}_1} e^{-|u|/\xi} \, du$$

$$= \frac{1}{\xi} \int_0^{+\infty} \omega_{n+1}\left(f; \frac{u}{\xi} \cdot \xi\right)_{\partial \mathbb{D}_1} e^{-u/\xi} \, du$$

$$\leq \omega_{n+1}(f; \xi)_{\partial \mathbb{D}_1} \frac{1}{\xi} \int_0^{+\infty} \left(1 + \frac{u}{\xi}\right)^{n+1} e^{-u/\xi} \, du$$

$$= \text{(reasoning exactly as in Gal [94], p. 254)}$$

$$= \sum_{k=0}^{n+1} \binom{n+1}{k} k! \omega_{n+1}(f; \xi)_{\partial \mathbb{D}_1}.$$

As above, we obtain

$$f(z) - W_{n,\xi}(f)(z) = \frac{1}{2C(\xi)} \int_{-\pi}^{\pi} (-1)^{n+1} \Delta_n^{n+1} f(e^{ix}) e^{-u^2/\xi^2} \, du,$$

which implies

$$|f(z) - W_{n,\xi}(f)(z)| \leq \frac{1}{C(\xi)} \int_0^{\pi} \omega_{n+1}(f; u)_{\partial \mathbb{D}_1} e^{-u^2/\xi^2} \, du$$

$$\leq \frac{1}{C(\xi)} \omega_{n+1}(f; \xi)_{\partial \mathbb{D}_1} \int_0^{\pi} \left[1 + \frac{u}{\xi}\right]^{n+1} e^{-u^2/\xi^2} \, du$$

$$\text{(reasoning exactly as in Gal [94], p. 260)}$$

$$\leq \frac{\int_0^{+\infty} [1 + u]^{n+1} e^{-u^2} \, du}{\int_0^{\pi} e^{-u^2} \, du} \cdot \omega_{n+1}(f; \xi)_{\partial \mathbb{D}_1}.$$

Similarly,

$$f(z) - W_{n,\xi}^*(f)(z) = \frac{1}{2C^*(\xi)} \int_{-\infty}^{+\infty} (-1)^{n+1} \Delta_u^{n+1} f(e^{ix}) e^{-u^2/\xi^2} \, du,$$

which implies as above

$$|f(z) - W_{n,\xi}^*(f)(z)| \leq \frac{1}{C^*(\xi)} \omega_{n+1}(f; \xi)_{\partial \mathbb{D}_1} \int_0^{\infty} \left[1 + \frac{u}{\xi}\right]^{n+1} e^{-u^2/\xi^2} \, du$$

$$\leq \frac{\int_0^{+\infty} [1 + u]^{n+1} e^{-u^2} \, du}{\int_0^{\infty} e^{-u^2} \, du} \omega_{n+1}(f; \xi)_{\partial \mathbb{D}_1}.$$

Finally, by the relation

$$f(z) - Q_{n,\xi}(f)(z) = \frac{1}{\frac{2}{\xi} \operatorname{arctg} \frac{\pi}{\xi}} \int_{-\pi}^{\pi} \frac{(-1)^{n+1} \Delta_u^{n+1} f(e^{ix})}{u^2 + \xi^2} \, du,$$

it follows (taking into account Anastassiou-Gal [18], p. 518 too)

$$|f(z) - Q_{n,\xi}(f)(z)| \leq \frac{\xi}{\operatorname{arctg} \frac{\pi}{\xi}} \int_0^{\pi} \frac{\omega_{n+1}(f; u)_{\partial \mathbb{D}_1}}{u^2 + \xi^2} \, du$$

$$\leq \frac{\xi}{\operatorname{arctg} \frac{\pi}{\xi}} \omega_{n+1}(f; \xi)_{\partial \mathbb{D}_1} \int_0^{\pi} \left[\frac{u}{\xi} + 1\right]^{n+1} \cdot \frac{1}{u^2 + \xi^2} \, du$$

$$= K(n, \xi) \omega_{n+1}(f; \xi)_{\partial \mathbb{D}_1}$$

which proves (ii).

(iii) Let $f(z) = \sum_{p=0}^{\infty} a_p z^p$, $z \in \mathbb{D}_1$. For fixed $z \in \mathbb{D}_1$, we can write $f(ze^{iku}) = \sum_{p=0}^{\infty} a_p e^{ikpu} z^p$ and since $|a_p e^{ikpu}| = |a_p|$ for all $u \in \mathbb{R}$ and the series $\sum_{p=0}^{\infty} a_p z^p$ is convergent, it follows that the series $\sum_{p=0}^{\infty} a_p e^{ikpu} z^p$ is uniformly convergent with respect to $u \in \mathbb{R}$. Therefore the series can be integrated term by term (with respect to u), that is

$$P_{n,\xi}(f)(z) = -\frac{1}{2\xi} \sum_{k=1}^{n+1} (-1)^k \binom{n+1}{k} \sum_{p=0}^{\infty} a_p z^p \int_{-\infty}^{+\infty} e^{ikpu} e^{-|u|/\xi}\, du.$$

But

$$-\frac{1}{2\xi} \int_{-\infty}^{+\infty} e^{ikpu} e^{-|u|/\xi}\, du$$

$$= -\frac{1}{2\xi} \int_{-\infty}^{\infty} [\cos(kpu) + i\sin(kpu)] e^{-|u|/\xi}\, du$$

$$= -\frac{1}{\xi} \int_{0}^{\infty} \cos(kpu) e^{-u/\xi}\, du$$

$$= -\frac{1}{\xi} \cdot \frac{e^{-u/\xi}\left[-\frac{1}{\xi}\cos(kpu) + k\sin(kpu)\right]}{k^2 p^2 + \frac{1}{\xi^2}}\Bigg|_{0}^{+\infty}$$

$$= \frac{1}{\xi} \cdot \frac{-\frac{1}{\xi}}{\xi^2 k^2 p^2 + 1} \cdot \xi^2 = -\frac{1}{\xi^2 k^2 p^2 + 1}.$$

Therefore we can write

$$P_{n,\xi}(f)(z) = \sum_{p=0}^{\infty} a_p z^p \cdot \left[-\sum_{k=1}^{n+1} (-1)^k \binom{n+1}{k} \cdot \frac{1}{\xi^2 k^2 p^2 + 1} \right] = \sum_{p=0}^{\infty} a_p b_{p,n}(\xi) z^p,$$

with

$$b_{p,n}(\xi) = \sum_{k=1}^{n+1} (-1)^{k+1} \binom{n+1}{k} \cdot \frac{1}{\xi^2 k^2 p^2 + 1},$$

for all $z \in \mathbb{D}_1$.

For the continuity property, let $z \in \overline{\mathbb{D}}_1$ and $z_m \in \overline{\mathbb{D}}_1$, $n \in \mathbb{N}$, with $\lim_{m \to \infty} z_m = z_0$. We have

$$|P_{n,\xi}(f)(z_m) - P_{n,\xi}(f)(z_0)| \leq \frac{1}{2\xi} \sum_{k=1}^{n+1} \binom{n+1}{k}$$

$$\cdot \int_{-\infty}^{+\infty} |f(z_m e^{iku}) - f(z_0 e^{iku})| e^{-|u|/\xi}\, du$$

$$\leq (2^{n+1} - 1)\omega_1(f; |z_m - z_0|)_{\overline{\mathbb{D}}_1}.$$

Passing to limit with $m \to \infty$, we get that $P_{n,\xi}(f)(z)$ is continuous on $\overline{\mathbb{D}}_1$.

The proofs for the other operators $W_{n,\xi}(f)(z)$, $W^*_{n,\xi}(f)(z)$ and $Q_{n,\xi}(f)(z)$ are similar. The formulas for $b_{1,n}(\xi)$, $c_{1,n}(\xi)$ and $d_{1,n}(\xi)$ are immediate from above.

Also, since $C^*(\xi) = \int_0^\infty e^{-u^2/\xi^2}\, du = \xi \int_0^\infty e^{-v^2}\, dv = \frac{\xi\sqrt{\pi}}{2}$ and

$$\int_0^\infty \cos(ku) e^{-u^2/\xi^2}\, du = \xi \int_0^\infty \cos(k\xi v) e^{-v^2}\, dv = \frac{\xi\sqrt{\pi} \cdot e^{-k^2\xi^2/4}}{2},$$

we get

$$c^*_{1,n}(\xi) = \frac{2}{\xi\sqrt{\pi}} \cdot \sum_{k=1}^{n+1} (-1)^{k+1} \binom{n+1}{k} \cdot \frac{\xi\sqrt{\pi}}{2} e^{-k^2\xi^2/4}$$

$$= \sum_{k=1}^{n+1} (-1)^{k+1} \binom{n+1}{k} e^{-k^2\xi^2/4}.$$

Now, by

$$0 = (-1+1)^{n+1} = \sum_{k=0}^{n+1} (-1)^k \binom{n+1}{k} = 1 + \sum_{k=1}^{n+1} (-1)^k \binom{n+1}{k},$$

it follows $\sum_{k=1}^{n+1} (-1)^{k+1} \binom{n+1}{k} = 1$.

Then, since

$$b_{1,n}(\xi) = \sum_{k=1}^{n+1} (-1)^{k+1} \binom{n+1}{k} \frac{1}{\xi^2 k^2 + 1}$$

and

$$c^*_{1,n}(\xi) = \sum_{k=1}^{n+1} (-1)^{k+1} \binom{n+1}{k} e^{-k^2\xi^2/4}$$

are obviously continuous functions of $\xi \in \mathbb{R}_+$ and

$$b_{1,n}(0) = c^*_{1,n}(0) = \sum_{k=1}^{n+1} (-1)^{k+1} \binom{n+1}{k} = 1,$$

there exists $\xi_n > 0$ such that $b_{1,n}(\xi) > 0$, $c^*_{1,n}(\xi) > 0$, $\forall \xi \in (0, \xi_n]$. Also, $c_{1,n}(\xi)$ and $d_{1,n}(\xi)$ are obviously continuous functions of $\xi \in \mathbb{R}_+ \setminus \{0\}$.

Since

$$C(\xi) = \int_0^\pi e^{-u^2/\xi^2}\, du = \xi \int_0^{\pi/\xi} e^{-v^2}\, dv$$

and

$$\int_0^\pi \cos(ku) e^{-u^2/\xi^2}\, du = \xi \int_0^{\pi/\xi} \cos(k\xi v) e^{-v^2}\, dv,$$

we get

$$\lim_{\xi\downarrow 0}\left[\int_0^\pi \cos(ku)e^{-u^2/\xi^2}\,du/C(\xi)\right]$$

$$=\lim_{\xi\downarrow 0}\int_0^{\pi/\xi}\cos(k\xi v)e^{-v^2}\,dv/\lim_{\xi\downarrow 0}\int_0^{\pi/\xi}e^{-v^2}\,dv$$

$$=\lim_{\xi\downarrow 0}\int_0^{\pi/\xi}\cos(k\xi v)e^{-v^2}\,dv/\int_0^{\infty}e^{-v^2}\,dv$$

$$=\frac{2}{\sqrt{\pi}}\lim_{\xi\downarrow 0}\int_0^{\pi/\xi}\cos(k\xi v)e^{-v^2}\,dv.$$

By the substitution $\xi v=u$ we get

$$\int_0^{\pi/\xi}[1-\cos(k\xi v)]e^{-v^2}\,dv=\frac{1}{\xi}\int_0^\pi [1-\cos(ku)]e^{-(u/\xi)^2}\,du,$$

that is

$$\left|\int_0^{\pi/\xi}\cos(k\xi v)e^{-v^2}\,dv-\int_0^\infty e^{-v^2}\,dv\right|$$

$$\le\left|\int_0^{\pi/\xi}\cos(k\xi v)e^{-v^2}\,dv-\int_0^{\pi/\xi}e^{-v^2}\,dv\right|$$

$$+\left|\int_0^{\pi/\xi}e^{-v^2}\,dv-\int_0^\infty e^{-v^2}\,dv\right|$$

$$\le\int_0^\pi |1-\cos(ku)|\xi^{-1}e^{-(u/\xi)^2}\,du+\left|\int_0^{\pi/\xi}e^{-v^2}\,dv-\int_0^\infty e^{-v^2}\,dv\right|.$$

Since

$$|1-\cos(ku)|=2\sin^2\frac{ku}{2}\le\frac{2k^2u^2}{4}=\frac{k^2u^2}{2},$$

we get

$$|1-\cos(ku)|\xi^{-1}e^{-(u/\xi)^2}\le\frac{k^2}{2}u^2\xi^{-1}e^{-(u/\xi)^2},$$

where

$$\lim_{\xi\downarrow 0}u^2\xi^{-1}e^{-(u/\xi)^2}=\lim_{\xi\downarrow 0}\frac{\frac{u^2}{\xi}}{e^{(u/\xi)^2}}=\lim_{\xi\downarrow 0}\frac{-\frac{u^2}{\xi^2}}{-\frac{2u^2}{\xi^3}e^{u^2/\xi^2}}=\lim_{\xi\downarrow 0}\frac{\xi}{e^{u^2/\xi^2}}=0,$$

that is $|1-\cos(ku)|\xi^{-1}e^{-(u/\xi)^2}\xrightarrow{\xi\to 0}0$, uniformly with respect to $u\in[0,\pi]$. (We applied the l'Hospital's rule). This immediately implies

$$\lim_{\xi\downarrow 0}\int_0^{\pi/\xi}\cos(ku)e^{-u^2/\xi^2}\,du/C(\xi)=\frac{2}{\sqrt{\pi}}\int_0^\infty e^{-v^2}\,dv=1.$$

Therefore,

$$\lim_{\xi \downarrow 0} c_{1,n}(\xi) = \sum_{k=1}^{n+1} (-1)^{k+1} \binom{n+1}{k} \lim_{\xi \downarrow 0} \frac{\int_0^\pi \cos(ku) e^{-u^2/\xi^2} \, du}{C(\xi)}$$

$$= \sum_{k=1}^{n+1} (-1)^{k+1} \binom{n+1}{k} = 1 > 0,$$

which implies that there exists $\xi_n > 0$ such that $c_{1,n}(\xi) > 0$, $\forall \xi \in (0, \xi_n]$.

In the case of $d_{1,n}(\xi)$, since

$$\frac{\xi}{\operatorname{arctg} \frac{\pi}{\xi}} = \frac{1}{\int_0^\pi \frac{du}{u^2 + u^2}} = \frac{1}{\frac{1}{\xi} \int_0^{\pi/\xi} \frac{dv}{v^2 + 1}}$$

and

$$\int_0^\pi \frac{\cos ku}{u^2 + \xi^2} \, du = \frac{1}{\xi} \int_0^{\pi/\xi} \frac{\cos(k\xi v)}{v^2 + 1} \, dv,$$

we get

$$\lim_{\xi \downarrow 0} \frac{\xi}{\operatorname{arctg} \frac{\pi}{\xi}} \cdot \int_0^\pi \frac{\cos ku}{u^2 + \xi^2} \, du = \lim_{\xi \downarrow 0} \frac{\int_0^{\pi/\xi} \frac{\cos(k\xi v)}{v^2 + 1} \, dv}{\int_0^{\pi/\xi} \frac{dv}{v^2 + 1}}$$

$$= \frac{\lim_{\xi \downarrow 0} \int_0^{\pi/\xi} \frac{\cos(k\xi v)}{v^2 + 1} \, dv}{\int_0^\infty \frac{dv}{v^2 + 1}}$$

$$= \frac{2}{\pi} \lim_{\xi \downarrow 0} \int_0^{\pi/\xi} \frac{\cos(k\xi v)}{v^2 + 1} \, dv$$

$$= \frac{2}{\pi} \cdot \int_0^\infty \frac{dv}{v^2 + 1} = 1.$$

Here, as in the above case, we write

$$\left| \int_0^{\pi/\xi} \frac{\cos(k\xi v)}{v^2 + 1} \, dv - \int_0^\infty \frac{dv}{v^2 + 1} \right|$$

$$\leq \left| \int_0^{\pi/\xi} \frac{\cos(k\xi v)}{v^2 + 1} \, dv - \int_0^{\pi/\xi} \frac{dv}{v^2 + 1} \right| + \left| \int_0^{\pi/\xi} \frac{dv}{v^2 + 1} - \int_0^\infty \frac{dv}{v^2 + 1} \right|$$

$$\leq \int_0^{\pi/\xi} \frac{[1 - \cos(k\xi v)]}{v^2 + 1} \, dv + \left| \int_0^{\pi/\xi} \frac{dv}{v^2 + 1} - \int_0^\infty \frac{dv}{v^2 + 1} \right|.$$

But

$$\int_0^{\pi/\xi} \frac{[1 - \cos(k\xi v)]}{v^2 + 1} \, dv = \frac{1}{\xi} \int_0^\pi \frac{(1 - \cos ku)}{1 + \left(\frac{u}{\xi}\right)^2} \, du = \frac{1}{\xi} \int_0^\pi \frac{2 \sin^2 \frac{ku}{2}}{1 + \left(\frac{u}{\xi}\right)^2} \, du$$

$$\leq \frac{k^2}{2} \int_0^\pi \frac{u^2}{\xi} \cdot \frac{1}{1 + \left(\frac{u}{\xi}\right)^2} \, du = \frac{k^2}{2} \int_0^\pi \xi \frac{u^2}{u^2 + \xi^2} \, du.$$

Denote $0 \le g_\xi(u) = \xi \frac{u^2}{u^2 + \xi^2} \le \xi$. We obviously have $\lim_{\xi \downarrow 0} g_\xi(u) = 0$, uniformly with respect to $u \in [0, \pi]$, which implies

$$\frac{2}{\pi} \lim_{\xi \downarrow 0} \int_0^{\pi/\xi} \frac{\cos(k\xi v)}{v^2 + 1} dv = \frac{2}{\pi} \int_0^\infty \frac{dv}{v^2 + 1} = 1.$$

Therefore,

$$\lim_{\xi \downarrow 0} d_{1,n}(\xi) = \sum_{k=1}^{n+1} (-1)^{k+1} \binom{n+1}{k} = 1 > 0,$$

which implies that there exists $\xi_n > 0$ such that $d_{1,n}(\xi) > 0$, for all $\xi \in (0, \xi_n]$.

Obviously, we can choose the same $\xi_n > 0$ for all the four operators $P_{n,\xi}(f)$, $W_{n,\xi}(f)$, $W_{n,\xi}^*(f)$ and $Q_{n,\xi}(f)$.

(iv) Analysing the proofs of the above points (i)-(iii), it easily follows that they hold by replacing everywhere $\mathbb{D}_1$ with $\mathbb{D}_r$, where $1 \le r < R$. Therefore the approximation error in (iii) can be expressed in terms of

$$\omega_{n+1}(f; \xi)_{\partial \mathbb{D}_r} = \sup\{|\Delta_u^{n+1} f(re^{it})|; |t| \le \pi, \ |u| \le \xi\},$$

with constants in front of ω_2 depending on $r \ge 1$.

By the mean value theorem for divided differences in Complex Analysis (see e.g. Stancu [172], p. 258, Exercise 4.20) we get

$$|\Delta_u^{n+1} f(re^{it})| \le |u|^{n+1} \|f^{(n+1)}\|_r,$$

where

$$\|f^{(n+1)}\|_r = \sup\{|f^{(n+1)}(z)|; |z| \le r\} \le \sum_{k=n+1}^\infty |a_k| k(k-1)...(k-n) r^{k-n-1}.$$

Then by (ii) for all $\xi \in (0, 1]$ it follows

$$\|P_{n,\xi}(f) - f\|_r \le C_{r,n}(f) \xi^{n+1},$$

$$\|W_{n,\xi}^*(f) - f\|_r \le C_{r,n}^*(f) \xi^{n+1}.$$

Now denoting by γ the circle of radius $r_1 > 1$ and center 0, since for any $|z| \le r$ and $v \in \gamma$, we have $|v - z| \ge r_1 - r$, by the Cauchy's formulas it follows that for all $|z| \le r$ and $\xi \in (0, 1]$, we have

$$|[W_{n,\xi}^*]^{(q)}(f)(z) - f^{(q)}(z)| = \frac{q!}{2\pi} \left| \int_\gamma \frac{W_{n,\xi}^*(f)(v) - f(v)}{(v-z)^{q+1}} dv \right|$$

$$\le C_{r_1,n}^*(f) \xi^{n+1} \frac{q!}{2\pi} \frac{2\pi r_1}{(r_1 - r)^{q+1}}$$

$$= C_{r_1,n}^*(f) \xi^{n+1} \frac{q! r_1}{(r_1 - r)^{q+1}},$$

which proves the upper estimate in approximation by $W_{n,\xi}^{(q)}(f)(z)$.

Similarly we get

$$|P_{n,\xi}^{(q)}(f)(z) - f^{(q)}(z)| \le C_{r_1,n}(f) \xi^{n+1} \frac{q! r_1}{(r_1 - r)^{q+1}},$$

which proves the theorem. $\qquad \square$

The geometric properties of the Jackson-type generalized complex convolutions are contained in the following results.

We present

Theorem 3.2.12. (Anastassiou-Gal [16]) *Let $f \in S_M$. Then, for all $\xi > 0$, $n \in \mathbb{N}$ we have*

$$\frac{1}{b_{1,n}(\xi)} P_{n,\xi}(S_M) \subset S_{M(2^{n+1}-1)/|b_{1,n}(\xi)|},$$

$$\frac{1}{c_{1,n}(\xi)} W_{n,\xi}(S_M) \subset S_{M(2^{n+1}-1)/|c_{1,n}(\xi)|},$$

$$\frac{1}{c_{1,n}^*(\xi)} W_{n,\xi}^*(S_M) \subset S_{M(2^{n+1}-1)/|c_{1,n}^*(\xi)|},$$

$$\frac{1}{d_{1,n}(\xi)} Q_{n,\xi}(S_M) \subset S_{M(2^{n+1}-1)/|d_{1,n}(\xi)|}.$$

Proof. Let

$$f(z) = z + \sum_{k=2}^{\infty} a_k z^k \in A^*(\overline{\mathbb{D}}_1), \quad |f'(z)| < M, \quad \forall z \in \mathbb{D}_1.$$

Since $a_0 = 0$, by Theorem 3.2.11 (iii) we get

$$P_{n,\xi}(f)(0) = W_{n,\xi}(f)(0) = W_{n,\xi}^*(f)(0) = Q_{n,\xi}(f)(0) = 0.$$

Also, since $a_1 = 1$, by Theorem 3.2.11 (iii) we get

$$\frac{1}{b_{1,n}(\xi)} \cdot P_{n,\xi}'(f)(0) = \frac{1}{c_{1,n}(\xi)} \cdot W_{n,\xi}'(f)(0)$$

$$= \frac{1}{c_{1,n}^*(\xi)} [W_{n,\xi}^*(f)]'(0) = \frac{1}{d_{1,n}(\xi)} \cdot Q_{n,\xi}'(f)(0) = 1,$$

which implies that

$$\frac{1}{b_{1,n}(\xi)} P_{n,\xi}(f), \ \frac{1}{c_{1,n}(\xi)} W_{n,\xi}(f), \ \frac{1}{c_{1,n}^*(\xi)} \cdot W_{n,\xi}^*(f), \ \frac{1}{d_{1,n}(\xi)} Q_{n,\xi}(f) \in A^*(\overline{\mathbb{D}}_1).$$

Also, by

$$P_{n,\xi}'(f)(z) = -\frac{1}{2\xi} \int_{-\infty}^{+\infty} e^{-|u|/\xi} \sum_{k=1}^{n+1} (-1)^k \binom{n+1}{k} f'(ze^{iku}) e^{iku} \, du,$$

we obtain

$$\left| \frac{1}{b_{1,n}(\xi)} P_{n,\xi}'(f)(z) \right| < \frac{M}{|b_{1,n}(\xi)|} \cdot \sum_{k=1}^{n+1} |(-1)^k| \binom{n+1}{k} = \frac{M(2^{n+1}-1)}{|b_{1,n}(\xi)|},$$

that is

$$\frac{1}{b_{1,n}(\xi)} \cdot P_{n,\xi}(f)(z) \in S_{M(2^{n+1}-1)/|b_{1,n}(\xi)|}.$$

The proofs for the other operators are similar, which proves the theorem. □

Remarks. 1) Recall that by e.g. Mocanu–Bulboacă–Sălăgean [138], p. 111, Exercise 5.4.1, $f \in S_M$, $M > 1$, implies that f is univalent in $\{z \in \mathbb{C}; |z| < \frac{1}{M}\} \subset \mathbb{D}_1$. Theorem 3.2.12 shows that $f \in S_M$ implies that $P_{n,\xi}(f)(z)$ is univalent in

$$\left\{z \in \mathbb{C}; |z| < \frac{|b_{1,n}(\xi)|}{M(2^{n+1}-1)}\right\} \subset \left\{z \in \mathbb{C}; |z| < \frac{1}{M}\right\} \subset \mathbb{D}_1,$$

since by Theorem 3.2.11, (iii), we have

$$|b_{p,n}(\xi)| \leq \sum_{k=1}^{n+1} \binom{n+1}{k} \cdot \frac{1}{\xi^2 k^2 p^2 + 1} < \sum_{k=1}^{n+1} \binom{n+1}{k} = 2^{n+1} - 1, \ \forall p = 0, 1, \dots.$$

For the operators $W_{n,\xi}(f)(z)$, $W_{n,\xi}^*(f)(z)$ and $Q_{n,\xi}(f)(z)$ similar conclusions hold by replacing above $b_{1,n}(\xi)$ by $c_{1,n}(\xi)$, $c_{1,n}^*(\xi)$ and $d_{1,n}(\xi)$, respectively.

2) For any fixed $n \in \mathbb{N}$, let us denote

$$B_{1,n} = \inf\{|b_{1,n}(\xi)|; \xi \in (0, \xi_n]\}, \ C_{1,n} = \inf\{|c_{1,n}(\xi)|; \xi \in (0, \xi_n]\},$$
$$C_{1,n}^* = \inf\{|c_{1,n}^*(\xi)|; \xi \in (0, \xi_n]\}, \ D_{1,n} = \inf\{|d_{1,n}(\xi)|; \xi \in (0, \xi_n]\}.$$

If $B_{1,n}$, $C_{1,n}$, $C_{1,n}^*$, $D_{1,n} > 0$, then by Theorem 3.2.12 the following properties hold :

$f \in S_M$ implies that $P_{n,\xi}(f)$ is univalent in $\{z \in \mathbb{C}; |z| < \frac{B_{1,n}}{M(2^{n+1}-1)}\}$, for all $x \in (0, \xi_n]$,

$f \in S_M$ implies that $W_{n,\xi}(f)$ is univalent in $\{z \in \mathbb{C}; |z| < \frac{C_{1,n}}{M(2^{n+1}-1)}\}$, for all $\xi \in (0, \xi_n]$,

$f \in S_M$ implies that $W_{n,\xi}^*(f)$ is univalent in $\{z \in \mathbb{C}; |z| < \frac{C_{1,n}^*}{M(2^{n+1}-1)}\}$, for all $\xi \in (0, \xi_n]$,

$f \in S_M$ implies that $Q_{n,\xi}(f)$ is univalent in $\{z \in \mathbb{C}; |z| < \frac{D_{1,n}}{M(2^{n+1}-1)}\}$, for all $\xi \in (0, \xi_n]$.

Therefore, it remains to calculate (for each fixed $n \in \mathbb{N}$), $B_{1,n}$, $C_{1,n}$, $C_{1,n}^*$, $D_{1,n}$, to check if $B_{1,n} > 0$, $C_{1,n} > 0$, $C_{1,n}^* > 0$, problems which are left to the reader as open questions.

3) It would be of interest to investigate for other geometric properties, for all the complex convolutions previously studied in Section 3.2.

At the end of this subsection firstly we obtain some applications to PDE of complex variables and secondly we extend the above results in this subsection from compact disks with centers in origin to Jordan domains of rectifiable boundaries.

Let $f \in A(\overline{\mathbb{D}}_1)$ and $t \in (0, +\infty)$. For reasons that come from physics, in the above expression of $W_\xi^*(f)(z)$ let us replace ξ by $2t$ and for the simplicity of notation, instead of $W_{2t}^*(f)(z)$ denote

$$W_t^*(f)(z) = \frac{1}{\sqrt{2\pi t}} \int_{-\infty}^{+\infty} f(ze^{-iu})e^{-u^2/(2t)}du, \ z \in \overline{\mathbb{D}}_1.$$

Indeed, the PDE in Theorem 3.2.13, (v) below represents the heat equation with complex spatial variable.

Our first goal is to show that this complex convolution defines a contraction semigroup on the Banach space $(A(\overline{\mathbb{D}}_1), \|\cdot\|_1)$. Then applications to PDE equations with complex spatial variables and real time variable are considered.

Theorem 3.2.13. (Gal-Gal-Goldstein [97]) *Let* $f \in A(\overline{\mathbb{D}}_1)$.
(i) For all $t > 0$, $W_t^*(f) \in A(\overline{\mathbb{D}}_1)$ *and*

$$W_t^*(f)(z) = \sum_{k=0}^{\infty} a_k d_k^*(t) z^k,$$

with

$$d_k^*(t) = \frac{1}{\sqrt{2\pi t}} \int_{-\infty}^{+\infty} e^{-u^2/(2t)} \cos ku\, du = e^{-k^2 t/2}, \; k \geq 1.$$

(ii) For all $z \in \overline{\mathbb{D}}_1$, $t > 0$, *the following estimate holds :*

$$|W_t^*(f)(z) - f(z)| \leq C\omega_1(f; \sqrt{t})_{\overline{\mathbb{D}}_1}.$$

Here $C > 0$ *is a constant independent of* t *and* f.
(iii) We have :

$$|W_t^*(f)(z) - W_s^*(f)(z)| \leq C_s|\sqrt{t} - \sqrt{s}|, \; \text{for all } z \in \overline{\mathbb{D}}_1, \; t \in V_s \subset (0, +\infty),$$

where $C_s > 0$ *is a constant depending on* f, *independent of* z *and* t *and* V_s *is any neighborhood of* s.
(iv) The operator W_t^* *is contractive, that is*

$$\|W_t^*(f)\|_1 \leq \|f\|_1, \quad \text{for all } t > 0, \; f \in A(\overline{\mathbb{D}}_1).$$

(v) $(W_t^*, t \geq 0)$ *is a* (C_0)-*contraction semigroup of linear operators on the Banach space* $(A(\overline{\mathbb{D}}_1), \|\cdot\|_1)$ *and the unique solution* $u(t, z)$ *(that belongs to* $A(\overline{\mathbb{D}}_1)$, *for each fixed* $t > 0$*) of the Cauchy problem*

$$\frac{\partial u}{\partial t}(t, z) = \frac{1}{2}\frac{\partial^2 u}{\partial \varphi^2}(t, z), \; (t, z) \in (0, +\infty) \times \mathbb{D}_1, \; z = re^{i\varphi}, \; z \neq 0,$$

$$u(0, z) = f(z), \; z \in \overline{\mathbb{D}}_1, \; f \in A(\overline{\mathbb{D}}_1),$$

is given by

$$u(t, z) = W_t^*(f)(z) = \frac{1}{\sqrt{2\pi t}} \int_{-\infty}^{+\infty} f(ze^{-iu})e^{-u^2/(2t)}\, du.$$

Proof.   (i) It is immediate by Theorem 3.2.8 (i).
(ii) We obtain

$$|W_t^*(f)(z) - f(z)| \leq \frac{1}{\sqrt{2\pi t}} \int_{-\infty}^{+\infty} |f(ze^{-iu}) - f(z)|e^{-u^2/(2t)}\, du$$

$$\leq \frac{1}{\sqrt{2\pi t}} \int_{-\infty}^{\infty} \omega_1(f; |1 - e^{-iu}|)_{\overline{D}} e^{-u^2/(2t)}\, du$$

$$= \frac{1}{\sqrt{2\pi t}} \int_{-\infty}^{+\infty} \omega_1\left(f; 2\left|\sin\frac{u}{2}\right|\right)_{\overline{\mathbb{D}}_1} e^{-u^2/(2t)}\, du$$

$$\leq \frac{1}{\sqrt{2\pi t}} \int_{-\infty}^{+\infty} \omega_1(f; |u|)_{\overline{\mathbb{D}}_1} e^{-u^2/(2t)}\, du$$

$$\leq \frac{1}{\sqrt{2\pi t}} \int_{-\infty}^{+\infty} \omega_1(f; \sqrt{t})_{\overline{\mathbb{D}}_1} \left(\frac{|u|}{\sqrt{t}} + 1\right) e^{-u^2/(2t)}\, du$$

$$= \omega_1(f; \sqrt{t})_{\overline{\mathbb{D}}_1} + \frac{\omega_1(f; \sqrt{t})_{\overline{\mathbb{D}}_1}}{\sqrt{t} \cdot \sqrt{2\pi t}} \int_0^{\infty} 2u e^{-u^2/(2t)}\, du.$$

Since $\int_0^{\infty} 2u e^{-u^2/(2t)} du = 2t \int_0^{\infty} e^{-v}\, dv = 2t$, we infer

$$|W_t^*(f)(z) - f(z)| \leq \omega_1(f; \sqrt{t})_{\overline{\mathbb{D}}_1} + \left[\omega_1(f; \sqrt{t})_{\overline{\mathbb{D}}_1}\right] \frac{2t}{t\sqrt{2\pi}} \leq C\omega_1(f; \sqrt{t})_{\overline{\mathbb{D}}_1}.$$

(iii) We have

$$|W_t^*(f)(z) - W_s^*(f)(z)| \leq \frac{\|f\|_1}{\sqrt{2\pi}} \int_{-\infty}^{+\infty} \left|\frac{e^{-u^2/t}}{\sqrt{t}} - \frac{e^{-u^2/s}}{\sqrt{s}}\right| du.$$

First, let us denote $\sqrt{t} = a$, $\sqrt{s} = b$. Applying now the mean value theorem, there exists a value $c \in (a, b)$, such that

$$\left|\frac{e^{-u^2/a^2}}{a} - \frac{e^{-u^2/b^2}}{b}\right| = |a - b| e^{-u^2/c^2} \left[\frac{2u^2}{c^4} - \frac{1}{c^2}\right],$$

which together with the fact that

$$\int_{-\infty}^{+\infty} e^{-u^2/(2c)} < \infty, \quad \int_{-\infty}^{+\infty} u^2 e^{-u^2/(2c)} < \infty,$$

it immediately implies the desired inequality for W_t^*.

(iv) Since $\frac{1}{\sqrt{2\pi t}} \int_{-\infty}^{+\infty} e^{-u^2/(2t)} du = 1$, we deduce

$$|W_t^*(f)(z)| \leq \frac{1}{\sqrt{2\pi t}} \int_{-\infty}^{+\infty} |f(ze^{-iu})| e^{-u^2/(2t)} du \leq \|f\|, \ z \in \overline{\mathbb{D}}_1,$$

which yields $\|W_t^*(f)\|_1 \leq \|f\|_1$.

(v) Let $f \in A(\overline{\mathbb{D}}_1)$, that is, $f(z) = \sum_{k=0}^{\infty} a_k z^k$, $z \in \mathbb{D}_1$. If $z \in \mathbb{D}_1$, $z = re^{i\varphi}$, $0 < r < 1$, then by (i), we can write $W_t^*(f)(z) = \sum_{k=0}^{\infty} a_k e^{-k^2 t/2} r^k e^{ki\varphi}$. It easily follows that $W_{t+s}^*(f)(z) = W_s^*[W_t^*(f)](z)$, for all t, $s > 0$. If z is on the boundary of $\mathbb{D}_1$, then we may take a sequence $(z_n)_{n\in\mathbb{N}}$ of points in $\mathbb{D}_1$ such that $lim_{n\to\infty} z_n = z$ and we apply the above relationship and the continuity property from (i). Furthermore, denoting $W_t^*(f)(z)$ by $T(t)(f)$, it is easy to check that the property $\lim_{t\searrow 0} T(t)(f) = f$, the continuity of $T(\cdot)$ and its contraction property follow from (ii), (iii) and (iv), respectively. Finally, all these facts together show that $(W_t^*, t \geq 0)$ is a (C_0)-contraction semigroup of linear operators on $A(\overline{\mathbb{D}}_1)$.

Consequently, since the above series representation for $W_t^*(f)(z)$ is uniformly convergent in any compact disk included in $\mathbb{D}_1$, it can be differentiated term by term, with respect to t and φ. We then easily obtain that $\frac{\partial W_t^*(f)(z)}{\partial t} = (1/2)\frac{\partial^2 W_t^*(f)(z)}{\partial \varphi^2}$. Finally, from the same series representation, it is easy to check that

$$W_0^*(f)(z) = f(z),\ z \in \overline{\mathbb{D}}_1.$$

We also note that in the differential equation we must take $z \neq 0$ simply because $z = 0$ has no polar representation, that is, $z = 0$ cannot be represented as function of φ. This completes the proof of the theorem. $\qquad\square$

The next result shows that the solution of the above Cauchy problem in Theorem 3.2.13, (v), preserve some interesting geometric properties of the boundary function.

Theorem 3.2.14. (Gal-Gal-Goldstein [97]) *Let $u(t,z)$ be the unique solution of the Cauchy problem in Theorem 3.2.13 (v).*

(i) As function of z, $u(t,z)$ has the following properties : if $f \in S_{3,\frac{1}{e^{1/2}}}$ then $u(t,z) \in S_3$, and if $f \in S_M$, $(M > 1)$ then $u(t,z)$ is univalent in $\{z \in \mathbb{C};\ |z| < \frac{1}{Me^{1/2}}\}$, for all $t \in (0, 1/2]$.

(ii) Assuming that f is analytic in an open set G including $\overline{\mathbb{D}}_1$, then for all $t \in (0, t_f]$ with sufficiently small $t_f > 0$ (depending on f), the solution $u(t,z)$ preserves as function of z, the starlikeness, convexity and spirallikeness of the boundary function f in $\overline{\mathbb{D}}_1$.

Proof.   (i) It is an immediate consequence of Corollary 3.2.10 for $\xi = 2t$.

(ii) From Theorem 3.2.8 (i) and (iii), it is readily seen that $W_t^*(f)(z)$ is analytic in G and $W_t^*(f)(z)$ converges to $f(z)$, (as $t \to 0$) uniformly in $\overline{\mathbb{D}}_1$. This convergence together the well-known Weierstrass's result yield that $[W_t^*(f)(z)]' \to f'(z)$ and $[W_t^*(f)(z)]'' \to f''(z)$, uniformly in $\overline{\mathbb{D}}_1$, as $t \to 0$. In all what follows in the proof, from now on, $P_t(f)(z)$ will denote $\frac{W_t^*(f)(z)}{b_1(t)}$, where $b_1(t)$ is the coefficient of z, in the Taylor series representation of the analytic function $W_t^*(f)(z)$. We will also denote by $P_t'(f)(z)$ and $P_t''(f)(z)$, the corresponding first and second-order derivatives of $P_t(f)(z)$ with respect to z.

If $f(0) = f'(0) - 1 = 0$, it is not difficult to observe that $P_t(f)(0) = \frac{f(0)}{b_1(t)} = 0$, $P_t'(f)(0) = \frac{W_t'(f)(0)}{b_1(t)} = 1$ and $b_1(t)$ converges to $f'(0) = 1$ as $t \to 0$. This obviously implies that, for $t \to 0$, we have $P_t(f)(z) \to f(z)$, $P_t'(f)(z) \to f'(z)$ and $P_t''(f)(z) \to f''(z)$, uniformly in $\overline{\mathbb{D}}_1$.

First, suppose that $f \in S^*(\overline{\mathbb{D}}_1)$, that is f is starlike (and univalent) in $\overline{\mathbb{D}}_1$. The univalence implies that $|f(z)| > 0$ for all $z \in \overline{\mathbb{D}}_1$, $z \neq 0$ and that we can write $f(z) = zg(z)$, with $g(z) \neq 0$, for all $z \in \overline{\mathbb{D}}_1$, where g is analytic in $\overline{\mathbb{D}}_1$. Write $P_t(f)(z)$ in the form $P_t(f)(z) = zR_t(f)(z)$. For $|z| = 1$, we have

$$|f(z) - P_t(f)(z)| = |z| \cdot |g(z) - R_t(f)(z)| = |g(z) - R_t(f)(z)|,$$

which, by the uniform convergence in $\overline{\mathbb{D}}_1$, of $P_t(f)$ to f, and by the maximum modulus principle, yields the uniform convergence in $\overline{\mathbb{D}}_1$, of $R_t(f)(z)$ to $g(z)$, as

$t \to 0$. Since g is continuous in $\overline{\mathbb{D}}_1$ and $|g(z)| > 0$ for all $z \in \overline{\mathbb{D}}_1$, then there exist an index $t_1 > 0$ and $a > 0$ depending on g (that is, on f), so that $|R_t(f)(z)| > a > 0$, for all $z \in \overline{\mathbb{D}}_1$ and all $t \in (0, t_1)$. For all $|z| = 1$, we also have

$$|f'(z) - P_t'(f)(z)| = |z[g'(z) - R_t'(f)(z)] + [g(z) - R_t(f)(z)]|$$
$$\geq ||z| \cdot |g'(z) - R_t'(f)(z)| - |g(z) - R_t(f)(z)||$$
$$= ||g'(z) - R_t'(f)(z)| - |g(z) - R_t(f)(z)||.$$

Clearly, the maximum modulus principle, the uniform convergence of $P_t'(f)$ to f' and of $R_t(f)$ to g, imply the uniform convergence of $R_t'(f)$ to g', as $t \to 0$. Then for $|z| = 1$, we obtain

$$\frac{zP_t'(f)(z)}{P_t(f)} = \frac{z[zR_t'(f)(z) + R_t(f)(z)]}{zR_t(f)(z)}$$
$$= \frac{zR_t'(f)(z) + R_t(f)(z)}{R_t(f)(z)} \to \frac{zg'(z) + g(z)}{g(z)}$$
$$= \frac{f'(z)}{g(z)} = \frac{zf'(z)}{f(z)}.$$

From this convergence and the maximum modulus principle, we infer

$$\frac{zP_t'(f)(z)}{P_t(f)} \to \frac{zf'(z)}{f(z)}, \text{ uniformly in } \overline{\mathbb{D}}_1.$$

Since $Re\left(\frac{zf'(z)}{f(z)}\right)$ is continuous in $\overline{\mathbb{D}}_1$, there exists $\alpha \in (0, 1)$, so that

$$Re\left(\frac{zf'(z)}{f(z)}\right) \geq \alpha, \text{ for all } z \in \overline{\mathbb{D}}_1.$$

Therefore

$$Re\left[\frac{zP_t'(f)(z)}{P_t(f)(z)}\right] \to Re\left[\frac{zf'(z)}{f(z)}\right] \geq \alpha > 0,$$

uniformly on $\overline{\mathbb{D}}_1$, that is, for any $0 < \beta < \alpha$ there is $t_f > 0$ so that for all $t \in (0, t_f)$, we have

$$Re\left[\frac{zP_t'(f)(z)}{P_t(f)(z)}\right] > \beta > 0, \text{ for all } z \in \overline{\mathbb{D}}_1.$$

Therefore, $P_t(f) \in S^*(\overline{\mathbb{D}}_1)$, for all $t \in (0, t_f)$ and since it only differs from $W_t^*(f)$ by a constant, this proves the starlikeness in $\overline{\mathbb{D}}_1$ of $W_t^*(f)(z)$, for all $t \in (0, t_f)$.

The proofs of the other cases, when f is convex or spirallike of order γ are similar and follow from the following uniform convergence on $\overline{\mathbb{D}}_1$:

$$Re\left[\frac{zP_t''(f)(z)}{P_t'(f)(z)}\right] + 1 \to Re\left[\frac{zf''(z)}{f'(z)}\right] + 1,$$

and

$$Re\left[e^{i\gamma}\frac{zP_{nt}'(f)(z)}{P_t(f)(z)}\right] \to Re\left[e^{i\gamma}\frac{zf'(z)}{f(z)}\right].$$

The proof is complete. $\qquad \square$

Replacing ξ by t (time) for reasons that come from physics, now we will consider the complex convolution operator of Poisson-Cauchy type

$$Q_t^*(f)(z) = \frac{t}{\pi} \int_{-\infty}^{+\infty} \frac{f(ze^{-iu})}{u^2 + t^2} \, du, \quad z \in \overline{\mathbb{D}}_1.$$

Theorem 3.2.15. (Gal-Gal-Goldstein [97]) *Let $f \in A(\overline{\mathbb{D}}_1)$.*
  (i) Then for all $t > 0$, $Q_t^*(f) \in A(\overline{\mathbb{D}}_1)$ and

$$Q_t^*(f)(z) = \sum_{k=0}^{\infty} a_k b_k^*(t) z^k,$$

with

$$b_k^*(t) = \frac{2t}{\pi} \int_0^{+\infty} \frac{1}{u^2 + t^2} \cos ku \, du = e^{-kt}, \, k \geq 1.$$

  (ii) For all $z \in \overline{\mathbb{D}}_1, t \in V_s \subset (0, +\infty)$,

$$|Q_t^*(f)(z) - Q_s^*(f)(z)| \leq C_s |t - s|,$$

where $C_s > 0$ is a constant independent of z, t and f, and V_s is any neighborhood of (fixed) $s > 0$.
  (iii) The operator Q_t^* is contractive, that is

$$\|Q_t^*(f)\|_1 \leq \|f\|_1, \text{ for all } t > 0, \, f \in A(\overline{\mathbb{D}}_1).$$

  (iv) $(Q_t^*, t \geq 0)$ *is a (C_0)-contraction semigroup of linear operators on the Banach space $(A(\overline{\mathbb{D}}_1), \|\cdot\|_1)$ and the unique solution $u(t, z)$ (that belongs to $A(\overline{\mathbb{D}}_1)$, for each fixed $t > 0$) of the Cauchy problem*

$$\frac{\partial^2 u}{\partial t^2}(t, z) + \frac{\partial^2 u}{\partial \varphi^2}(t, z) = 0, \, (t, z) \in \mathbb{D}_1 \times (0, +\infty), \, z = re^{i\varphi}, \, z \neq 0,$$

$$u(0, z) = f(z), \, z \in \overline{\mathbb{D}}_1, \, f \in A(\overline{\mathbb{D}}_1),$$

is given by

$$u(t, z) = Q_t^*(f)(z) = \frac{t}{\pi} \int_{-\infty}^{+\infty} f(ze^{-iu}) \frac{du}{t^2 + u^2}.$$

Proof.   (i) It is exactly Theorem 3.2.5 (i).
  (ii) We have

$$|Q_t^*(f)(z) - Q_s^*(f)(z)| \leq \frac{\|f\|_1}{\pi} \left| \int_{\infty}^{+\infty} \left[\frac{t}{t^2 + u^2} - \frac{s}{s^2 + u^2} \right] du \right|.$$

However, since

$$\frac{t}{t^2 + u^2} - \frac{s}{s^2 + u^2} = \frac{(t - s)(u^2 - ts)}{(t^2 + u^2)(s^2 + u^2)},$$

integrating this relation with respect to u, for all $t \in V_s$, we obtain

$$\left| \int_{-\infty}^{+\infty} \frac{u^2 - ts}{(t^2 + u^2)(s^2 + u^2)} du \right|$$

$$= \left| \int_{-\infty}^{+\infty} \frac{1}{s^2 + u^2} du - (t^2 + ts) \int_{-\infty}^{+\infty} \frac{du}{(u^2 + t^2)(s^2 + u^2)} \right| \leq C_s < +\infty,$$

since the integrals $\int_{-\infty}^{+\infty} \frac{1}{s^2 + u^2} du$ and $\int_{-\infty}^{+\infty} \frac{du}{(u^2 + t^2)(s^2 + u^2)}$ are finite. This immediately implies (ii).

(iii) Since $\frac{t}{\pi} \int_{-\infty}^{+\infty} \frac{1}{t^2 + u^2} du = 1$, we deduce the following estimate :

$$|Q_t^*(f)(z)| \leq \frac{t}{\pi} \int_{-\infty}^{+\infty} |f(ze^{-iu})| \frac{1}{t^2 + u^2} du \leq \|f\|_1, \ z \in \overline{\mathbb{D}}_1.$$

It is readily seen that $\|Q_t^*(f)\|_1 \leq \|f\|_1$.

(iv) The property $Q_{t+s}^*(f)(z) = Q_t^*[Q_s^*(f)](z)$, for all t, $s > 0$, $f \in A(\overline{\mathbb{D}}_1)$ and $z \in \overline{\mathbb{D}}_1$ is immediate from the representation in (i). To prove that $\lim_{t \searrow 0} Q_t^*(f) = f$, for any $f \in A(\overline{\mathbb{D}}_1)$, let $f = U + iV$, $z = re^{ix}$ be fixed and denote $F(v) = U[r\cos(v), r\sin(v)]$, $G(v) = V[r\cos(v), r\sin(v)]$. We can write

$$Q_t^*(f)(z) = \frac{t}{\pi} \int_{-\infty}^{+\infty} F(x - u) \frac{1}{t^2 + u^2} du + i \frac{t}{\pi} \int_{-\infty}^{+\infty} G(x - u) \frac{1}{t^2 + u^2} du.$$

From the maximum modulus principle, when estimating the quantity $|Q_t^*(f)(z) - f(z)|$, we may take $r = |z| = 1$, that is, $z = e^{ix}$ in the above formulas for $F(v)$ and $G(v)$. Therefore, passing to limit as $t \searrow 0$ and taking into account the above property (see e.g. Goldstein [100], Exercise 2.18.8), we find

$$\lim_{t \searrow 0} |Q_t^*(f)(z) - f(z)| \leq \lim_{t \searrow 0} \left| \frac{t}{\pi} \int_{-\infty}^{+\infty} F(x - u) \frac{1}{t^2 + u^2} du - F(x) \right|$$

$$+ \lim_{t \searrow 0} \left| \frac{t}{\pi} \int_{-\infty}^{+\infty} G(x - u) \frac{1}{t^2 + u^2} du - G(x) \right| = 0,$$

which holds for $z \in \overline{\mathbb{D}}_1$.

Combining this with the properties in (ii) and (iii), the first assertion of (iv) follows. Finally, from the series representation of $Q_t^*(f)(z)$ (cf. (i)) and reasoning exactly as in Theorem 3.2.13 (v), we can easily check the second assertion in (iv). This completes the proof of the theorem. $\square$

Remark. The property $\lim_{t \searrow 0} |Q_t^*(f)(z) = f(z)$ for all $z \in \overline{\mathbb{D}}_1$ proved in the above proof of Theorem 3.2.15 (iv), in fact holds uniformly with respect to $z \in \overline{\mathbb{D}}_1$. Indeed, since F and G in the proof of Theorem 3.2.15 (iv) are continuous and 2π-periodic on $\mathbb{R}$, for all $x \in \mathbb{R}$ by Anghelutza [30] we have

$$\left| \frac{t}{\pi} \int_{-\infty}^{+\infty} F(x - u) \frac{1}{t^2 + u^2} du - F(x) \right| \leq C\omega_1(F; \xi) \log \left(\frac{1}{\omega_1(F; \xi)} \right),$$

and a similar estimate for G holds. Here log denotes the natural logarithm and $C > 0$ is independent of x and ξ. We can suppose that f is not a constant, that is that F and G are not constant (otherwise the trivial case $Q_t^*(f)(z) = f(z)$ for all z holds). Therefore, there exists a $\xi_0 > 0$ such that $\omega_1(F; \xi) > 0$ for all $0 < \xi \leq \xi_0$. To simplify the notation denote $H(\xi) = \omega_1(F; \xi)$, $0 \leq \xi \leq \xi_0$. Clearly $H(0) = 0$, $H(\xi) > 0$ for $\xi \neq 0$ and H is continuous and increasing on $[0, \xi_0]$. Therefore obviously that we can choose $\xi_0 > 0$ such that in addition, $H(\xi) < 1$ for all $\xi \in [0, \xi_0]$.

We will show that $\lim_{\xi \to 0} H(\xi)[- \log H(\xi)] = 0$, which will imply the uniform convergence with respect to $z \in \overline{\mathbb{D}}_1$.

For this purpose, let us suppose that would exists a sequence $(\xi_n)_n$, $\xi_n \in (0, \xi_0]$, such that $\xi_n \to 0$ as $n \to \infty$ and $\lim_{n \to \infty} H(\xi_n)[- \log H(\xi_n)] = a > 0$. This immediately implies that for any $0 < \varepsilon < a$, there exists an $n_0 \in \mathbb{N}$ such that

$$|a - H(\xi_n)[- \log H(\xi_n)]| < \varepsilon, \text{ for all } n \geq n_0,$$

that is $0 < \rho = a - \varepsilon < H(\xi_n)[- \log H(\xi_n)]$ for all $n \geq n_0$. It follows

$$\rho \frac{1}{H(\xi_n)} \leq \log \frac{1}{H(\xi_n)}, \text{ for all } n \geq n_0,$$

or denoting $h_n = H(\xi_n)$ we get $h_n \to \infty$ and

$$\rho \leq \frac{\log h_n}{h_n}, \text{ for all } n \geq n_0,$$

with $\rho > 0$ independent of n. But passing here to limit as $n \to \infty$, since $\lim_{n \to \infty} \frac{\log h_n}{h_n} = 0$, we obtain the contradiction $0 < \rho = 0$.

In conclusion, for all the sequences $(\xi_n)_n$, $\xi_n \in (0, \xi_0]$, such that $\xi_n \to 0$ as $n \to \infty$, we necessarily have $\lim_{n \to \infty} H(\xi_n)[- \log H(\xi_n)] = 0$, which proves our assertion.

The next result shows that the solution of the above Cauchy problem preserves some interesting geometric properties of the boundary function.

Theorem 3.2.16. (Gal-Gal-Goldstein [97]) *Let $u(t, z)$ be the unique solution of the Cauchy problem in Theorem 3.2.15 (iv).*

(i) As function of z, $u(t, z)$ has the following properties : if $f \in S_{3, \frac{1}{e}}$, then $u(t, z) \in S_3$, and, if $f \in S_M$, $(M > 1)$ then $u(t, z)$ is univalent in $\{z \in \mathbb{C}; |z| < \frac{1}{eM}\}$, for all $t \in (0, 1]$

(ii) Assuming that f is analytic in an open set G including $\overline{\mathbb{D}}_1$, then, for all $t \in (0, t_f]$ with a sufficiently small $t_f > 0$ (depending on f), the solution $u(t, z)$ preserves as function of z, the starlikeness, convexity and spirallikeness of the boundary function f in $\overline{\mathbb{D}}_1$.

Proof. (i) Both assertions are immediate consequences of Theorem 3.2.6 (iii).

(ii) The proof is similar to that of Theorem 3.2.14 (ii), as a consequence of Theorem 3.2.5 (i) and (iii). We leave the rigorous details to the reader. □

In what follows we will extend some of the above results to more general domains in $\mathbb{C}$. We will consider here the cases of the complex operators $P_\xi(f)(z)$ in Theorem 3.2.1, $Q_\xi^*(f)(z)$ in Theorem 3.2.5 and $W_\xi^*(f)(z)$ in Theorem 3.2.8, which can be generalized for approximation of $f \in A(\overline{G})$ in Jordan domains $G \subset \mathbb{C}$, as follows.

Keeping the notations in Section 1.0, let $f \in A(\overline{G})$ and $F \in A(\overline{\mathbb{D}}_1)$ be such that $f = T[F]$. Here T denotes the Faber operator which is supposed to be continuous as mapping from $A(\overline{\mathbb{D}}_1)$ to $A(\overline{G})$.

For $\xi > 0$, $k \in \mathbb{N} \bigcup \{0\}$ and $\rho_{k,\xi} \in (0,1]$, attach to $F \in A(\overline{\mathbb{D}}_1)$, $F(w) = \sum_{k=0}^\infty a_k w^k, w \in \overline{\mathbb{D}}_1$, the complex operators

$$O_\xi(F)(w) = \sum_{k=0}^\infty a_k \rho_{k,\xi} w^k.$$

Then, suggested by the Remark after Theorem 1.0.12, attach to $f = T[F]$ the complex operator $L_\xi(f) = T[O_\xi(F)]$ given by

$$L_\xi(f)(z) = \sum_{k=0}^\infty a_k \rho_{k,\xi} F_k(z), \ z \in \overline{G},$$

where $F_k(z)$ denotes the Faber polynomial of degree k attached to G.

But according to Theorem 1.0.11, we have $a_k = a_k(f)$-the Faber coefficients of f on G and according to Definition 1.0.10, (ii), we can write

$$a_k(f) = \frac{1}{2\pi i} \int_{|u|=1} \frac{f[\Psi(u)]}{u^{k+1}} du = \frac{1}{2\pi i} \int_{-\pi}^\pi f[\Psi(e^{it})] e^{-ikt} dt, k \in \mathbb{N} \cup \{0\}.$$

Therefore, the complex operators attached to an $f \in A(\overline{G})$ can be formally written as

$$L_\xi(f)(z) = \sum_{k=0}^\infty a_k(f) \rho_{k,\xi} F_k(z), \ z \in \overline{G}.$$

Remark. Of course that of interest are the cases when $L_\xi(f)(z)$ is well defined in $\overline{G}$ and belongs to $A(\overline{G})$. For example, let us suppose that $f \in A(\overline{G})$, where G is a bounded simply connected domain with the boundary Γ a regular analytic curve, that is Γ is an analytic mapping and its derivative never vanishes. In this case, by Theorem 1 and its proof, p. 51-52 in the book of Suetin [186] it follows that $f(z) = \sum_{k=0}^\infty a_k(f) F_k(z)$, where the series is uniformly and absolutely convergent in any compact subset K of G. More exactly, applying the ideas in the proof of the general Theorem 3, p. 54 in Suetin [186], we easily obtain that there exists $d \in (0,1)$ such that $|a_k F_k(z)| \leq Cd^k$, for all $k = 0, 1, 2, ...$ and $z \in K$, which obviously implies that the series $\sum_{k=0}^\infty a_k(f) F_k(z)$ is absolutely (and uniformly) convergent in K. As an immediate consequence, it follows that the series defining $L_\xi(f)(z)$ also is absolutely and uniformly convergent in K, that is $L_\xi(f)(z)$ is analytic in G.

The two theorems below presents some approximation properties of $L_\xi(f)$, $\xi \in (0,1]$ corresponding to the extensions of $P_\xi(f)$, $Q_\xi^*(f)$ and $W_\xi^*(f)$ when $f \in A(\overline{G})$, in the light of the ideas mentioned in the above remark.

Theorem 3.2.17. *Let $G \subset \mathbb{C}$ be a bounded simply connected domain whose boundary Γ is a regular and analytic curve and for $f \in A(\overline{G})$ define $L_\xi(f)(z) = \sum_{k=0}^{\infty} a_k(f)\rho_{k,\xi}F_k(z)$, where $F_k(z)$ are the Faber polynomials attached to G, $a_k(f)$ are the Faber coefficients of f and $\rho_{k,\xi} \in (0,1]$ for all $\xi \in (0,1]$, $k \in \mathbb{N}$. Also, for any $r > 0$ define as G_r the interior of the (closed) level curve Γ_r given by*

$$\Gamma_r = \{z; |\Phi(z)| = r\} = \{\Psi(w); |w| = r\}.$$

(i) Then $L_\xi(f)(z)$ is analytic in G and if f is not a constant function on G and $\rho_{k,\xi} \neq 1$ for all $\xi \in (0,1]$ and $k \in \mathbb{N}$, then there exists $0 < \beta_0 < 1$ such that for all $\beta_0 < \beta < 1$ we have the saturation result

$$\max_{z \in \overline{G_\beta}} |f(z) - L_\xi(f)(z)| \neq o(\min_{1 \leq k} |1 - \rho_{k,\xi}|), \ \xi \in (0,1].$$

Recall here that by definition $a_\xi = o(b_\xi)$ if $\lim_{\xi \to 0} \frac{a_\xi}{b_\xi} = 0$.

(ii) Take $\rho_{k,\xi} = \frac{1}{1+\xi^2 k^2}$, that is $L_\xi(f)$ is the generalization of $P_\xi(f)$. If f is not a constant function on G then there exists $0 < \beta_0 < 1$ such that for all $\beta_0 < \beta < 1$ we have

$$\max_{z \in \overline{G_\beta}} |f(z) - L_\xi(f)(z)| \sim \xi^2, \ \xi \in (0,1],$$

where the constants in the equivalence depend only on f, β and G_β (but are independent of ξ).

(iii) Take $\rho_{k,\xi} = e^{-k\xi}$, that is $L_\xi(f)$ is the generalization of $Q_\xi^*(f)$. If f is not a constant function on G then there exists $0 < \beta_0 < 1$ such that for all $\beta_0 < \beta < 1$ we have

$$\max_{z \in \overline{G_\beta}} |f(z) - L_\xi(f)(z)| \sim \xi, \ \xi \in (0,1],$$

where the constants in the equivalence depend only on f, β and G_β.

(iv) Take $\rho_{k,\xi} = e^{-k^2 \xi/4}$, that is $L_\xi(f)$ is the generalization of $W_\xi^*(f)$. If f is not a constant function on G then there exists $0 < \beta_0 < 1$ such that for all $\beta_0 < \beta < 1$ we have

$$\max_{z \in \overline{G_\beta}} |f(z) - L_\xi(f)(z)| \sim \xi, \ \xi \in (0,1],$$

where the constants in the equivalence depend only on f, β and G_β.

Proof. (i) First we use some ideas in the proof of Theorem 1, p. 51-52 in the book of Suetin [186]. Thus, since Γ is a regular analytic curve, then the mapping $\Phi(z)$ (and consequently its inverse $\Psi(z)$) can be analytically and univalently continued across the boundary Γ in the domain G. More exactly, there exists $0 < \beta_0 < 1$ such that $\Psi(z)$ is analytic and univalent in Ext_{β_0} (excepting the point ∞), where $Ext_{\beta_0} = \tilde{\mathbb{C}} \setminus \overline{G_{\beta_0}}$, with G_{β_0} denoting the interior of the (closed) level curve Γ_{β_0} given by

$$\Gamma_{\beta_0} = \{z; |\Phi(z)| = \beta_0\} = \{\Psi(w); |w| = \beta_0\}.$$

Let $\beta_0 < r < \beta < 1$.

Now, by Definition 1.0.10 (ii) we have $a_k(f) = \frac{1}{2\pi i} \int_{|u|=1} \frac{f(\Psi(u))}{u^{k+1}} du$. Also, by Kövari-Pommerenke [120], p. 198, Lemma 2, we have that the integral $\frac{1}{2\pi i} \int_{|u|=1} \frac{F_k(\Psi(u))}{u^{m+1}} du$ is equal to zero for $m \neq k$ and equal to 1 for $m = k$. But taking into account the Cauchy's result, clearly we can write $a_k(f) = \frac{1}{2\pi i} \int_{|u|=\beta} \frac{f(\Psi(u))}{u^{k+1}} du$ and that $\frac{1}{2\pi i} \int_{|u|=\beta} \frac{F_k(\Psi(u))}{u^{m+1}} du$ is equal to zero for $m \neq k$ and equal to 1 for $m = k$.

Let $L_\xi(f)(z) = \sum_{k=0}^\infty a_k(f)\rho_{k,\xi} F_k(z)$, $z \in \overline{G_r}, \xi \in (0,1]$, with $\rho_{k,\xi} \in (0,1]$, for all $\xi \in (0,1]$ and $k \in \mathbb{N} \bigcup \{0\}$.

By the proof of Theorem 1, p. 52 in Suetin [186] we have

$$|F_k(z)| \leq C(r)r^k, \text{ for all } z \in \overline{G_r}.$$

Also, by the above formula for $a_k(f)$ we easily obtain $|a_k(f)| \leq \frac{C(\beta,f)}{\beta^k}$. Note here that $C(r), C(\beta,f) > 0$ are independent of k. It follows

$$|a_k(f)\rho_{k,\xi} F_k(z)| \leq C(r,\beta,f)\left(\frac{r}{\beta}\right)^k \text{ for all } z \in \overline{G_r}, k = 0, 1, 2, \ldots.$$

Since $\frac{r}{\beta} < 1$, it follows that the series which defines $L_\xi(f)(z)$ is absolutely and uniformly convergent in the compact $\overline{G_r}$ since it is majorized by a geometric progression. Therefore $L_\xi(f)$ is analytic in any $\overline{G_r}$ with $\beta_0 < r < 1$, which immediately implies its analyticity in G. Note that by the above reasonings does not follow that $L_\xi(f)(z)$ is continuous on $\overline{G}$.

On the other hand we immediately obtain

$$\frac{1}{2\pi i} \int_{|u|=\beta} \frac{f(\Psi(u)) - L_\xi(f)(\Psi(u))}{u^{m+1}} du = (1 - \rho_{m,\xi})a_m(f), m \geq 0.$$

This implies

$$|1 - \rho_{m,\xi}| \cdot |a_m(f)| \leq \max_{z \in \overline{G_\beta}} |f(z) - L_\xi(f)(z)|.$$

Now, let us assume that the conclusion of Theorem 3.2.17 is false, that is there exists $0 < \beta < 1$ such that

$$\max_{z \in \overline{G_\beta}} |f(z) - L_\xi(f)(z)| = o(\min_{1 \leq k} |1 - \rho_{k,\xi}|).$$

It implies

$$|1 - \rho_{m,\xi}| \cdot |a_m(f)| = o(\min_{1 \leq k} |1 - \rho_{k,\xi}|), \text{ for all } m \geq 1$$

and therefore

$$\lim_{\xi \searrow 0} \frac{|1 - \rho_{m,\xi}| \cdot |a_m(f)|}{\min_{1 \leq k} |1 - \rho_{k,\xi}|} = 0,$$

which means $a_m(f) = 0$, for all $m \geq 1$. Then, by Theorem 1, p. 44 in Gaier [76] we get $f(z) = a_0(f)$ for all $z \in \overline{G_\beta}$. By the identity theorem on the analytic functions it follows that f is constant in $G := G_1$, a contradiction.

(ii) By Theorem 3.2.1, (i) we get $\rho_{k,\xi} = \frac{1}{1+\xi^2 k^2}$ and

$$\min_{1\leq k}(1 - \rho_{k,\xi}) = \frac{\xi^2}{1+\xi^2} \geq \frac{\xi^2}{2}, \text{ for all } \xi \in (0,1].$$

Denote $V_\xi = \inf_{1\leq p}(\frac{\xi^2 p^2}{1+\xi^2 p^2})$. We get $V_\xi = \frac{\xi^2}{1+\xi^2}$ and for all $\xi \in (0,1]$ it follows $V_\xi \geq \xi^2/2$.

By the following lower estimate from the above point (i),

$$|1 - \rho_{m,\xi}| \cdot |a_m(f)| \leq \max_{z\in\overline{G_\beta}} |f(z) - L_\xi(f)(z)|,$$

for all $p \geq 1$ and $\xi \in (0,1]$ it follows

$$\frac{2\max_{z\in\overline{G_\beta}} |f(z) - L_\xi(f)(z)|}{\xi^2} \geq \frac{\max_{z\in\overline{G_\beta}} |f(z) - L_\xi(f)(z)|}{V_\xi}$$

$$\geq \frac{\max_{z\in\overline{G_\beta}} |f(z) - L_\xi(f)(z)|}{\frac{\xi^2 p^2}{1+\xi^2 p^2}} \geq |a_p(f)|.$$

This implies that if there exists a subsequence $(\xi_k)_k$ in $(0,1]$ with $\lim_{k\to\infty}\xi_k = 0$ and such that $\lim_{k\to\infty} \frac{\max_{z\in\overline{G_\beta}}|f(z)-L_\xi(f)(z)|}{\xi_k^2} = 0$ then $a_p = 0$ for all $p \geq 1$, that is f is constant on $\overline{G_\beta}$.

Therefore, if f is not a constant then $\inf_{\xi\in(0,1]} \frac{\max_{z\in\overline{G_\beta}}|f(z)-L_\xi(f)(z)|}{\xi^2} > 0$, which implies that there exists a constant $C_\beta(f) > 0$ such that

$$\frac{\max_{z\in\overline{G_\beta}} |f(z) - L_\xi(f)(z)|}{\xi^2} \geq C_\beta(f) \text{ for all } \xi \in (0,1],$$

that is

$$\max_{z\in\overline{G_\beta}} |f(z) - L_\xi(f)(z)| \geq C_\beta(f)\xi^2, \text{ for all } \xi \in (0,1].$$

For the upper estimate, first by Theorem 1, p. 51 in Suetin [186], we have $f(z) = \sum_{k=0}^\infty a_k(f)F_k(z)$, for all $z \in G$, where the series is absolutely and uniformly convergent inside G. Taking into account the estimate for $|a_k(f)F_k(z)|$, $z \in \overline{G_r}$ from the above point (i) and denoting $d = \frac{r}{\beta} < 1$, for all $z \in \overline{G_r}$ we obtain

$$|f(z) - L_\xi(f)(z)| \leq \sum_{k=0}^\infty (1 - \rho_{k,\xi})|a_k(f)F_k(z)| = \sum_{k=1}^\infty \frac{\xi^2 k^2}{1+\xi^2 k^2}|a_k(f)F_k(z)|$$

$$\leq \xi^2 \sum_{k=1}^\infty k^2 |a_k(f)F_k(z)| \leq C\xi^2 \sum_{k=0}^\infty k^2 d^k.$$

But taking into account that $\sum_{k=0}^\infty d^k = \frac{1}{1-d}$, by differentiation with respect to d we easily obtain $\sum_{k=1}^\infty kd^{k-1} = \frac{1}{(1-d)^2}$ and $\sum_{k=2}^\infty k(k-1)d^{k-2} = \frac{2}{(1-d)^3}$, which immediately implies $\sum_{k=0}^\infty k^2 d^k < +\infty$ and therefore

$$|f(z) - L_\xi(f)(z)| \leq M\xi^2, \text{ for all } z \in \overline{G_r}, \xi \in (0,1].$$

Since $r > \beta_0$ is arbitrary this proves (ii).

(iii) By Theorem 3.2.5, (i) and (iv) we get $\rho_{k,\xi} = e^{-k\xi}$ and

$$\min_{1 \le k}(1 - \rho_{k,\xi}) = 1 - e^{-\xi} \ge \xi/3, \text{ for all } \xi \in (0,1].$$

Denote $V_\xi = \inf_{1 \le p}(1 - e^{-p\xi})$. We get $V_\xi = 1 - e^{-\xi}$ and by the mean value theorem applied to $h(x) = e^{-x}$ on $[0, \xi]$ there exists $\eta \in (0, \xi)$ such that for all $\xi \in (0, 1]$ we have

$$V_\xi = h(0) - h(\xi) = (-\xi)h'(\eta) = (\xi)e^{-\eta} \ge (\xi)e^{-\xi} \ge e^{-1}\xi \ge \xi/3.$$

By the lower estimate

$$|1 - \rho_{m,\xi}| \cdot |a_m(f)| \le \max_{z \in \overline{G_\beta}} |f(z) - L_\xi(f)(z)|,$$

for all $p \ge 1$ and $\xi \in (0, 1]$ it follows

$$\frac{3 \max_{z \in \overline{G_\beta}} |f(z) - L_\xi(f)(z)|}{\xi} \ge \frac{\max_{z \in \overline{G_\beta}} |f(z) - L_\xi(f)(z)|}{V_\xi}$$

$$\ge \frac{\max_{z \in \overline{G_\beta}} |f(z) - L_\xi(f)(z)|}{1 - e^{-p\xi}} \ge |a_p|.$$

Reasoning exactly as at the above point (ii), it follows that if f is not a constant then there exists a constant $C_\beta(f) > 0$ such that

$$\max_{z \in \overline{G_\beta}} |f(z) - L_\xi(f)(z)| \ge C_\beta \xi, \text{ for all } \xi \in (0, 1].$$

For the upper estimate we reason as at the above point (ii). By applying the mean value theorem to $1 - e^{-k\xi} = g(0) - g(k)$ with $g(x) = e^{-\xi x}$, there exists $p \in (0, k)$ such that $1 - e^{-k\xi} = k\xi e^{-p\xi} \le k\xi$, which implies

$$|f(z) - L_\xi(f)(z)| \le \sum_{k=0}^{\infty}(1 - e^{-k\xi})|a_k(f)F_k(z)| \le \sum_{k=1}^{\infty} k\xi|a_k(f)F_k(z)|$$

$$\le \xi^2 \sum_{k=1}^{\infty} k|a_k(f)F_k(z)| \le C\xi \sum_{k=0}^{\infty} kd^k,$$

where $\sum_{k=0}^{\infty} kd^k < +\infty$.

(iv) By Theorem 3.2.8, (i) and (iv) we get $\rho_{k,\xi} = e^{-k^2\xi/4}$ and

$$\min_{1 \le k}(1 - \rho_{k,\xi}) = 1 - e^{-\xi/4} \ge \xi/8, \text{ for all } \xi \in (0, 1].$$

Denote $V_\xi = \inf_{1 \le p}(1 - e^{-p^2\xi/4})$. We get $V_\xi = 1 - e^{-\xi/4}$ and by the mean value theorem applied to $h(x) = e^{-x/4}$ on $[0, \xi]$ there exists $\eta \in (0, \xi)$ such that for all $\xi \in (0, 1]$ we have

$$V_\xi = h(0) - h(\xi) = (-\xi)h'(\eta) = (\xi/4)e^{-\eta/4} \ge (\xi/4)e^{-\xi/4} \ge \frac{e^{-1/4}}{4}\xi \ge \xi/8.$$

As at the above points (ii) and (iii), for all $p \geq 1$ and $\xi \in (0, 1]$ it follows

$$\frac{8 \max_{z \in \overline{G_\beta}} |f(z) - L_\xi(f)(z)|}{\xi} \geq \frac{\max_{z \in \overline{G_\beta}} |f(z) - L_\xi(f)(z)|}{V_\xi}$$

$$\geq \frac{\max_{z \in \overline{G_\beta}} |f(z) - L_\xi(f)(z)|}{1 - e^{-p^2 \xi/4}} \geq |a_p|.$$

Again reasoning as the above points (ii) and (iii) it follows that if f is not a constant then there exists a constant $C_\beta(f) > 0$ such that

$$\max_{z \in \overline{G_\beta}} |f(z) - L_\xi(f)(z)| \geq C_\beta \xi, \text{ for all } \xi \in (0, 1].$$

For the upper estimate, we reason as at the above point (iii). By applying the mean value theorem to $1 - e^{-k^2 \xi/4} = g(0) - g(k)$ with $g(x) = e^{-\xi x^2/4}$, there exists $p \in (0, k)$ such that $1 - e^{-k^2 \xi/4} = (-k) \left(-\frac{2p\xi}{4} \right) e^{-p^2 \xi/4} \leq \frac{k^2 \xi}{2}$, which implies

$$|f(z) - L_\xi(f)(z)| \leq \sum_{k=0}^{\infty} (1 - e^{-k^2 \xi/4}) |a_k(f) F_k(z)| \leq \sum_{k=1}^{\infty} \frac{k^2 \xi}{2} |a_k(f) F_k(z)|$$

$$\leq \frac{\xi}{2} \sum_{k=1}^{\infty} k^2 |a_k(f) F_k(z)| \leq C\xi \sum_{k=0}^{\infty} k^2 d^k,$$

where $\sum_{k=0}^{\infty} k^2 d^k < +\infty$. Theorem 3.2.17 is proved. $\qquad \square$

The results in Theorem 3.2.17 can be improved if we suppose that G is a continuum (that is a compact connected subset of $\mathbb{C}$), without any requirement on its boundary. More exactly the following result holds.

Theorem 3.2.18. *Let G be a continuum and suppose that f is analytic in G, that is there exists $R > 1$ such that f is analytic in G_R (which includes G). Here recall that G_R denotes the interior of the closed level curve Γ_R given by $\Gamma_R = \{z; |\Phi(z)| = R\} = \{\Psi(w); |w| = R\}$. Attach to f the operators $L_\xi(f)(z) = \sum_{k=0}^{\infty} a_k(f) \rho_{k,\xi} F_k(z)$, $\xi \in (0, 1]$, where $F_k(z)$ are the Faber polynomials attached to G, $a_k(f)$ are the Faber coefficients of f and $\rho_{k,\xi} \in (0, 1]$ for all $\xi \in (0, 1]$, $k \in \mathbb{N}$.*

(i) Then $L_\xi(f)(z)$ is analytic in G and if f is not a constant function on G and $\rho_{k,\xi} \neq 1$ for all $\xi \in (0, 1]$ and $k \in \mathbb{N}$, then for all $1 < \beta < R$ the saturation result

$$\max_{z \in \overline{G_\beta}} |f(z) - L_\xi(f)(z)| \neq o(\min_{1 \leq k} |1 - \rho_{k,\xi}|), \ \xi \in (0, 1],$$

holds.

(ii) Take $\rho_{k,\xi} = \frac{1}{1 + \xi^2 k^2}$, that is $L_\xi(f)$ is the generalization of $P_\xi(f)$. If f is not a constant function on G then for all $1 < \beta < R$ we have

$$\max_{z \in \overline{G_\beta}} |f(z) - L_\xi(f)(z)| \sim \xi^2, \ \xi \in (0, 1],$$

where the constants in the equivalence depend only on f, β and G_β (but are independent of ξ).

(iii) Take $\rho_{k,\xi} = e^{-k\xi}$, that is $L_\xi(f)$ is the generalization of $Q^*_\xi(f)$. If f is not a constant function on G then for all $1 < \beta < R$ we have

$$\max_{z \in \overline{G_\beta}} |f(z) - L_\xi(f)(z)| \sim \xi, \ \xi \in (0,1],$$

where the constants in the equivalence depend only on f, β and G_β.

(iv) Take $\rho_{k,\xi} = e^{-k^2\xi/4}$, that is $L_\xi(f)$ is the generalization of $W^*_\xi(f)$. If f is not a constant function on G then we have

$$\max_{z \in G} |f(z) - L_\xi(f)(z)| \sim \xi, \ \xi \in (0,1],$$

where the constants in the equivalence depend only on f, β and G_β.

Proof. (i) First we use some ideas in the proof of Theorem 3, p. 54 in the book of Suetin [186]. Since there exists $R > 1$ such that f is analytic G_R and since $G \subset G_R$, there exists r with $1 < r < R$ such that $G \subset \overline{G_r}$. Let β satisfy the conditions $1 < r < \beta < R$.

Now, by Definition 1.0.10 (ii) we have $a_k(f) = \frac{1}{2\pi i} \int_{|u|=1} \frac{f(\Psi(u))}{u^{k+1}} du$. Also, by Kövari-Pommerenke [120], p. 198, Lemma 2, we have that the integral $\frac{1}{2\pi i} \int_{|u|=1} \frac{F_k(\Psi(u))}{u^{m+1}} du$ is equal to zero for $m \neq k$ and equal to 1 for $m = k$. But taking into account the Cauchy's result, clearly we can write $a_k(f) = \frac{1}{2\pi i} \int_{|u|=\beta} \frac{f(\Psi(u))}{u^{k+1}} du$ and that $\frac{1}{2\pi i} \int_{|u|=\beta} \frac{F_k(\Psi(u))}{u^{m+1}} du$ is equal to zero for $m \neq k$ and equal to 1 for $m = k$.

Let $L_\xi(f)(z) = \sum_{k=0}^{\infty} a_k(f)\rho_{k,\xi}F_k(z)$, $z \in \overline{G_r}, \xi \in (0,1]$, with $\rho_{k,\xi} \in (0,1]$, for all $\xi \in (0,1]$ and $k \in \mathbb{N} \bigcup \{0\}$.

By the inequality (13), p. 44 in Suetin [186] we have

$$|F_k(z)| \leq C(r)r^k, \text{ for all } z \in \overline{G_r}.$$

Also, by the above formula for $a_k(f)$ we easily obtain $|a_k(f)| \leq \frac{C(\beta,f)}{\beta^k}$. Note here that $C(r), C(\beta, f) > 0$ are independent of k. It follows

$$|a_k(f)\rho_{k,\xi}F_k(z)| \leq C(r,\beta,f) \left(\frac{r}{\beta}\right)^k \text{ for all } z \in \overline{G_r}, \ k = 0, 1, 2, ..., .$$

Since $\frac{r}{\beta} < 1$, it follows that the series which defines $L_\xi(f)(z)$ is absolutely and uniformly convergent in the compact $\overline{G_r}$ since it is majorized by a geometric progression. Therefore $L_\xi(f)$ is analytic in G.

On the other hand we immediately obtain

$$\frac{1}{2\pi i} \int_{|u|=\beta} \frac{f(\Psi(u)) - L_\xi(f)(\Psi(u))}{u^{m+1}} du = (1 - \rho_{m,\xi})a_m(f), m \geq 0.$$

This implies

$$|1 - \rho_{m,\xi}| \cdot |a_m(f)| \leq \max_{z \in \overline{G_\beta}} |f(z) - L_\xi(f)(z)|.$$

Now, let us assume that the conclusion of Theorem 3.2.17 is false, that is there exists $1 < \beta < R$ such that

$$\max_{z \in \overline{G_\beta}} |f(z) - L_\xi(f)(z)| = o(\min_{1 \leq k} |1 - \rho_{k,\xi}|).$$

It implies

$$|1 - \rho_{m,\xi}| \cdot |a_m(f)| = o(\min_{1 \le k} |1 - \rho_{k,\xi}|), \text{ for all } m \ge 1$$

and therefore

$$\lim_{\xi \searrow 0} \frac{|1 - \rho_{m,\xi}| \cdot |a_m(f)|}{\min_{1 \le k} |1 - \rho_{k,\xi}|} = 0,$$

which means $a_m(f) = 0$, for all $m \ge 1$. Then, by Theorem 1, p. 44 in Gaier [76] we get $f(z) = a_0(f)$ for all $z \in \overline{G_\beta}$. Since $G \subset \overline{G_\beta}$ it follows that f is constant in G, a contradiction.

(ii) By Theorem 3.2.1, (i) we get $\rho_{k,\xi} = \frac{1}{1+\xi^2 k^2}$ and

$$\min_{1 \le k}(1 - \rho_{k,\xi}) = \frac{\xi^2}{1 + \xi^2} \ge \frac{\xi^2}{2}, \text{ for all } \xi \in (0, 1].$$

Denote $V_\xi = \inf_{1 \le p}(\frac{\xi^2 p^2}{1+\xi^2 p^2})$. We get $V_\xi = \frac{\xi^2}{1+\xi^2}$ and for all $\xi \in (0, 1]$ it follows $V_\xi \ge \xi^2/2$.

By the following lower estimate from the above point (i),

$$|1 - \rho_{m,\xi}| \cdot |a_m(f)| \le \max_{z \in \overline{G_\beta}} |f(z) - L_\xi(f)(z)|,$$

for all $p \ge 1$ and $\xi \in (0, 1]$ it follows

$$\frac{2 \max_{z \in \overline{G_\beta}} |f(z) - L_\xi(f)(z)|}{\xi^2} \ge \frac{\max_{z \in \overline{G_\beta}} |f(z) - L_\xi(f)(z)|}{V_\xi}$$

$$\ge \frac{\max_{z \in \overline{G_\beta}} |f(z) - L_\xi(f)(z)|}{\frac{\xi^2 p^2}{1+\xi^2 p^2}} \ge |a_p(f)|.$$

This implies that if there exists a subsequence $(\xi_k)_k$ in $(0, 1]$ with $\lim_{k \to \infty} \xi_k = 0$ and such that $\lim_{k \to \infty} \frac{\max_{z \in \overline{G_\beta}} |f(z) - L_\xi(f)(z)|}{\xi_k^2} = 0$ then $a_p = 0$ for all $p \ge 1$, that is f is constant on $\overline{G_\beta}$.

Therefore, if f is not a constant then $\inf_{\xi \in (0,1]} \frac{\max_{z \in \overline{G_\beta}} |f(z) - L_\xi(f)(z)|}{\xi^2} > 0$, which implies that there exists a constant $C_\beta(f) > 0$ such that

$$\frac{\max_{z \in \overline{G_\beta}} |f(z) - L_\xi(f)(z)|}{\xi^2} \ge C_\beta(f) \text{ for all } \xi \in (0, 1],$$

that is

$$\max_{z \in \overline{G_\beta}} |f(z) - L_\xi(f)(z)| \ge C_\beta(f)\xi^2, \text{ for all } \xi \in (0, 1].$$

For the upper estimate, first by Theorem 2, p. 52 in Suetin [186] we have $f(z) = \sum_{k=0}^{\infty} a_k(f) F_k(z)$, for all $z \in G$, where the series is absolutely and uniformly convergent on the whole G. Taking into account the estimate for $|a_k(f)F_k(z)|$, $z \in \overline{G_r}$ from the above point (i) and denoting $d = \frac{r}{\beta} < 1$, for all $z \in \overline{G_r}$ we obtain

$$|f(z) - L_\xi(f)(z)| \leq \sum_{k=0}^{\infty} (1 - \rho_{k,\xi})|a_k(f)F_k(z)| = \sum_{k=1}^{\infty} \frac{\xi^2 k^2}{1 + \xi^2 k^2}|a_k(f)F_k(z)|$$

$$\leq \xi^2 \sum_{k=1}^{\infty} k^2 |a_k(f)F_k(z)| \leq C\xi^2 \sum_{k=0}^{\infty} k^2 d^k.$$

But taking into account that $\sum_{k=0}^{\infty} d^k = \frac{1}{1-d}$, by differentiation with respect to d we easily obtain $\sum_{k=1}^{\infty} kd^{k-1} = \frac{1}{(1-d)^2}$ and $\sum_{k=2}^{\infty} k(k-1)d^{k-2} = \frac{2}{(1-d)^3}$, which immediately implies $\sum_{k=0}^{\infty} k^2 d^k < +\infty$ and therefore

$$|f(z) - L_\xi(f)(z)| \leq M\xi^2, \text{ for all } z \in \overline{G_r}, \xi \in (0,1].$$

Since r with $1 < r < R$ is arbitrary, this proves (ii).

(iii) The proof follows word for word that for Theorem 3.2.17, (iii), with the unique difference that here $\beta > 1$.

(iv) The proof follows word for word that for Theorem 3.2.17, (iv), with the unique difference that here $\beta > 1$. Theorem 3.2.18 is proved. $\square$

Remark. It is natural to ask for extensions of the Cauchy problems in Theorem 3.2.13 (v) and in Theorem 3.2.15 (iv) for domains in $\mathbb{C}$. For example, let us suppose that $f \in A(\overline{G})$, where G is a bounded simply connected domain with the boundary Γ a regular analytic curve, that is Γ is an analytic mapping and its derivative never vanishes. In this case, by Theorem 1 and its proof, p. 51-52 in the book of Suetin [186] it follows that $f(z) = \sum_{k=0}^{\infty} a_k(f)F_k(z)$, where the series is uniformly and absolutely convergent in any compact subset K of G. Alternatively, we can suppose here that G is a continuum (that is a compact connected subset of $\mathbb{C}$), without any requirement on its boundary. In this case too, from Theorem 2 in Suetin [186], p. 52, it follows $f(z) = \sum_{k=0}^{\infty} a_k(f)F_k(z)$, where the series is uniformly convergent in G.

For this extension we need a suitable concept of derivative called *Faber derivative* and introduced in Bruj-Schmieder [48], Definition 3, p. 165. Taking into account the above hypothesis on G, it easily follows that its Faber derivative (of arbitrary order r) introduced in Bruj-Schmieder [48], Definition 3 and denoted by $F^{(r)}(f)(z)$ satisfies $\int_{-\pi}^{\pi} F^{(r)}[\psi(e^{it})]dt = 0$ and can be written as

$$F^{(r)}(f)(z) = \sum_{k=0}^{\infty} a_k(F^{(r)})F_k(z).$$

Therefore, taking into account Theorem 5, p. 166 in the same paper of Bruj-Schmieder [48], for the Faber derivative of order r of f formally it easily follows the relationship

$$F^{(r)}(f)(z) = \sum_{k=0}^{\infty} (ik)^r a_k(f)F_k(z),$$

for all z in any compact set included in G. Here $i^2 = -1$.

The following two corollaries are immediate.

Corollary 3.2.19. *Let us suppose that $f \in A(\overline{G})$, where $G \subset \mathbb{C}$ is, for example a continuum (or a bounded simply connected domain with the boundary Γ a regular analytic curve). Then $W_t^*(f)(z) = \sum_{k=0}^{\infty} a_k(f)e^{-k^2 t/2}F_k(z) := u(f)(t,z)$, $z \in \overline{G}$ is a semigroup of linear operators on the Banach space $(A(\overline{G}), \|\cdot\|_{C(G)})$ and is solution (that belongs to $A(\overline{G})$, for each fixed $t > 0$) of the "Cauchy problem"*

$$\frac{\partial u}{\partial t}(t,z) = \frac{1}{2}F''(u(f))(t,z), \ (t,z) \in (0,+\infty) \times G, \ z = re^{i\varphi}, \ z \neq 0,$$

$$u(0,z) = f(z), \ z \in \overline{G}, \ f \in A(\overline{G}).$$

Corollary 3.2.20. *Let us suppose that $f \in A(\overline{G})$, where $G \subset \mathbb{C}$ is, for example a continuum (or a bounded simply connected domain with the boundary Γ a regular analytic curve). Then $Q_t^*(f)(z) = \sum_{k=0}^{\infty} a_k(f)e^{-kt}F_k(z) := u(f)(t,z)$, $z \in \overline{G}$ is a semigroup of linear operators on the Banach space $(A(\overline{G}), \|\cdot\|_{C(G)})$ and is solution (that belongs to $A(\overline{G})$, for each fixed $t > 0$) of the "Cauchy problem"*

$$\frac{\partial u}{\partial t}(t,z) + F''(u(f))(t,z) = 0, \ (t,z) \in (0,+\infty) \times G, \ z = re^{i\varphi}, \ z \neq 0,$$

$$u(0,z) = f(z), \ z \in \overline{G}, \ f \in A(\overline{G}).$$

Remark. For $\overline{G} = \overline{\mathbb{D}}_1$, these equations become those in Theorem 3.2.13 (v) and Theorem 3.2.15 (iv).

3.2.2 Complex q-Picard and q-Gauss-Weierstrass Singular Integrals

In this subsection we extend the results in the case of classical complex Picard and Gauss-Weierstrass singular integrals proved in the previous subsection, to their q-analogues.

For $f(z) = \sum_{k=0}^{\infty} a_k z^k$, $z \in \mathbb{D}_R$, $\lambda \in \mathbb{R}$, $\lambda > 0$, $0 < q < 1$, $r \in \mathbb{N} \bigcup \{0\}$ and $z \in \overline{\mathbb{D}}_R$, let us define the q-complex singular integrals

$$P_{r\lambda}(f; q, z) \equiv P_{r\lambda}(f; z)$$

$$:= -\frac{(1-q)}{2[\lambda]_q \ln q^{-1}} \sum_{k=1}^{r+1} (-1)^k \frac{q^{(r-k+1)(r-k)/2}}{q^{(r+1)r/2}} \begin{bmatrix} r+1 \\ k \end{bmatrix}_q \int_{-\infty}^{\infty} \frac{f\left(ze^{i[k]_q t}\right)}{E_q\left(\frac{(1-q)|t|}{[\lambda]_q}\right)} dt$$

and

$$W_{r\lambda}(f; q, z) \equiv W_{r\lambda}(f; z)$$

$$:= -\frac{1}{\pi \sqrt{[\lambda]_q}\, (q^{1/2}; q)_{1/2}}$$

$$\cdot \sum_{k=1}^{r+1} (-1)^k \frac{q^{(r-k+1)(r-k)/2}}{q^{(r+1)r/2}} \begin{bmatrix} r+1 \\ k \end{bmatrix}_q \int_{-\infty}^{\infty} \frac{f\left(ze^{i[k]_q t}\right)}{E_q\left(\frac{t^2}{[\lambda]_q}\right)} dt$$

called as the complex q- Jackson type generalization of the q-Picard and q-Gauss-Weierstrass singular integrals, respectively. For $r = 0$ we denote these singular integrals by $P_\lambda(f; q, z) \equiv P_\lambda(f; z)$ and $W_\lambda(f; q, z) \equiv W_\lambda(f; z)$, respectively.

First we present the approximation properties.

Theorem 3.2.21. (Aral-Gal [32]) *Let* $f(z) = \sum_{k=0}^{\infty} a_k z^k, z \in \mathbb{D}_R$ *be with* $a_0 = 0, a_1 = 1$ *and* $\lambda > 0, 0 < q < 1$. *We have :*

(i) $P_\lambda(f; q, z) := P_\lambda(f; z)$ *is continuous in* $\overline{\mathbb{D}}_R$, *analytic in* $\mathbb{D}_R$ *so that*

$$P_\lambda(f; z) = \sum_{k=0}^{\infty} a_k c_k(\lambda, q) z^k, z \in \mathbb{D}_R, P_\lambda(f; 0) = 0 \text{ and}$$

$$c_k(\lambda, q) = \frac{(1-q)}{[\lambda]_q \ln q^{-1}} \int_0^{\infty} \frac{cos(ku)}{E_q\left(\frac{(1-q)u}{[\lambda]_q}\right)} du, k = 0, 1, \dots$$

Also, there exists $\widehat{q} \in (0, 1)$ *such that for all* $q \in (\widehat{q}, 1)$ *we have* $c_1(\lambda, q) > 0$ *and if we choose* q_λ *such that* $0 < q_\lambda < 1$ *and* $q_\lambda \to 1$ *as* $\lambda \to 0$, *then we have* $\lim_{\lambda \to 0} c_1(\lambda, q_\lambda) = 1$;

(ii) $|P_\lambda(f; z) - f(z)| \leq (R + 1)(1 + \frac{1}{q})\omega_1(f; [\lambda]_q)_{\overline{\mathbb{D}}_R}$, *for all* $z \in \overline{\mathbb{D}}_R$, *where*

$$\omega_1(f; \delta)_{\overline{\mathbb{D}}_R} = \sup\{|f(z_1) - f(z_2)|; z_1, z_2 \in \overline{\mathbb{D}}_R, |z_1 - z_2| \leq \delta\}.$$

Proof.   (i) Let $z_0, z_n \in \overline{\mathbb{D}}_R$ be with $\lim_{n \to \infty} z_n = z_0$. Since $|e^{iu}| = 1$, we get

$$|P_\lambda(f; z_n) - P_\lambda(f; z_0)|$$

$$\leq \frac{(1-q)}{2[\lambda]_q \ln q^{-1}} \int_{-\infty}^{+\infty} |f(z_n e^{iu}) - f(z_0 e^{iu})| \cdot \frac{1}{E_q\left(\frac{(1-q)|u|}{[\lambda]_q}\right)} du$$

$$\leq \frac{(1-q)}{2[\lambda]_q \ln q^{-1}} \int_{-\infty}^{+\infty} \omega_1(f; |z_n - z_0|)_{\overline{\mathbb{D}}_R} \cdot \frac{1}{E_q\left(\frac{(1-q)|u|}{[\lambda]_q}\right)} du$$

$$= \omega_1(f; |z_n - z_0|)_{\overline{\mathbb{D}}_R}.$$

Passing to limit with $n \to \infty$, it follows that $P_\lambda(f; z)$ is continuous at $z_0 \in \overline{\mathbb{D}}_R$, since f is continuous on $\overline{\mathbb{D}}_R$. It remains to prove that $P_\lambda(f; z)$ is analytic in $\mathbb{D}_R$. We can write $f(z) = \sum_{k=0}^{\infty} a_k z^k$, $z \in \mathbb{D}_R$. For fixed $z \in \mathbb{D}_R$, we get $f(ze^{iu}) = \sum_{k=0}^{\infty} a_k e^{iku} z^k$ and since $|a_k e^{iku}| = |a_k|$, for all $u \in \mathbb{R}$ and the series $\sum_{k=0}^{\infty} a_k z^k$ is absolutely convergent, it follows that the series $\sum_{k=0}^{\infty} a_k e^{iku} z^k$ is uniformly convergent with respect to $u \in \mathbb{R}$. This immediately implies that the series can be integrated term by term, i.e.

$$P_\lambda(f;z) = \frac{(1-q)}{2\,[\lambda]_q \ln q^{-1}} \sum_{k=0}^\infty a_k z^k \left(\int_{-\infty}^\infty e^{iku} \cdot \frac{1}{E_q\left(\frac{(1-q)|u|}{[\lambda]_q} \right)} du \right)$$

$$= \sum_{k=0}^\infty a_k c_k(\lambda,q) z^k, \text{ where } c_k(\lambda,q)$$

$$= \frac{(1-q)}{[\lambda]_q \ln q^{-1}} \int_0^\infty \frac{\cos(ku)}{E_q\left(\frac{(1-q)u}{[\lambda]_q} \right)} du.$$

Since $a_0 = 0$, we get $P_\lambda(f;0) = 0$.

Then we have

$$c_1(\lambda,q) = \frac{(1-q)}{[\lambda]_q \ln q^{-1}} \int_0^\infty \frac{\cos(u)}{E_q\left(\frac{(1-q)u}{[\lambda]_q} \right)} du = \frac{(1-q)}{\ln q^{-1}} \int_0^\infty \frac{\cos([\lambda]_q u)}{E_q\left((1-q)u \right)} du.$$

Now, if we choose $q_\lambda \to 1$ as $\lambda \to 0$, then we get $[\lambda]_{q_\lambda} \to 0$ (see Aral [31]). Since $\lim_{q\to 1^-} E_q\left((1-q)t \right) = e^t$ (see Gasper-Rahman [99], p. 9, (1.3.16)) and $\lim_{q\to 1^-} [\lambda]_q = \lambda$, by Lebesgue's Dominated Convergence theorem, we obtain

$$\lim_{\lambda\to 0} c_1(\lambda,q_\lambda) = \int_0^\infty e^{-t} du = 1 \text{ and}$$

$$\lim_{q\to 1^-} c_1(\lambda,q) = \int_0^\infty \frac{\cos(\lambda u)}{e^u} du > \text{ (by e.g. Anastassiou-Gal [18], p. 4)} > 0.$$

Thus, there exists $\widehat{q} \in (0,1)$ such that for all $q \in (\widehat{q},\,1)$ we have $c_1(\lambda,q) > 0$.

(ii) By the Maximum Modulus Principle, it suffices to take $|z| = R$. Since $|e^{iu} - 1| \le 2|\sin\frac{u}{2}| \le |u|$ for all $u \in \mathbb{R}$, we easily get

$$|P_\lambda(f;z) - f(z)| \le \frac{(1-q)}{2\,[\lambda]_q \ln q^{-1}} \int_{-\infty}^\infty \omega_1(f;|ze^{iu}-z|)_{\overline{\mathbb{D}}_R} \cdot \frac{1}{E_q\left(\frac{(1-q)|u|}{[\lambda]_q} \right)} du$$

$$\le \frac{(1-q)}{2\,[\lambda]_q \ln q^{-1}} \int_{-\infty}^\infty \omega_1(f;R|u|)_{\overline{\mathbb{D}}_R} \cdot \frac{1}{E_q\left(\frac{(1-q)|u|}{[\lambda]_q} \right)} du$$

$$\le \omega_1(f;[\lambda]_q)_{\overline{\mathbb{D}}_R}(R+1) \frac{(1-q)}{2\,[\lambda]_q \ln q^{-1}} \int_{-\infty}^\infty \left(1 + \frac{|u|}{[\lambda]_q} \right) \cdot \frac{1}{E_q\left(\frac{(1-q)|u|}{[\lambda]_q} \right)} du$$

$$\le \text{(by Aral [31])} \le (R+1)\left(1 + \frac{1}{q} \right) \omega_1(f;[\lambda]_q)_{\overline{\mathbb{D}}_R}. \qquad \square$$

Theorem 3.2.22. (Aral-Gal [32]) *(i) If $f(z) = \sum_{k=0}^\infty a_k z^k$ is analytic in $\mathbb{D}_R$, then for all $\lambda > 0$, $0 < q < 1$, $W_\lambda(f;q,z) := W_\lambda(f;z)$ is analytic in $\mathbb{D}_R$ and we have in $\mathbb{D}_R$*

$$W_\lambda(f;z) = \sum_{k=0}^\infty a_k d_k(\lambda,q) z^k,$$

where

$$d_k(\lambda, q) = \frac{2}{\pi\sqrt{[\lambda]_q}\,(q^{1/2};\,q)_{1/2}} \int_0^\infty \frac{\cos(ku)}{E_q\left(\frac{u^2}{[\lambda]_q}\right)}\,du.$$

Also, there exists $\widehat{q} \in (0,1)$ such that for all $q \in (\widehat{q},\ 1)$ we have $d_1(\lambda, q) > 0$ and if we choose q_λ such that $0 < q_\lambda < 1$ and $q_\lambda \to 1$ as $\lambda \to 0$, then we have $\lim_{\lambda\to 0} d_1(\lambda, q_\lambda) = 1$.

In addition, if f is continuous on $\overline{\mathbb{D}}_R$ then $W_\lambda(f; z)$ is continuous on $\overline{\mathbb{D}}_R$.

(ii)

$$|W_\lambda(f; z) - f(z)| \le (R+1)\left(1 + \sqrt{q^{-1/2}(1 - q^{1/2})}\right)\omega_1\left(f; \sqrt{[\lambda]_q}\right)_{\overline{\mathbb{D}}_R},$$

for all $z \in \overline{\mathbb{D}}_R$.

Proof.   (i) Reasoning as for the $P_\lambda(f)$ operator, we easily deduce

$$W_\lambda(f; z) = \frac{1}{\pi\sqrt{[\lambda]_q}\,(q^{1/2};\,q)_{1/2}} \int_{-\infty}^{+\infty} \sum_{k=0}^\infty a_k z^k e^{iuk} \cdot \frac{1}{E_q\left(\frac{u^2}{[\lambda]_q}\right)}\,du$$

$$= \sum_{k=0}^\infty a_k d_k(\lambda, q)z^k,\ \text{where } d_k(\lambda, q) = \frac{2}{\pi\sqrt{[\lambda]_q}\,(q^{1/2};\,q)_{1/2}} \int_0^{+\infty} \frac{cos(ku)}{E_q\left(\frac{u^2}{[\lambda]_q}\right)}.$$

Similar results with those for $c_1(\lambda, q)$ (in Theorem 3.2.21), can be obtained for $d_1(\lambda, q)$ too. Indeed, if we choose q_λ such that $0 < q_\lambda < 1$ and $q_\lambda \to 1$ as $\lambda \to 0$, then from Lebesgue's Dominated Convergence theorem, we get

$$\lim_{\lambda\to 0} d_1(\lambda, q_\lambda) = \lim_{\lambda\to 0} \frac{2}{\pi\sqrt{[\lambda]_q}\,(q^{1/2};\,q)_{1/2}} \int_0^\infty \frac{\cos(u)}{E_q\left(\frac{u^2}{[\lambda]_q}\right)}\,du$$

$$= \lim_{\lambda\to 0} \frac{2}{\pi\,(q^{1/2};\,q)_{1/2}} \int_0^\infty \frac{\cos(\sqrt{[\lambda]_q}u)}{E_q(u^2)}\,du = 1,$$

(Here, for the last equality see e.g. Alvarez-Nodarse-Atakishiyeva-Atakishiyev [6], p. 132).

Similarly we can see that $\lim_{q\to 1^-} d_1(\lambda, q) > 0$, which implies that there exists $\widehat{q} \in (0,1)$ such that for all $q \in (\widehat{q},\ 1)$ we have $d_1(\lambda, q) > 0$.

The proof of continuity of $W_\lambda(f; z)$ is similar to that for $P_\lambda(f; z)$.

(ii) Reasoning as in the case of $P_\lambda(f; z)$, we can write

$$|W_\lambda(f; z) - f(z)| \le \frac{1}{\pi\sqrt{[\lambda]_q}\,(q^{1/2};\,q)_{1/2}} \int_{-\infty}^{+\infty} |f(ze^{-iu}) - f(z)|\frac{1}{E_q\left(\frac{u^2}{[\lambda]_q}\right)}\,du$$

$$\le \omega_1(f; \sqrt{[\lambda]_q})_{\overline{\mathbb{D}}_R}(R+1)\frac{1}{\pi\sqrt{[\lambda]_q}\,(q^{1/2};\,q)_{1/2}}$$

$$\cdot \int_{-\infty}^{+\infty}\left(1 + \frac{|u|}{\sqrt{[\lambda]_q}}\right)\frac{1}{E_q\left(\frac{u^2}{[\lambda]_q}\right)}\,du$$

$$\le (R+1)\left(1 + \sqrt{q^{-1/2}(1 - q^{1/2})}\right)\omega_1\left(f; \sqrt{[\lambda]_q}\right)_{\overline{\mathbb{D}}_R}.$$

(For the last inequality see Aral [31].) □

Theorem 3.2.23. (Aral-Gal [32]) *Let $R > 0$, $z \in \overline{\mathbb{D}}_R$, $\lambda \in (0, 1]$, $0 < q < 1$ and $r \in \mathbb{N}$. For f analytic in $\mathbb{D}_R$ and continuous on $\overline{\mathbb{D}}_R$ we have*

$$|P_{r\lambda}(f; z) - f(z)| \leq \frac{1}{q^{(r+1)r/2}} \sum_{k=0}^{r+1} \binom{r+1}{k} \frac{[k]_q!}{q^{\frac{k(k+1)}{2}}} \omega_{r+1,q} \left(f; [\lambda]_q\right)_{\partial \mathbb{D}_R},$$

$$|W_{(2r-1)\lambda}(f; z) - f(z)| \leq 2^{2r-1} \left(1 + q^{-\frac{r^2}{2}} \left(q^{1/2}; q\right)_r\right) \omega_{2r,q} \left(f; \sqrt{[\lambda]_q}\right)_{\partial \mathbb{D}_R}$$

where

$$\omega_{r,q}(f; \delta)_{\partial \mathbb{D}_R} = \sup\{|\Delta_u^r f(Re^{ix})|; |x| \leq \pi, |u| \leq \delta\}.$$

Proof. Let $z \in \overline{\mathbb{D}}_R$, $|z| = R$ be fixed. Because of the Maximum Modulus Principle, it suffices to estimate $|P_{r\lambda}(f; z) - f(z)|$, for this $|z| = R$, $z = Re^{ix}$. Reasoning now exactly as in the proof of Theorem 3.2.21, we get

$$f(z) - P_{r\lambda}(f; z) = \frac{(1-q)}{2[\lambda]_q \ln q^{-1}} \frac{(-1)^{r+1}}{q^{(r+1)r/2}} \int_{-\infty}^{\infty} \frac{\Delta_{q,t}^{r+1} f(Re^{ix})}{E_q\left(\frac{(1-q)|t|}{[\lambda]_q}\right)} dt,$$

which implies

$$|f(z) - P_{r\lambda}(f; z)| \leq \frac{(1-q)}{2[\lambda]_q \ln q^{-1}} \frac{1}{q^{(r+1)r/2}} \int_{-\infty}^{\infty} \frac{\omega_{r+1,q}(f; |t|)_{\partial \mathbb{D}_R}}{E_q\left(\frac{(1-q)|t|}{[\lambda]_q}\right)} dt$$

$$\leq \omega_{r+1,q} \left(f; [\lambda]_q\right)_{\partial \mathbb{D}_R} \frac{1}{q^{(r+1)r/2}} \sum_{k=0}^{r+1} \binom{r+1}{k} \frac{[k]_q!}{q^{\frac{k(k+1)}{2}}}.$$

The proof in the case of $W_{(2r-1)\lambda}(f; z)$ is similar. □

Remark. Since f is supposed to be analytic, by Theorems 3.2.21 and 3.2.22 it is immediate that the orders of approximation by $P_\lambda(f; q, z)$ and $W_\lambda(f; q, z)$ are $O([\lambda]_q)$ and $O(\sqrt{[\lambda]_q})$, respectively. Also, from Theorem 3.2.23 it follows that for any $r \in \mathbb{N}$, the orders of approximation by $P_{r\lambda}(f; q; z)$ and $W_{(2r-1)\lambda}(f; q; z)$ are $O([\lambda]_q^{r+1})$ and $O([\lambda]_q^r)$, respectively.

The geometric properties are consequences of Theorems 3.2.21 and 3.2.22 and are expressed by the following.

Theorem 3.2.24. (Aral-Gal [32]) *Let us suppose that $G \subset \mathbb{C}$ is open, such that $\overline{\mathbb{D}}_1 \subset G$ and $f : G \to \mathbb{C}$ is analytic in G. Denote by $(B_\lambda(f)(z))_{\lambda>0}$ any from $(P_\lambda(f; q, z))_{\lambda>0}$, $(W_\lambda(f; q, z))_{\lambda>0}$, where we choose $q := q_\lambda$ such that $0 < q_\lambda < 1$ and $q_\lambda \to 1$ as $\lambda \to 0$.*

(i) If f is univalent in $\overline{\mathbb{D}}_1$, then there exists $\lambda_0 > 0$ sufficiently small (depending on f), such that for all $\lambda \in (0, \lambda_0)$, $B_\lambda(f)(z)$ are univalent in $\overline{\mathbb{D}}_1$.

(ii) Let $\gamma \in (-\pi/2, \pi/2)$. *If* $f(0) = f'(0) - 1 = 0$ *(and* $f(z) \neq 0$*, for all* $z \in \overline{\mathbb{D}}_1 \setminus \{0\}$ *in the case of spirallikeness of order* γ*) and* f *is starlike (convex, spirallike of order* γ*, respectively) in* $\overline{\mathbb{D}}_1$*, that is for all* $z \in \overline{\mathbb{D}}_1$

$$Re\left(\frac{zf'(z)}{f(z)}\right) > 0 \left(Re\left(\frac{zf''(z)}{f'(z)}\right) + 1 > 0, Re\left(e^{i\gamma}\frac{zf'(z)}{f(z)}\right) > 0, resp.\right),$$

then there exists $\lambda_0 > 0$ *sufficiently small (depending on* f*, and on* f *and* γ *in the case of spirallikeness), such that for all* $\lambda \in (0, \lambda_0)$*,* $B_\lambda(f)(z)$ *are starlike (convex, spirallike of order* γ*, respectively) in* $\overline{\mathbb{D}}_1$*.*

If $f(0) = f'(0) - 1 = 0$ *(and* $f(z) \neq 0$*, for all* $z \in \mathbb{D}_1 \setminus \{0\}$ *in the case of spirallikeness of order* γ*) and* f *is starlike (convex, spirallike of order* γ*, respectively) only in* $\mathbb{D}_1$ *(that is the corresponding inequalities hold only in* $\mathbb{D}_1$*), then for any disk of radius* $0 < \rho < 1$ *and center* 0 *denoted by* $\mathbb{D}_\rho$ *, there exists* $\lambda_0 > 0$ *sufficiently small (depending on* f *and* $\mathbb{D}_\rho$*, and in addition on* γ *for spirallikeness), such that for all* $\lambda \in (0, \lambda_0)$*,* $B_\lambda(f)(z)$ *are starlike (convex, spirallike of order* γ*, respectively) in* $\overline{\mathbb{D}}_\rho$ *(that is, the corresponding inequalities hold in* $\overline{\mathbb{D}}_\rho$*).*

Proof. (i) Reasoning as in Aral [31], Theorem 2.3, we get uniform convergence (as $\lambda \to 0$) in the Theorems 3.2.21 and 3.2.22, which together with a well-known results concerning sequences of analytic functions converging locally uniformly to an univalent function (see e.g. Kohr-Mocanu [118], p. 130, Theorem 4.1.17) implies the univalence of $B_\lambda(f)(z)$ for sufficiently small λ.

For the proof of the conclusions in (ii), let us make some general useful considerations. By Theorems 3.2.21 and 3.2.22 (reasoning again as in Aral [31], Theorem 2.3), it follows that for $\lambda \to 0$, we have $B_\lambda(f)(z) \to f(z)$, uniformly in any compact disk included in G. By the well-known Weierstrass' result (see e.g. Kohr-Mocanu [118], p. 18, Theorem 1.1.6), this implies that $B'_\lambda(f)(z) \to f'(z)$ and $B''_\lambda(f)(z) \to f''(z)$, uniformly in any compact disk in G and therefore in $\overline{\mathbb{D}}_1$ too, when $\lambda \to 0$. In all what follows, denote $P_\lambda(f)(z) = \frac{B_\lambda(f)(z)}{b_1(\lambda, q_\lambda)}$, where $b_1(\lambda, q_\lambda) > 0$ (for λ sufficiently small) is the coefficient of z in the Taylor series representing the analytic function $B_\lambda(f)(z)$.

If $f(0) = f'(0) - 1 = 0$, then we get $P_\lambda(f)(0) = \frac{f(0)}{b_1(\lambda, q_\lambda)} = 0$ and $P'_\lambda(f)(0) = \frac{B'_\lambda(f)(0)}{b_1(\lambda, q_\lambda)} = 1$. Also, if $f(0) = 0$ and $f'(0) = 1$, then $b_1(\lambda, q_\lambda)$ converges to $f'(0) = 1$ as $\lambda \to 0$, which obviously implies that for $\lambda \to 0$, we have $P_\lambda(f)(z) \to f(z)$, $P'_\lambda(f)(z) \to f'(z)$ and $P''_\lambda(f)(z) \to f''(z)$, uniformly in $\overline{\mathbb{D}}_1$.

(ii) Suppose first that f is starlike in $\overline{\mathbb{D}}_1$. By hypothesis we get $|f(z)| > 0$ for all $z \in \overline{\mathbb{D}}_1$ with $z \neq 0$, which from the univalence of f in $\mathbb{D}_1$, implies that we can write $f(z) = zg(z)$, with $g(z) \neq 0$, for all $z \in \overline{\mathbb{D}}_1$, where g is analytic in $\mathbb{D}_1$ and continuous in $\overline{\mathbb{D}}_1$.

Write $P_\lambda(f)(z)$ in the form $P_\lambda(f)(z) = zQ_\lambda(f)(z)$. For $|z| = 1$ we have

$$|f(z) - P_\lambda(f)(z)| = |z| \cdot |g(z) - Q_\lambda(f)(z)| = |g(z) - Q_\lambda(f)(z)|,$$

which by the uniform convergence in $\overline{\mathbb{D}}_1$ of $P_\lambda(f)$ to f and by the maximum modulus principle, implies the uniform convergence in $\overline{\mathbb{D}}_1$ of $Q_\lambda(f)(z)$ to $g(z)$, as $\lambda \to 0$.

Since g is continuous in $\overline{\mathbb{D}}_1$ and $|g(z)| > 0$ for all $z \in \overline{\mathbb{D}}_1$, there exist an index $\lambda_0 > 0$ and $a > 0$ depending on g, such that $|Q_\lambda(f)(z)| > a > 0$, for all $z \in \overline{\mathbb{D}}_1$ and all $\lambda \in (0, \lambda_0)$. Also, for all $|z| = 1$, we have

$$
\begin{aligned}
|f'(z) - P'_\lambda(f)(z)| &= |z[g'(z) - Q'_\lambda(f)(z)] + [g(z) - Q_\lambda(f)(z)]| \\
&\geq |\ |z| \cdot |g'(z) - Q'_\lambda(f)(z)| - |g(z) - Q_\lambda(f)(z)|\ | \\
&= |\ |g'(z) - Q'_\lambda(f)(z)| - |g(z) - Q_\lambda(f)(z)|\ |,
\end{aligned}
$$

which from the maximum modulus principle, the uniform convergence of $P'_\lambda(f)$ to f' and of $Q_\lambda(f)$ to g, evidently implies the uniform convergence of $Q'_\lambda(f)$ to g', as $\lambda \to 0$. Then, for $|z| = 1$, we get

$$
\begin{aligned}
\frac{zP'_\lambda(f)(z)}{P_\lambda(f)} &= \frac{z[zQ'_\lambda(f)(z) + Q_\lambda(f)(z)]}{zQ_\lambda(f)(z)} \\
&= \frac{zQ'_\lambda(f)(z) + Q_\lambda(f)(z)}{Q_\lambda(f)(z)} \to \frac{zg'(z) + g(z)}{g(z)} \\
&= \frac{f'(z)}{g(z)} = \frac{zf'(z)}{f(z)},
\end{aligned}
$$

which again from the maximum modulus principle, implies

$$
\frac{zP'_\lambda(f)(z)}{P_\lambda(f)} \to \frac{zf'(z)}{f(z)}, \text{ uniformly in } \overline{\mathbb{D}}_1.
$$

Since $Re\left(\frac{zf'(z)}{f(z)}\right)$ is continuous in $\overline{\mathbb{D}}_1$, there exists $\alpha \in (0, 1)$, such that

$$
Re\left(\frac{zf'(z)}{f(z)}\right) \geq \alpha, \text{ for all } z \in \overline{\mathbb{D}}_1.
$$

Therefore

$$
Re\left[\frac{zP'_\lambda(f)(z)}{P_\lambda(f)(z)}\right] \to Re\left[\frac{zf'(z)}{f(z)}\right] \geq \alpha > 0,
$$

uniformly on $\overline{\mathbb{D}}_1$, i.e. for any $0 < \beta < \alpha$, there is $\lambda_0 > 0$ such that for all $\lambda \in (0, \lambda_0)$ we have

$$
Re\left[\frac{zP'_\lambda(f)(z)}{P_\lambda(f)(z)}\right] > \beta > 0, \text{ for all } z \in \overline{\mathbb{D}}_1.
$$

Since $P_\lambda(f)(z)$ differs from $B_\lambda(f)(z)$ only by a constant, this proves the starlikeness in $\overline{\mathbb{D}}_1$.

If f is supposed to be starlike only in $\mathbb{D}_1$, the proof is identical, with the only difference that instead of $\overline{\mathbb{D}}_1$, we reason for $\overline{\mathbb{D}}_\rho$.

The proofs in the cases when f is convex or spirallike of order γ are similar and follows from the following uniform convergence (on $\overline{\mathbb{D}}_1$ or on $\overline{\mathbb{D}}_\rho$)

$$
Re\left[\frac{zP''_\lambda(f)(z)}{P'_\lambda(f)(z)}\right] + 1 \to Re\left[\frac{zf''(z)}{f'(z)}\right] + 1,
$$

and

$$
Re\left[e^{i\gamma}\frac{zP'_{n\lambda}(f)(z)}{P_\lambda(f)(z)}\right] \to Re\left[e^{i\gamma}\frac{zf'(z)}{f(z)}\right].
$$

The proof is complete. $\qquad\qquad\qquad\qquad\qquad\qquad\qquad\qquad\qquad\square$·

Remark. By using Theorem 3.2.23 and reasoning as above, it is not difficult to prove that the geometric properties in Theorem 3.2.24 remain valid for $P_{r\lambda}(f;z)$ and $W_{r\lambda}(f;z)$ too.

3.2.3 *Post-Widder Complex Convolution*

In the case of real functions, the Post-Widder operator is given by (see e.g. Ditzian [63])

$$P_n(f)(x) = \frac{1}{n!}\left(\frac{n}{x}\right)^{n+1}\int_0^\infty e^{-nu/x}u^n f(u)\,du, \quad x > 0,\ f \in C[0,+\infty).$$

Making the change of variable $\frac{1}{x} = y$, we obtain

$$P_n(f)\left(\frac{1}{y}\right) = \frac{1}{n!}n^{n+1}y^{n+1}\int_0^\infty e^{-nuy}u^n f(u)\,du = (\text{by } uy = v)$$

$$= \frac{1}{n!}n^{n+1}y^{n+1}\int_0^\infty e^{-nv}\frac{v^n}{y^n}f\left(\frac{v}{y}\right)\frac{1}{y}\,dy$$

$$= \frac{n^{n+1}}{n!}\int_0^\infty e^{-nv}v^n f\left(\frac{v}{y}\right)dv.$$

Denoting now $\frac{1}{y} = w$, we can write

$$P_n(f)(w) = \frac{n^{n+1}}{n!}\int_0^\infty e^{-nv}v^n f(vw)\,dv, \quad w > 0.$$

Now, let $\mathbb{D}_1 = \{z \in \mathbb{C};\ |z| < 1\}$ and suppose

$$f \in A(\overline{\mathbb{D}}_1) = \{f;\ \text{analytic on } \mathbb{D}_1 \text{ and continuous on } \overline{\mathbb{D}}_1\}.$$

Suggested by the above form we propose the following complex Post-Widder operator by

$$P_{n,\alpha}(f)(z) = \frac{n^{n+1}}{n!}\int_0^\infty e^{-nv}v^n f(ze^{iv/\alpha_n})\,dv, \quad z \in \overline{\mathbb{D}}_1,$$

where $\alpha = (\alpha_n)_n$, $\alpha_n \nearrow +\infty$, is arbitrary, fixed.

First a Jackson-type estimate in approximation of $f \in A(\overline{\mathbb{D}}_1)$ by $P_{n,\alpha}(f)(z)$ and a global smoothness preservation property are proved.

Secondly we prove some geometric properties of $P_{n,\alpha}(f)(z)$, in the sense that it preserves some sufficient conditions for starlikeness and univalence satisfied by $f \in A^*(\overline{\mathbb{D}}_1)$.

Concerning the approximation properties of $P_{n,\alpha}(f)(z)$ we present the following result. Note that Theorem 3.2.25 (i),(ii) and (iii) were proved in Anastassiou-Gal [17], while Theorem 3.2.25 (iv) is new.

Theorem 3.2.25. *Let* $f\colon \overline{\mathbb{D}}_1 \to \mathbb{C}$ *be continuous on* $\overline{\mathbb{D}}_1$. *For all* $\delta \geq 0$, $z \in \overline{\mathbb{D}}_1$ *and* $n \in \mathbb{N}$ *we have:*

(i) $\omega_1(P_{n,\alpha}(f);\delta)_{\overline{\mathbb{D}}_1} \leq \omega_1(f;\delta)_{\overline{\mathbb{D}}_1}.$

(ii) $|P_{n,\alpha}(f)(z) - f(z)| \leq 3\omega_1\left(f; \frac{1}{\alpha_n}\right)_{\overline{\mathbb{D}}_1}.$

(iii) If $f(z) = \sum_{k=0}^{\infty} a_k z^k$ is analytic in $\mathbb{D}_1$ and continuous in $\overline{\mathbb{D}}_1$, then $P_{n,\alpha}(f)$ is analytic in $\mathbb{D}_1$ and continuous in $\overline{\mathbb{D}}_1$ for all $n \in \mathbb{N}$ and $\alpha = (\alpha_n)_n$. Also, we can write

$$P_{n,\alpha}(f)(z) = \sum_{k=0}^{\infty} B_{k,\alpha} z^k, \quad \forall z \in \mathbb{D}_1,$$

where $B_{k,\alpha} = a_k A_{k,\alpha}$ with

$$A_{k,\alpha} = \frac{n^{n+1}}{n!} \int_0^\infty e^{-nv} v^n e^{ikv/\alpha_n} \, dv.$$

If $\alpha_n = n$, $\forall n \in \mathbb{N}$, then

$$|A_{1,\alpha}| := |A_{1,n}| = \frac{1}{\sqrt{2}} \cdot \frac{(n!)^2}{\prod\limits_{k=1}^{n} \sqrt{k^4 + 1}},$$

that is $A_{1,n} \neq 0$, for all $n \in \mathbb{N}$.

(iv) In addition, let us suppose that $f(z) = \sum_{k=0}^{\infty} a_k z^k$ for all $z \in \mathbb{D}_R$, $R > 1$. For all $1 \leq r < r_1 < R$, $n \in \mathbb{N}$ and $q \in \mathbb{N} \bigcup \{0\}$ we have

$$\|P_{n,\alpha}^{(q)}(f) - f^{(q)}\|_r \leq \frac{C}{\alpha_n}, n \in \mathbb{N},$$

where the constant C depends only on f, q, r, r_1.

Proof. (i) Let $|z_1 - z_2| \leq \delta$, $z_1, z_2 \in_{\overline{\mathbb{D}}_1}$. We have

$$|z_1 e^{iv/\alpha_n} - z_2 e^{iv/\alpha_n}| = |z_1 - z_2|,$$

which implies

$$|P_{n,\alpha}(f)(z_1) - P_{n,\alpha}(f)(z_2)|$$
$$\leq \frac{n^{n+1}}{n!} \int_0^\infty e^{-nv} v^n |f(z_1 e^{iv/\alpha_n}) - f(z_2 e^{iv/\alpha_n})| dv$$
$$\leq \left[\frac{n^{n+1}}{n!} \int_0^\infty e^{-nv} v^n \, dv \right] \omega_1(f; \delta)_{\overline{\mathbb{D}}_1}.$$

Passing to supremum with $|z_1 - z_2| \leq \delta$, it follows

$$\omega_1(P_{n,\alpha}(f); \delta)_{\overline{\mathbb{D}}_1} \leq \omega_1(f; \delta)_{\overline{\mathbb{D}}_1}, \quad \forall \delta \geq 0.$$

(ii) Since $\frac{n^{n+1}}{n!} \int_0^\infty e^{-nv} v^n \, dv = 1$, we get

$$|P_{n,\alpha}(f)(z) - f(z)| \leq \frac{n^{n+1}}{n!} \int_0^\infty e^{-nv} v^n |f(z e^{iv/\alpha_n}) - f(z)| \, dv$$
$$\leq \frac{n^{n+1}}{n!} \int_0^\infty e^{-nv} v^n \omega_1(f; |z| \cdot |e^{iv/\alpha_n} - 1|)_{\overline{\mathbb{D}}_1} \, dv.$$

Combined with the inequality

$$|z| \cdot |e^{iv/\alpha_n} - 1| \leq 2 \left| \sin \frac{v}{2\alpha_n} \right| \leq \frac{v}{\alpha_n}, \quad \forall v > 0,$$

it follows

$$|P_{n,\alpha}(f)(z) - f(z)| \leq \frac{n^{n+1}}{n!} \int_0^\infty e^{-nv} v^n \omega_1 \left(f; \frac{v}{\alpha_n} \right)_{\overline{\mathbb{D}}_1} dv$$

$$\leq \omega_1 \left(f; \frac{1}{\alpha_n} \right)_{\overline{\mathbb{D}}_1} \cdot \frac{n^{n+1}}{n!} \int_0^\infty e^{-nv} v^n (v+1) \, dv.$$

Since

$$\int_0^\infty e^{-nv} v^{n+1} \, dv = \frac{(n+1)!}{n^{n+2}},$$

we arrive at

$$|P_{n,\alpha}(f)(z) - f(z)| \leq \omega_1 \left(f; \frac{1}{\alpha_n} \right)_{\overline{\mathbb{D}}_1} \left[1 + \frac{n+1}{n} \right] \leq 3\omega_1 \left(f; \frac{1}{\alpha_n} \right)_{\overline{\mathbb{D}}_1}.$$

(iii) Let $z_0, z_m \in \overline{\mathbb{D}}_1$ be with $\lim_{m \to \infty} z_m = z_0$. We have

$$|P_{n,\alpha}(f)(z_m) - P_{n,\alpha}(f)(z_0)|$$

$$\leq \frac{n^{n+1}}{n!} \int_0^\infty e^{-nv} v^n |f(z_m e^{iv/\alpha_n}) - f(z_0 e^{iv/\alpha_n})| \, dv$$

$$\leq \omega_1(f; |z_m - z_0|)_{\overline{\mathbb{D}}_1} \cdot \frac{n^{n+1}}{n!} \int_0^\infty e^{-nv} v^n \, dv = \omega_1(f; |z_m - z_0|)_{\overline{\mathbb{D}}_1}.$$

Passing to the limit with $m \to \infty$, it follows the continuity of $P_{n,\alpha}(f)$ at $z_0 \in \overline{\mathbb{D}}_1$.

Now, let $f(z) = \sum_{k=0}^\infty a_k z^k$, $z \in \mathbb{D}_1$, be analytic in $\mathbb{D}_1$. For fixed $z \in \mathbb{D}_1$, we get

$$f(z e^{iv/\alpha_n}) = \sum_{k=0}^\infty a_k e^{ikv/\alpha_n} z^k$$

and since $|a_k e^{ikv/\alpha_n}| = |a_k|$, for all $v \geq 0$ and the series $\sum_{k=0}^\infty a_k z^k$ is convergent, it follows that the series $\sum_{k=0}^\infty a_k e^{ikv/\alpha_n} z^k$ is uniformly convergent with respect to $v \geq 0$. This immediately implies that the series can be integrated term by term, that is

$$P_{n,\alpha}(f)(z) = \sum_{k=0}^\infty a_k z^k \frac{n^{n+1}}{n!} \int_0^\infty e^{-nv} v^n e^{-ikv/\alpha_n} \, dv = \sum_{k=0}^\infty a_k A_{k,\alpha} z^k,$$

where

$$A_{k,\alpha} = \frac{n^{n+1}}{n!} \int_0^\infty e^{-nv} v^n e^{-ikv/\alpha_n} \, dv, \quad k = 0, 1, \ldots .$$

By using the substitution $-nv = x$ and taking $\alpha_n = n$, $\forall n \in \mathbb{N}$, we get

$$A_{1,\alpha} := A_{1,n} = \frac{1}{n!} \int_0^\infty e^{-x} x^n e^{-ix/n^2} \, dx = \frac{1}{n!} \int_0^\infty x^n e^{-x(1+i/n^2)} \, dx.$$

Denoting the last integral (without $\frac{1}{n!}$) by I_n and integrating by parts, we immediately obtain

$$I_n = \frac{x^n \cdot e^{-x(1+i/n^2)}}{-(1+i/n^2)} \Big|_0^\infty + \frac{n}{(1+i/n^2)} \cdot I_{n-1} = \frac{n}{(1+i/n^2)} I_{n-1},$$

for all $n = 1, 2, \ldots$. Taking the modulus, it follows

$$|I_n| = \frac{n^3}{\sqrt{n^4 + 1}} \cdot |I_{n-1}|, \quad \forall n = 1, 2, \ldots.$$

Here

$$I_0 = \int_0^\infty e^{-x(1+i)} \, dx = \frac{e^{-x(1+i)}}{-(1+i)} \Big|_0^\infty = \frac{1}{1+i},$$

which implies $|I_0| = \frac{1}{\sqrt{2}}$.

Therefore $|I_1| = \frac{1}{\sqrt{2}} \cdot \frac{1}{\sqrt{2}} = \frac{1}{2}$. Writing

$$|I_1| = \frac{1}{2}$$

$$|I_2| = \frac{2^3}{\sqrt{2^4 + 1}} \cdot |I_1|$$

$$\vdots$$

$$|I_n| = \frac{n^3}{\sqrt{n^4 + 1}} \cdot |I_{n-1}|,$$

and taking the products of both members, we arrive at

$$|I_n| = \frac{1}{2} \cdot \frac{(n!)^3}{\prod\limits_{k=1}^n \sqrt{k^4 + 1}} \cdot \sqrt{2}.$$

It follows

$$|A_{1,\alpha}| = |A_{1,n}| = \frac{1}{n!} \cdot |I_n| = \frac{1}{\sqrt{2}} \cdot \frac{(n!)^2}{\prod\limits_{k=1}^n \sqrt{k^4 + 1}} \neq 0, \quad \text{for all } n \in \mathbb{N},$$

that is $A_{1,n} \neq 0$, $\forall n \in \mathbb{N}$.

(iv) Analysing the proofs of the above points (i)-(iii), it easily follows that they hold by replacing everywhere $\mathbb{D}_1$ with $\mathbb{D}_r$, where $1 \leq r < R$. Therefore the approximation error in (ii) can be expressed in terms of

$$\omega_1(f; \delta)_{\overline{\mathbb{D}}_r} = \sup\{|f(u) - f(v)|; |u - v| \leq \delta, u, v \in \overline{\mathbb{D}}_r\},$$

with constants in front of ω_1 depending on $r \geq 1$.

By the mean value theorem for divided differences in Complex Analysis (see e.g. Stancu [172], p. 258, Exercise 4.20) we get

$$\omega_1(f;\delta)_{\overline{\mathbb{D}}_r} \leq \|f'\|_r \delta,$$

where

$$\|f'\|_r = \sup\{|f'(z)|; |z| \leq r\} \leq \sum_{k=1}^{\infty} |a_k| k r^{k-1}.$$

Then by (ii) for all $n \in \mathbb{N}$ it follows

$$\|P_{n,\alpha}(f) - f\|_r \leq \frac{C_r(f)}{\alpha_n}.$$

Now denoting by γ the circle of radius $r_1 > 1$ and center 0, since for any $|z| \leq r$ and $v \in \gamma$, we have $|v - z| \geq r_1 - r$, by the Cauchy's formulas it follows that for all $|z| \leq r$ and $\xi \in (0,1]$, we have

$$\begin{aligned}
|P_{n,\alpha}^{(q)}(f)(z) - f^{(q)}(z)| &= \frac{q!}{2\pi} \left| \int_{\gamma} \frac{P_{n,\alpha}(f)(v) - f(v)}{(v-z)^{q+1}} dv \right| \\
&\leq \frac{C_{r_1}(f)}{\alpha_n} \cdot \frac{q!}{2\pi} \frac{2\pi r_1}{(r_1 - r)^{q+1}} \\
&= \frac{C_{r_1}(f)}{\alpha_n} \cdot \frac{q! r_1}{(r_1 - r)^{q+1}},
\end{aligned}$$

which proves the upper estimate in approximation by $P_{n,\alpha}^{(q)}(f)(z)$. $\square$

Remark. If we take, for example, $\alpha_n = n^p$, then by Theorem 3.2.25 (ii), we get that the order of approximation by $P_{n,\alpha}(f)(z)$ will be $\omega_1\left(f; \frac{1}{n^p}\right)_{\overline{\mathbb{D}}_1}$.

Concerning the geometric properties of $P_{n,\alpha}(f)(z)$ the following result holds.

Theorem 3.2.26. (Anastassiou-Gal [17]) *(i)* $|B_{k,\alpha}| \leq |a_k|, \quad \forall k = 0, 1, \ldots$.
(ii) $P_{n,\alpha}(\mathcal{P}) \subset \mathcal{P}$, $\frac{1}{A_{1,\alpha}} P_{n,\alpha}(S_{3,A_{1,\alpha}}) \subset S_3$ *and*

$$\frac{1}{A_{1,\alpha}} P_{n,\alpha}(S_M) \subset S_{M/|A_{1,\alpha}|},$$

for $\alpha = (\alpha_n)_n$, $\alpha_n = n$, $\forall n = 1, 2, \ldots$, *where*

$$S_{3,a} = \left\{ f \in S_3; |f''(z)| \leq |a| \right\} \subset S_3,$$

for $|a| < 1$ *and*

$$S_B = \left\{ f \in A(\overline{\mathbb{D}}_1); |f'(z)| < B, \forall z \in \mathbb{D}_1 \right\}.$$

Proof. (i) We have $|B_{k,\alpha}| = |a_k A_{k,\alpha}| = |a_k| \cdot |A_{k,\alpha}| \leq |a_k|$, since $|A_{k,\alpha}| \leq 1$, $\forall k = 0, 1, 2, \ldots$.

(ii) By Theorem 3.2.25 (iii) we get

$$|A_{k,\alpha}| \leq \frac{n^{n+1}}{n!} \int_0^{\infty} e^{-nv} v^n \, dv = 1, \quad \forall k = 0, 1, 2, \ldots,$$

and $|A_{0,\alpha}| = |a_0|$.

Let $f \in \mathcal{P}$, $f = U + iV$, $U > 0$ on $\mathbb{D}_1$. Since $a_0 = f(0) = 1$, we obtain $P_{n,\alpha}(f)(0) = a_0 A_{0,\alpha} = 1$. Also, by

$$P_{n,\alpha}(f)(z) = \frac{n^{n+1}}{n!} \int_0^\infty e^{-nv} v^n \big[U(r\cos(x + v/\alpha_n), r\sin(x + v/\alpha_n)) $$
$$+ iV(r\cos(x + v/\alpha_n), r\sin(x + v/\alpha_n)) \big]\, dv, \quad \forall z = re^{ix} \in \mathbb{D}_1,$$

it follows

$$\mathrm{Re}[P_{n,\alpha}(f)(z)] = \frac{n^{n+1}}{n!} \int_0^\infty e^{-nv} v^n U(r\cos(x + v/\alpha_n), r\sin(x + v/\alpha_n))\, dv > 0,$$

that is $P_{n,\alpha}(f) \in \mathcal{P}$.

Let $f \in A^*(\overline{\mathbb{D}}_1)$. It follows $a_0 = 0$, $a_1 = 1$, which by Theorem 3.2.25 (iii) implies

$$\frac{1}{A_{1,\alpha}} P_{n,\alpha}(f)(0) = 0, \quad \frac{1}{A_{1,\alpha}} P'_{n,\alpha}(f)(0) = \frac{A_{1,\alpha}}{A_{1,\alpha}} = 1.$$

If $|f''(z)| < |A_{1,\alpha}|$ then

$$\left| \frac{1}{A_{1,\alpha}} P''_{n,\alpha}(f)(z) \right| \leq \frac{1}{|A_{1,\alpha}|} \cdot \frac{n^{n+1}}{n!} \int_0^\infty e^{-nv} v^n |e^{2iv/\alpha_n}|$$
$$\cdot |f''(ze^{iv/\alpha_n})|\, dv \leq 1,$$

which means that $\frac{1}{A_{1,\alpha}} P_{n,\alpha}(f) \in S_3$. Also, if $|f'(z)| < M$, $\forall z \in \mathbb{D}_1$, then we obtain

$$\left| \frac{1}{A_{1,\alpha}} \cdot P'_{n,\alpha}(f)(z) \right| \leq \frac{1}{|A_{1,\alpha}|} \cdot \frac{n^{n+1}}{n!} \int_0^\infty e^{-nv} v^n |e^{iv/\alpha_n}| \cdot |f'(ze^{iv/\alpha_n})|\, dv$$
$$< \frac{M}{|A_{1,\alpha}|} \cdot \frac{n^{n+1}}{n!} \int_0^\infty e^{-nv} v^n\, dv = \frac{M}{|A_{1,\alpha}|},$$

that is $\frac{1}{A_{1,\alpha}} P_{n,\alpha}(f) \in S_{M/|A_{1,\alpha}|}$. Since $|A_{1,\alpha}| \leq 1$, it follows that $\frac{M}{|A_{1,\alpha}|} \geq M$. $\qquad\square$

3.2.4 *Rotation-Invariant Complex Convolutions*

For $f \in A(\overline{\mathbb{D}}_1)$, $z \in \overline{\mathbb{D}}_1$, let us consider the complex rotation-invariant integral operators given by

$$Y_k(f)(z) = 2^k \int_{-\infty}^{+\infty} f(ze^{iv}) \varphi(-2^k v)\, dv,$$

and the generalized complex rotation-invariant integral operators given by

$$L_{k,j}(f)(z) = \frac{2^k}{j} \int_{-\infty}^{+\infty} \ell_k(f)(2^k ze^{iv}) \varphi\left(-\frac{2^k}{j} v\right) dv, \quad k \in \mathbb{Z},\ j \in \mathbb{N}.$$

Here $i^2 = -1$, φ is a real-valued function of compact support $\subseteq [-a, a]$, $a > 0$, $\varphi(x) \geq 0$, $\int_{-\infty}^{+\infty} \varphi(x - u)\, du = 1$, $\forall x \in \mathbb{R}$, and $\{\ell_k\}_{k \in \mathbb{Z}}$ is a sequence of linear operators from $A(\overline{\mathbb{D}}_1)$ to $A(\overline{\mathbb{D}}_1)$ defined by recurrence as $\ell_k(f)(z) = \ell_0(f_k)(z)$, $z \in \overline{D}$, where $f_k(z) = f\left(\frac{z}{2^k}\right)$, $z \in \overline{D}$ and $\ell_0 \colon A(\overline{\mathbb{D}}_1) \to A(\overline{\mathbb{D}}_1)$ is a linear operator.

Also, let us consider the Jackson-type generalization of $L_{k,j}(f)(z)$ given by

$$I_{k,q}(f)(z) = -\sum_{j=1}^{q}(-1)^j\binom{q}{j}L_{k,j}(f)(z), \quad \forall k \in \mathbb{Z}, \ q \in \mathbb{N}.$$

Note that the real variants (for real-valued functions of a real variable) of these operators were studied in Anastassiou-Gonska [21] and Anastassiou–Gal [12], [13].

In this subsection approximation and shape preserving properties (in geometric function theory) for the above complex rotation-invariant integral operators are presented.

First the approximation properties are presented. Note that the results in Theorem 3.2.27 (i)-(v) were obtained in Anastassiou-Gal [19], while Theorem 3.2.27 (vi) is new.

Theorem 3.2.27. *Let $f \in A(\overline{\mathbb{D}}_1)$.*
(i) For all $z \in \overline{\mathbb{D}}_1$ and $k \in \mathbb{Z}$, we have

$$|f(z) - Y_k(f)(z)| \leq 3\omega_1\left(f; \frac{a}{2^k}\right)_{\overline{\mathbb{D}}_1};$$

(ii) For all $z \in \overline{\mathbb{D}}_1$, $k \in \mathbb{Z}$, $j \in \mathbb{N}$, we have

$$|f(z) - L_{k,j}(f)(z)| \leq \omega_1\left(f; \frac{mja+n}{2^{k+r}}\right)_{\overline{\mathbb{D}}_1},$$

where for fixed $a > 0$ it is assumed that

$$\sup_{\substack{z,y\in\overline{\mathbb{D}}_1 \\ |z-y|\leq a}}|\ell_0(f)(z) - f(y)| \leq \omega_1\left(f; \frac{ma+n}{2^r}\right)_{\overline{\mathbb{D}}_1};$$

(iii) For the hypothesis in (ii) and $q \in \mathbb{N}$, we have

$$|f(z) - I_{k,q}(f)(z)| \leq (2^q - 1)\omega_1\left(f; \frac{mqa+n}{2^{k+r}}\right)_{\overline{\mathbb{D}}_1};$$

(iv)

$$\omega_1(Y_k(f); \delta)_{\overline{\mathbb{D}}_1} \leq \omega_1(f; \delta)_{\overline{\mathbb{D}}_1}, \quad \delta > 0, \ k \in \mathbb{Z},$$
$$\omega_1(L_{k,j}(f); \delta)_{\overline{\mathbb{D}}_1} \leq \omega_1(f; \delta)_{\overline{\mathbb{D}}_1}, \quad \delta > 0, \ k \in \mathbb{Z}, \ j \in \mathbb{N},$$
$$\omega_1(I_{k,q}(f); \delta)_{\overline{\mathbb{D}}_1} \leq (2^q - 1)\omega_1(f; \delta)_{\overline{\mathbb{D}}_1}, \quad \delta > 0, \ k \in \mathbb{Z}, \ q \in \mathbb{N},$$

in the hypothesis

$$|\ell_0(f)(x - u + h) - \ell_0(f)(x - u)| \leq \omega_1(f; h)_{\overline{\mathbb{D}}_1}, \quad \forall h > 0,$$

$\forall x, u \in \overline{\mathbb{D}}_1$ *with $x - u$, $x - u + h \in \overline{\mathbb{D}}_1$.*
(v) If $f(z) = \sum_{p=0}^{\infty} a_p z^p$ is analytic in $\mathbb{D}_1$ and continuous in $\overline{\mathbb{D}}_1$, then $Y_k(f)(z)$,
$L_{k,j}(f)(z)$ *and $I_{k,q}(f)(z)$ are analytic in $\mathbb{D}_1$ and continuous in $\overline{\mathbb{D}}_1$. The analyticity of $L_{k,j}(f)(z)$ and $I_{k,q}(f)(z)$ is proved here only for $\ell_0(f) \equiv f$.*

Also we can write

$$Y_k(f)(z) = \sum_{p=0}^{\infty} a_p b_{p,k} z^p, \quad z \in \mathbb{D}_1,$$

where

$$b_{p,k} = \int_{-\infty}^{+\infty} \cos\left(\frac{pu}{2^k}\right) \varphi(u)\, du, \quad p = 0, 1, \ldots, \quad k \in \mathbb{Z}.$$

If $\ell_0(f) \equiv f$ then

$$L_{k,j}(f)(z) = \sum_{p=0}^{\infty} a_p b_{p,k,j} z^p, \quad z \in \mathbb{D}_1, \quad k \in \mathbb{Z}, \quad j \in \mathbb{N}$$

with

$$b_{p,k,j} = \int_{-\infty}^{\infty} \cos\left(\frac{pju}{2^k}\right) \varphi(u)\, du = b_{pj,k},$$

and

$$I_{k,q}(f)(z) = \sum_{p=0}^{\infty} a_p c_{p,k,q} z^p, \quad z \in \mathbb{D}_1, \quad k \in \mathbb{Z}, \quad q \in \mathbb{N}$$

with

$$c_{p,k,q} = \sum_{j=1}^{q} (-1)^{j+1} \binom{q}{j} b_{p,k,j}.$$

If $\varphi(x) = 1 - x$, $x \in [0,1]$, $\varphi(x) = 1 + x$, $x \in [-1, 0]$, $\varphi(x) = 0$, $x \in \mathbb{R} \setminus (0, 1)$, then

$$b_{1,k} = 2^{2k+1}\left(1 - \cos\frac{1}{2^k}\right) > 0, \quad \forall k \in \mathbb{Z},$$

$$b_{j,k} = b_{1,k,j} = \frac{2^{2k+1}}{j^2}\left(1 - \cos\frac{j}{2^k}\right) > 0, \quad \forall k \in \mathbb{Z}, \quad j \in \mathbb{N},$$

$$c_{1,k,q} = \sum_{j=1}^{q} (-1)^{j+1} \binom{q}{j} \frac{2^{2k+1}}{j^2}\left(1 - \cos\frac{j}{2^k}\right), \quad \forall k \in \mathbb{Z}, \quad q \in \mathbb{N},$$

and

$$c_{p,k,q} = \sum_{j=1}^{q} (-1)^{j+1} \binom{q}{j} \frac{2^{2k+1}}{p^2 j^2}\left(1 - \cos\frac{pj}{2^k}\right), \quad \forall k \in \mathbb{Z}, \quad q \in \mathbb{N}, \quad p = 0, 1, 2, \ldots.$$

(vi) Let us suppose that $\ell_0(f) \equiv f$, the hypothesis in (ii) is satisfied for all $z \in \overline{\mathbb{D}}_r$ and that $f(z) = \sum_{k=0}^{\infty} a_k z^k$ for all $z \in \mathbb{D}_R$, $R > 1$. For all $1 \le r < r_1 < R$, $k \in \mathbb{Z}$, $j, q \in \mathbb{N}$ and $s \in \mathbb{N} \bigcup \{0\}$ we have

$$\|Y_k^{(s)}(f) - f^{(s)}\|_r \le \frac{C_1}{2^k},$$

$$\|L_{k,j}^{(s)}(f) - f^{(s)}\|_r \le \frac{C_2}{2^{k+r}},$$

$$\|I_{k,q}^{(s)}(f) - f^{(s)}\|_r \le \frac{C_3}{2^{k+r}},$$

where C_1 depends on f, s, r, r_1 and a, C_2 depends on f, s, r, r_1, a, m and n, and C_3 depends on f, s, r, r_1, a, m, q and n.

Proof.   (i) Since

$$2^k \int_{-\infty}^{+\infty} \varphi(-2^k v)\, dv = \int_{-\infty}^{+\infty} \varphi(u)\, du = 1$$

we obtain

$$|f(z) - Y_k(f)(z)| = \left| 2^k \int_{-\infty}^{+\infty} [f(z) - f(ze^{iv})]\varphi(-2^k v)\, dv \right|$$

$$\le 2^k \int_{-\infty}^{+\infty} |f(z) - f(ze^{iv})|\varphi(-2^k v)\, dv$$

$$\le 2^k \int_{-\infty}^{+\infty} \omega_1(f; |z| \cdot |1 - e^{iv}|)_{\overline{\mathbb{D}}_1}\varphi(-2^k v)\, dv$$

$$\le 2^k \int_{-\infty}^{+\infty} \omega_1\left(f; 2\left|\sin\frac{v}{2}\right|\right)_{\overline{\mathbb{D}}_1}\varphi(-2^k v)\, dv$$

$$\le 2^k \int_{-\infty}^{+\infty} \omega_1(f; |v|)_{\overline{\mathbb{D}}_1}\varphi(-2^k v)\, dv$$

$$\le \omega_1\left(f; \frac{a}{2^k}\right)_{\overline{\mathbb{D}}_1} \cdot 2^k \int_{-\infty}^{+\infty} \left[\frac{2^k}{a}|v| + 1\right]\varphi(-2^k v)\, dv$$

$$= \omega_1\left(f; \frac{a}{2^k}\right)_{\overline{\mathbb{D}}_1} \cdot \left[1 + \frac{2^k \cdot 2^k}{a}\int_{-\infty}^{+\infty} |v|\varphi(-2^k v)\, dv\right].$$

But

$$\frac{2^k \cdot 2^k}{a}\int_{-\infty}^{+\infty} |v|\varphi(-2^k v)\, dv = (\text{by } u = -2^k v)$$

$$= \frac{2^k \cdot 2^k}{a} \cdot \int_{-\infty}^{+\infty} \frac{|u|}{2^k} \cdot \varphi(u) \cdot \frac{du}{2^k} = \frac{1}{a}\int_{-\infty}^{+\infty} |u|\varphi(u)\, du$$

$$= \frac{1}{a}\int_{-a}^{a} |u|\varphi(u)\, du \le \frac{1}{a}\int_{-a}^{a} |u|\, du = \frac{2}{a} \cdot a = 2,$$

which immediately proves (i).

   (ii) By

$$\frac{2^k}{j}\int_{-\infty}^{+\infty} \varphi\left(-\frac{2^k}{j}v\right)\, dv = \int_{-\infty}^{+\infty} \varphi(u)\, du = 1,$$

we get

$$L_{k,j}(f)(z) - f(z) = \frac{2^k}{j} \int_{-\infty}^{+\infty} [\ell_k(f)(2^k z e^{iv}) - f(z)] \varphi\left(-\frac{2^k}{j}v\right) dv$$

$$= \frac{2^k}{j} \int_{-\infty}^{+\infty} [\ell_0(f_k)(2^k z e^{iv}) - f_k(2^k z)] \varphi\left(-\frac{2^k}{j}v\right) dv$$

$$\left(\text{by } -\frac{2^k}{j}v = u\right)$$

$$\leq \int_{-\infty}^{+\infty} \left|\ell_0(f_k)\left(2^k z e^{i\left(-\frac{j}{2^k}u\right)}\right) - f_k(2^k z)\right| \varphi(u)\, du$$

$$= \int_{-a}^{a} \left|\ell_0(f_k)\left(2^k z e^{i\left(-\frac{j}{2^k}u\right)}\right) - f_k(2^k z)\right| \varphi(u)\, du.$$

But

$$\left|2^k z e^{i\left(-\frac{j}{2^k}u\right)} - 2^k z\right| \leq 2^k \cdot 2 \sin\left|\frac{j}{2 \cdot 2^k}u\right| \leq 2^k \cdot \frac{j|u|}{2^k} = j|u| \leq ja,$$

for all $|z| \leq 1$, $k \in \mathbb{Z}$, $j \in \mathbb{N}$, which implies (reasoning as in Anastassiou–Gal [12], p. 9)

$$\int_{-a}^{a} \left|\ell_0(f_k)\left(2^k z e^{i\left(-\frac{j}{2^k}u\right)}\right) - f_k(2^k z)\right| \varphi(u)\, du$$

$$\leq \int_{-a}^{a} \sup\{|\ell_0(f_k)(w) - f_k(y)|; |w - y| \leq ja\} \varphi(u)\, du$$

$$\leq \omega_1\left(f; \frac{mja + n}{2^{k+r}}\right)_{\overline{\mathbb{D}}_1},$$

which proves (ii).

(iii) By the relation $-\sum_{j=1}^{q}(-1)^j \binom{q}{j} = 1$, we get

$$|I_{k,q}(f)(z) - f(z)| = \left|-\sum_{j=1}^{q}(-1)^j \binom{q}{j} L_{k,j}(f)(z) - \left(-\sum_{j=1}^{q}(-1)^j \binom{q}{j}\right) f(z)\right|$$

$$= \left|\sum_{j=1}^{q}(-1)^j \binom{q}{j} [L_{k,j}(f)(z) - f(z)]\right|$$

$$\leq \sum_{j=1}^{q} \binom{q}{j} \cdot |L_{k,j}(f)(z) - f(z)|$$

$$\leq \sum_{j=1}^{q} \binom{q}{j} \omega_1\left(f; \frac{mja + n}{2^{k+r}}\right)_{\overline{\mathbb{D}}_1}$$

$$\leq (2^q - 1)\omega_1\left(f; \frac{mqa + n}{2^{k+r}}\right)_{\overline{\mathbb{D}}_1},$$

which proves (iii) too.

(iv) Let $|z_1 - z_2| \leq \delta$, $z_1, z_2 \in \overline{\mathbb{D}}_1$. We get

$$|Y_k(f)(z_1) - Y_k(f)(z_2)| \leq 2^k \int_{-\infty}^{+\infty} |f(z_1 e^{iv}) - f(z_2 e^{iv})| \varphi(-2^k v) \, dv$$

$$\leq \omega_1(f; |z_1 - z_2|)_{\overline{\mathbb{D}}_1} \leq \omega_1(f; \delta)_{\overline{\mathbb{D}}_1},$$

where from passing to supremum with $|z_1 - z_2| \leq \delta$, we obtain

$$\omega_1(Y_k(f) : \delta)_{\overline{\mathbb{D}}_1} \leq \omega(f; \delta)_{\overline{\mathbb{D}}_1}, \quad \forall \delta > 0, \ k \in \mathbb{Z}.$$

Then,

$$|L_{k,j}(f)(z_1) - L_{k,j}(f)(z_2)|$$

$$\leq \frac{2^k}{j} \int_{-\infty}^{+\infty} |\ell_k(f)(2^k z_1 e^{iv}) - \ell_k(f)(2^k z_2 e^{iv})| \varphi\left(-\frac{2^k}{j} v\right) dv$$

$$\left(\text{by } -\frac{2^k}{j} v = u\right)$$

$$\leq \int_{-\infty}^{+\infty} |\ell_0(f_k)\left(2^k z_1 e^{i\left(-\frac{j}{2^k} u\right)}\right) - \ell_0(f_k)\left(2^k z_2 e^{i\left(-\frac{j}{2^k} u\right)}\right)| \varphi(u) \, du$$

$$\leq \omega_1(f; |z_1 - z_2|)_{\overline{\mathbb{D}}_1} \leq \omega_1(f; \delta)_{\overline{\mathbb{D}}_1},$$

where from passing to supremum with $|z_1 - z_2| \leq \delta$, we obtain

$$\omega_1(L_{k,j}(f); \delta)_{\overline{\mathbb{D}}_1} \leq \omega_1(f; \delta)_{\overline{\mathbb{D}}_1}.$$

The inequality

$$\omega_1(I_{k,q}(f); \delta)_{\overline{\mathbb{D}}_1} \leq (2^q - 1)\omega_1(f; \delta)_{\overline{\mathbb{D}}_1}$$

follows immediately from the above inequality for $L_{k,j}$ and from the relation

$$\sum_{j=1}^{q} \binom{q}{j} = 2^q - 1,$$

which proves (iv).

(v) Let $z_0, z_n \in \overline{\mathbb{D}}_1$ be with $\lim_{n \to \infty} z_n = z_0$. We get (as in the proof of (iv))

$$|Y_k(f)(z_n) - Y_k(f)(z_0)| \leq \omega_1(f; |z_n - z_0|)_{\overline{\mathbb{D}}_1},$$

$$|L_{k,j}(f)(z_n) - L_{k,j}(f)(z_0)| \leq \omega_1(f; |z_n - z_0|)_{\overline{\mathbb{D}}_1},$$

$$|I_{k,q}(f)(z_n) - I_{k,q}(f)(z_0)| \leq \omega_1(f; |z_n - z_0|)_{\overline{\mathbb{D}}_1},$$

which proves the continuity of these operators in $\overline{\mathbb{D}}_1$.

It remains to prove that $Y_k(f)(z)$, $L_{k,j}(f)(z)$ and $I_{k,q}(f)(z)$ are analytic in $\mathbb{D}_1$. By hypothesis we have $f(z) = \sum_{p=0}^{\infty} a_p z^p$, $z \in \mathbb{D}_1$. Let $z \in \mathbb{D}_1$ be fixed. We get

$$f(ze^{iv}) = \sum_{p=0}^{\infty} a_p e^{ipv} z^p$$

and since $|a_p e^{ipv}| = |a_p|$ for all $v \in \mathbb{R}$ and the series $\sum_{p=0}^{\infty} a_p z^k$ is convergent, it follows that the series $\sum_{p=0}^{\infty} a_p e^{ipv} z^p$ is uniformly convergent with respect to $v \in \mathbb{R}$. This immediately implies that the series can be integrated term by term, that is

$$Y_k(f)(z) = \int_{-\infty}^{+\infty} f\left(ze^{i\left(-\frac{u}{2^k}\right)}\right) \varphi(u)\, du$$

$$= \sum_{p=0}^{\infty} a_p \left[\int_{-\infty}^{+\infty} e^{i\left(-\frac{pu}{2^k}\right)} \varphi(u)\, du\right] z^p$$

$$= \sum_{p=0}^{\infty} a_p \left[\int_{-\infty}^{+\infty} \cos\left(-\frac{pu}{2^k}\right) \varphi(u)\, du\right] z^p = \sum_{p=0}^{\infty} a_p b_{p,k} z^k,$$

since cos is even function.

If $\ell_0(f) \equiv f$ then $\ell_k(f)(2^k z e^{iv}) = f(z e^{iv})$ and we obtain

$$L_{k,j}(f)(z) = \frac{2^k}{j} \int_{-\infty}^{+\infty} f(z e^{iv}) \varphi\left(-\frac{2^k v}{j}\right) dv$$

and reasoning as for $Y_k(f)(z)$ we immediately obtain

$$L_{k,j}(f)(z) = \sum_{p=0}^{\infty} a_p b_{p,k,j} z^p, \quad z \in \mathbb{D}_1,$$

with

$$b_{k,p,j} = \int_{-\infty}^{+\infty} \cos\left(\frac{pju}{2^k}\right) \varphi(u)\, du.$$

The development for $I_{k,q}(f)(z)$ follows easily from above, which proves (v). For the particular choice of $\varphi(x)$, we have :

$$b_{j,k} = b_{1,k,j} = \int_{-1}^{0} \cos\left(\frac{ju}{2^k}\right) \cdot (1+u)\, du + \int_{0}^{1} \cos\left(\frac{ju}{2^k}\right) \cdot (1-u)\, du$$

$$= 2\int_{0}^{1} (1-u) \cos\frac{ju}{2^k}\, du = 2\left[\sin\left(\frac{ju}{2^k}\right) \cdot \frac{2^k}{j}\right]\Bigg|_{0}^{1} - 2\int_{0}^{1} u \cos\frac{ju}{2^k}\, du$$

$$= \frac{2^{k+1}}{j} \sin\frac{j}{2^k} - 2\left[\frac{2^{2k}}{j^2} \cos\frac{ju}{2^k} + \frac{2^k}{j} u \sin\frac{ju}{2^k}\right]\Bigg|_{0}^{1}$$

$$= \frac{2^{2k+1}}{j^2}\left(1 - \cos\frac{j}{2^k}\right) > 0, \quad \forall k \in \mathbb{Z},\ j \in \mathbb{N}.$$

For $j = 1$ we get $b_{1,k,1} := b_{1,k} = 2^{2k+1}\left(1 - \cos\frac{1}{2^k}\right) > 0$. Therefore,

$$c_{1,k,q} = \sum_{j=1}^{q} (-1)^{j+1} \binom{q}{j} b_{1,k,j} = 2^{2k+1} \sum_{j=1}^{q} (-1)^{j+1} \binom{q}{j} \frac{1}{j^2}\left(1 - \cos\frac{j}{2^k}\right),$$

and

$$c_{p,k,q} = \sum_{j=1}^{q} (-1)^{j+1} \binom{q}{j} b_{p,k,j} = \sum_{j=1}^{q} (-1)^{j+1} \binom{q}{j} b_{pj,k}$$

$$= \sum_{j=1}^{q} (-1)^{j+1} \binom{q}{j} \frac{2^{2k+1}}{p^2 j^2} \left(1 - \cos \frac{pj}{2^k}\right).$$

(vi) Analysing the proofs of the above points (i)-(v), it easily follows that they hold by replacing everywhere $\mathbb{D}_1$ with $\mathbb{D}_r$, where $1 \le r < R$. Therefore the approximation errors in (i)-(iii) can be expressed in terms of

$$\omega_1(f;\delta)_{\overline{\mathbb{D}}_r} = \sup\{|f(u) - f(v)|; |u - v| \le \delta\},$$

with constants in front of ω_1 depending on $r \ge 1$.

By the mean value theorem for divided differences in Complex Analysis (see e.g. Stancu [172], p. 258, Exercise 4.20) we get

$$\omega_1(f;\delta)_{\overline{\mathbb{D}}_r} \le \delta \|f'\|_r,$$

where

$$\|f'\|_r = \sup\{|f'(z)|; |z| \le r\} \le \sum_{k=1}^{\infty} |a_k| k r^{k-n-1}.$$

Then by (i), (ii) and (iii) it follows

$$\|Y_k(f) - f\|_r \le \frac{C_1}{2^k},$$

$$\|L_{k,j}(f) - f\|_r \le \frac{C_2}{2^{k+r}},$$

$$\|I_{k,j}(f) - f\|_r \le \frac{C_3}{2^{k+r}},$$

respectively.

Now denoting by γ the circle of radius $r_1 > 1$ and center 0, since for any $|z| \le r$ and $v \in \gamma$, we have $|v - z| \ge r_1 - r$, by the Cauchy's formulas it follows that for all $|z| \le r$ we have

$$|Y_k^{(s)}(f)(z) - f^{(q)}(z)| = \frac{s!}{2\pi} \left| \int_{\gamma} \frac{Y_k(f)(v) - f(v)}{(v - z)^{s+1}} dv \right|$$

$$\le \frac{C_1}{2^k} \cdot \frac{s!}{2\pi} \frac{2\pi r_1}{(r_1 - r)^{s+1}}$$

$$= \frac{C_1}{2^k} \cdot \frac{s! r_1}{(r_1 - r)^{s+1}},$$

which proves the upper estimate in approximation by $Y_k^{(s)}(f)(z)$.

Similarly we get the upper estimates in approximation by $L_{k,j}^{(s)}(f)(z)$ and $I_{k,q}^{(s)}(f)(z)$, which proves the theorem. $\qquad\square$

The geometric properties are expressed by the following results.

Theorem 3.2.28. (Anastassiou-Gal [19]) *It holds that $Y_k(\mathcal{P}) \subset \mathcal{P}$, $\forall k \in \mathbb{N}$,*

$$\frac{1}{b_{1,k}} Y_k(S_{3,b_{1,k}}) \subset S_3, \quad \frac{1}{b_{1,k}} Y_k(S_M) \subset S_{M/|b_{1,k}|} \quad \forall k \in \mathbb{Z}.$$

If $\ell_0(f) \equiv f$ then

$$L_{k,j}(\mathcal{P}) \subset \mathcal{P}, \quad \frac{1}{b_{1,k,j}} L_{k,j}(S_{3,b_{1,k,j}}) \subset S_3,$$

$$\frac{1}{b_{1,k,j}} L_{k,j}(S_M) \subset S_{M/|b_{1,k,j}|}, \quad \forall k \in \mathbb{Z}, \; j \in \mathbb{N}.$$

Here in all the cases we take $\varphi(x) = 1 - x$, $x \in [0,1]$, $\varphi(x) = 1 + x$, $x \in [-1,0]$, $\varphi(x) = 0$, $x \in \mathbb{R} \setminus (0,1)$.

Proof. Since from Theorem 3.2.27 (v) we have

$$b_{0,k} = b_{0,k,j} = \int_{-\infty}^{+\infty} \varphi(u) \, du = 1, \quad \forall k \in \mathbb{Z}, \; j \in \mathbb{N},$$

it follows $Y_k(f)(0) = L_{k,j}(f)(0) = a_0$, that is if $f \in \mathcal{P}$, $f = U + iV$ then $a_0 = 1$ and $U > 0$ on $\mathbb{D}_1$, which implies $Y_k(f)(0) = L_{k,j}(f)(0) = 1$,

$$\mathrm{Re}[Y_k(f)(z)] = 2^k \int_{-\infty}^{+\infty} U(r\cos(x+v), r\sin(x+v))\varphi(-2^k v) \, dv > 0,$$

$\forall z = re^{ix} \in \mathbb{D}_1$, and for $\ell_0(f) \equiv f$,

$$\mathrm{Re}[L_{k,j}(f)(z)] = \frac{2^k}{j} \int_{-\infty}^{+\infty} U(r\cos(x+v), r\sin(x+v))\varphi\left(-\frac{2^k v}{j}\right) dv > 0,$$

for all $z = re^{ix} \in \mathbb{D}_1$, that is $Y_k(\mathcal{P})$, $L_{k,j}(\mathcal{P}) \subset \mathcal{P}$.

Let $f(0) = f'(0) - 1 = 0$. From Theorem 3.2.27, (v) we get

$$\frac{1}{b_{1,k}} \cdot Y_k(f)(0) = \frac{1}{b_{1,k}} B'_k(f)(0) - 1 = 0$$

and if $\ell_0(f) \equiv f$ then

$$\frac{1}{b_{1,k,j}} L_{k,j}(f)(0) = \frac{1}{b_{1,k,j}} \cdot L'_{k,j}(f)(0) - 1 = 0.$$

Also, for $f \in S_{3,b_{1,k}}$ we get

$$\left| \frac{1}{b_{1,k}} Y''_k(f)(z) \right| \leq \frac{1}{|b_{1,k}|} 2^k \int_{-\infty}^{+\infty} |f''(ze^{iv}) e^{2iv}| \varphi(-2^k v) \, dv$$

$$\leq 2^k \int_{-\infty}^{+\infty} \varphi(-2^k v) \, dv = 1,$$

that is $\frac{1}{b_{1,k}} Y_k(f) \in S_3$, then for $f \in S_M$ it follows

$$\left| \frac{1}{b_{1,k}} Y'_k(f)(z) \right| \leq \frac{1}{|b_{1,k}|} 2^k \int_{-\infty}^{+\infty} |f'(ze^{iv}) e^{iv}| \varphi(-2^k v) \, dv < \frac{M}{|b_{1,k}|}, \quad z \in \mathbb{D}_1,$$

that is $\frac{1}{b_{1,k}} Y_k(f) \in S_{M/|b_{1,k}|}$.

The proof in the case of $L_{k,j}$ is similar, which proves the theorem. $\square$

Remarks. 1) From the proof of Theorem 3.2.28 we obtain the following geometric properties: if $f \in S_{3,b_{1,k}}$ then $Y_k(f)$ is starlike (and univalent) on $\mathbb{D}_1$, if $f \in S_M$ then $Y_k(f)$ is univalent in

$$\left\{ z \in \mathbb{C}; \ |z| < \frac{|b_{1,k}|}{M} \right\} \subset \left\{ z \in \mathbb{C}; \ |z| < \frac{1}{M} \right\},$$

and by Theorem 3.2.27 (v)

$$|b_{1,k}| \leq \int_{-\infty}^{+\infty} \left| \cos\left(\frac{u}{2^k} \right) \right| \varphi(u) \, du \leq \int_{-\infty}^{+\infty} \varphi(u) \, du = 1;$$

if $\ell_0(f) \equiv f$ then $f \in S_{3,b_{1,k,j}}$ implies that $L_{k,j}(f)$ is starlike (and univalent) on $\mathbb{D}_1$ and $f \in S_M$ implies that $L_{k,j}(f)$ is univalent in

$$\left\{ z \in \mathbb{C}; \ |z| < \frac{|b_{1,k,j}|}{M} \right\} \subset \left\{ z \in \mathbb{C}; \ |z| < \frac{1}{M} \right\},$$

since by Theorem 3.2.27 (v)

$$|b_{1,k,j}| \leq \int_{-\infty}^{+\infty} \left| \cos\left(\frac{pju}{2^k} \right) \right| \varphi(u) \, du \leq \int_{-\infty}^{+\infty} \varphi(u) \, du = 1.$$

2) Let $\ell_0(f) \equiv f$. If $c_{1,k,q} \neq 0$ then similarly we get

$$\frac{1}{c_{1,k,q}} I_{k,q}(S_M) \subset S_{M(2^q-1)/|c_{1,k,q}|},$$

that if $f \in S_M$ implies $I_{k,q}(f)$ is univalent in

$$\left\{ z \in \mathbb{C}; \ |z| < \frac{|c_{1,k,q}|}{M(2^q - 1)} \right\} \subset \left\{ z \in \mathbb{C}; \ |z| < \frac{1}{M} \right\},$$

since by Theorem 3.2.27 (v)

$$|c_{1,k,q}| = \left| \sum_{j=1}^{q} (-1)^{j+1} \binom{q}{j} b_{1,k,j} \right|$$

$$\leq \sum_{j=1}^{q} \binom{q}{j} |b_{1,k,j}| \leq \sum_{j=1}^{q} \binom{q}{j} = 2^q - 1.$$

3) For $\varphi(x) = 1 - x$, $\forall x \in [0,1]$, $\varphi(x) = 1 + x$, $\forall x \in [-1,0]$, $\varphi(x) = 0$, $x \in \mathbb{R} \backslash (0,1)$, let us consider

$$b_1 = \inf\{|b_{1,k}|; k \in \mathbb{N}\} = \inf \left\{ 2^{2k+1} \left(1 - \cos\frac{1}{2^k} \right); k \in \mathbb{N} \right\},$$

$$b_1^* = \inf\{|b_{1,k,j}|; k, j \in \mathbb{N}, j \leq 2^{k+1}\} \text{ and } c_{1,q} = \inf\{|c_{1,k,q}|; k \in \mathbb{N}\}.$$

We have:

$$|b_{1,k}| = 2^{2k+1} \left(1 - \cos\frac{1}{2^k} \right) = 2^{2k+2} \sin^2 \frac{1}{2^{k+1}} = \left(2^{k+1} \sin\frac{1}{2^{k+1}} \right)^2,$$

$$|b_{1,k,j}| = \frac{2^{2k+1}}{j^2} \left(1 - \cos\frac{j}{2^k} \right) = \frac{2^{2k+2}}{j^2} \sin^2 \frac{j}{2^{k+1}} = \left(\frac{2^{k+1}}{j} \cdot \sin\frac{j}{2^{k+1}} \right)^2,$$

which by the fact that $f(t) = t \sin \frac{1}{t}$ is increasing for $t \geq 1$, $f(1) = \sin 1$, implies

$$0 < b_1 = \left(4 \sin \frac{1}{4}\right)^2 = 16 \sin^2 \frac{1}{4}.$$

Also, since $1 \leq \frac{2^{k+1}}{j}$, $j = \overline{1, 2^{k+1}}$, we get $b_1^* = \sin^2 1$. Therefore, it is immediate the following.

Corollary 3.2.29. (Anastassiou-Gal [19]) (i) *If*

$$f \in A^*(\overline{\mathbb{D}}_1) = \{f \in A(\overline{\mathbb{D}}_1); f(0) = f'(0) - 1 = 0\},$$

$|f''(z)| \leq 16 \sin^2 \frac{1}{4}$, $\forall z \in \mathbb{D}_1$ *then* $Y_k(f) \in S_3$, *for all* $k \in \mathbb{N}$ *and if* $f \in S_M$, $M > 1$, *then* $Y_k(f)$ *is univalent in* $\left\{z \in \mathbb{C}; |z| < \frac{16 \sin^2 \frac{1}{4}}{M}\right\}$, *for all* $k \in \mathbb{N}$;
  (ii) *If* $f \in A^*(\overline{\mathbb{D}}_1)$, $|f''(z)| \leq \sin^2 1$, $\forall z \in \mathbb{D}_1$, *then* $L_{k,j}(f) \in S_3$ *and if* $f \in S_M$, $M > 1$, *then* $L_{k,j}(f)$ *is univalent in* $\left\{z \in \mathbb{C}; |z| < \frac{\sin^2 1}{M}\right\}$, *for all* $k, j \in \mathbb{N}$, $j \leq 2^{k+1}$.

Remarks. 1) It would be of interest to find other geometric properties of the operators Y_k, $L_{k,j}$ and $I_{k,q}$.
  2) Let $f \in A^*(\overline{\mathbb{D}}_1)$ and define $f_\alpha(z) := f(\alpha z)$ for all $\alpha, z \in \overline{\mathbb{D}}_1$. The operator Φ is called *rotation invariant* iff $\Phi(f_\alpha) = (\Phi(f))_\alpha$. We assume that

$$\ell_0(f(2^{-k}\bullet))(az) = \ell_0(f(2^{-k}\alpha\bullet))(z), \quad k \in \mathbb{Z},$$

a condition fulfilled trivially by Y_k operators, case of $\ell_0(f) = f$. Then easily one proves that $\ell_k(f_\alpha) = (\ell_k(f))_\alpha$ and $Y_k(f_\alpha) = (Y_k f)_\alpha$,

$$(L_{k,j}(f_\alpha)) = (L_{k,j}(f))_\alpha, \quad I_{k,q}(f_\alpha) = (I_{k,q}(f))_\alpha.$$

So all operators we are dealing with here are *rotation invariant*.

3.2.5 *Sikkema Complex Convolutions*

In a series of papers, Sikkema [164], [165], [166], [167] and Totik [193] studied the approximation properties of the convolution integral operators of real variable (for $\rho \to +\infty$),

$$U_\rho(f)(x) = \frac{1}{I_\rho} \int_{-\infty}^{+\infty} f(x - t)\beta^\rho(t)\, dt, \quad I_\rho = \int_{-\infty}^{+\infty} \beta^\rho(t)\, dt,$$

where $f, \beta : \mathbb{R} \to \mathbb{R}$ satisfy some suitable properties.
  In this subsection we study approximation and geometric properties of the complexified version of the above operators, given by

$$L_\rho(f)(z) = \frac{1}{I_\rho} \int_{-\infty}^{+\infty} f(ze^{-it})\beta^\rho(t)\, dt, \quad z \in \overline{\mathbb{D}}_r \quad f \in A(\overline{\mathbb{D}}_r), \quad \rho \geq 1, \quad r \geq 1,$$

where recall that

$$A(\overline{\mathbb{D}}_r) = \{f : \overline{\mathbb{D}}_r \to \mathbb{C}; f \text{ is analytic in } \mathbb{D}_r \text{ and continuous in } \overline{\mathbb{D}}_r\}.$$

Let us suppose that $\beta : \mathbb{R} \to \mathbb{R}$ satisfies the following five properties:

a) $\beta(t) \geq 0$, $\forall t \in \mathbb{R}$, $\beta(0) = 1$;
b) $\forall \delta > 0$, $\sup\{\beta(t); |t| \geq \delta\} < 1$;
c) $\beta(t)$ is continuous at 0;
d) $t^2\beta(t)$ is Lebesgue integrable over $\mathbb{R}$;
e) β is even on $\mathbb{R}$, i.e. $\beta(-t) = \beta(t)$, $\forall t \in \mathbb{R}$.

First we present:

Theorem 3.2.30. (Anastassiou-Gal [20]) *If $f(z) = \sum_{k=0}^{\infty} a_k z^k$ is analytic in $\mathbb{D}_1$ and continuous in $\overline{\mathbb{D}}_1$, then $L_\rho(f)$ is analytic in $\mathbb{D}_1$ and continuous in $\overline{\mathbb{D}}_1$, for all $\rho \geq 1$. Also, we can write*

$$L_\rho(f)(z) = \sum_{k=0}^{\infty} A_k(\rho)z^k, \quad z \in \mathbb{D}_1,$$

where

$$A_k(\rho) = \frac{a_k}{I_\rho} \cdot \int_{-\infty}^{+\infty} \cos(kt)\beta^\rho(t)\,dt, \quad k = 0, 1, 2, \ldots,$$

and

$$|A_k(\rho)| \leq |a_k|, \quad k = 0, 1, 2, \ldots.$$

Proof. Let $z_0, z_n \in \overline{\mathbb{D}}_1$ be such that $\lim_{n \to \infty} z_n = z_0$. We have

$$|L_\rho(f)(z_n) - L_\rho(f)(z_0)| \leq \frac{1}{I_\rho} \int_{-\infty}^{+\infty} |f(z_n e^{-it}) - f(z_0 e^{-it})|\beta^\rho(t)\,dt$$

$$\leq \frac{1}{I_\rho} \int_{-\infty}^{+\infty} \omega_1\big(f; |e^{-it}| \cdot |z_n - z_0|\big)_{\overline{\mathbb{D}}_1} \beta^\rho(t)\,dt$$

$$= \omega_1\big(f; |z_n - z_0|\big)_{\overline{\mathbb{D}}_1},$$

which proves the continuity of $L_\rho(f)$ in $\overline{\mathbb{D}}_1$.

Now, let $f(z) = \sum_{k=0}^{\infty} a_k z^k$, $z \in \mathbb{D}_1$, be analytic in $\mathbb{D}_1$. For fixed $z \in \mathbb{D}_1$, we get $f(ze^{-it}) = \sum_{k=0}^{\infty} a_k e^{-ikt} z^k$, and since $|a_k e^{-ikt}| = |a_k|$, for all $t \in \mathbb{R}$ and the series $\sum_{k=0}^{\infty} a_k z^k$ is convergent, it follows that the series $\sum_{k=0}^{\infty} a_k e^{-ikt} z^k$ is uniformly convergent with respect to $t \in \mathbb{R}$. This immediately implies that the series can be integrated term by term, that is

$$L_\rho(f)(z) = \sum_{k=0}^{\infty} a_k z^k \cdot \frac{1}{I_\rho} \int_{-\infty}^{+\infty} [\cos(kt) - i\sin(kt)]\beta^\rho(t)\,dt$$

$$= \sum_{k=0}^{\infty} a_k \left(\frac{1}{I_\rho} \int_{-\infty}^{+\infty} \cos(kt)\beta^\rho(t)\,dt\right) z^k.$$

Then, it is immediate that

$$|A_k| \leq |a_k| \cdot \frac{1}{I_\rho} \int_{-\infty}^{+\infty} |\cos(kt)|\beta^\rho(t)\,dt \leq |a_k|, \quad k = 0, 1, \ldots \qquad \square$$

Remark. In the rest of the section those particular choices of $\beta(t)$ for which $\int_{-\infty}^{+\infty}(\cos t)\beta^\rho(t)\,dt \neq 0$ will be important.

Concerning the approximation properties, we have

Theorem 3.2.31. (Anastassiou-Gal [20]) *If $f \in A(\overline{\mathbb{D}}_1)$ then*

$$|L_\rho(f)(z) - f(z)| \leq 2\big(1 + B_\rho(\delta)\big)\omega_1(f;\delta)_{\partial\mathbb{D}_1}, \quad \forall z \in \overline{\mathbb{D}}_1,\ \delta > 0,\ \rho \geq 1,$$

where

$$B_\rho(\delta) = \frac{1}{I_\rho} \int_{-\infty}^{+\infty} |t|\delta^{-1}\beta^\rho(t)\,dt.$$

Proof. By Theorem 3.2.30, $L_\rho(f)$ is analytic in $\mathbb{D}_1$ and continuous on $\overline{\mathbb{D}}_1$, $\forall \rho \geq 1$, $f \in A(\overline{\mathbb{D}}_1)$, so from the Maximum Modulus Principle, for the estimate $|L_\rho(f)(z) - f(z)|$ it suffices to take $|z| = 1$.

For $f \in A(\overline{\mathbb{D}}_1)$ and $|z| = 1$, we can write

$$f(z) = U(\cos u, \sin u) + iV(\cos u, \sin u), \quad \forall z = e^{iu} \in \partial\mathbb{D}_1.$$

Denoting $F(u) = U(\cos u, \sin u)$, $G(u) = V(\cos u, \sin u)$, $u \in \mathbb{R}$, by Sikkema [164], p. 356, Theorem 2, we get

$$|U_\rho(F)(u) - F(u)| \leq \big(1 + B_\rho(\delta)\big)\omega_1(F;\delta)_{\mathbb{R}},$$
$$|U_\rho(G)(u) - G(u)| \leq \big(1 + B_\rho(\delta)\big)\omega_1(G;\delta)_{\mathbb{R}}, \quad \forall u \in \mathbb{R},\ \delta > 0,\ \rho \geq 1.$$

But, for $|z| = 1$, $z = e^{iu}$, we have $L_\rho(f)(z) = U_\rho(F)(u) + iU_\rho(G)(u)$ and for

$$\omega_1(f;\delta)_{\partial\mathbb{D}_1} = \sup\{|f(e^{iu}) - f(e^{iv})|; u, v \in \mathbb{R}, |u - v| \leq \delta\},$$

it is easy to check the inequalities

$$\omega_1(F;\delta)_{\mathbb{R}} \leq \omega_1(f;\delta)_{\partial\mathbb{D}_1}, \quad \omega_1(G;\delta)_{\mathbb{R}} \leq \omega_1(f;\delta)_{\partial\mathbb{D}_1}, \quad \delta > 0,$$

since

$$|F(u) - F(v)| \leq |f(e^{iu}) - f(e^{iv})|, \quad |G(u) - G(v)| \leq |f(e^{iu}) - f(e^{iv})|.$$

In conclusion, for $|z| = 1$ we get

$$|L_\rho(f)(z) - f(z)| \leq 2\big(1 + B_\rho(\delta)\big)\omega_1(f;\delta)_{\partial\mathbb{D}_1}, \quad \forall \delta > 0,\ \rho \geq 1,$$

which proves the theorem. $\square$

Remarks. 1) In the upper estimate in Theorem 3.2.31 it is important to make the best choice (depending of course on β) for $\delta := \delta(\rho)$, in such a way that if $\rho \to \infty$ then $\delta(\rho) \to 0$ in the fastest possible way and at the same time $B_\rho(\delta)$ to remain bounded with respect to $\rho > 0$. Some well-known examples for $\beta(t)$ are as follows : $\beta(t) = e^{-t^2}$ (in this case $L_\rho(f)(z)$ is exactly $W_\xi^*(f)(z)$ studied by Theorem 3.2.8), $\beta(t) = e^{-|t|}$ (in this case $L_\rho(f)(z)$ is exactly $Q_\xi^*(f)(z)$ studied by Theorem 3.2.1), $\beta(t) = \cos(\pi t/2)$ if $|t| \leq 1$, $\beta(t) = 0$ if $|t| > 1$ (in this case $L_\rho(f)(z)$ is exactly $P_n(f)(z)$ studied by Theorems 3.1.1–3.1.3 and by Corollary 3.1.4) and

$\beta(t) = 1 - |t|^{2p}$ if $|t| \leq 1$, $\beta(t) = 0$ if $|t| > 1$ (in this case we would obtain a complex version of the Landau operator). Also, for other choices for $\beta(t)$ see e.g. Sikkema [164], pp. 358-359.

2) The lower estimate for approximation of f by $L_\rho(f)(z)$ can be obtained in a similar manner with the cases of the complex convolutions $W_\xi^*(f)(z)$ (in the proof of Theorem 3.2.8) or $Q_\xi^*(f)(z)$ (in the proof of Theorem 3.2.1). Let us sketch here the corresponding reasonings. Firstly let us observe that denoting $d_k(\rho) = \frac{1}{T_\rho} \int_{-\infty}^{\infty} \cos(kt) \cdot \beta^\rho(t)dt$ and writing $f(z) = \sum_{k=0}^{\infty} a_k z^k$ we get

$$L_\rho(f)(z) = \sum_{k=0}^{\infty} a_k d_k(\rho) z^k.$$

Now take $z = re^{i\varphi}$ and $p \in \mathbb{N} \bigcup \{0\}$. We have

$$\frac{1}{2\pi}[f(z) - L_\rho(f)(z)]e^{-ip\varphi} = \frac{1}{2\pi} \sum_{k=0}^{\infty} a_k r^k e^{i\varphi(k-p)}[1 - d_k(\rho)].$$

Integrating from $-\pi$ to π we obtain

$$\frac{1}{2\pi} \int_{-\pi}^{\pi} [f(z) - L_\rho(f)(z)]e^{-ip\varphi}d\varphi = a_p r^p[1 - d_p(\rho)].$$

Passing now to absolute value we easily obtain

$$|a_p|r^p|1 - d_p(\rho)| \leq \|f - L_\rho(f)\|_r.$$

Denote $V_\rho = \inf_{1 \leq p} |1 - d_p(\rho)|$. By the above lower estimate for $\|L_\rho(f) - f\|_r$, for all $p \geq 1$ and $\rho > 0$ it follows

$$\frac{\|L_\rho(f) - f\|_r}{V_\rho} \geq \frac{\|L_\rho(f) - f\|_r}{|1 - d_p(\rho)|} \geq |a_p|r^p.$$

This implies that if there exists a subsequence $(\rho_k)_k$ with $\lim_{k \to \infty} \rho_k = +\infty$ and such that $\lim_{k \to \infty} \frac{\|L_{\rho_k}(f) - f\|_r}{V_{\rho_k}} = 0$ then $a_p = 0$ for all $p \geq 1$, that is f is constant on $\overline{\mathbb{D}}_r$.

Therefore, if f is not a constant then $\inf_{\rho > 0} \frac{\|L_\rho(f) - f\|_r}{V_\rho} > 0$, which implies that there exists a constant $C_r(f) > 0$ such that $\frac{\|L_\rho(f) - f\|_r}{V_\rho} \geq C_r(f)$, for all $\rho > 0$, that is

$$\|L_\rho(f) - f\|_r \geq C_r(f)V_\rho, \text{ for all } \rho > 0.$$

Now, concerning the geometric properties, first we present

Theorem 3.2.32. (Anastassiou-Gal [20]) *Let* $f \in A(\overline{\mathbb{D}}_1)$ *and* $L_\rho(f)(z) = \sum_{k=0}^{\infty} A_k(\rho)z^k$. *Suppose that* $\beta(t)$ *is chosen such that* $A_1(\rho) \neq 0$, *for all* $\rho \geq 1$. *Then, for all* $\rho \geq 1$ *we have*

$$\frac{1}{A_1(\rho)}L_\rho\big(S_{3,A_1(\rho)}\big) \subset S_3, \quad \frac{1}{A_1(\rho)}L_\rho(S_M) \subset S_{M/|A_1(\rho)|},$$

where $M > 1$ and recall that

$$S_3 = \{f \in A(\overline{\mathbb{D}}_1); |f''(z)| \le 1, \ \forall z \in \mathbb{D}_1\},$$
$$S_{3,A_1(\rho)} = \{f \in A(\overline{\mathbb{D}}_1); |f''(z)| \le |A_1(\rho)|, \ \forall z \in \mathbb{D}_1\},$$
$$S_B = \{f \in A(\overline{\mathbb{D}}_1); |f'(z)| < B, \ \forall z \in \mathbb{D}_1\}.$$

Proof. Let $f \in A(\overline{\mathbb{D}}_1)$, $f(z) = \sum_{k=0}^{\infty} a_k z^k$, $z \in \mathbb{D}_1$. It follows $a_0 = 0$, $a_1 = 1$, which by Theorem 3.2.30 immediately implies

$$\frac{1}{A_1(\rho)} \cdot L_\rho(f)(0) = 0, \quad \frac{1}{A_1(\rho)} L'_\rho(f)(0) = \frac{A_1(\rho)}{A_1(\rho)} = 1,$$

i.e.

$$\frac{1}{A_1(\rho)} L_\rho(f) \in A(\overline{\mathbb{D}}_1).$$

Then, by

$$\frac{1}{A_1(\rho)} L'_\rho(f)(z) = \frac{1}{I_\rho} \int_{-\infty}^{+\infty} e^{-it} f'(ze^{-it}) \beta^\rho(t) \, dt, \quad z \in \mathbb{D}_1,$$

and

$$\frac{1}{A_1(\rho)} L''_\rho(f)(z) = \frac{1}{I_\rho} \int_{-\infty}^{+\infty} e^{-2it} f''(ze^{-it}) \beta^\rho(t) \, dt, \quad z \in \mathbb{D}_1,$$

we get:

$f \in S_{3,A_1(\rho)}$ implies $|f''(z)| \le |A_1(\rho)|$, $\forall z \in \mathbb{D}_1$, i.e.

$$\frac{1}{|A_1(\rho)|} \cdot |L''_\rho(f)(z)| \le \frac{1}{|A_1(\rho)|} \frac{1}{I_\rho} \int_{-\infty}^{+\infty} |f''(ze^{-it}) e^{-2it}| \beta^\rho(t) \, dt \le 1, \ \forall z \in \mathbb{D}_1$$

and

$f \in S_M$ implies $|f'(z)| < M$, $\forall z \in \mathbb{D}_1$, i.e.

$$\frac{1}{|A_1(\rho)|} \cdot |L'_\rho(f)(z)| \le \frac{1}{|A_1(\rho)|} \frac{1}{I_\rho} \int_{-\infty}^{+\infty} |f'(ze^{-it}) e^{-it}| \beta^\rho(t) \, dt < \frac{M}{|A_1(\rho)|},$$

which proves the theorem. □

Remarks. 1) It is known (see e.g. Obradovici [145]) that $f \in S_3$ implies that f is starlike (and univalent) in $\mathbb{D}_1$ and that $f \in S_M$ implies that f is univalent in $\{z \in \mathbb{C}; |z| < \frac{1}{M}\} \subset \mathbb{D}_1$ (see e.g. Mocanu-Bulboacă-Sălăgean [138], p. 111, Exercise 5.4.1).

Since by Theorem 3.2.30 we have $|A_1(\rho)| \le |a_1| = 1$, it follows that $S_{3,A_1(\rho)} \subset S_3$ and $\frac{M}{|A_1(\rho)|} \ge M > 1$, i.e. if $f \in S_{3,A_1(\rho)}$ then $L_\rho(f)$ remains starlike (and univalent) in $\mathbb{D}_1$ and if $f \in S_M$ then $L_\rho(f)$ is univalent in

$$\left\{z \in \mathbb{C}; |z| < \frac{|A_1(\rho)|}{M}\right\} \subset \left\{z \in \mathbb{C}; |z| < \frac{1}{M}\right\}.$$

2) Denote $A = \inf\{|A_1(\rho)|; \rho \geq 1\}$. If $A > 0$, then by Theorem 3.2.32 we get the following invariant geometric properties: if $f \in S_{3,A}$ then $L_\rho(f) \in S_3$ for all $\rho \geq 1$ and if $f \in S_M$, $M > 1$, then $L_\rho(f)$ is univalent in

$$\left\{ z \in \mathbb{C}; |z| < \frac{A}{M} \right\}, \quad \forall \rho \geq 1.$$

Therefore, would remain to calculate A and to check that $A > 0$ for various choices of $\beta(t)$, problems which are left to the reader as open questions.

The second result concerning geometric properties of $L_\rho(f)$ is the following.

Theorem 3.2.33. (Anastassiou-Gal [20]) *Let $f \in A(\overline{\mathbb{D}_r})$, $r > 1$, and suppose that $\beta(t)$ is such that for any bounded $g : \mathbb{R} \to \mathbb{R}$, we have $\lim_{\rho\to\infty} U_\rho(g) = g$, uniformly in any compact interval of $\mathbb{R}$.*

(i) If f is starlike in $\overline{\mathbb{D}}_1$ (that is, $Re\left(\frac{zf'(z)}{f(z)}\right) > 0$, for all $z \in \overline{\mathbb{D}}_1$), then there exists $\rho_0 > 0$ (depending on f), such that for all $\rho \geq \rho_0$, $L_\rho(f)(z)$ are starlike in $\overline{\mathbb{D}}_1$.

If f is starlike only in $\mathbb{D}_1$, then for any disk of radius $0 < \lambda < 1$ denoted by $\mathbb{D}_\lambda$, there exists ρ_0 (depending on f and $\mathbb{D}_\lambda$), such that for all $\rho \geq \rho_0$, $L_\rho(f)(z)$ are starlike in $\overline{\mathbb{D}}_\lambda$ (that is, $Re\left(\frac{zL'_\rho(f)(z)}{L_\rho(f)(z)}\right) > 0$, for all $z \in \overline{\mathbb{D}}_\lambda$).

(ii) If f is convex in $\overline{\mathbb{D}}_1$ (that is, $Re\left(\frac{zf''(z)}{f'(z)}\right) + 1 > 0$, for all $z \in \overline{\mathbb{D}}_1$), then there exists ρ_0 (depending on f), such that for all $\rho \geq \rho_0$, $L_\rho(f)(z)$ are convex in $\overline{\mathbb{D}}_1$.

If f is convex only in $\mathbb{D}_1$, then for any disk of radius $0 < \lambda < 1$ denoted by $\mathbb{D}_\lambda$, there exists ρ_0 (depending on f and $\mathbb{D}_\lambda$), such that for all $\rho \geq \rho_0$, $L_\rho(f)(z)$ are convex in $\overline{\mathbb{D}}_\lambda$ (that is, $Re\left(\frac{zL''_\rho(f)(z)}{L'_\rho(f)(z)}\right) + 1 > 0$, for all $z \in \overline{\mathbb{D}}_\lambda$).

Proof. First let us make some general useful considerations. By hypothesis, it follows that for $\rho \to \infty$, we have $L_\rho(f)(z) \to f(z)$, uniformly in any compact disk included in $\mathbb{D}_r$, that is in $\overline{\mathbb{D}}_1$ too. Indeed, this is immediate from the relationship $L_\rho(f)(z) = U_\rho(F_\lambda)(u) + iU_\rho(G_\lambda)(u)$, where $F_\lambda(u) = U(\lambda cos(u), \lambda sin(u))$, $G_\lambda(u) = V(\lambda cos(u), \lambda sin(u))$, $f(z) = U(x, y) + iV(x, y)$, $|z| = \lambda \in [0, r), z = \lambda e^{iu} = x + iy$, $u \in [0, 2\pi]$.

By the well-known Weierstrass' result, this implies that $L'_\rho(f)(z) \to f'(z)$ and $L''_\rho(f)(z) \to f''(z)$, uniformly in any compact disk in $\mathbb{D}_r$ and therefore in $\overline{\mathbb{D}}_1$ too, when $\rho \to \infty$.

Then, denoting by $A_1(\rho)$ the coefficient of z in the Taylor series in Theorem 3.2.30 representing the analytic function $L_\rho(f)(z)$, since $A_1(\rho) = L'_\rho(0)$ and $\lim_{\rho\to\infty} L'_\rho(0) = f'(0) = 1$, it follows that $\lim_{\rho\to\infty} A_1(\rho) = 1$ and for all $\rho \geq \rho_0$ we have $A_1(\rho) > 0$.

Let us denote $P_\rho(f)(z) = \frac{L_\rho(f)(z)}{A_1(\rho)}$, for all $\rho \geq \rho_0$.

By $f(0) = f'(0) - 1 = 0$ we get $P_\rho(f)(0) = \frac{f(0)}{A_1(\rho)} = 0$ and $P'_\rho(f)(0) = \frac{L'_\rho(f)(0)}{A_1(\rho)} = 1$. Also, we obviously have $P_\rho(f)(z) \to f(z)$, $P'_\rho(f)(z) \to f'(z)$ and $P''_\rho(f)(z) \to f''(z)$, uniformly in $\overline{\mathbb{D}}_1$.

(i) By hypothesis we get $|f(z)| > 0$ for all $z \in \overline{\mathbb{D}}_1$ with $z \neq 0$, which from the univalence of f in $\mathbb{D}_1$, implies that we can write $f(z) = zg(z)$, with $g(z) \neq 0$, for all $z \in \overline{\mathbb{D}}_1$, where g is analytic in $\mathbb{D}_1$ and continuous in $\overline{\mathbb{D}}_1$.

Write $P_\rho(f)(z)$ in the form $P_\rho(f)(z) = zQ_\rho(f)(z)$.

Let $|z| = 1$. We have

$$|f(z) - P_\rho(f)(z)| = |z| \cdot |g(z) - Q_\rho(f)(z)| = |g(z) - Q_\rho(f)(z)|,$$

which by the uniform convergence in $\overline{\mathbb{D}}_1$ of $P_\rho(f)$ to f and by the Maximum Modulus Principle, implies the uniform convergence in $\overline{\mathbb{D}}_1$ of $Q_\rho(f)(z)$ to $g(z)$, as $\rho \to \infty$.

Since g is continuous in $\overline{\mathbb{D}}_1$ and $|g(z)| > 0$ for all $z \in \overline{\mathbb{D}}_1$, there exist ρ_0 and $a > 0$ depending on g, such that $|Q_\rho(f)(z)| > a > 0$, for all $z \in \overline{\mathbb{D}}_1$ and all $\rho \geq \rho_0$.

Also, for all $|z| = 1$, we have

$$
\begin{aligned}
|f'(z) - P'_\rho(f)(z)| &= |z[g'(z) - Q'_\rho(f)(z)] + [g(z) - Q_\rho(f)(z)]| \\
&\geq |\ |z| \cdot |g'(z) - Q'_\rho(f)(z)| - |g(z) - Q_\rho(f)(z)|\ | \\
&= |\ |g'(z) - Q'_\rho(f)(z)| - |g(z) - Q_\rho(f)(z)|\ |,
\end{aligned}
$$

which from the Maximum Modulus Principle, the uniform convergence of $P'_\rho(f)$ to f' and of $Q_\rho(f)$ to g, evidently implies the uniform convergence of $Q'_\rho(f)$ to g', as $\rho \to \infty$.

Then, for $|z| = 1$, we get

$$
\begin{aligned}
\frac{zP'_\rho(f)(z)}{P_\rho(f)} &= \frac{z[zQ'_\rho(f)(z) + Q_\rho(f)(z)]}{zQ_\rho(f)(z)} \\
&= \frac{zQ'_\rho(f)(z) + Q_\rho(f)(z)}{Q_\rho(f)(z)} \to \frac{zg'(z) + g(z)}{g(z)} \\
&= \frac{f'(z)}{g(z)} = \frac{zf'(z)}{f(z)},
\end{aligned}
$$

which again from the Maximum Modulus Principle, implies

$$\frac{zP'_\rho(f)(z)}{P_\rho(f)} \to \frac{zf'(z)}{f(z)}, \text{ uniformly in } \overline{\mathbb{D}}_1.$$

Since $Re\left(\frac{zf'(z)}{f(z)}\right)$ is continuous in $\overline{\mathbb{D}}_1$, there exists $\alpha \in (0,1)$, such that

$$Re\left(\frac{zf'(z)}{f(z)}\right) \geq \alpha, \text{ for all } z \in \overline{\mathbb{D}}_1.$$

Therefore

$$Re\left[\frac{zP'_\rho(f)(z)}{P_\rho(f)(z)}\right] \to Re\left[\frac{zf'(z)}{f(z)}\right] \geq \alpha > 0,$$

uniformly in $\overline{\mathbb{D}}_1$, i.e. for any $0 < \beta < \alpha$, there is ρ_0 such that for all $\rho \geq \rho_0$ we have

$$Re\left[\frac{zP'_\rho(f)(z)}{P_\rho(f)(z)}\right] > \beta > 0, \text{ for all } z \in \overline{\mathbb{D}}_1.$$

Since $P_\rho(f)(z)$ differs from $L_\rho(f)(z)$ only by a constant, this proves the first part in (i).

For the second part, the proof is identical with the first part, with the only difference that instead of $\overline{\mathbb{D}}_1$, we reason for $\overline{\mathbb{D}}_\lambda$.

(ii) For the first part, by hypothesis there is $\alpha \in (0,1)$, such that

$$Re\left[\frac{zf''(z)}{f'(z)}\right] + 1 \ge \alpha > 0,$$

uniformly in $\overline{\mathbb{D}}_1$. It is not difficult to show that this is equivalent with the fact that for any $\beta \in (0, \alpha)$, the function $zf'(z)$ is starlike of order β in $\overline{\mathbb{D}}_1$ (see e.g. Mocanu-Bulboacă-Sălăgean [138], p. 77), which implies $f'(z) \ne 0$, for all $z \in \overline{\mathbb{D}}_1$, i.e. $|f'(z)| > 0$, for all $z \in \overline{\mathbb{D}}_1$. Also, by the same type of reasonings as those from the above point (i), we get

$$Re\left[\frac{zP_\rho''(f)(z)}{P_\rho'(f)(z)}\right] + 1 \to Re\left[\frac{zf''(z)}{f'(z)}\right] + 1 \ge \alpha > 0,$$

uniformly in $\overline{\mathbb{D}}_1$. As a conclusion, for any $0 < \beta < \alpha$, there is $\rho_0 > 0$ depending on f, such that for all $\rho \ge \rho_0$ we have

$$Re\left[\frac{zP_\rho''(f)(z)}{P_\rho'(f)(z)}\right] + 1 > \beta > 0, \text{ for all } z \in \overline{\mathbb{D}}_1.$$

The proof of second part in (ii) is similar, which proves the theorem. $\qquad\square$

3.3 Nonlinear Complex Convolutions

Geometric properties of nonlinear integral transforms generated by the (nonlinear) Hadamard product (of $\frac{z^2}{f(z)}$) with special functions of hypergeometric or of Hurwitz-type, were obtained in several papers, see e.g. Ponnusamy-Singh-Vasundhra [153], Ponnusamy-Vasundhra [154].

In this section we move in a different direction. Thus, we present convergence, shape preserving results and rate of approximation of analytic functions by some nonlinear complex integral convolution operators, related with the linear convolutions studied in the previous two sections.

More exactly, we present approximation and shape preserving properties (that is preservation of univalence, starlikeness, convexity, etc) for the family of nonlinear complex convolution operators of the form

$$T_\xi(f)(z) = \int_{-\pi}^{\pi} L_\xi(t) H_\xi[f(ze^{-it})]dt, \xi > 0, z \in \overline{\mathbb{D}}_1,$$

where $i^2 = -1$, f is analytic in the open unit disk $\mathbb{D}_1$ and continuous in $\overline{\mathbb{D}}_1$, $L_\xi : \mathbb{R} \to \mathbb{R}$ and $H_\xi : \mathbb{C} \to \mathbb{C}$.

The results in this section extends those of the real case in Angeloni-Vinti [28], [29], with the addition that while the idea of shape preservation is absent in the real case, in the complex case it is present.

First we recall some notations. For $\mathbb{D}_1 = \{\in \mathbb{C}; |z| < 1\}$, $\mathbb{D}_r = \{z \in \mathbb{C}; |z| < r\}$, $C_r = \{z \in \mathbb{C}; |z| = r\}$, $0 < r < 1$, let us consider

$$A(\overline{\mathbb{D}}_1) = \{f : \overline{\mathbb{D}}_1 \to \mathbb{C}; f \text{ is continuous in } \overline{\mathbb{D}}_1 \text{ and analytic in } \mathbb{D}_1\},$$

and the total variation of $f \in A(\overline{\mathbb{D}}_1)$ on C_r, by $V_r(f) = V_{[-\pi,\pi]}(F_r)$, where $F_r(u) = f(re^{iu})$, $u \in [-\pi, \pi]$ and $V_{[-\pi,\pi]}$ denotes the (usual) total variation of F_r on $[-\pi, \pi]$.

Since for $f \in A(\overline{\mathbb{D}}_1)$ and $r \in (0,1)$, the modulus of derivative $f'(z)$ is bounded on $\overline{\mathbb{D}}_r$, by the classical mean value theorem (inequality) in complex analysis it easily follows that $V_r(f) < \infty$, for any $r \in (0,1)$.

Denote by $L^1_{2\pi}$, the class of all 2π-periodic functions $f : \mathbb{R} \to \mathbb{R}$, which are Lebesgue integrable over $[-\pi, \pi]$. Everywhere in the section let us consider the family of kernel functions $K_\xi : \mathbb{R} \times \mathbb{C} \to \mathbb{C}$, $\xi > 0$, of the form $K_\xi(t, u) = L_\xi(t) H_\xi(u)$, with $H_\xi : \mathbb{C} \to \mathbb{C}$ and $L_\xi : \mathbb{R} \to \mathbb{R}$ are 2π-periodic functions, such that, in addition they satisfy the following properties :

1) L_ξ is Lebesgue measurable, $L_\xi \in L^1_{2\pi}$, $\|L_\xi\|_{L^1} \le A$ for all $\xi > 0$ and $\lim_{\xi \to \infty} \int_{-\pi}^{\pi} L_\xi(t) dt = 1$;

2) $\lim_{\xi \to \infty} \int_{\delta \le |t| \le \pi} |L_\xi(t)| dt = 0$, for any fixed $\delta > 0$;

3) $H_\xi(z), \xi > 0$, are entire functions with $H_\xi(0) = 0$, for all $\xi > 0$ and $\lim_{\xi \to \infty} H_\xi(u) = u$, for all $u \in \mathbb{C}$;

4) For every $M > 0$, there exists $K_M > 0$ (independent of ξ), such that such that

$$|H_\xi(u) - H_\xi(v)| \le K_M |u - v|, \text{ for all } u, v \in \overline{\mathbb{D}}_M, \xi > 0,$$

where $\mathbb{D}_M = \{z \in \mathbb{C}; |z| \le M\}$.

Remark. A simple example for $H_\xi(u)$ satisfying the above conditions is given by $H_\xi(u) = u + \frac{1}{\xi+1} g(u)$, where $g : \mathbb{C} \to \mathbb{C}$ is an entire function satisfying $g(0) = 0$. Indeed, since g is entire function, for any $M > 0$ it has bounded derivative on $\overline{\mathbb{D}}_M$ and let us denote that bound by B_M. By the classical mean value inequality in complex analysis it follows

$$|H_\xi(u) - H_\xi(v)| \le |u - v| + \frac{1}{1 + \xi} |g(u) - g(v)|$$

$$\le (1 + B_M)|u - v|, \text{ for all } u, v \in \overline{\mathbb{D}}_M, \xi > 0,$$

which means that condition 4) is satisfied.

Also, since $H_\xi(u) - u = \frac{1}{\xi+1} g(u)$, this immediately implies 3) too.

The first result shows the invariance property of the nonlinear convolutions $T_\xi, \xi > 0$.

Theorem 3.3.1. (Gal [88]) *Suppose that the above conditions 1)-4) are fulfilled. If $f \in A(\overline{\mathbb{D}}_1)$ then $T_\xi(f) \in A(\overline{\mathbb{D}}_1)$, for all $\xi > 0$.*

Proof. Fix $\xi > 0$. First we show that $T_\xi(f)$ is differentiable in any $z_0 \in \mathbb{D}_1$. Indeed, it is clear that there exists $r < 1$ such that $z_0 \in \mathbb{D}_r$. Let $z_n \to z_0$, as

$n \to \infty$, $z_n \neq z_0$, for all $n \in \mathbb{N}$. Obviously, without loss of generality, we can suppose that $z_n \in \mathbb{D}_r$, for all $n \in \mathbb{N}$. We get

$$\frac{T_\xi(f)(z_n) - T_\xi(f)(z_0)}{z_n - z_0} = \int_{-\pi}^{\pi} L_\xi(t) \left\{ \frac{H_\xi[f(z_n e^{-it})] - H_\xi[f(z_0 e^{-it})]}{z_n - z_0} \right\} dt.$$

Denoting now

$$F_n(t) = \frac{H_\xi[f(z_n e^{-it})] - H_\xi[f(z_0 e^{-it})]}{z_n - z_0},$$

since H is entire function and $f \in A(\overline{\mathbb{D}}_1)$, it is immediate that for each $t \in [-\pi, \pi]$, we have

$$\lim_{n \to \infty} F_n(t) = H_\xi'[f(z_0 e^{-it})] \cdot f'(z_0 e^{-it})[e^{-it}].$$

This show that pointwise for $t \in [-\pi, \pi]$, we have

$$\lim_{n \to \infty} L(t)_\xi F_n(t) = L_\xi(t) H_\xi'[f(z_0 e^{-it})] \cdot f'(z_0 e^{-it})[e^{-it}].$$

On the other hand, by the property 4) and by the classical mean value inequality in complex analysis, there is $\eta \in \overline{\mathbb{D}}_r$, such that

$$|L_\xi(t) \cdot F_n(t)| \leq |L_\xi(t)| \cdot \left| \frac{H_\xi[f(z_n e^{-it})] - H_\xi[f(z_0 e^{-it})]}{z_n - z_0} \right| \leq$$

$$|L_\xi(t)| \cdot M_f \cdot \left| \frac{f(z_n e^{-it}) - f(z_0 e^{-it})}{z_n - z_0} \right| \leq |L_\xi(t)| \cdot M_f \cdot |f'(\eta)| \leq |L_\xi(t)| \cdot M_f \cdot \|f'\|_{\overline{\mathbb{D}}_r}.$$

By the Lebesgue's dominated convergence theorem, we can pass to limit under the integral sign, which shows that $\lim_{n \to \infty} \frac{T_\xi(f)(z_n) - T_\xi(f)(z_0)}{z_n - z_0}$ exists, i.e. $T_\xi(f)(z)$ is analytic in $\mathbb{D}_1$.

For the continuity at $z_0 \in \overline{\mathbb{D}}_1$, let $z_n \in \overline{\mathbb{D}}_1$, with $z_n \to z_0$ as $n \to \infty$. We have

$$|T_\xi(f)(z_n) - T_\xi(f)(z_0)| = \left| \int_{-\pi}^{\pi} L_\xi(t)[H_\xi(f(z_n e^{-it})) - H_\xi(f(z_0 e^{-it}))] dt \right|$$

$$\leq \int_{-\pi}^{\pi} |L_\xi(t)| \cdot |f(z_n e^{-it}) - f(z_0 e^{-it})| dt,$$

which by the continuity of f at $z_0 \in \overline{\mathbb{D}}_1$, immediately implies the continuity of $T_\xi(f)$ too at z_0. $\qquad \square$

The next result represent an estimation in variation for the nonlinear convolutions T_ξ, $\xi > 0$.

Theorem 3.3.2. (Gal [88]) *Suppose that the above conditions 1)-4) are fulfilled. Then, for any $f \in A(\overline{\mathbb{D}}_1)$, there exists a constant $C_f > 0$ (depending only on f), such that*

$$V_r[T_\xi(f)] \leq C_f V_r[f], \text{ for all } 0 < r < 1, \xi > 0.$$

Proof. For $0 < r < 1$, $f \in A(\overline{\mathbb{D}}_1)$ and $\xi > 0$, denote $T_{r,\xi}(u) = T_\xi(f)(re^{iu})$. Also, let $s_0 = -\pi < s_1 < ... < s_n = \pi$ be a partition of $[-\pi, \pi]$ and denote $z_j = re^{is_j}, j = 0, ..., n$.

Denoting by $M_f > 0$, the smallest positive constant such that $f[\overline{\mathbb{D}}_1] \subset \overline{\mathbb{D}}_{M_f}$ and applying property 4), we get

$$\sum_{j=1}^n |T_{r,\xi}(s_j) - T_{r,\xi}(s_{j-1})|$$

$$\leq \sum_{j=1}^n \int_{-\pi}^\pi |L_\xi(t)| \cdot |H_\xi[f(z_j e^{-it})] - H_\xi[f(z_{j-1}e^{-it})]| dt$$

$$\leq M_f \int_{-\pi}^\pi |L_\xi(t)| \sum_{j=1}^n |f(z_j e^{-it}) - f(z_{j-1}e^{-it})| dt \leq A \cdot M_f V_r(f),$$

which proves the theorem. $\qquad\qquad\qquad\qquad\qquad\qquad\qquad\qquad\qquad\qquad$ $\square$

Concerning the convergence, it holds

Theorem 3.3.3. (Gal [88]) *Suppose that the above conditions 1)-4) are fulfilled and* $f \in A(\overline{\mathbb{D}}_1)$. *Then* $\lim_{\xi \to \infty} T_\xi(f)(z) = f(z)$, *uniformly in any compact disk included in* $\mathbb{D}_1$.

Proof. We can write

$$T_\xi(f)(z) - f(z) = \int_{-\pi}^\pi L_\xi(t) H_\xi[f(ze^{-it})] dt - f(z)$$

$$= \int_{-\pi}^\pi L_\xi(t)\{H_\xi[f(ze^{-it})] - H_\xi[f(z)]\} dt$$

$$+ \int_{-\pi}^\pi L_\xi(t)[H_\xi[f(z)] - f(z)] dt$$

$$+ f(z) \left[\int_{-\pi}^\pi L_\xi(t) dt - 1 \right] := E_1 + E_2 + E_3.$$

We will estimate all $E_k, k = 1, 2, 3$. For this purpose, let $z \in \mathbb{D}_1$ and take $\varepsilon > 0$ arbitrary small, fixed. First, we note that there exists $r < 1$ such that $z \in \mathbb{D}_r$.

We have

$$|E_3| \leq |f(z)| \cdot \left| \int_{-\pi}^\pi L_\xi(t) dt - 1 \right|,$$

which by condition 1) implies that there exist $\xi_3 > 0$, such that

$$|E_3| \leq \varepsilon \cdot \|f\|_{\overline{\mathbb{D}}_1}, \text{ for all } \xi > \xi_3.$$

Also,

$$|E_2| \leq \int_{-\pi}^\pi |L_\xi(t)| \cdot |H_\xi[f(z)] - f(z)| dt,$$

which by the conditions 3) and 1), implies that there exists $\xi_2 > 0$, such that for all $\xi > \xi_2$ we have

$$|E_2| \leq \varepsilon \cdot \int_{-\pi}^{\pi} |L_\xi(t)| \leq \varepsilon A.$$

Finally, by the conditions 4) and 2) we get

$$|E_1| \leq \int_{-\pi}^{\pi} |L_\xi(t)| \cdot |H_\xi[f(ze^{-it})] - H_\xi[f(z)]| dt$$

$$\leq M_f \int_{-\pi}^{\pi} |L_\xi(t)| \cdot |f(ze^{it}) - f(z)| dt$$

$$\leq M_f \int_{-\pi}^{\pi} |L_\xi(t)| \omega_1(f; |z| \cdot |e^{it} - 1|)_{\overline{\mathbb{D}}_r} dt$$

$$\leq M_f \int_{-\pi}^{\pi} |L_\xi(t)| \omega_1(f; 2|z| \cdot |sin(t/2)|)_{\overline{\mathbb{D}}_r} dt$$

$$\leq M_f \int_{-\pi}^{\pi} |L_\xi(t)| \omega_1(f; |t|)_{\overline{\mathbb{D}}_r} dt$$

$$\leq M_f \int_{-\varepsilon}^{\varepsilon} |L_\xi(t)| \cdot \omega_1(f; \varepsilon)_{\overline{\mathbb{D}}_r} dt + M_f \int_{\varepsilon \leq |t| \leq \pi} |L_\xi(t)| \cdot 2\|f\|_{\overline{\mathbb{D}}_1} dt$$

$$\leq M_f \cdot A \cdot \omega_1(f; \varepsilon)_{\overline{\mathbb{D}}_r} + 2M_f \cdot \|f\|_{\overline{\mathbb{D}}_1} \int_{\varepsilon \leq |t| \leq \pi} |L_\xi(t)| dt$$

$$\leq M_f \cdot A \cdot \varepsilon \|f'\|_{\overline{\mathbb{D}}_r} + 2M_f \|f\|_{\overline{\mathbb{D}}_1} \varepsilon, \text{ for all } \xi > \xi_1.$$

Here

$$\omega_1(f; \varepsilon)_{\overline{\mathbb{D}}_r} := sup\{|f(u) - f(v)|; u, v \in \overline{\mathbb{D}}_r, |u - v| \leq \varepsilon\},$$

and $\| \cdot \|_{\overline{\mathbb{D}}_M}$ denotes the uniform norm in $\overline{\mathbb{D}}_M$. Collecting all the above estimates and denoting $\xi_0 = max\{\xi_1, \xi_2, \xi_3\}$, for all $\xi > \xi_0$ we get

$$|T_\xi(f)(z) - f(z)| \leq |E_1| + |E_2| + |E_3|$$
$$\leq M_f \cdot A \cdot \varepsilon \|f'\|_{\overline{\mathbb{D}}_r} + 2M_f \|f\|_{\overline{\mathbb{D}}_1} \varepsilon + \varepsilon A + \varepsilon \cdot \|f\|_{\overline{\mathbb{D}}_1},$$

which implies that for all $z \in \mathbb{D}_1$, we have (pointwise)

$$\lim_{\xi \to \infty} T_\xi(f)(z) = f(z).$$

On the other hand, we will show that $(T_\xi(f)(z))_{\xi > 0}$ is uniformly bounded (that is independent of ξ) on each $\overline{\mathbb{D}}_r$. Indeed, taking into account that $H_\xi(0) = 0$ and taking $v = 0$ in condition 4), for all $\xi > 0$ and $|z| \leq r$ we obtain

$$|T_\xi(f)(z)| \leq$$

$$\int_{-\pi}^{\pi} |L_\xi(t)| \cdot |H_\xi[f(ze^{-it})]| dt \leq K_f \int_{-\pi}^{\pi} |L_\xi(t)| \cdot |f(ze^{-it})| dt \leq K_f \cdot A \cdot \|f\|_{\overline{\mathbb{D}}_r}.$$

Since by Theorem 3.3.1, it follows $T_\xi(f) \in A(\overline{\mathbb{D}}_1)$ for all $\xi > 0$, applying the classical Vitali's result it follows the uniform convergence on each $\overline{\mathbb{D}}_r$, with $0 < r < 1$. $\square$

Corollary 3.3.4. (Gal [88]) *Suppose that the conditions 1)-4) are fulfilled. If $f \in A(\overline{\mathbb{D}}_R)$ with $R > 1$, then $T_\xi(f) \in A(\overline{\mathbb{D}}_R)$, for all $\xi > 0$ and $\lim_{\xi \to \infty} T_\xi(f)(z) = f(z)$, uniformly in any compact disk included in $\mathbb{D}_R$, that is in $\mathbb{D}_1$ too.*

Proof. The proof follows exactly the lines of the proofs for Theorem 3.3.1 and Theorem 3.3.3, but in this case for $f \in A(\overline{\mathbb{D}}_R)$ with $R > 1$. □

The following result gives an estimate for the convergence result in Corollary 3.3.4 and it is a consequence of the estimates obtained in the proof of Theorem 3.3.3.

For this purpose, we need some additional requirements, described below (see e.g. Angeloni-Vinti [28]).

Let $h : \mathbb{R}_0^+ \to \mathbb{R}_0^+$ with the properties : h is continuous at 0, $h(0) = 0$ and $h(u) > 0$ for $u > 0$. One says that $L_\xi(t)$ is a h-singular kernel, if $|\int_{-\pi}^{\pi} L_\xi(t)dt - 1| = O[h(1/\xi)]$ and for every $\delta > 0$,

$$\int_{\delta \leq |t| \leq \pi} |L_\xi(t)|dt = O[h(1/\xi)], \text{ as } \xi \to \infty.$$

Corollary 3.3.5. (Gal [88]) *Suppose that in addition to the above conditions 1)-4), the following conditions are fulfilled :*

a) $L_\xi(t)$ are h-singular kernels, for all $\xi > 0$;

b) $|H_\xi(u) - u| = O[h(1/\xi)]$, as $\xi \to \infty$, on each compact disk in $\mathbb{C}$;

c) there exists $\delta > 0$, such that

$$\int_{0 \leq |t| \leq \delta} |L_\xi(t)| \cdot |t|dt = O[h(1/\xi)], \text{ as } \xi \to \infty.$$

Then, for any $f \in A(\overline{\mathbb{D}}_R)$ with $R > 1$, and for sufficiently large ξ, we have

$$\|T_\xi(f) - f\|_{\overline{\mathbb{D}}_1} = O[h(1/\xi)].$$

Proof. From the proof of Theorem 3.3.3, for all $|z| \leq 1$, it follows

$$|T_\xi(f)(z) - f(z)| \leq |E_1| + |E_2| + |E_3|,$$

and (taking into account on the hypothesis too)

$$|E_3| \leq |f(z)| \cdot \left| \int_{-\pi}^{\pi} L_\xi(t)dt - 1 \right| \leq \|f\|_{\overline{\mathbb{D}}_1} \cdot \left| \int_{-\pi}^{\pi} L_\xi(t)dt - 1 \right| = O[h(1/\xi)],$$

$$|E_2| \leq \int_{-\pi}^{\pi} |L_\xi(t)| \cdot |H_\xi[f(z)] - f(z)|dt = O[h(1/\xi)],$$

$$|E_1| \leq M_f \int_{-\pi}^{\pi} |L_\xi(t)| \cdot \omega_1(f; |t|)_{\overline{\mathbb{D}}_1} dt$$

$$= M_f \int_{\delta \leq |t| \leq \pi} |L_\xi(t)| \cdot \omega_1(f; |t|)_{\overline{\mathbb{D}}_1} dt + M_f \int_{0 \leq |t| \leq \delta} |L_\xi(t)| \omega_1(f; |t|)_{\overline{\mathbb{D}}_1} dt$$

$$\leq 2M_f \|f\|_{\overline{\mathbb{D}}_1} \int_{\delta \leq |t| \leq \pi} |L_\xi(t)|dt + \int_{0 \leq |t| \leq \delta} |L_\xi(t)| \cdot \|f'\|_{\overline{\mathbb{D}}_1} \cdot |t|dt = O[h(1/\xi)].$$

All these imply the conclusion in Corollary 3.3.5. □

Remark. Concrete examples for various $L_\xi(t)$ satisfying the above results, can be found in e.g. Angeloni-Vinti [28], [29].

In what follows we present the geometric properties of the nonlinear complex operators $T_\xi(f)(z), \xi > 0$.

Theorem 3.3.6. (Gal [88]) *Assume that the above conditions 1)-4) are fulfilled and let us suppose that $f : \overline{\mathbb{D}}_R \to \mathbb{C}$ is analytic in $\mathbb{D}_R$, where $R > 1$.*

If $f(0) = f'(0) - 1 = 0$ and f is starlike (convex, spirallike of type η, respectively) in $\overline{\mathbb{D}}_1$, that is for all $z \in \overline{\mathbb{D}}_1$ (see e.g. Mocanu-Bulboacă-Sălăgean [138])

$$Re\left(\frac{zf'(z)}{f(z)}\right) > 0 \left(Re\left(\frac{zf''(z)}{f'(z)}\right) + 1 > 0, Re\left(e^{i\eta}\frac{zf'(z)}{f(z)}\right) > 0, resp.\right),$$

then there exists ξ_0 depending on f (and on η for spirallikeness), such that for all $\xi \geq \xi_0$, $T_\xi(f)(z)$, are starlike (convex, spirallike of type η, respectively) in $\overline{\mathbb{D}}_1$.

If $f(0) = f'(0) - 1 = 0$ and f is starlike (convex, spirallike of type η, respectively) only in $\mathbb{D}_1$ (that is the corresponding inequalities hold only in $\mathbb{D}_1$), then for any disk of radius $0 < r < 1$ and center 0 denoted by $\mathbb{D}_r$, there exists $\xi_0 = \xi_0(f, \mathbb{D}_r)$ (ξ_0 depends on η too in the case of spirallikeness), such that for all $\xi \geq \xi_0$, $T_\xi(f)(z)$, are starlike (convex, spirallike of type η, respectively) in $\overline{\mathbb{D}}_r$ (that is, the corresponding inequalities hold in $\overline{\mathbb{D}}_r$).

Proof. By Corollary 3.3.4, it follows that for $\xi \to \infty$, we have $T_\xi(f)(z) \to f(z)$, uniformly for $|z| \leq 1$, which by the well-known Weierstrass's theorem implies $[T_\xi(f)]'(z) \to f'(z)$ and $[T_\xi(f)]''(z) \to f''(z)$, for $\xi \to \infty$, uniformly in $\overline{\mathbb{D}}_1$. In all what follows, denote $P_\xi(f)(z) = \frac{T_\xi(f)(z)}{[T_\xi(f)]'(0)}$, well defined for sufficiently large ξ. We easily get $P_\xi(f)(0) = 0$, $P'_\xi(f)(0) = 1$, $P_\xi(f)(z) \to f(z)$, $P'_\xi(f)(z) \to f'(z)$ and $P''_\xi(f)(z) \to f''(z)$, uniformly in $\overline{\mathbb{D}}_1$.

Suppose first that f is starlike in $\overline{\mathbb{D}}_1$. Then, by hypothesis we get $|f(z)| > 0$ for all $z \in \overline{\mathbb{D}}_1$ with $z \neq 0$, which from the univalence of f in $\mathbb{D}_1$, implies that we can write $f(z) = zg(z)$, with $g(z) \neq 0$, for all $z \in \overline{\mathbb{D}}_1$, where g is analytic in $\mathbb{D}_1$ and continuous in $\overline{\mathbb{D}}_1$.

Write $P_\xi(f)(z)$ in the form $P_\xi(f)(z) = zQ_\xi(f)(z)$. For $|z| = 1$ we have

$$|f(z) - P_\xi(f)(z)| = |z| \cdot |g(z) - Q_\xi(f)(z)| = |g(z) - Q_\xi(f)(z)|,$$

which by the uniform convergence in $\overline{\mathbb{D}}_1$ of $P_\xi(f)$ to f and by the maximum modulus principle, implies the uniform convergence in $\overline{\mathbb{D}}_1$ of $Q_\xi(f)(z)$ to $g(z)$.

Since g is continuous in $\overline{\mathbb{D}}_1$ and $|g(z)| > 0$ for all $z \in \overline{\mathbb{D}}_1$, there exist $\xi_1 > 0$ and $a > 0$ depending on g, such that $|Q_\xi(f)(z)| > a > 0$, for all $z \in \overline{\mathbb{D}}_1$ and all $\xi \geq \xi_0$. Also, for all $|z| = 1$, we have

$$\begin{aligned}
|f'(z) - P'_\xi(f)(z)| &= |z[g'(z) - Q'_\xi(f)(z)] + [g(z) - Q_\xi(f)(z)]| \\
&\geq |\ |z| \cdot |g'(z) - Q'_\xi(f)(z)| - |g(z) - Q_\xi(f)(z)|\ | \\
&= |\ |g'(z) - Q'_\xi(f)(z)| - |g(z) - Q_\xi(f)(z)|\ |,
\end{aligned}$$

which from the maximum modulus principle, the uniform convergence of $P'_\xi(f)$ to f' and of $Q_\xi(f)$ to g, evidently implies the uniform convergence of $Q'_\xi(f)$ to g'.

Then, for $|z| = 1$, we get

$$
\begin{aligned}
\frac{z P'_\xi(f)(z)}{P_\xi(f)} &= \frac{z[z Q'_\xi(f)(z) + Q_\xi(f)(z)]}{z Q_\xi(f)(z)} \\
&= \frac{z Q'_\xi(f)(z) + Q_\xi(f)(z)}{Q_\xi(f)(z)} \to \frac{z g'(z) + g(z)}{g(z)} \\
&= \frac{f'(z)}{g(z)} = \frac{z f'(z)}{f(z)},
\end{aligned}
$$

which again from the maximum modulus principle, implies

$$
\frac{z P'_\xi(f)(z)}{P_\xi(f)} \to \frac{z f'(z)}{f(z)}, \text{ uniformly in } \overline{\mathbb{D}}_1.
$$

Since $Re\left(\frac{z f'(z)}{f(z)}\right)$ is continuous in $\overline{\mathbb{D}}_1$, there exists $\varepsilon \in (0,1)$, such that

$$
Re\left(\frac{z f'(z)}{f(z)}\right) \geq \varepsilon, \text{ for all } z \in \overline{\mathbb{D}}_1.
$$

Therefore

$$
Re\left[\frac{z P'_\xi(f)(z)}{P_\xi(f)(z)}\right] \to Re\left[\frac{z f'(z)}{f(z)}\right] \geq \varepsilon > 0
$$

uniformly on $\overline{\mathbb{D}}_1$, i.e. for any $0 < \rho < \varepsilon$, there is ξ_0 such that for all $\xi \geq \xi_0$, we have

$$
Re\left[\frac{z P'_\xi(f)(z)}{P_\xi(f)(z)}\right] > \rho > 0, \text{ for all } z \in \overline{\mathbb{D}}_1.
$$

Since $P_\xi(f)(z)$ differs from $T_\xi(f)(z)$ only by a constant, this proves the starlikeness of $T_\xi(f)(z)$, for sufficiently large ξ.

If f is supposed to be starlike only in $\mathbb{D}_1$, the proof is identical, with the only difference that instead of $\overline{\mathbb{D}}_1$, we reason for $\overline{\mathbb{D}}_r$ with $r < 1$.

The proofs in the cases when f is convex or spirallike of order η are similar and follow from the following uniform convergence (on $\overline{\mathbb{D}}_1$ or on $\overline{\mathbb{D}}_r$) as $\xi \to \infty$

$$
Re\left[\frac{z P''_\xi(f)(z)}{P'_\xi(f)(z)}\right] + 1 \to Re\left[\frac{z f''(z)}{f'(z)}\right] + 1
$$

and

$$
Re\left[e^{i\eta} \frac{z P'_\xi(f)(z)}{P_\xi(f)(z)}\right] \to Re\left[e^{i\eta} \frac{z f'(z)}{f(z)}\right].
$$

$\square$

3.4   Bibliographical Notes and Open Problems

Theorems 3.1.1, 3.1.2, 3.1.3, Corollary 3.1.4, Theorem 3.1.5, Corollary 3.1.11, Theorems 3.1.12 and 3.1.13, Corollary 3.1.14, Theorems 3.1.15-3.1.16, 3.2.1 (iv), 3.2.5 (iv), 3.2.8 (iv), 3.2.11 (iv), 3.2.17, 3.2.18, Corollaries 3.2.19, 3.2.20, Theorems 3.2.25 (iv), 3.2.27 (iv) are new and appear for the first time here.

Open Problem 3.4.1. Let $f : \mathbb{D}_R \to \mathbb{C}$ be analytic on $\mathbb{D}_R$ with $R > 1$ and suppose that $1 \le r < R$ and $n \in \mathbb{N}$ are fixed. Then for all $\xi \in (0, 1]$ we have

$$\|Q_\xi(f) - f\|_r \le C_r(f)\xi, \text{ from Theorem 3.2.5, (iii)},$$

$$\|W_\xi(f) - f\|_r \le C_r(f)\sqrt{\xi}, \text{ from Theorem 3.2.8, (iii)},$$

$$\|W_{n,\xi}(f) - f\|_r \le C_{n,r}(f)\xi^{n+1}, \text{ from Theorem 3.2.11, (ii)}.$$

The exact order of approximations (with respect to ξ) remain as open questions.

Open Problem 3.4.2. It remains to prove the lower estimates for $\|P_{n,\xi}^{(q)}(f) - f^{(q)}\|_r$ and $\|[W_{n,\xi}^*]^{(q)}(f) - f^{(q)}\|_r$ (for the upper estimates see Theorem 3.2.11, (iv)).

Open Problem 3.4.3. It remains to prove the lower estimate for $\|P_{n,\alpha}^{(q)}(f) - f^{(q)}\|_r$ (for the upper estimate see Theorem 3.2.25, (iv)).

Open Problem 3.4.4. It remains to prove the lower estimates for $\|Y_k^{(s)}(f) - f^{(s)}\|_r$, $\|L_{k,j}^{(s)}(f) - f^{(s)}\|_r$, $\|I_{k,q}^{(s)}(f) - f^{(s)}\|_r$, (for the upper estimate see Theorem 3.2.27, (vi)).

Open Problem 3.4.5. It would be of interest to study the complex Sikkema-type operators, for various kernels (different from those classical already mentioned in Subsection 3.2.4, Remark 1 after the proof of Theorem 3.2.31) introduced in the Sikkema's papers [164], [165], [166], [167].

Open Problem 3.4.6. Taking into account the Remark after the proof of Theorem 3.2.23, it is an open question if for $q_n \to 1$, $0 < q_n < 1$, the approximation order by $P_{1/n}(f; q_n; z)$ and $W_{1/n}(f; q_n; z)$ are exactly $\frac{1}{[n]_{q_n}}$ and $\frac{1}{\sqrt{[n]_{q_n}}}$, respectively.

Chapter 4

Appendix : Related Topics

This chapter is a collection of several distinct directions of research generalizing and/or being related with those in the previous ones : approximation by quaternion Bernstein polynomials and approximation of vector-valued functions by Bernstein and convolution type operators.

4.1 Bernstein Polynomials of Quaternion Variable

As it is well-known, the field of complex numbers can be extended to more general algebraic structures (with several complex units) called hypercomplex numbers. These structures can be divided in two main classes : commutative hypercomplex structures which are rings with divisors of zeros and noncommutative hypercomplex structures which are fields (so without divisors of zero), the most known being the so-called quaternion numbers and Clifford algebras.

Therefore, it is natural to see for extensions of the approximation results in the previous sections, to approximation by Bernstein-type operators of hypercomplex variables. In this section we limit our consideration to the case of Bernstein polynomials of quaternion variable. For this purpose first we make a short introduction.

The quaternion field is defined by

$$\mathbb{H} = \{q = x_1 + x_2 i + x_3 j + x_4 k; x_1, x_2, x_3, x_4 \in \mathbb{R}\},$$

where the complex units $i, j, k \notin \mathbb{R}$ satisfy

$$i^2 = j^2 = k^2 = -1, \ ij = -ji = k, \ jk = -kj = i, \ ki = -ik = j.$$

It is a noncommutative field and since obviously $\mathbb{C} \subset \mathbb{H}$, it extends the class of complex numbers. On $\mathbb{H}$ can be defined the norm $\|q\| = \sqrt{x_1^2 + x_2^2 + x_3^3 + x_4^2}$, for $q = x_1 + x_2 i + x_3 j + x_4 k$.

If $G \subset \mathbb{H}$ then a function $f : G \to \mathbb{H}$ can be written in the form

$$f(q) =$$

$$f_1(x_1, x_2, x_3, x_4) + f_2(x_1, x_2, x_3, x_4)i + f_3(x_1, x_2, x_3, x_4)j + f_4(x_1, x_2, x_3, x_4)k,$$

$q = x_1 + x_2 i + x_3 j + x_4 k \in G$, where f_i are real valued functions, $i = 1, 2, 3, 4$.

It is well-known the fact that a direct attempt to generalize the concept of differentiability for f as

$$\lim_{q \to q_0} (q - q_0)^{-1} [f(q) - f(q_0)] \in \mathbb{H}, q_0 \in G,$$

or as

$$\lim_{q \to q_0} [f(q) - f(q_0)](q - q_0)^{-1} \in \mathbb{H}, q_0 \in G,$$

necessarily implies that f is of the form $f(q) = Aq + B$ (see Mejlihzon [135]).

For this reason, the theory of holomorphic functions of quaternion variable can be constructed in several other ways, by producing different classes of holomorphic functions. We mention below only two ways. The first one is given by the following.

Definition 4.1.1. (Moisil [139]) Let $f = f_1 + f_2 i + f_3 j + f_4 k$ be such that each f_i has continuous partial derivatives of order one, $i = 1, 2, 3, 4$. Define $F = \frac{\partial}{\partial x_1} + \frac{\partial}{\partial x_2} i + \frac{\partial}{\partial x_3} j + \frac{\partial}{\partial x_4} k$. One says that f is left differentiable (monogenic) at q_0 if $F f(q_0) = 0$. In this case, the derivative of f at q_0 will be given $f'(q_0) = \overline{D}(f)(q_0)$, where $\overline{F} = \frac{\partial}{\partial x_1} - \frac{\partial}{\partial x_2} i - \frac{\partial}{\partial x_3} j - \frac{\partial}{\partial x_4} k$.

If f is monogenic at each q, then it is called holomorphic.

Remarks. 1) In the case of complex variable, the differential operator F one reduces to the areolar derivative of f and the operator $\overline{F}$ becomes the derivative of f.

2) f will be called right differentiable at q_0 if $f F(q_0) = 0$.

The second kind of definition for holomorphy is suggested by the Weierstrass's idea in the case of complex variable.

Definition 4.1.2. Denoting $\mathbb{D}_R = \{q \in \mathbb{H}; \|q\| < R\}$, one says that $f : \mathbb{D}_R \to \mathbb{H}$ is left analytic in $\mathbb{D}_R$ if $f(q) = \sum_{k=0}^{\infty} c_k q^k$, for all $q \in \mathbb{D}_R$, where $c_k \in \mathbb{H}$ for all $k = 0, 1, 2, ...,$. Also, f is called right analytic in $\mathbb{D}_R$ if $f(q) = \sum_{k=0}^{\infty} q^k c_k$, for all $q \in \mathbb{D}_R$.

Remark. While in the case of complex variable the two concepts in Definitions 4.1.1. and 4.1.2 coincide, in the case of quaternion variable this does not happen. The most suitable concept for our purpose is that in Definition 4.1.2.

Concerning the Bernstein polynomials, due to non-commutativity, for $R > 1$ to a function $f : \mathbb{D}_R \to \mathbb{H}$, three distinct Bernstein polynomials can be attached, as follows :

$$B_n(f)(z) = \sum_{l=0}^{n} f\left(\frac{l}{n}\right) \binom{n}{l} z^l (1 - z)^{n-l}, z \in \mathbb{H},$$

$$B_n^*(f)(z) = \sum_{l=0}^{n} \binom{n}{l} z^l (1 - z)^{n-l} f\left(\frac{l}{n}\right), z \in \mathbb{H},$$

$$B_n^{**}(f)(z) = \sum_{l=0}^{n} \binom{n}{l} z^l f\left(\frac{l}{n}\right) (1-z)^{n-l}, z \in \mathbb{H}.$$

We may call them as the left Bernstein polynomials, right Bernstein polynomials and middle Bernstein polynomials, respectively.

It is easy to show by a simple example that these kinds of Bernstein polynomials do not converge for any continuous function f. Indeed, if we take $f(z) = izi$, then we easily get

$$|B_n(f)(z) - izi| = |B_n^{*}(f)(z) - izi| = |B_n^{**}(f)(z) - izi| = |-z - izi| =$$

$$|-iz + zi| > 0, \quad \text{for all } z \neq i.$$

However, for each kind of Bernstein polynomial there exists a suitable class of functions for which the convergence holds. To prove that we need some auxiliary results.

Theorem 4.1.3. *Suppose that $f : \mathbb{D}_R \to \mathbb{H}$ has the property that $f(z) \in \mathbb{R}$ for all $z \in [0,1]$. Then we have the representation formula*

$$B_n(f)(z) = \sum_{m=0}^{n} \binom{n}{m} \Delta_{1/n}^m f(0) z^m, \text{ for all } z \in \mathbb{H}.$$

Proof. Because of the hypothesis on f, the values $f\left(\frac{l}{n}\right)$ commutes with the other terms in the expression of $B_n(f)(z)$, so that taking into account that $z^{n+m} = z^n z^m = z^m z^n$ (from associativity), $z^l(1-z)^{n-l} = (1-z)^{n-l}z^l$, $\alpha z = z\alpha$, for all $\alpha \in \mathbb{R}$, $z \in \mathbb{H}$ and that

$$(1-z)^{n-l} = \sum_{s=0}^{n-l} (-1)^s \binom{n-l}{s} z^s,$$

reasoning exactly as in the case of Bernstein polynomials of real variable (see e.g. Lorentz [125], p. 13) we obtain

$$B_n(f)(z) = \sum_{l=0}^{n} f\left(\frac{l}{n}\right) \binom{n}{l} z^l \sum_{s=0}^{n-l} (-1)^s \binom{n-l}{s} z^s = \sum_{m=0}^{n} \binom{n}{m} \Delta_{1/n}^m f(0) z^m,$$

which proves the theorem. $\qquad\square$

Remark. It is clear that Theorem 4.1.3 holds for the right and middle Bernstein polynomials, $B_n^{*}(f)(z)$ and $B_n^{**}(f)(z)$ too.

Now we are in position to state the following approximation result.

Theorem 4.1.4. *Suppose that $f : \mathbb{D}_R \to \mathbb{H}$ is left analytic in $\mathbb{D}_R$, i.e. $f(z) = \sum_{p=0}^{\infty} c_p z^p$, for all $z \in \mathbb{D}_R$, where $c_p \in \mathbb{H}$ for all $p = 0, 1, 2, ...,$. Then for all $1 \leq r < R$, $\|z\| \leq r$ and $n \in \mathbb{N}$ we have*

$$\|B_n(f)(z) - f(z)\| \leq \frac{2}{n} \sum_{p=2}^{\infty} \|c_p\| p(p-1) r^p.$$

Proof.   First it is easy to see that $B_n(f)(0) = f(0)$ and $B_n(f)(1) = f(1)$. Then, for any $f(z) = e_p(z) = z^p$, taking into account Theorem 4.1.3 too we get

$$B_n(e_p)(z) = \sum_{l=0}^{n} \binom{n}{l} \Delta_{1/n}^l e_p(0) z^l,$$

and $B_n(e_p)(1) = 1 = \sum_{l=0}^{n} \binom{n}{l} \Delta_{1/n}^l e_p(0)$, where since e_p is convex of any order it follows that $\binom{n}{l} \Delta_{1/n}^l e_p(0) \geq 0$ for all $0 \leq l \leq n$.

Moreover, taking into account the formula between the finite differences and divided differences, we can write

$$B_n(e_p)(z) = \sum_{l=0}^{n} \binom{n}{l} l! \frac{1}{n^l} [0, 1/n, ..., l/n; e_p] z^l$$

$$= \sum_{l=0}^{n} \frac{n(n-1)...(n-l+1)}{n^l} [0, 1/n, ..., l/n; e_p] z^l.$$

On the other hand, $B_n(f)$ has the properties

$$B_n(f+g) = B_n(f) + B_n(g), \ B_n(\alpha f) = \alpha B_n(f), \alpha \in \mathbb{H},$$

$$B_n(f\alpha) \neq \alpha B_n(f), \alpha \in \mathbb{H} \setminus \mathbb{R}.$$

Now let us prove the relationship $B_n(f)(z) = \sum_{k=0}^{\infty} c_k B_n(e_k)(z)$. Denoting $f_m(z) = \sum_{j=0}^{m} c_j z^j$, $\|z\| \leq r$, $m \in \mathbb{N}$, since from the above linearity of B_n we obviously have $B_n(f_m)(z) = \sum_{k=0}^{m} c_k B_n(e_k)(z)$, it suffices to prove that for any fixed $n \in \mathbb{N}$ and $\|z\| \leq r$ with $r \geq 1$, we have $\lim_{m \to \infty} B_n(f_m)(z) = B_n(f)(z)$. But this is immediate from $\lim_{m \to \infty} \|f_m - f\|_r = 0$ (where $\|f\|_r := \sup\{\|f(z)\|; \|z\| \leq r\}$ and from the inequality

$$\|B_n(f_m)(z) - B_n(f)(z)\| \leq \sum_{k=0}^{n} \binom{n}{k} \|z^k(1-z)^{n-k}\| \cdot \|f_m - f\|_r$$

$$\leq M_{r,n} \|f_m - f\|_r,$$

valid for all $\|z\| \leq r$.

Therefore we immediately get that

$$\|B_n(f)(z) - f(z)\| \leq \sum_{p=0}^{\infty} \|c_p\| \cdot \|B_n(e_p)(z) - e_p(z)\|.$$

To estimate $\|B_n(e_p) - e_p\|$ two possibilities exist : 1) $0 \leq p \leq n$; 2) $p > n$.

Case 1). We get $B_n(e_p)(z) = \sum_{l=0}^{p} \frac{n(n-1)...(n-l+1)}{n^l} [0, 1/n, ..., l/n; e_p] z^l$ and denoting $C_{n,l} = \frac{n(n-1)...(n-l+1)}{n^l} [0, 1/n, ..., l/n; e_p]$ it follows

$$B_n(e_p)(z) - e_p(z) = \left(1 - \frac{n(n-1)...(n-(p-1))}{n^p}\right) e_p(z) + \sum_{l=0}^{p-1} C_{n,l} z^l.$$

Passing to the norm $\|\cdot\|$ with $\|z\| \leq r$ and taking into account an inequality in the proof of Theorem 1.2.1, (ii), we obtain

$$\|B_n(e_p)(z) - e_p(z)\| \leq \left|1 - \frac{n(n-1)...(n-(p-1))}{n^p}\right| r^p + \sum_{l=0}^{p-1} C_{n,l} r^p$$

$$= 2\left|1 - \frac{n(n-1)...(n-(p-1))}{n^p}\right| r^p$$

$$\leq \frac{p(p-1)}{n} r^p.$$

Case 2). For all $\|z\| \leq r$, $r \geq 1$, $z \in \mathbb{H}$, $p > n \geq 1$ we get

$$\|B_n(e_p)(z)\| \leq \sum_{l=0}^{p} \frac{n(n-1)...(n-l+1)}{n^l} [0, 1/n, ..., l/n; e_p] \cdot \|z^l\| \leq r^n,$$

and therefore

$$\|B_n(e_p)(z) - e_p(z)\| \leq r^n + r^p \leq 2r^p \leq 2r^p n \leq 2r^p \frac{p(p-1)}{n}.$$

In conclusion, combining both Cases 1 and 2 we obtain

$$\|B_n(f)(z) - f(z)\| \leq \sum_{p=0}^{\infty} \|c_p\| \cdot \|B_n(e_p)(z) - e_p(z)\| \leq \frac{2}{n} \sum_{p=0}^{\infty} \|c_p\| p(p-1),$$

which proves the theorem.
$\square$

In a similar manner we obtain the following.

Corollary 4.1.5. *Suppose that $f : \mathbb{D}_R \to \mathbb{H}$ is right analytic in $\mathbb{D}_R$, i.e. $f(z) = \sum_{p=0}^{\infty} z^p c_p$, for all $z \in \mathbb{D}_R$, where $c_p \in \mathbb{H}$ for all $p = 0, 1, 2, ...,$. Then for all $1 \leq r < R$, $\|z\| \leq r$ and $n \in \mathbb{N}$ we have*

$$\|B_n^*(f)(z) - f(z)\| \leq \frac{2}{n} \sum_{p=2}^{\infty} \|c_p\| p(p-1) r^p.$$

Remark. It is not difficult to see that in the case of Bernstein-type polynomials $B_n^{**}(f)(z)$, an estimate of the form in Theorem 4.1.4 cannot be obtained, because in general we cannot write a formula of the type $B_n^{**}(f)(z) = \sum_{p=0}^{\infty} c_p B_n^{**}(e_p)(z)$ for f left analytic or of the type $B_n^{**}(f)(z) = \sum_{p=0}^{\infty} B_n^{**}(e_p)(z) c_p$ for f right analytic.

4.2 Approximation of Vector-Valued Functions

By using a nice and powerful method based on a classical result in Functional Analysis, in this section we study the approximation of vector-valued functions of real variable and of complex variable, by the corresponding Bernstein-type or convolution-type operators.

4.2.1 *Real Variable Case*

In this subsection we present some results concerning the approximation of functions $f : I \to X$, where I is a subinterval of the real numbers $\mathbb{R}$ and X is a normed space over $\mathbb{R}$..

The case of functions of one real variable is of intrinsic value and gives the main ideas of the method. Because of these reasons it is the first considered below in full details. Also, because we can take as X the space of all complex numbers $\mathbb{C}$, it follows that this case one frames into the title of the present book.

In essence, in this subsection we prove basic results in the approximation of vector-valued functions of real variable by polynomials with coefficients in normed spaces, called here generalized polynomials. Thus we obtain estimates in terms of Ditzian-Totik L^p-moduli of smoothness and inverse theorems for approximation by Bernstein, Bernstein-Kantorovich, Szász-Mirakjan, Baskakov generalized operators and their Kantorovich analogues, Post-Widder, Jackson-type generalized operators, etc. Some applications to approximation of random functions and of fuzzy-number-valued functions are given.

We need the following useful concepts.

Let $(X, \| \cdot \|)$ be a normed space over $\mathbb{R}$. Similar to the case of real-valued functions, the following concepts in the Definitions 4.2.1-4.2.3 can be introduced.

Definition 4.2.1. (i) A generalized algebraic polynomial of degree $\leq n$, with coefficients in X will be an expression of the form $P_n(x) = \sum_{k=0}^{n} c_k x^k$, where $c_k \in X, k = 0, ..., n$ and $x \in [a, b]$.

A generalized trigonometric polynomial of degree $\leq n$ with coefficients in X will be an expression of the form $T_n(x) = a_0 + \sum_{k=1}^{n} [a_k \cos(kx) + b_k \sin(kx)]$, where $a_0, a_k, b_k \in X, k = 1, ..., n$ and $x \in \mathbb{R}$.

(ii) Denote by $\mathcal{P}_n[a, b]$, $\mathcal{T}_n$ the sets of all generalized algebraic and trigonometric polynomials of degree $\leq n$ with coefficients in X, respectively, $\|f\|_\infty = \sup_x \{\|f(x)\|\}$, $\|f\|_p = (\int_a^b \|f(x)\|^p dx)^{1/p}$, if $f : [a, b] \to X$, $\|f\|_p = (\int_0^{2\pi} \|f(x)\|^p dx)^{1/p}$, if $f : \mathbb{R} \to X$ is 2π-periodic, $1 \leq p < \infty$.

Also, if $\|f\|_\infty < \infty$ then we write that $f \in C([a, b]; X)$ (or $f \in C_{2\pi}(\mathbb{R}; X)$) and if $\|f\|_p < \infty$, $1 \leq p < \infty$, we write $f \in L^p([a, b]; X)$ (or $f \in L^p_{2\pi}(\mathbb{R}; X)$), depending if f is defined on $[a, b]$ or 2π-periodic on $\mathbb{R}$, respectively.

Definition 4.2.2. $f : [a, b] \to X$ will be called Riemann integrable on $[a, b]$, if there exists an element $I \in X$ denoted by $\int_a^b f(x)dx$, with the following property : for any $\varepsilon > 0$, there exists $\delta > 0$, such that for any division of $[a, b]$, $d : a = x_0 < ... < x_n = b$ with the norm $\nu(d) < \delta$ and any intermediary points $\xi_i \in [x_i, x_{i+1}]$, we have $\|S(f; d, \xi_i) - I\| < \varepsilon$, where $S(f; d, \xi_i) = \sum_{i=0}^{n-1} f(\xi_i)(x_{i+1} - x_i)$.

Also, denote by $L^p([a, b]; X) = \{f : [a, b] \to X; f$ is pth Bochner-Lebesgue integrable and $\int_a^b \|f(x)\|^p dx < +\infty\}, 1 \leq p < \infty$, where the equality between two functions in $L^p([a, b]; X)$ is considered in the almost everywhere sense. For $p = +\infty$ we consider $L^p([a, b]; X) = C([a, b]; X)$.

Definition 4.2.3. (i) For $f : [a,b] \to X$, $f \in L^p([a,b]; X)$, the kth L^p-modulus of smoothness of f on $[a,b]$ will be given by

$$\omega_k(f;\delta)_p = \sup\left\{\left(\int_a^{b-kh} \|\Delta_h^k f(x)\|^p dx\right)^{1/p} ; 0 \le h \le \delta\right\}, \text{ if } 1 \le p < +\infty,$$

and

$$\omega_k(f;\delta)_\infty = \sup_{0 \le h \le \delta} \{\sup\{\|\Delta_h^k f(x)\|; x, x+kh \in [a,b]\}\}.$$

For $f : \mathbb{R} \to X$, 2π-periodic, $f \in L^p([0, 2\pi]; X)$ one define

$$\omega_k(f;\delta)_p = \sup\left\{\left(\int_0^{2\pi} \|\Delta_h^k f(x)\|^p dx\right)^{1/p} ; 0 \le h \le \delta\right\}, \text{ if } 1 \le p < +\infty,$$

and

$$\omega_k(f;\delta)_\infty = \sup_{0 \le h \le \delta} \{\sup\{\|\Delta_h^k f(x)\|; x \in \mathbb{R}\}\}.$$

Here $\Delta_h^k f(x) = \sum_{j=0}^k (-1)^{k-j} \binom{k}{j} f(x+jh)$.

(ii) Let $f : I \to X$, $f \in L^p(I; X)$, where I is a subinterval of $\mathbb{R}$. The kth Ditzian-Totik L^p-modulus of smoothness will be given by

$$\omega_\phi^k(f;\delta)_{L^p} := \omega_\phi^k(f;\delta)_p = \sup_{0 \le h \le \delta} \|\overline{\Delta}_{h\phi(x)}^k f(x)\|_{L^p(I)},$$

$1 \le p \le +\infty$, where $\phi(x)$ is a suitable step-weight attached to I and $\overline{\Delta}_h^k f(x) = \sum_{j=0}^k (-1)^j \binom{k}{j} f(x+kh/2-jh)$, if $x, x \pm kh/2 \in I$, $\overline{\Delta}_h^k f(x) = 0$, otherwise. Here $\|f\|_p := \|f\|_{L^p(I)} = (\int_I \|f(x)\|^p dx)^{1/p}$, if $1 \le p < +\infty$ and $\|f\|_\infty = \sup\{\|f(x)\|; x \in I\}$.

In particular, for $[a,b] = [0,1]$ and $p = \infty$, we have

$$\omega_\phi^2(f;\delta) = \sup\{\sup\{\|f(x+h\phi(x)) - 2f(x) + f(x-h\phi(x))\|; x \in I_{2,h}\},$$

$$h \in [0,\delta]\},$$

respectively, where $I_{2,h} = \left[-\frac{1-h^2}{1+h^2}, \frac{1-h^2}{1+h^2}\right]$, $\phi(x) = \sqrt{x(1-x)}$, $\delta \le 1$.

Remark. In the applications we will encounter the following step-weight functions ϕ :

$$\phi^2(x) = (1-x^2), \text{ if } I = [-1,1],$$

$$\phi^2(x) = x(1-x), \text{ if } I = [0,1],$$

$$\phi^2(x) = x, \phi^2(x) = x^2, \text{ or } \phi^2(x) = x(1+x), \text{ if } I = [0,+\infty).$$

The main tool used in our proofs is represented by the following well-known result in Functional Analysis.

Theorem 4.2.4. (see e.g. Muntean [141], p. 183) *Let $(X, \| \cdot \|)$ be a normed space over the real or complex numbers and denote by X^* the conjugate space of X. Then, $\|x\| = \sup\{|x^*(x)| : x^* \in X^*, \|\|x^*\|\| \leq 1\}$, for all $x \in X$.*

Let $f : [0,1] \to X$ be continuous on $[0,1]$. The generalized Bernstein and Bernstein-Kantorovich polynomials attached to f can be defined by

$$B_n(f)(x) = \sum_{k=0}^{n} p_{n,k}(x) f\left(\frac{k}{n}\right),$$

and

$$K_n(f)(x) = (n+1) \sum_{k=0}^{n} p_{n,k}(x) \int_{k/(n+1)}^{(k+1)/(n+1)} f(t)dt,$$

respectively, where $p_{n,k}(x) = \binom{n}{k} x^k (1-x)^{n-k}$ and the integral $\int_a^b f(t)dt$ is defined as the limit for $m \to \infty$ in the norm $\| \cdot \|$, of the all (classical defined) Riemann sums $\sum_{i=0}^{m}(x_{i+1} - x_i)f(\xi_i)$.

Also, the generalized trigonometric polynomials of Jackson-type attached to a 2π-periodic continuous function $f : \mathbb{R} \to X$, can be defined by

$$J_n(f)(x) = \frac{3}{2\pi n(2n^2 + 1)} \int_{-\pi}^{\pi} K_n(x-t)f(t)dt,$$

where $K_n(t) = \left(\frac{\sin(nt/2)}{\sin(t/2)}\right)^4$.

We present

Theorem 4.2.5. *(Gal [92]) Let $f : [0,1] \to X$ be continuous on $[0,1]$. There exist the absolute constants C_1, C_2, such that for all $n \in \mathbb{N}$ we have :*
  (i)

$$C_1 \omega_\phi^2(f; \frac{1}{\sqrt{n}})_\infty \leq \|B_n(f) - f\|_\infty \leq C_2 \omega_\phi^2(f; \frac{1}{\sqrt{n}})_\infty,$$

where $\|f\|_\infty = \sup\{\|f(x)\|; x \in [0,1]\}$;
  (ii)

$$C_1[\omega_\phi^2(f; \frac{1}{\sqrt{n}})_\infty + \omega_1(f; \frac{1}{n})_\infty] \leq \|K_n(f) - f\|_\infty$$

$$\leq C_2[\omega_\phi^2(f; \frac{1}{\sqrt{n}})_\infty + \omega_1(f; \frac{1}{n})_\infty];$$

  (iii)

$$\|B_n(f)(x) - f(x)\| \leq M \left[\frac{x(1-x)}{n}\right]^{\alpha/2} \forall x \in [0,1], \text{ if and only if}$$

$$\omega_2(f; \delta)_\infty = O(\delta^\alpha),$$

where $\alpha \leq 2$;

(iv) If, in addition, $f : \mathbb{R} \to X$ is continuous and 2π-periodic then

$$\|J_n(f) - f\|_\infty \leq C_1 \omega_2(f; \frac{1}{n})_\infty.$$

Proof. For $x^* \in X^*, 0 < \|\|x^*\|\| \leq 1$ let us define the function $g : [0,1] \to K, g(x) = x^*(f(x))$. Obviously g is continuous on $[0,1]$.

(i) By Knoop-Zhou [116], Totik [192] we have

$$C_1 \omega_\phi^2(g; \frac{1}{\sqrt{n}})_\infty \leq \|B_n(g) - g\|_\infty \leq C_2 \omega_\phi^2(g; \frac{1}{\sqrt{n}})_\infty,$$

where $\|g\|_\infty = \sup\{|g(x)|; x \in [0,1]\}$.

By the linearity of x^* we get

$$\omega_\phi^2(g; \frac{1}{\sqrt{n}})_\infty$$
$$= \sup\{\sup\{|x^*[f(x + h\phi(x)) - 2f(x) + f(x - h\phi(x))]|; x \in I_{2,h}\},$$
$$\quad h \in [0, 1/\sqrt{n}]\}$$
$$\leq \sup\{\sup\{\|\|x^*\|\| \cdot \|f(x + h\phi(x)) - 2f(x) + f(x - h\phi(x))\|; x \in I_{2,h}\},$$
$$\quad h \in [0, 1/\sqrt{n}]\}$$
$$\leq \omega_\phi^2(f; \frac{1}{\sqrt{n}})_\infty.$$

Also, from the linearity of x^* it easily follows

$$B_n(g)(x) - g(x) = x^*[B_n(f)(x) - f(x)]$$

and from the right-hand side of the inequalities for g, we get for all $x \in [0,1]$

$$|x^*[B_n(f)(x) - f(x)]| \leq C_2 \omega_\phi^2(f; \frac{1}{\sqrt{n}})_\infty.$$

Passing to supremum with x^* and taking into account Theorem 4.2.4, it follows

$$\|B_n(f)(x) - f(x)\| \leq C_2 \omega_\phi^2(f; \frac{1}{\sqrt{n}})_\infty,$$

for all $x \in [0,1]$, i.e.

$$\|B_n(f) - f\|_\infty \leq C_2 \omega_\phi^2(f; \frac{1}{\sqrt{n}})_\infty.$$

On the other hand, by the left-hand side of the inequalities for g, i.e.

$$C_1 \omega_\phi^2(g; \frac{1}{\sqrt{n}})_\infty \leq \|B_n(g) - g\|_\infty,$$

we get

$$C_1 \omega_\phi^2(g; \frac{1}{\sqrt{n}})_\infty \leq \sup\{|x^*[B_n(f)(x) - f(x)]|; x \in [0,1]\}$$
$$\leq \sup\{\|\|x^*\|\| \cdot \|B_n(f)(x) - f(x)\|; x \in [0,1]\}$$
$$\leq \|B_n(f) - f\|_\infty.$$

Now, for fixed $x \in I_{2,h}$ and $h \in [0, 1/\sqrt{n}]$, we have

$$C_1 |x^*[f(x + h\phi(x)) - 2f(x) + f(x - h\phi(x))]| \leq \omega_\phi^2(g; \frac{1}{\sqrt{n}})_\infty \leq \|B_n(f) - f\|_\infty,$$

wherefrom passing to supremum with x^* and taking into account Theorem 4.2.4, we get

$$C_1 \|f(x + h\phi(x)) - 2f(x) + f(x - h\phi(x))\| \leq \|B_n(f) - f\|_\infty.$$

Passing now to supremum with $x \in I_{2,h}$ and $h \in [0, 1/\sqrt{n}]$ we obtain

$$C_1 \omega_\phi^2(f; \frac{1}{\sqrt{n}})_\infty \leq \|B_n(f) - f\|_\infty,$$

which proves the point (i).

(ii) By Gonska-Zhou [104] we have

$$C_1 [\omega_\phi^2(g; \frac{1}{\sqrt{n}})_\infty + \omega_1(g; \frac{1}{n})_\infty] \leq \|K_n(g) - g\|_\infty \leq C_2 [\omega_\phi^2(g; \frac{1}{\sqrt{n}})_\infty + \omega_1(g; \frac{1}{n})_\infty].$$

From the linearity and the continuity of x^* we easily get $K_n(g)(x) - g(x) = x^*[K_n(f)(x) - f(x)]$. Also, since

$$\omega_1(g; \frac{1}{n})_\infty = \sup\{|x^*[f(v) - f(w)]|; v, w \in [0, 1], |v - w| \leq \frac{1}{n}\}$$

$$\leq \sup\{\|\|x^*\|\| \cdot \|f(v) - f(w)\|; v, w \in [0, 1], |v - w| \leq \frac{1}{n}\}$$

$$\leq \omega_1(f; \frac{1}{n})_\infty,$$

reasoning exactly as for the point (i), we immediately get

$$\|K_n(f) - f\|_\infty \leq C_2 [\omega_\phi^2(f; \frac{1}{\sqrt{n}})_\infty + \omega_1(f; \frac{1}{n})_\infty].$$

On the other hand, by the left-hand side of the inequalities for g, i.e.

$$C_1 [\omega_\phi^2(g; \frac{1}{\sqrt{n}})_\infty + \omega_1(g; \frac{1}{n})_\infty] \leq \|K_n(g) - g\|_\infty,$$

we get

$$C_1 [\omega_\phi^2(g; \frac{1}{\sqrt{n}})_\infty + \omega_1(g; \frac{1}{n})_\infty] \leq \sup\{|x^*[K_n(f)(x) - f(x)]|; x \in [0, 1]\}$$

$$\leq \sup\{\|\|x^*\|\| \cdot \|K_n(f)(x) - f(x)\|; x \in [0, 1]\}$$

$$\leq \|K_n(f) - f\|_\infty.$$

Now, for any fixed $x \in I_{2,h}$, $h \in [0, 1/\sqrt{n}]$, $v, w \in [0, 1]$ with $|v - w| \leq \frac{1}{n}$, we have

$$C_1 [|x^*(f(x + h\phi(x)) - 2f(x) + f(x - h\phi(x)))| + |x^*(f(v) - f(w))|]$$

$$\leq C_1 [\omega_\phi^2(g; \frac{1}{\sqrt{n}})_\infty + \omega_1(g; \frac{1}{n})_\infty] \leq \|K_n(f) - f\|_\infty,$$

wherefrom passing to supremum with x^* we get

$$C_1 \sup\{|x^*(f(x + h\phi(x)) - 2f(x) + f(x - h\phi(x)))|; \|\|x^*\|\| \le 1\}$$
$$\le C_1 \sup\{|x^*(f(x + h\phi(x)) - 2f(x) + f(x - h\phi(x)))| + |x^*(f(v) - f(w))|;$$
$$\|\|x^*\|\| \le 1\} \le \|K_n(f) - f\|_\infty,$$

and

$$C_1 \sup\{|x^*(f(v) - f(w))|; \|\|x^*\|\| \le 1\}$$
$$\le C_1 \sup\{|x^*(f(x + h\phi(x)) - 2f(x) + f(x - h\phi(x)))| + |x^*(f(v) - f(w))|;$$
$$\|\|x^*\|\| \le 1\} \le \|K_n(f) - f\|_\infty.$$

By Theorem 4.2.4 we obtain

$$C_1 \|f(x + h\phi(x)) - 2f(x) + f(x - h\phi(x))\| \le \|K_n)f) - f\|_\infty$$

and

$$C_1 \|f(v) - f(w)\| \le \|K_n(f) - f\|_\infty.$$

Passing now to supremum with $x \in I_{2,h}$ and $h \in [0, 1/\sqrt{n}]$ and with $|v - w| \le \frac{1}{n}$, respectively, we obtain

$$C_1 \omega_\phi^2(f; \frac{1}{\sqrt{n}})_\infty \le \|K_n(f) - f\|_\infty,$$

and

$$C_1 \omega_1(f; \frac{1}{n})_\infty \le \|K_n(f) - f\|_\infty,$$

implying

$$\frac{C_1}{2}[\omega_\phi^2(f; \frac{1}{\sqrt{n}})_\infty + \omega_1(f; \frac{1}{n})_\infty] \le \|K_n(f) - f\|_\infty,$$

which proves the point (ii) too.

(iii) By Berens-Lorentz [37] we have

$$|B_n(g)(x) - g(x)| \le M \left[\frac{x(1-x)}{n}\right]^{\alpha/2} \forall x \in [0, 1], \text{ if and only if}$$

$$\omega_2(g; \delta)_\infty = O(\delta^\alpha).$$

First, let us suppose that

$$\|B_n(f)(x) - f(x)\| \le M \left[\frac{x(1-x)}{n}\right]^{\alpha/2} \forall x \in [0, 1].$$

We get

$$|B_n(g)(x) - g(x)| = |x^*[B_n(f)(x) - f(x)]|$$
$$\le \|\|x^*\|\| \cdot \|B_n(f) - f\|$$
$$\le M \left[\frac{x(1-x)}{n}\right]^{\alpha/2} \forall x \in [0, 1].$$

This implies $\omega_2(g;\delta)_\infty \leq C\delta^\alpha$, that is for any fixed x and h satisfying $x, x + h, x - h \in [0,1]$, $h \in [0,\delta]$ we get $|x^*[f(x+h) - 2f(x) + f(x-h)]| \leq C\delta^\alpha$.

Passing to supremum with x^* and taking into account Theorem 4.2.4, it follows $\|f(x+h) - 2f(x) + f(x-h)\| \leq C\delta^\alpha$.

Passing now to supremum with x and h as above, we obtain $\omega_2(f;\delta)_\infty \leq C\delta^\alpha$.

Conversely, let us suppose that $\omega_2(f;\delta)_\infty \leq C\delta^\alpha$. For x and h satisfying $x, x + h, x - h \in [0,1]$, $h \in [0,\delta]$ we have

$$
\begin{aligned}
|g(x+h) - 2g(x) + g(x-h)| &= |x^*[f(x+h) - 2f(x) + f(x-h)]| \\
&\leq |\|x^*\|| \cdot \|f(x+h) - 2f(x) + f(x-h)\| \\
&\leq \|f(x+h) - 2f(x) + f(x-h)\| \\
&\leq \omega_2(f;\delta)_\infty \leq C\delta^\alpha.
\end{aligned}
$$

Passing to supremum with x and h as above, it follows $\omega_2(g;\delta)_\infty \leq C\delta^\alpha$, which implies

$$
|B_n(g)(x) - g(x)| \leq M \left[\frac{x(1-x)}{n}\right]^{\alpha/2} \forall x \in [0,1].
$$

This implies

$$
|x^*[B_n(f)(x) - f(x)]| \leq M \left[\frac{x(1-x)}{n}\right]^{\alpha/2} \forall x \in [0,1].
$$

Passing to supremum with x^* and taking into account Theorem 4.2.4 we immediately obtain

$$
\|B_n(f)(x) - f(x)\| \leq M \left[\frac{x(1-x)}{n}\right]^{\alpha/2} \forall x \in [0,1],
$$

which proves (iii).

(iv) By Lorentz [126], p. 56 we have

$$
\|J_n(g) - g\|_\infty \leq C\omega_2(g; \frac{1}{n})_\infty.
$$

Reasoning exactly as in the first part of the proof of (ii), we immediately get the desired conclusion. $\qquad\square$

Remark. In the recent paper Anastassiou-Gal [15], results concerning best approximation by generalized polynomials with coefficients in vector spaces over $\mathbb{R}$ or $\mathbb{C}$ were obtained.

In what follows, for $f : [0,+\infty) \to X$ continuous and bounded on $[0,+\infty)$, we can attach the following operators :

$$
S_n(f)(x) = \sum_{k=0}^\infty s_{n,k}(x)f(k/n),
$$

$$
V_n(f)(x) = \sum_{k=0}^\infty v_{n,k}(x)f(k/n),
$$

$$P_n(f)(x) = \frac{(n/x)^n}{(n-1)!} \int_0^\infty e^{-nu/x} u^{n-1} f(u) du,$$

where $s_{n,k}(x) = e^{-nx}(nx)^k/k!$, $v_{n,k}(x) = \binom{n+k-1}{k} x^k(1-x)^{-n-k}$, called the generalized Szász-Mirakjan, Baskakov and Post-Widder operators, respectively.

Denote by $L_n(f)(x)$ any from the above operators (including $K_n(f)(x)$) and by $L_{n,r}(f)(x)$ the combination of the form (see Ditzian-Totik [64], p. 116) $L_{n,r}(f)(x) = \sum_{i=0}^{r-1} C_i(n) L_{n_i}(f)(x)$, where n_i and $C_i(n)$ satisfy the relations (9.27), a), b), c) and d) in Ditzian-Totik [64], p. 116.

The main result of this section is formally exactly the same as that for real-valued functions (see Ditzian-Totik [64], p. 117, Theorem 9.3.2) and can be stated as follows.

Theorem 4.2.6. (Anastassiou-Gal [14]) *Let $(X, \|\cdot\|)$ be a real normed space and $f : I \to X$ be continuous and bounded on I, where $I \subset \mathbb{R}$ is a subinterval of $\mathbb{R}$. Then*

$$\|L_{n,r}(f) - f\|_B \le M[\omega_\phi^{2r}(f; \frac{1}{\sqrt{n}})_B + n^{-r}\|f\|_B],$$

for all $n, r \in \mathbb{N}$ and for $\alpha < 2r$

$$\|L_{n,r}(f) - f\|_B \le C_1 n^{-\alpha/2} \text{ iff } \omega_\phi^{2r}(f; h)_B \le C_2 h^\alpha.$$

Here $\omega_\phi^{2r}(f; h)$ is given by Definition 4.2.3, (ii) and :
1) $I = [0, 1]$, $\phi^2(x) = x(1-x)$, $B = L^p[0,1]$, $1 \le p < \infty$, $\|f\|_B = (\int_0^1 \|f(x)\|^p dx)^{1/p}$, if $L_n(f) := K_n(f)$;
2) $I = [0, +\infty)$, $\phi^2(x) = x$, $B = C_b[0, +\infty)$, ($\|f\|_B = \sup\{\|f(x)\|; x \in [0, +\infty)\}$ $< +\infty$ iff $f \in C_b[0, +\infty))$, for $L_n(f) := S_n(f)$;
3) $I = [0, +\infty)$, $\phi^2(x) = x(1+x)$, $B = C_b[0, +\infty)$, if $L_n(f) := V_n(f)$;
4) $I = [0, +\infty)$, $\phi^2(x) = x^2$, $B = C_b[0, +\infty)$, for $L_n(f) := P_n(f)$.

Proof. For $f : I \to X$, X^* the conjugate of X and $x^* \in B_1 = \{x^* \in X^*; \|x^*\| \le 1\}$, let us define $g : I \to \mathbb{R}$ by $g(x) = x^*(f(x))$. Since x^* is linear and continuous, it commutes with $\sum$ and integral $\int$ and therefore we easily get $L_n(g)(x) = x^*[L_n(f)(x)]$ and $L_{n,r}(g)(x) = x^*[L_{n,r}(f)(x)]$.

We will prove only the cases 1) and 3) in the statement, since the other cases are similar.

Case 1). Let $1 \le p < \infty$. By Ditzian-Totik [64], p. 117, Theorem 9.3.2 we have

$$\|K_{n,r}(g) - g\|_{L^p[0,1]} \le M[\omega_\phi^{2r}(g; \frac{1}{\sqrt{n}})_{L^p[0,1]} + n^{-r}\|g\|_{L^p[0,1]}],$$

for all $n, r \in \mathbb{N}$ and for $\alpha < 2r$

$$\|K_{n,r}(g) - g\|_{L^p[0,1]} \le C_1 n^{-\alpha/2} \text{ iff } \omega_\phi^{2r}(g; h)_{L^p[0,1]} \le C_2 h^\alpha.$$

But $K_{n,r}(g)(x) - g(x) = x^*[K_{n,r}(f)(x) - f(x)]$ and

$$\|\overline{\Delta}^{2r}_{\delta\phi(x)} g(x)\|_{L^p[0,1]} = \left(\int_0^1 |\overline{\Delta}^{2r}_{\delta\phi(x)} g(x)|^p dx \right)^{1/p}$$

$$= \left(\int_0^1 |x^*(\overline{\Delta}^{2r}_{\delta\phi(x)} g(x))|^p dx \right)^{1/p}$$

$$\leq \left(\int_0^1 [\|\|x^*\|\| \cdot \|\overline{\Delta}^{2r}_{\delta\phi(x)} f(x)\|]^p dx \right)^{1/p}$$

$$\leq \left(\int_0^1 [\|\overline{\Delta}^{2r}_{\delta\phi(x)} f(x)\|]^p dx \right)^{1/p},$$

which immediately implies $\omega^{2r}_\phi(g; \frac{1}{\sqrt{n}})_{L^p[0,1]} \leq \omega^{2r}_\phi(f; \frac{1}{\sqrt{n}})_{L^p[0,1]}$.

Similarly, $\|g\|_{L^p[0,1]} = (\int_0^1 |x^*(f(x))|^p dx)^{1/p} \leq \|f\|_{L^p[0,1]}$.

Therefore, for all $x^* \in B_1$ we get

$$\|K_{n,r}(g) - g\|_{L^p[0,1]} \leq M[\omega^{2r}_\phi(f; \frac{1}{\sqrt{n}})_{L^p[0,1]} + n^{-r}\|f\|_{L^p[0,1]}],$$

i.e.

$$\left(\int_0^1 |x^*(f(x) - K_{n,r}(f)(x))|^p dx \right)^{1/p} \leq M[\omega^{2r}_\phi(f; \frac{1}{\sqrt{n}})_{L^p[0,1]} + n^{-r}\|f\|_{L^p[0,1]}].$$

In what follows, we need the following equality : for any $F : [0,1] \to X$ with $\|F\|_{L^p} < +\infty$, we have

$$\int_0^1 \sup\{|x^*(F(x))|^p; x^* \in B_1\}dx = \sup\{\int_0^1 |x^*(F(x))|^p dx; x^* \in B_1\}.$$

Indeed, by $|x^*(F(x))|^p \leq \sup\{|x^*(F(x))|^p; x \in B_1\}, \forall x \in [0,1]$, integrating and then passing to supremum with $x^* \in B_1$, we immediately get

$$\sup\{\int_0^1 |x^*(F(x))|^p dx; x^* \in B_1\} \leq \int_0^1 \sup\{|x^*(F(x))|^p; x^* \in B_1\}dx.$$

To prove the converse inequality, from the definition of supremum, for each $\varepsilon > 0$, there exists $y^* \in B_1$ (depending on ε), such that

$$\sup\{|x^*(F(x))|^p; x^* \in B_1\} \leq |y^*(F(x))|^p + \varepsilon, \forall x \in [0,1].$$

Integrating, we easily obtain

$$\int_0^1 \sup\{|x^*(F(x))|^p; x^* \in B_1\}dx \leq \int_0^1 |y^*(F(x))|^p dx + \varepsilon$$

$$\leq \sup\{\int_0^1 |x^*(F(x))|^p dx; x^* \in B_1\} + \varepsilon.$$

Passing with $\varepsilon \to 0$, we get the converse inequality too, which proves the claimed equality.

Consequently, we obtain

$$\int_0^1 \sup\{|x^*(f(x) - K_{n,r}(f)(x))|^p; x^* \in B_1\}dx$$

$$= \sup\{\int_0^1 |x^*(f(x) - K_{n,r}(f)(x))|^p dx; x^* \in B_1\}$$

$$\leq \{M[\omega_\phi^{2r}(f; \frac{1}{\sqrt{n}})_{L^p[0,1]} + n^{-r}\|f\|_{L^p[0,1]}]\}^p.$$

Since Theorem 4.2.4 immediately implies $\|x\|^p = \sup\{|x^*(x)|^p; x^* \in B_1\}, 0 < p < \infty$, from the previous relations we get

$$\|K_{n,r}(f) - f\|_{L^p[0,1]} \leq M[\omega_\phi^{2r}(f; \frac{1}{\sqrt{n}})_{L^p[0,1]} + n^{-r}\|f\|_{L^p[0,1]}].$$

Now, in order to prove the converse result for $K_{n,r}(f)(x)$, firstly let us suppose that $\|K_{n,r}(f) - f\|_{L^p[0,1]} \leq C_1 n^{-\alpha/2}$. We have

$$\|K_{n,r}(g) - g\|_{L^p[0,1]} = \left(\int_0^1 |x^*(K_{n,r}(f)(x) - f(x))|^p dx\right)^{1/p}$$

$$\leq \left(\int_0^1 \||x^*|\|^p\|(K_{n,r}(f)(x) - f(x)\|^p dx\right)^{1/p}$$

$$\leq \|K_{n,r}(f) - f\|_{L^p[0,1]} \leq C_1 n^{-\alpha/2},$$

for all $x^* \in B_1$.

From Ditzian-Totik [64], p. 117, Theorem 9.3.2, it follows that

$$\omega_\phi^{2r}(g; h)_{L^p[0,1]} \leq C_2 h^\alpha,$$

where C_2 depends only on C_1 (i.e. C_2 does not depend on $x^* \in B_1$). Then, for fixed $0 \leq \delta \leq h, x \in [0,1]$, we have

$$\|\overline{\Delta}_{\delta\phi(x)}^{2r} g(x)\|_{L^p[0,1]} = \left(\int_0^1 |x^*(\overline{\Delta}_{\delta\phi(x)}^{2r} f(x))|^p dx\right)^{1/p} \leq C_2 h^\alpha.$$

Passing to supremum with $x^* \in B_1$ and taking again into account that the integral commutes with supremum, we get

$$\left(\int_0^1 \|\overline{\Delta}_{\delta\phi(x)}^{2r} f(x)\|^p dx\right)^{1/p} \leq C_2 h^\alpha, \forall \delta \in (0, h].$$

Passing now to supremum with δ, we get $\omega_\phi^{2r}(f; h)_{L^p[0,1]} \leq C_2 h^\alpha$.

Conversely, let us suppose that $\omega_\phi^{2r}(f; h)_{L^p[0,1]} \leq C_2 h^\alpha$. We get

$$\omega_\phi^{2r}(g; h)_{L^p[0,1]} \leq \omega_\phi^{2r}(f; h)_{L^p[0,1]} \leq C_2 h^\alpha,$$

which by Ditzian-Totik [64], p. 117, Theorem 9.3.2 implies the inequality $\|K_{n,r}(g) - g\|_{L^p[0,1]} \leq C_1 n^{-\alpha/2}$, i.e.

$$\left(\int_0^1 |x^*[K_{n,r}(f)(x) - f(x)]|^p dx\right)^{1/p} \leq C_1 n^{-\alpha/2}, \forall x^* \in B_1.$$

Passing to supremum with $x^* \in B_1$ and reasoning as for the above lines, we finally arrive at

$$\left(\int_0^1 \|K_{n,r}(f)(x) - f(x)\|^p dx\right)^{1/p} \leq C_1 n^{-\alpha/2},$$

which proves the Case 1.

Case 3. The proof is a little simpler than in the Case 1. Let $B = C_b[0, \infty)$. First from $|x^*(u)| \leq \||x^*\|| \cdot \|u\| \leq \|u\|, \forall x^* \in B_1$, we immediately get $\omega_\phi^{2r}(g; \frac{1}{\sqrt{n}})_B \leq \omega_\phi^{2r}(f; \frac{1}{\sqrt{n}})_B, \|g\|_B \leq \|f\|_B$, where $\|f\|_B = \sup\{\|f(x)\|; x \in [0, +\infty)\}$.

From Ditzian-Totik [64], p.117, Theorem 9.3.2 this implies

$$\|V_{n,r}(g) - g\|_B \leq M[\omega_\phi^{2r}(f; \frac{1}{\sqrt{n}})_B + n^{-r}\|f\|_B].$$

Let $x^* \in B_1, x \in [0, +\infty)$ be fixed. From the above inequality we obtain

$$|x^*(V_{n,r}(f)(x) - f(x))| \leq M[\omega_\phi^{2r}(f; \frac{1}{\sqrt{n}})_B + n^{-r}\|f\|_B].$$

Passing to supremum with $x^* \in B_1$ and taking into account Theorem 4.2.4 we get

$$\|V_{n,r}(f)(x) - f(x)\| \leq M[\omega_\phi^{2r}(f; \frac{1}{\sqrt{n}})_B + n^{-r}\|f\|_B],$$

i.e. passing to supremum with $x \in [0, \infty)$ it follows

$$\|V_{n,r}(f) - f\|_B \leq M[\omega_\phi^{2r}(f; \frac{1}{\sqrt{n}})_B + n^{-r}\|f\|_B].$$

Now, let us suppose

$$\|V_{n,r}(f) - f\|_B \leq C_1 n^{-\alpha/2}.$$

It follows

$$|V_{n,r}(g)(x) - g(x)| = |x^*(V_{n,r}(f)(x) - f(x))|$$
$$\leq \||x^*\|| \cdot \|V_{n,r}(f)(x) - f(x)\| \leq C_1 n^{-\alpha/2},$$

i.e. $\|V_{n,r}(f) - f\|_B \leq C_1 n^{-\alpha/2}$.

Then, from the same Theorem 9.3.2 in Ditzian-Totik [64] it follows $\omega_\phi^{2r}(g; h)_B \leq C_2 h^\alpha$. Let $0 \leq \delta \leq h$ and x be fixed. We get

$$|\overline{\Delta}_{\delta\phi(x)}^{2r} g(x)| = |x^*(\overline{\Delta}_{\delta\phi(x)}^{2r} f(x))| \leq \omega_\phi^{2r}(g; h)_B \leq C_2 h^\alpha,$$

for all $x^* \in B_1$. Passing to supremum with $x^* \in B_1$, by Theorem 4.2.4 it follows $\|\overline{\Delta}_{\delta\phi(x)}^{2r} f(x)\| \leq C_2 h^\alpha$, and passing to supremum with δ and x, we get $\omega_\phi^{2r}(f; h)_B \leq C_2 h^\alpha$.

Conversely, let us suppose $\omega_\phi^{2r}(f; h)_B \leq C_2 h^\alpha$. It follows $\omega_\phi^{2r}(g; h)_B \leq C_2 h^\alpha$, which from the same Theorem 9.3.2 in Ditzian-Totik [64] implies $\|V_{n,r}(g) - g\|_B \leq C_1 n^{-\alpha/2}$. This means that for all $x^* \in B_1$ and $x \in [0, \infty)$, we have

$$|x^*(V_{n,r}(f)(x) - f(x))| \leq C_1 n^{-\alpha/2}.$$

Passing to supremum first with $x^* \in B_1$, by Theorem 4.2.4 and then passing to supremum with $x \in [0, +\infty)$, finally we arrive at $\|V_{n,r}(f) - f\|_B \leq C_1 n^{-\alpha/2}$, which proves the theorem. $\qquad \square$

Some applications of the above results to the approximation of random functions by random polynomials and of fuzzy-number-valued functions by fuzzy polynomials can be obtained.

First, let us recall that if (S, B, P) is a probability space (P is the probability), then the set of almost sure (a.s.) finite real random variables is denoted by $L(S, B, P)$ and it is a normed (Banach) space with respect to the norm $\|g\| = \int_S |g(t)| dP(t)$. Here, for $g_1, g_2 \in L(S, B, P)$ we consider $g_1 = g_2$ if $g_1(t) = g_2(t)$, a.s. $t \in S$.

A random function defined on $[0, 1]$ is a mapping $f : [0, 1] \to L(S, B, P)$ and we denote $f(x)(t) \in \mathbb{R}$ by $f(x, t)$. For this kind of f and the Bernstein random polynomials defined by $B_n(f)(x, t) = \sum_{k=0}^{n} p_{n,k}(x) f(\frac{k}{n}, t)$, a direct consequence of Theorem 4.2.5, (i) is the following.

Corollary 4.2.7. (Gal [92]) *If $f : [0, 1] \to L(S, B, P)$ is continuous on $[0, 1]$, then*

$$C_1 \omega_\phi^2(f; \frac{1}{\sqrt{n}})_\infty \leq \|B_n(f) - f\|_\infty \leq C_2 \omega_\phi^2(f; \frac{1}{\sqrt{n}})_\infty,$$

where $\|f\|_\infty = \sup\{\|f(x)\|; x \in [0, 1]\} = \sup\{\int_S |f(x, t)| dP(t); x \in [0, 1]\}$.

Now, for the random function $f : I \to L(S, B, P)$, continuous and bounded on I, where $I = [0, +\infty)$, let us consider the random operators given by $L_{n,r}(f)(x, t)$, where $L_{n,r}$ are defined by Theorem 4.2.6 and $L_{n,r}(f)(x, t)$ means that the usual $L_{n,r}$ is applied to the random function $f(x, t)$ considered as function of x only ($t \in S$ is fixed, arbitrary).

A direct consequence of Theorem 4.2.6 is the following (keeping the notations there).

Corollary 4.2.8. (Anastassiou-Gal [14]) *If $f : [0, +\infty) \to L(S, B, P)$ is continuous and bounded on $[0, +\infty)$, then*

$$\|L_{n,r}(f) - f\|_B \leq M[\omega_\phi^{2r}(f; \frac{1}{\sqrt{n}})_B + n^{-r}\|f\|_B],$$

for all $n, r \in \mathbb{N}$ and for $\alpha < 2r$

$$\|L_{n,r}(f) - f\|_B \leq C_1 n^{-\alpha/2} \text{ iff } \omega_\phi^{2r}(f; h)_B \leq C_2 h^\alpha.$$

Here

$$\|f\|_B = \sup\{\|f(x)\|; x \in [0, +\infty)\} = \sup\{\int_S |f(x, t)| dP(t); x \in [0, +\infty)\}$$

and f is called bounded on $[0, +\infty)$ if $\|f\|_B < +\infty$.

In what follows, we present some applications to the approximation of fuzzy-number-valued functions by fuzzy polynomials.

First let us recall a few facts concerning the fuzzy-number valued functions.

Given a set $X \neq \emptyset$, a fuzzy subset of X is a mapping $u : X \to [0, 1]$ and obviously any classical subset A of X can be considered as a fuzzy subset of X defined by $\chi_A : X \to [0, 1]$, $\chi_A(x) = 1$, if $x \in A$, $\chi_A(x) = 0$ if $x \in X \setminus A$. (see e.g. Zadeh [204][11]).

Let us denote by $\mathbb{R}_{\mathcal{F}}$ the class of fuzzy subsets of real axis $\mathbb{R}$ (i.e. $u : \mathbb{R} \to [0,1]$), satisfying the following properties:

(i) $\forall u \in \mathbb{R}_{\mathcal{F}}$, u is normal i.e. $\exists x_u \in \mathbb{R}$ with $u(x_u) = 1$;

(ii) For all $u \in \mathbb{R}_{\mathcal{F}}$, u is convex fuzzy set, i.e.

$$u(tx + (1-t)y) \geq \min\{u(x), u(y)\},$$

$\forall t \in [0,1]$, $x, y \in \mathbb{R}$;

(iii) $\forall u \in \mathbb{R}_{\mathcal{F}}$, u is upper semi-continuous on $\mathbb{R}$;

(iv) $\overline{\{x \in \mathbb{R} : u(x) > 0\}}$ is compact, where $\overline{A}$ denotes the closure of A.

Then $\mathbb{R}_{\mathcal{F}}$ is called the space of fuzzy real numbers (see e.g. Dubois-Prade [66]).

Remark. Obviously $\mathbb{R} \subset \mathbb{R}_{\mathcal{F}}$, because any real number $x_0 \in \mathbb{R}$, can be described as the fuzzy number whose value is 1 for $x = x_0$ and 0 otherwise.

For $0 < r \leq 1$ and $u \in \mathbb{R}_{\mathcal{F}}$ define $[u]^r = \{x \in \mathbb{R}; u(x) \geq r\}$ and $[u]^0 = \overline{\{x \in \mathbb{R}; u(x) > 0\}}$. Then it is well known that for each $r \in [0,1]$, $[u]^r$ is a bounded closed interval. For $u, v \in \mathbb{R}_{\mathcal{F}}$ and $\lambda \in \mathbb{R}$, we have the sum $u \oplus v$ and the product $\lambda \odot u$ defined by $[u \oplus v]^r = [u]^r + [v]^r$, $[\lambda \odot u]^r = \lambda [u]^r$, $\forall r \in [0,1]$, where $[u]^r + [v]^r$ means the usual addition of two intervals (as subsets of $\mathbb{R}$) and $\lambda [u]^r$ means the usual product between a scalar and a subset of $\mathbb{R}$ (see e.g. Dubois-Prade [66], Congxin-Zengtai [201]).

Define $D : \mathbb{R}_{\mathcal{F}} \times \mathbb{R}_{\mathcal{F}} \to \mathbb{R}_+ \cup \{0\}$ by

$$D(u, v) = \sup_{r \in [0,1]} \max\{|u_-^r - v_-^r|, |u_+^r - v_+^r|\},$$

where $[u]^r = [u_-^r, u_+^r]$, $[v]^r = [v_-^r, v_+^r]$.

The following properties are known (Dubois-Prade [66]):

$D(u \oplus w, v \oplus w) = D(u, v)$, $\forall u, v, w \in \mathbb{R}_{\mathcal{F}}$

$D(k \odot u, k \odot v) = |k| D(u, v)$, $\forall u, v \in \mathbb{R}_{\mathcal{F}}, \forall k \in \mathbb{R}$;

$D(u \oplus v, w \oplus e) \leq D(u, w) + D(v, e)$, $\forall u, v, w, e \in \mathbb{R}_{\mathcal{F}}$ and $(\mathbb{R}_{\mathcal{F}}, D)$ is a complete metric space.

Also, we need the following Riemann integral, as particular case of the Henstock integral introduced by Congxin-Zengtai [201].

A function $f : [a, b] \to \mathbb{R}_{\mathcal{F}}$, $[a, b] \subset \mathbb{R}$ is called Riemann integrable on $[a, b]$, if there exists $I \in \mathbb{R}_{\mathcal{F}}$, with the property: $\forall \epsilon > 0$, $\exists \delta > 0$, such that for any division of $[a, b]$, $d : a = x_0 < ... < x_n = b$ of norm $\nu(d) < \delta$, and for any points $\xi_i \in [x_i, x_{i+1}]$, $i = \overline{0, n-1}$, we have

$$D\left(\sum_{i=0}^{n-1}{}^* f(\xi_i) \odot (x_{i+1} - x_i), I\right) < \epsilon,$$

where $\sum^*$ means sum with respect to $\oplus$. Then we denote $I = \int_a^b f(x)\,dx$.

A crucial result for our reasonings will be the following known result.

Theorem 4.2.9. (see e.g. Congxin-Zengtai [201]) $\mathbb{R}_{\mathcal{F}}$ *can be embedded in* $\mathbb{B} = \bar{C}[0,1] \times \bar{C}[0,1]$, *where* $\bar{C}[0,1]$ *is the class of all real valued bounded functions* $f :$

$[0,1] \to \mathbb{R}$ *such that f is left continuous for any $x \in (0,1]$, f has right limit for any $x \in [0,1)$ and f is right continuous at 0. With the norm $\|\cdot\| = \sup_{x \in [0,1]} |f(x)|$, $\bar{C}[0,1]$ is a Banach space. Denote $\|\cdot\|_{\mathbb{B}}$ the usual product norm i.e. $\|(f,g)\|_{\mathbb{B}} = \max\{\|f\|, \|g\|\}$. Let us denote the embedding by $j : \mathbb{R}_{\mathcal{F}} \to \mathbb{B}$, $j(u) = (u_-, u_+)$. Then $j(\mathbb{R}_{\mathcal{F}})$ is a closed convex cone in $\mathbb{B}$ and j satisfies the following properties:*

(i) $j(s \odot u \oplus t \odot v) = s \cdot j(u) + t \cdot j(v)$ for all $u, v \in \mathbb{R}_{\mathcal{F}}$ and $s, t \geq 0$ (here ".") and "+" denote the scalar multiplication and addition in $\mathbb{B}$);

(ii) $D(u,v) = \|j(u) - j(v)\|_{\mathbb{B}}$ (i.e. j embeds $\mathbb{R}_{\mathcal{F}}$ in $\mathbb{B}$ isometrically.)

Now, for $f : [0,1] \to \mathbb{R}_{\mathcal{F}}$ a fuzzy-number-valued function, the generalized Bernstein and Bernstein-Kantorovich polynomials attached to f can be defined by

$$B_n(f)(x) = \sum_{k=0}^{n} p_{n,k}(x) f\left(\frac{k}{n}\right),$$

and

$$K_n(f)(x) = (n+1) \sum_{k=0}^{n} p_{n,k}(x) \odot \int_{k/(n+1)}^{(k+1)/(n+1)} f(t)dt,$$

respectively, where $p_{n,k}(x) = \binom{n}{k} x^k (1-x)^{n-k}$ and the integral $\int_a^b f(t)dt$ is defined as the limit for $m \to \infty$ in the distance D, of the all (classical defined) Riemann sums $\sum_{i=0}^{m} (x_{i+1} - x_i) \odot f(\xi_i)$. (Here all the sums are with respect to the operation $\oplus$.)

Also, let us define the following moduli of continuity and smoothness of f :

$$\omega_1(f; \delta)_\infty = \sup\{D(f(x+h), f(x)); x, x+h \in [0,1], 0 \leq h \leq \delta\},$$

$$\omega_\phi^2(f; \delta)_\infty = \sup\{D[f(x+h\phi(x)) \oplus f(x - h\phi(x)), 2 \odot f(x)];$$

$$x, x+h\phi(x), x - h\phi(x) \in [0,1], 0 \leq h \leq \delta\}.$$

Here $\phi^2(x) = x(1-x)$.

We present

Theorem 4.2.10. (Anastassiou-Gal [14]) *Let $f : [0,1] \to \mathbb{R}_{\mathcal{F}}$ be continuous on $[0,1]$. There exist the absolute constants C_1, C_2, such that for all $n \in \mathbb{N}$ we have :*

(i)

$$C_1 \omega_\phi^2\left(f; \frac{1}{\sqrt{n}}\right)_\infty \leq \sup\{D[B_n(f)(x), f(x)]; x \in [0,1]\} \leq C_2 \omega_\phi^2\left(f; \frac{1}{\sqrt{n}}\right)_\infty.$$

(ii)

$$C_1\left[\omega_\phi^2\left(f; \frac{1}{\sqrt{n}}\right)_\infty + \omega_1\left(f; \frac{1}{n}\right)_\infty\right] \leq$$

$$\sup\{D[K_n(f)(x), f(x)]; x \in [0,1]\} \leq C_2\left[\omega_\phi^2\left(f; \frac{1}{\sqrt{n}}\right)_\infty + \omega_1\left(f; \frac{1}{n}\right)_\infty\right].$$

Proof. Define $g : [0,1] \to X$ by $g(x) = j[f(x)], x \in [0,1]$, where j is given by Theorem 4.2.9 and $X = \bar{C}[0,1] \times \bar{C}[0,1]$ endowed with the norm in Theorem 4.2.9, denoted by $\| \cdot \|$.

(i) According to Theorem 4.2.5, there exist the absolute constants C_1, C_2, such that for all $n \in \mathbb{N}$ we have

$$C_1 \omega_\phi^2(g; \frac{1}{\sqrt{n}})_\infty \leq \|B_n(g) - g\|_\infty \leq C_2 \omega_\phi^2(g; \frac{1}{\sqrt{n}})_\infty,$$

where $\|f\|_\infty = \sup\{\|f(x)\|; x \in [0,1]\}$.

By Theorem 4.2.9, (ii), we notice

$$\|g(x+h) - g(x)\| = \|j[f(x+h)] - j[f(x)]\| = D[f(x+h), f(x)],$$

$$\|g(x + h\phi(x)) + g(x - h\phi(x)) - 2g(x)\|$$
$$= \|j[f(x + h\phi(x)) \oplus f(x - h\phi(x))] - j[2 \odot f(x)]\|$$
$$= D[f(x + h\phi(x)) \oplus f(x - h\phi(x)), 2 \odot f(x)],$$

which immediately implies $\omega_1(g; \delta)_\infty = \omega_1(f; \delta)_\infty$ and $\omega_\phi^2(g; \delta)_\infty = \omega_\phi^2(f; \delta)_\infty$, for all $\delta \geq 0$.

Since on the other hand, from the linearity of j over the positive scalars we have

$$\|g(x) - B_n(g)(x)\| = \|j[f(x)] - j[B_n(f)(x)]\| = D[f(x), B_n(f)(x)],$$

we easily arrive at the estimates in the statement.

(ii) Since j is linear over the positive scalars and j commutes with the integral, we get $K_n(g)(x) = j[K_n(f)(x)]$ and by similar reasoning as for the above point (i), we get the desired estimates. The theorem is proved. $\qquad \square$

4.2.2   *Complex Variable Case*

First we recall some known concepts and results.

Definition 4.2.11. (see e.g. Hille-Phillips [109], p. 92–93) Let $(X, \| \cdot \|)$ be a complex Banach space, $R > 1$ and $f \colon \overline{\mathbb{D}}_R \to X$. We say that f is holomorphic on $\mathbb{D}_R$ if for any $x^* \in B_1 = \{x^* \colon X \to \mathbb{C}; \ x^* \text{ linear and continuous}, \|\|x^*\|\| \leq 1\}$, the function $g \colon \overline{\mathbb{D}}_R \to \mathbb{C}$ given by $g(z) = x^*[f(z)]$, is holomorphic on $\mathbb{D}_R$. (Here $\|\| \cdot \|\|$ represents the usual norm in the dual space X^*.)

We denote by $A(\overline{\mathbb{D}}_R; X)$ the space of all functions $f \colon \overline{\mathbb{D}}_R \to X$ which are continuous on $\overline{\mathbb{D}}_R$ and holomorphic on $\mathbb{D}_R$. It is a Banach space with respect to the norm $\|f\|_R = \max\{\|f(z)\|; z \in \overline{\mathbb{D}}_R\}$

Note that everywhere in this subsection $(X, \| \cdot \|)$ will be a complex Banach space.

Theorem 4.2.12. (see e.g. Hille-Phillips [109], p. 93) *If $f \colon \mathbb{D}_R \to X$ is holomorphic on $\mathbb{D}_R$, then $f(z)$ is continuous (as mapping between two metric spaces) and differentiable in the sense that exists $f'(z) \in \mathbb{C}$ given by*

$$\lim_{h \to 0} \left\| \frac{f(z+h) - f(z)}{h} - f'(z) \right\| = 0,$$

uniformly with respect to z in any compact subset of $\mathbb{D}_R$.

Theorem 4.2.13. *(see e.g. Hille-Phillips [109], p. 97) If $f \colon \mathbb{D}_R \to X$ is holomorphic on $\mathbb{D}_R$, then we have the Taylor expansion*

$$f(z) = \sum_{n=0}^{\infty} \frac{f^{(n)}(0)}{n!} z^n, \quad z \in \mathbb{D}_R,$$

where the series converges uniformly on any compact subset of $\mathbb{D}_R$.

For the beginning, to $f \colon \mathbb{D}_R \to X$ holomorphic on $\mathbb{D}_R$ with $R > 1$, let us attach the Bernstein operator of complex variable

$$B_n(f)(z) = \sum_{k=0}^{n} \binom{n}{k} z^k (1-z)^{n-k} f(k/n), |z| \le R.$$

Theorem 4.2.14. *Let $(X, \|\cdot\|)$ be a complex Banach space. Suppose that $R > 1$ and $f \colon \mathbb{D}_R \to X$ is holomorphic in $\mathbb{D}_R$, that is $f(z) = \sum_{k=0}^{\infty} c_k z^k$, for all $z \in \mathbb{D}_R$, with $c_k \in X$, for all k.*

(i) Let $1 \le r < R$ be arbitrary fixed. For all $|z| \le r$ and $n \in \mathbb{N}$, we have

$$\|B_n(f)(z) - f(z)\| \le \frac{K_r(f)}{n},$$

where $0 < K_r(f) = \frac{3r(1+r)}{2} \sum_{j=2}^{\infty} j(j-1)\|c_j\| r^{j-2} < \infty$.

(ii) If $1 \le r < r_1 < R$ are arbitrary fixed, then for all $|z| \le r$ and $n, p \in \mathbb{N}$,

$$\|B_n^{(p)}(f)(z) - f^{(p)}(z)\| \le \frac{K_{r_1}(f) p! r_1}{n(r_1 - r)^{p+1}},$$

where $K_{r_1}(f)$ is given as at the above point (i).

(iii) Let $r \in [1, R)$. Then for all $n \in \mathbb{N}, |z| \le r$, we have

$$\left\| B_n(f)(z) - f(z) - \frac{z(1-z)}{2n} f''(z) \right\| \le \frac{5(1+r)^2}{2n} \cdot \frac{M_r(f)}{n},$$

where $M_r(f) = \sum_{k=3}^{\infty} \|c_k\| k(k-1)(k-2)^2 r^{k-2} < \infty$.

Proof. Let $x^* \in X^*$ be with $\|\|x^*\|\| \le 1$ and define $g \colon\colon \mathbb{D}_R \to \mathbb{C}$ by $g(z) = x^*[f(z)]$, for all $z \in \mathbb{D}_R$. By Definition 4.2.11 g is holomorphic (analytic) in $\mathbb{D}_R$.

(i) Applying to g Theorem 1.1.2, (i), for all $|z| \le r$ and $n \in \mathbb{N}$ it follows

$$|B_n(g)(z) - g(z)| \le \frac{C_r(g)}{n},$$

where $C_r(g) = \frac{3r(1+r)}{2} \sum_{j=2}^{\infty} j(j-1)|a_j| r^{j-2}$, and a_j are the coefficients in the series representation of the analytic function $g(z) = x^*[f(z)]$. But the linearity and continuity of x^* implies $g(z) = \sum_{j=0}^{\infty} x^*(c_j) z^j$ and $|a_j| = |x^*(c_j)| \le \|\|x^*\|\| \cdot \|c_j\| \le \|c_j\|$, for all j. Therefore, since by the linearity of x^* we have $|B_n(g)(z) - g(z)| = |x^*[B_n(f)(z) - f(z)]|$, it follows

$$|x^*[B_n(f)(z) - f(z)]| \le \frac{K_r(f)}{n},$$

with $K_r(f) = \frac{3r(1+r)}{2} \sum_{j=2}^{\infty} j(j-1)\|c_j\|r^{j-2}$.

Passing above to supremum with $\|x^*\| \leq 1$ and taking into account Theorem 4.2.4, we obtain

$$\|B_n(f)(z) - f(z)\| \leq \frac{K_r(f)}{n},$$

for all $|z| \leq r$ and $n \in \mathbb{N}$.

(ii) Applying now to g Theorem 1.1.2, (ii) and taking into account that $g^{(p)}(z) = x^*[f^{(p)}(z)]$, similar reasonings with those from the above point (i) prove the desired estimate.

(iii) It is immediate by applying Theorem 1.1.3, (ii) to g and by using similar reasoning with those from the above points (i) and (ii). $\qquad\qquad\square$

As in the case of complex-valued functions, we can prove that in fact the order of approximation in Theorem 4.2.14 is exactly $\frac{1}{n}$. In this sense we present

Theorem 4.2.15. *Let $(X, \|\cdot\|)$ be a complex Banach space. Suppose that $R > 1$ and $f : \mathbb{D}_R \to X$ is holomorphic in $\mathbb{D}_R$, that is $f(z) = \sum_{k=0}^{\infty} c_k z^k$, for all $z \in \mathbb{D}_R$, with $c_k \in X$, for all k. If f is not of the form $f(z) = c_0 + c_1 z$, with $c_0, c_1 \in X$, then for any $r \in [1, R)$ we have*

$$\|B_n(f) - f\|_r \geq \frac{K_r(f)}{n}, n \in \mathbb{N},$$

where $\|f\|_r = \max\{\|f(z)\|; |z| \leq r\}$ and the constant $K_r(f)$ depends only on f and r.

Proof. By the hypothesis on f, there exists at least one $c_s \neq 0$ with $s \geq 2$. As a consequence of Hahn-Banach theorem, there exists $y^* \in X^*$, with $\||y^*\|| = 1$ and $y^*(c_s) = \|c_s\| \neq 0$.

Define $g : \mathbb{D}_R \to \mathbb{C}$ by $g(z) = y^*[f(z)]$, for all $z \in \mathbb{D}_R$. It follows that $g(z)$ cannot be of the form $g(z) = az + b$ with $a, b \in \mathbb{C}$. Indeed, supposing the contrary, from the linearity and continuity of y^* we obtain $g(z) = \sum_{j=0}^{\infty} y^*(c_j)z^j = az + b$, which implies $y^*(c_j) = 0$, for all $j \geq 2$, a contradiction. Applying now Theorem 1.1.4 to g (see better the proof of Theorem 1.5.3), there exists an index n_0 such that for all $n \geq n_0$ and $|z| \leq r$ we have

$$|B_n(g)(z) - g(z)|_r \geq \frac{1}{n}\left\|\frac{e_1(1 - e_1)g''}{4}\right\|_r,$$

where $\left\|\frac{e_1(1-e_1)g''}{4}\right\|_r > 0$.

This is equivalent to

$$|y^*[B_n(f)(z) - f(z)]| \geq \frac{1}{n}\left\|y^*\left[\frac{e_1(1 - e_1)f''}{4}\right]\right\|_r.$$

But since

$$|y^*[B_n(f)(z) - f(z)]| \leq \||y^*\|| \cdot \|B_n(f)(z) - f(z)\| \leq \|B_n(f)(z) - f(z)\|,$$

it follows that for all $|z| \leq r$ and $n \geq n_0$ we have

$$\|B_n(f)(z) - f(z)\| \geq \frac{1}{n} \left\| y^* \left[\frac{e_1(1-e_1)f''}{4} \right] \right\|_r.$$

For $1 \leq n \leq n_0 - 1$ we reason as in the proof of Theorem 1.1.4, so that the conclusion in the statement is immediate. □

If for $f : \mathbb{D}_R \to X$ holomorphic in $\mathbb{D}_R$ we consider the Butzer's linear combination of complex Bernstein polynomials defined by the recurrence

$$L_n^{[0]}(f)(z) = B_n(f)(z), (2^q - 1)L_n^{[2q]}(f)(z) = 2^q L_{2n}^{[2q-2]}(f)(z) - L_n^{[2q-2]}(f)(z),$$

where $z \in \mathbb{C}, q = 1, 2, ...,$ then taking into account the Remarks 2), 3), 4), and 5) after the proof of Theorem 1.4.1 and Theorem 1.4.2, by similar reasonings as above we obtain the following.

Theorem 4.2.16. *Let $(X, \| \cdot \|)$ be a complex Banach space. Suppose that $R > 1$, q is a given natural number and $f : \mathbb{D}_R \to X$ is holomorphic in $\mathbb{D}_R$. If f is not of the form $f(z) = c_0 + c_1 z + ... + c_q z^q$, with $c_0, c_1, ..., c_q \in X$, then for any $r \in [1, R)$ and $n \in \mathbb{N}$ we have*

$$\left\| L_n^{[2q-2]}(f) - f \right\|_r \sim \frac{1}{n^q},$$

where $\|f\|_r = \max\{\|f(z)\|; |z| \leq r\}$.

Remarks. 1) By using the above method, the results in Chapter 1 can be extended in a similar manner to the case of vector-valued Bernstein-type operators of a complex variable.

2) The method in this section can be used for other type too of vector-valued operators of a complex variable. In what follows we illustrate this method for some convolution-kind operators.

For $f : \overline{\mathbb{D}}_R \to X$ holomorphic on $\mathbb{D}_R$, $f(z) = \sum_{k=0}^{\infty} c_j z^j$, $c_j \in X$ for all j, let us attach the convolution operators of complex variable of de la Vallée Poussin kind and of Riesz-Zygmund kind, given by

$$P_n(f)(z) = \frac{1}{2\pi} \int_{-\pi}^{\pi} f(ze^{iu})K_n(u)du = \frac{1}{\binom{2n}{n}} \sum_{j=0}^{n} c_j \binom{2n}{n+j} z^j$$

$$= \sum_{j=0}^{n} c_j \frac{(n!)^2}{(n-j)!(n+j)!} z^j,$$

and

$$R_{n,s}(f)(z) = \sum_{k=0}^{n-1} c_k \left[1 - \left(\frac{k}{n} \right)^s \right] z^k, n \in \mathbb{N},$$

respectively, where $s \in \mathbb{N}$.

Let $x^* \in X^*$ be with $\|\|x^*\|\| = 1$ and for $f \in A(\overline{\mathbb{D}}_R; X)$ define $g : \overline{\mathbb{D}}_R \to \mathbb{C}$ by $g(z) = x^*[f(z)]$. Taking into account Theorems 3.2.1, 3.2.2 and 3.2.3 applied to

$P_n(g)(z)$ and Theorems 3.2.6, 3.2.8 applied to $R_{n,s}(g)(z)$ and using similar reasonings with those in the proofs of Theorems 4.2.14 and 4.2.15, we immediately obtain the following two results.

Theorem 4.2.17. *Let $(X, \|\cdot\|)$ be a complex Banach space. Suppose that $R > 1$ and $f : \mathbb{D}_R \to X$ is holomorphic in $\mathbb{D}_R$, that is $f(z) = \sum_{k=0}^{\infty} c_k z^k$, for all $z \in \mathbb{D}_R$, with $c_k \in X$, for all k.*

(i) Let $r \in [1, R)$ be arbitrary fixed. For all $|z| \le r$ and $n \in \mathbb{N}$ we have

$$\|P_n(f)(z) - f(z)\| \le \frac{M_r(f)}{n},$$

where $M_r(f) = \sum_{k=1}^{\infty} \|c_k\| k^2 r^k < \infty$.

(ii) If $1 \le r < r_1 < R$ and $p \in \mathbb{N}$ then for all $|z| \le r$ and $n \in \mathbb{N}$ we have

$$\|P_n^{(p)}(f)(z) - f^{(p)}(z)\| \le \frac{r_1 p! M_{r_1}(f)}{(r_1 - r)^{p+1} n},$$

where $M_{r_1}(f)$ is given at the point (i).

(iii) Let $r \in [1, R)$. For all $|z| \le r$ and $n \in \mathbb{N}$ we have

$$\left\| P_n(f)(z) - f(z) + \frac{zf''(z)}{n} + \frac{zf'(z)}{n} \right\| \le \frac{A_r(f)}{n^2},$$

where $A_r(f) = \sum_{k=1}^{\infty} \|c_k\| k^4 r^k < \infty$.

(iv) If f is not of the form $f(z) = c_0$ for all z, then for any $r \in [1, R)$ we have

$$\|P_n(f) - f\|_r \ge \frac{K_r(f)}{n}, n \in \mathbb{N},$$

where $\|f\|_r = \max\{\|f(z)\|; |z| \le r\}$ and the constant $K_r(f)$ depends only on f and r.

Theorem 4.2.18. *Let $(X, \|\cdot\|)$ be a complex Banach space. Suppose that $R > 1$ and $f : \mathbb{D}_R \to X$ is holomorphic in $\mathbb{D}_R$, that is $f(z) = \sum_{k=0}^{\infty} c_k z^k$, for all $z \in \mathbb{D}_R$, with $c_k \in X$, for all k.*

(i) If $1 \le r < r_1 < R$ and $s, p \in \mathbb{N}$ then for all $|z| \le r$ and $n \in \mathbb{N}$ we have

$$\|R_{n,s}^{(p)}(f)(z) - f^{(p)}(z)\| \le \frac{r_1 p! C_{r_1,s}(f)}{(r_1 - r)^{p+1} n^s},$$

where $C_{r_1,s}(f) = \sum_{k=0}^{\infty} \|c_k\| k^s r^k < \infty$.

(ii) Let $r \in [1, R)$ and $s \in \mathbb{N}$. For all $|z| \le r$ and $n \in \mathbb{N}$ we have

$$\left\| R_{n,s}(f)(z) - f(z) + \frac{1}{n^s} \sum_{j=1}^{s} \alpha_{j,s} z^j f^{(j)}(z) \right\| \le \frac{A_{r,s}(f)}{n^{s+1}},$$

where $A_{r,s}(f) = \sum_{k=1}^{\infty} \|c_k\| k^{s+1} r^k < \infty$ and the real numbers $\alpha_{j,s}$ are defined by Lemma 3.2.7.

For other related results let us mention the following two results.

Theorem 4.2.19. (Anastassiou-Gal [16]) *Let $(X, \|\cdot\|)$ be a complex normed space and $f \in A(\overline{\mathbb{D}}_1; X)$.*

(i) Define

$$J_n(f)(z) = \frac{1}{2\pi n'[2(n')^2 + 1]} \int_{-\pi}^{\pi} f(ze^{iu}) K_n(u)\, du,$$

where $K_n(u) = [\sin(n'u/2)/\sin(u/2)]^4$, $n' = [n/2] + 1$, and $\int_{-\pi}^{\pi}$ is the classical Riemann integral for vector-valued functions. Then $J_n(f)(z)$ is a polynomial in z of degree $\leq n$, with coefficients in X, which satisfies the estimate

$$\|f(z) - J_n(f)(z)\| \leq C\omega_2\left(f; \frac{1}{n}\right)_{\partial \mathbb{D}_1}, \quad \forall z \in \overline{\mathbb{D}}_1,$$

where

$$\omega_p(f; \delta)_{\partial \mathbb{D}_1} = \sup\{\|\Delta_u^p f(e^{ix})\|; |x| \leq \pi, |u| \leq \delta\}, \quad p = 2, 3, \ldots;$$

(ii) Define the polynomials in z (with coefficients in X)

$$I_{n,p}(f)(z) = -\int_{-\pi}^{\pi} K_{n,r}(u) \sum_{k=1}^{p+1} (-1)^k \binom{p+1}{k} f(ze^{iku})\, du,$$

where $K_{n,r}(u) = \lambda_{n,r}[\sin(nu/2)/\sin(u/2)]^{2r}$, r is the smallest integer which satisfies $r \geq (p+2)/2$ and the constants $\lambda_{n,r}$ are chosen such that $\int_{-\pi}^{\pi} K_{n,r}(u)\, du = 1$. Then we have

$$\|I_{n,p}(f)(z) - f(z)\| \leq C_p \omega_{p+1}\left(f; \frac{1}{n}\right)_{\partial \mathbb{D}_1}, \quad \forall z \in \overline{\mathbb{D}}_1;$$

(iii) Define

$$V_n(f)(z) = 2L_{2n}(f)(z) - L_n(f)(z),$$

where

$$L_n(f)(z) = \frac{1}{2n\pi} \int_{-\pi}^{\pi} f(ze^{iu}) F_n(u)\, du,$$

$F_n(u) = [\sin(nu/2)/\sin(u/2)]^2$. *Then $V_n(f)(z)$ is a polynomial of degree $\leq 2n - 1$ in z, with coefficients in X which satisfies the estimate*

$$\|f(z) - V_n(f)(z)\| \leq 4E_n(f)_\infty(\overline{\mathbb{D}}_1), \quad \forall z \in \overline{\mathbb{D}}_1,$$

where $E_n(f)_\infty(\overline{\mathbb{D}}_1) = \inf\{\|f - P\|_{\overline{\mathbb{D}}_1}; P$ polynomial of degree $\leq n$ in z, with coefficients in $X\}$, $\|f\|_{\overline{\mathbb{D}}_1} = \sup\{\|f(z)\|; z \in \overline{\mathbb{D}}_1\}$.

(iv) For $f \in A(\overline{\mathbb{D}}_1; X)$ define $Q_{n,\xi}(f)(z)$ and $W_{n,\xi}(f)(z)$ by

$$Q_{n,\xi}(f)(z) = -\frac{1}{\frac{2}{\xi}\arctan\left(\frac{\pi}{\xi}\right)} \cdot \sum_{k=1}^{n+1} (-1)^k \binom{n+1}{k} \int_{-\pi}^{\pi} \frac{f(ze^{iku})}{u^2 + \xi^2}\, du,$$

$$W_{n,\xi}(f)(z) = -\frac{1}{2C(\xi)} \cdot \sum_{k=1}^{n+1} (-1)^k \binom{n+1}{k} \int_{-\pi}^{\pi} f(ze^{iku}) e^{-u^2/\xi^2}\, du.$$

Then we have

$$\|Q_{n,\xi}(f)(z) - f(z)\| \leq K(n,\xi)\omega_{n+1}(f;\xi)_{\partial \mathbb{D}_1}, \quad \forall z \in \overline{\mathbb{D}}_1, \xi \in (0,1], \ n \in \mathbb{N}$$

and

$$\|W_{n,\xi}(f)(z) - f(z)\| \leq C_n\omega_{n+1}(f;\xi)_{\partial \mathbb{D}_1}, \quad \forall z \in \overline{\mathbb{D}}_1, \ \xi \in (0,1], \ n \in \mathbb{N},$$

where

$$K(n,\xi) = \frac{\int_0^{\pi/\xi} \frac{(u+1)^{n+1}}{u^2+1} du}{\tan^{-1}\left(\frac{\pi}{\xi}\right)}, \ C_n = \frac{\int_0^\infty (1+u)^{n+1} e^{-u^2} du}{\int_0^\pi e^{-u^2} du}.$$

(v) If for $f \in A(\overline{\mathbb{D}}_1; X)$ we consider the operators

$$Q_\xi(f)(z) = \frac{\xi}{\pi}\int_{-\pi}^{\pi} \frac{f(ze^{iu})}{u^2+\xi^2} du, \quad z \in \overline{\mathbb{D}}_1, \ \xi > 0,$$

$$W_\xi(f)(z) = \frac{1}{\sqrt{\pi\xi}}\int_{-\pi}^{\pi} f(ze^{iu})e^{-u^2/\xi} du, \quad z \in \overline{\mathbb{D}}_1, \ \xi > 0,$$

then we have

$$\|Q_\xi(f)(z) - f(z)\| \leq C\frac{\omega_2(f;\xi)_{\partial \mathbb{D}_1}}{\xi}, \quad \forall z \in \overline{\mathbb{D}}_1, \ \xi \in (0,1]$$

and

$$\|W_\xi(f)(z) - f(z)\| \leq C\frac{\omega_2(f;\xi)_{\partial \mathbb{D}_1}}{\xi} + C\sqrt{\xi}\|f\|_{\overline{\mathbb{D}}_1}, \quad \forall z \in \overline{\mathbb{D}}_1, \ \xi \in (0,1],$$

where $\||f\|\|_{\overline{\mathbb{D}}_1} = \sup\{\|f(z)\|; |z| \leq 1\}$.

Remark. For the proof of Theorem 4.2.19, as above is used the x^*-functional method to the corresponding results for complex-valued functions in the Sections 3.1 and 3.2.

Theorem 4.2.20. (Aral-Gal [32]) *Let $f \in A(\overline{\mathbb{D}}_R; X)$ where $(X, \|\cdot\|)$ is a complex normed space. If for $\lambda > 0$, $0 < q < 1$, we consider the q-operators*

$$P_\lambda(f;q,z) \equiv P_\lambda(f;z) := \frac{(1-q)}{2[\lambda]_q \ln q^{-1}}\int_{-\infty}^{\infty} \frac{f(ze^{it})}{E_q\left(\frac{(1-q)|t|}{[\lambda]_q}\right)} dt,$$

$$W_\lambda(f;q,z) \equiv W_\lambda(f;z) := \frac{1}{\pi\sqrt{[\lambda]_q}(q^{1/2};q)_{1/2}}\int_{-\infty}^{\infty} \frac{f(ze^{it})}{E_q\left(\frac{t^2}{[\lambda]_q}\right)} dt,$$

then we have

$$\|P_\lambda(f;z) - f(z)\| \leq (R+1)(1 + \frac{1}{q})\omega_1(f;[\lambda]_q)_{\overline{\mathbb{D}}_R},$$

$$\|W_\lambda(f;z) - f(z)\| \leq (R+1)\left(1 + \sqrt{q^{-1/2}(1-q^{1/2})}\right)\omega_1\left(f;\sqrt{[\lambda]_q}\right)_{\overline{\mathbb{D}}_R},$$

for all $z \in \overline{\mathbb{D}}_R$, where $\omega_1(f; \delta)_{\overline{\mathbb{D}}_R} = \sup\{\|f(z_1) - f(z_2)\|; z_1, z_2 \in \overline{\mathbb{D}}_R, |z_1 - z_2| \leq \delta\}$.

Proof. Let $x^* \in B_1$ and define $g(z) = x^*[f(z)]$, $g: \overline{\mathbb{D}}_R \to \mathbb{C}$. By Theorem (?, I don't know yet its number) we have $|P_\lambda(g; z) - g(z)| \leq 2(1 + \frac{1}{q})\omega_1(g; [\lambda]_q)_{\overline{\mathbb{D}}_R}$, for all $z \in \overline{\mathbb{D}}_R$, where

$$\omega_1(g; \delta)_{\overline{\mathbb{D}}_R} = \sup\{|x^*[f(z_1) - f(z_2)]|; z_1, z_2 \in \overline{\mathbb{D}}_R, |z_1 - z_2| \leq \delta\}$$
$$\leq \sup\{\|f(z_1) - f(z_2)\|; z_1, z_2 \in \overline{\mathbb{D}}_R, |z_1 - z_2| \leq \delta\} = \omega_1(f; \delta)_{\overline{\mathbb{D}}_R}.$$

Therefore, we obtain $|x^*[P_\lambda(f; z) - f(z)]| \leq 2(1 + \frac{1}{q})\omega_1(f; [\lambda]_q)_{\overline{\mathbb{D}}_R}$, for all $x^* \in B_1$, and passing here to supremum, according to Theorem 4.2.4 it follows the required estimate. The proof in the case of $W_\lambda(f; z)$ is similar. $\qquad\square$

4.3 Strong Approximation by Complex Taylor Series

In this section we show that some classical results in the strong approximation by Fourier series can be extended to complex approximation by Taylor series in the unit disk.

Let $g(x)$ be a continuous and 2π periodic function and let

$$g(x) \sim \frac{a_0}{2} + \sum_{k=1}^{\infty}(a_k \cos kx + b_k \sin kx)$$

be its Fourier series. Denote by $s_n(f)(x) = \frac{a_0}{2} + \sum_{k=1}^{n}(a_k \cos kx + b_k \sin kx)$ and $E_n(g)$ the best approximation of g in the uniform norm by trigonometric polynomials of degree $\leq n$. Also, for $m \leq n$ let us define the de la Vallée Poussin sum by $\sigma_{n,m}(g)(x) = \frac{1}{m+1}\sum_{k=n-m}^{n} s_k(g)(x)$. Concerning $\sigma_{n,m}(g)(x)$, in Stechkin [185] the general result

$$\|g - \sigma_{n,m}(g)\| \leq C \sum_{k=0}^{n} \frac{E_{n-m+k}(g)}{n+k+1}, 0 \leq m \leq n, m = 0, 1, ...,$$

was obtained, where $\|\cdot\|$ denotes the uniform norm in the space of all continuous and 2π-periodic functions and $C > 0$ is an absolute constant independent of g, n and m.

Defining the strong de la Vallée Poussin means by

$$V_{n,m}(g)(x) = \frac{1}{m+1}\sum_{k=n-m}^{n} |s_k(g)(x) - g(x)|,$$

the above inequality was extended to $V_{n,m}(g)$ in Leindler [122] as follows

$$\|V_{n,m}(g)\| \leq C \sum_{k=0}^{n} \frac{E_{n-m+k}(g)}{n+k+1}, 0 \leq m \leq n, m = 0, 1, ...,$$

where again $C > 0$ is an absolute constant independent of g, n and m.

Moreover, this last inequality was generalized in Leindler-Meir [123] to the strong de la Vallée Poussin means with exponent $p \geq 1$,

$$V_{n,m}^{(p)}(g)(x) = \left\{ \frac{1}{m+1} \sum_{k=n-m}^{n} |s_k(g)(x) - g(x)|^p \right\}^{1/p}$$

by the inequality

$$\|V_{n,m}^{(p)}(g)\| \leq C(\log n)^{1-(1/p)} \left(\sum_{k=0}^{n} \frac{E_{n-m+k}^p(g)}{n+k+1} \right)^{1/p}, 0 \leq m \leq n, n \geq 2,$$

where $C > 0$ is independent of g, m, n and p. For $p = 1$ this result becomes that in Leindler [122], which clearly implies the inequality in Stechkin [185].

Now, for $\overline{\mathbb{D}}_1 = \{z \in \mathbb{C}; |z| \leq 1\}$ denote

$$\overline{A}(\overline{\mathbb{D}}_1) = \{f : \overline{\mathbb{D}}_1 \to \mathbb{C}; f \text{ is analytic in } \overline{\mathbb{D}}_1\},$$

endowed with the uniform norm $\|f\|_{\overline{\mathbb{D}}_1} = \max\{|f(z)|; |z| \leq 1\}$. Clearly $f \in \overline{A}(\overline{\mathbb{D}}_1)$ means that there exists $R > 1$ such that f is analytic in $\mathbb{D}_R = \{z \in \mathbb{C}; |z| < R\}$.

For $f \in \overline{A}(\overline{\mathbb{D}}_1)$ and $0 \leq m \leq n$, let us define the de la Vallée Poussin mean

$$\sigma_{n,m}(f)(z) = \frac{1}{m+1} \sum_{k=n-m}^{n} T_k(f)(z),$$

where $T_k(f)(z) = \sum_{j=0}^{k} \frac{f^{(j)}(0)}{j!} z^j$ represents the kth Taylor partial sum of f.

In Gal [95], the proof of Theorem 2.1 (see also the proof of Theorem 3.2.1 in the book Gal [77]) we pointed out the validity of the following inequality analogous to that in Stechkin [185], given by

$$\|f - \sigma_{n,m}(f)\|_{\overline{\mathbb{D}}_1} \leq C \sum_{k=0}^{n} \frac{E_{n-m+k}(f)}{n+k+1}, 0 \leq m \leq n, m = 0, 1, ...,$$

where $E_n(f) = \inf\{\|f - P\|_{\overline{\mathbb{D}}_1}; P \text{ is complex polynomial of degree } \leq n\}$ and $C > 0$ is an absolute constant independent of f, n and m.

The next theorem gives a positive answer to a problem raised in the book Gal [77] (see there the Remark 3 after the proof of Theorem 3.2.1, (iii), pp. 229-230) concerning the complex analogues of the two above mentioned generalizations of the Stechkin's inequality, obtained for Fourier series in Leindler [122] and Leindler-Meir [123].

Theorem 4.3.1. (Gal [96]) *For $f \in \overline{A}(\overline{\mathbb{D}}_1)$ let us consider the strong de la Vallée Poussin means with exponent $p \geq 1$,*

$$\Delta_{n,m}^{(p)}(f)(z) = \left(\frac{1}{m+1} \sum_{k=n-m}^{n} |T_k(f)(z) - f(z)|^p \right)^{1/p}.$$

For all $0 \leq m \leq n$ and $n \geq 2$ we have

$$\|\Delta_{n,m}^{(p)}(f)\|_{\overline{\mathbb{D}}_1} \leq C[\log(n)]^{1-(1/p)} \left(\sum_{k=0}^{n} \frac{E_{n-m+k}^p(f)}{m+k+1} \right)^{1/p},$$

where $C > 0$ is independent of f, m, n and p.

Proof. First denoting $z = re^{it} \in \overline{\mathbb{D}}_1$ and $f(z) = f(ze^{it}) = \sum_{k=0}^{\infty} c_k r^k e^{ikt}$, we get

$$\frac{1}{2\pi} \int_{-\pi}^{\pi} f(re^{it}) e^{-ikt} dt = c_k r^k, \text{ for all } k = 0, 1, ..., .$$

Then, for any $k = 1, 2, ...$, we observe that it follows

$$c_k r^k e^{ik\theta} = \frac{e^{ik\theta}}{2\pi} \int_{-\pi}^{\pi} f(re^{it}) e^{-ikt} dt + \frac{e^{-ik\theta}}{2\pi} \int_{-\pi}^{\pi} f(re^{it}) e^{ikt} dt$$

$$= \frac{1}{2\pi} \int_{-\pi}^{\pi} f(re^{it}) [e^{ik(\theta-t)} + e^{ik(t-\theta)}] dt$$

$$= \frac{1}{\pi} \int_{-\pi}^{\pi} f(re^{it}) \cos[k(t - \theta)] dt$$

$$= \frac{1}{\pi} \int_{-\pi}^{\pi} f(re^{i(t+\theta)}) \cos(kt) dt.$$

This immediately implies

$$T_n(f)(re^{i\theta}) = \frac{1}{\pi} \int_{-\pi}^{\pi} f(re^{i(t+\theta)}) \left(1 + \sum_{k=1}^{n} \cos(kt)\right) dt$$

$$= \frac{1}{\pi} \int_{-\pi}^{\pi} f(re^{i(t+\theta)}) D_n(t) dt.$$

By the maximum modulus principle, in all the estimates we can take $|z| = 1$, that is $z = e^{i\theta}$. Therefore, denoting $f(e^{i\theta}) = F[\cos(\theta), \sin(\theta)] + iG[\cos(\theta), \sin(\theta)]$, or for the simplicity of notations, $f(e^{i\theta}) := F(\theta) + iG(\theta)$, we can write

$$T_n(f)(e^{i\theta}) = s_n(F)(\theta) + is_n(G)(\theta), \text{ for all } \theta \in [-\pi, \pi].$$

By the Minkowski's inequality and by applying (3), for all $\theta \in [-\pi, \pi]$ we immediately obtain

$$|\Delta_{n,m}^{(p)}(f)(e^{i\theta})| \leq |V_{n,m}^{(p)}(F)(\theta)| + |V_{n,m}^{(p)}(G)(\theta)| \leq \|V_{n,m}^{(p)}(F)\| + \|V_{n,m}^{(p)}(G)\|$$

$$\leq C(\log n)^{1-(1/p)} \left[\left(\sum_{k=0}^{n} \frac{E_{n-m+k}^p(F)}{n+k+1}\right)^{1/p} + \left(\sum_{k=0}^{n} \frac{E_{n-m+k}^p(G)}{n+k+1}\right)^{1/p}\right].$$

Now, let $E_k(f) = \|f - P^*\|_{\overline{\mathbb{D}}_1}$ where $P^*(z)$ is the polynomial of best approximation of degree $\leq k$. By the maximum modulus principle we get

$$E_k(f)$$
$$= \max_{\theta \in [-\pi,\pi]} \{|f(e^{i\theta}) - P^*(e^{i\theta})|\} = \max_{\theta \in [-\pi,\pi]} \{|F(\theta) - P_1^*(\theta) + i[G(\theta) - P_2^*(\theta)]|\}$$
$$= \max_{\theta \in [-\pi,\pi]} \sqrt{[F(\theta) - P_1^*(\theta)]^2 + [G(\theta) - P_2^*(\theta)]^2},$$

where $P^*(e^{i\theta}) = P_1^*(\theta) + iP_2^*(\theta)$ with P_1^* and P_2^* trigonometric polynomials of degree $\leq k$.

Since

$$E_k(F) \leq \max_{\theta \in [-\pi, \pi]} \{|F(\theta) - P_1^*(\theta)|\}$$

and

$$E_k(G) \leq \max_{\theta \in [-\pi, \pi]} \{|G(\theta) - P_2^*(\theta)|\},$$

by the above inequalities we easily get $E_k(F) \leq E_k(f)$, $E_k(G) \leq E_k(f)$ and the estimate in the statement. $\qquad\square$

Remark. For $p = 1$ it is clear that the estimate in Theorem 4.3.1 also implies the estimate for $\|f - \sigma_{n,m}(f)\|$ obtained in the proof of Theorem 3.2.1 in the book Gal [77], p. 227 (see also the proof of Theorem 2.1 in Gal [95]).

4.4   Bibliographical Notes and Open Problems

Theorems 4.1.3, 4.1.4 and Corollary 4.1.5, Theorems 4.2.14, 4.2.15, 4.2.16, 4.2.17 and 4.2.18 appear for the first time here.

Open Problem 4.4.1. Let us suppose that $f \in A(\overline{G})$ (that is f is analytic on G), where G is a continuum in $\mathbb{C}$ (that is a compact connected subset of $\mathbb{C}$). If we denote by $F_k(z)$, $k = 0, 1, 2, ...$, the Faber polynomials attached to G, then from Theorem 2 in Suetin [186], p. 52, it is known that $f(z) = \sum_{k=0}^{\infty} a_k(f)F_k(z)$, where the series is uniformly convergent in G. Also, for $P_n(z) = \sum_{k=0}^{n} a_k(f)F_k(z)$, let us define the de la Vallée Poussin-type means

$$\sigma_{n,m}(f; G)(z) = \frac{1}{m+1} \sum_{k=n-m}^{n} P_k(f)(z), \, z \in G, \, m \leq n$$

and the strong de la Vallée Poussin-type means with exponent $p \geq 1$,

$$\Delta_{n,m}^{(p)}(f : G)(z) = \left(\frac{1}{m+1} \sum_{k=n-m}^{n} |P_k(f)(z) - f(z)|^p \right)^{1/p}, \, z \in G, \, m \leq n.$$

It is an open question if the following three inequalities can hold

$$\|f - \sigma_{n,m}(f; G)\|_G \leq C \sum_{k=0}^{n} \frac{E_{n-m+k}(f; G)}{n+k+1}, m \leq n, \, m = 0, 1, ...,$$

$$\|\Delta_{n,m}(f; G)\|_G \leq C \sum_{k=0}^{n} \frac{E_{n-m+k}^p(f; G)}{m+k+1}, m \leq n, \, n \geq 2,$$

$$\|\Delta_{n,m}^{(p)}(f; G)\|_G \leq C[\log(n)]^{1-(1/p)} \left(\sum_{k=0}^{n} \frac{E_{n-m+k}^p(f; G)}{m+k+1} \right)^{1/p}, m \leq n, n \geq 2,$$

where $\| \cdot \|_G$ denotes the uniform norm on $C(G)$, $C > 0$ is an absolute constant independent of f, n, m and p but dependent on G and

$$E_n(f; G) = \inf\{\|f - P\|_G; P \text{ is complex polynomial of degree } \leq n\}.$$

Open Problem 4.4.2. A famous result of Bernstein [44], [45] states that for any $\lambda > 0$, λ not even integer, there exists finite the limit $\lim_{n \to \infty} n^\lambda E_n(|x|^\lambda)_\infty > 0$, where

$$E_n(|x|^\lambda)_\infty = \inf\{\max\{|P(x) - |x|^\lambda|; x \in [-1, 1]\}; P \in \mathcal{P}_n(\mathbb{R})\},$$

and $\mathcal{P}_n(\mathbb{R})$ denotes the set of all real polynomials (that is with real coefficients) of degree $\leq n$ (for details and generalizations of this result to weighted approximation see the recent book of Ganzburg [98]).

On the other hand, taking into account that $|x|^\lambda$ is convex for $\lambda \geq 1$ and concave for $0 < \lambda < 1$, denoting

$$E_n^{(+2)}(f)_\infty = \inf\{\|f - P\|_\infty; P \in \mathcal{P}_n(\mathbb{R}), P''(x) \geq 0, \forall x \in [-1, 1]\},$$

and

$$E_n^{(-2)}(f)_\infty = \inf\{\|f - P\|_\infty; P \in \mathcal{P}_n(\mathbb{R}), P''(x) \leq 0, \forall x \in [-1, 1]\},$$

it has sense to consider here the open problem if there exists finite the limits $\lim_{n \to \infty} n^\lambda E_n^{(-2)}(|x|^\lambda)_\infty$ for $0 < \lambda < 1$ and $\lim_{n \to \infty} n^\lambda E_n^{(+2)}(|x|^\lambda)_\infty$ for $\lambda \geq 1$?

Note that since by Kopotun-Leviatan-Shevchuk [119], for f convex (or concave, respectively) on $[-1, 1]$ and $\lambda > 0$ we have

$$E_n(f)_\infty = \mathcal{O}(n^{-\lambda}) \quad \text{iff } E_n^{(\pm 2)}(f)_\infty = \mathcal{O}(n^{-\lambda}), \quad n \to \infty,$$

it easily follows that the sequences $(n^\lambda E_n^{(-2)}(|x|^\lambda)_\infty)_{n \in \mathbb{N}}$ for $0 < \lambda < 1$ and $(n^\lambda E_n^{(+2)}(|x|^\lambda)_\infty)_{n \in \mathbb{N}}$ for $\lambda \geq 1$, are bounded.

More general, we can consider the following open question concerning the shape preserving limit results. Thus, taking into account that by Leviatan-Shevchuk [124], for a convex function f we have $E_n^{(+2)}(f)_{L_p} \leq E_{n-2}(f'')_{L_p}$, for all $n \geq 2$ and $0 \leq p \leq \infty$ and this estimate cannot be improved, it is an open question if for f convex on $[-1, 1]$ and the sequence $(\lambda_n)_n$ satisfying the conditions in Ganzburg [98], p. 3, we have

$$\lim_{n \to \infty} E_n^{(+2)}\left(f; L_\infty\left[-\frac{1 - \lambda_n}{n\sigma}, \frac{1 - \lambda_n}{n\sigma}\right]\right) < \infty(\text{ or } \leq A_\sigma^{(+2)}(f, L_\infty(\mathbb{R})) < \infty) \qquad ?$$

Here $A_\sigma^{(+2)}(f, L_\infty(\mathbb{R})) = \inf\{\|f - g\|_{L_\infty(\mathbb{R})}; g \in B_\sigma, g \text{ convex on } \mathbb{R}\}$ and B_σ denotes the class of all entire functions of exponential type σ.

All the above questions could be considered in complex setting too. Thus, denoting $\|f\|_{\overline{\mathbb{D}}_1} = \max\{|f(z)|; z \in \overline{\mathbb{D}}_1\}$ and for $f \in C(\overline{\mathbb{D}}_1)$,

$$E_n(f)(\overline{\mathbb{D}}_1) = \inf\{\|P - f\|_{\overline{\mathbb{D}}_1}; P \in \mathcal{P}_n(\mathbb{C})\},$$

with $\mathcal{P}_n(\mathbb{C})$ representing the set of all complex polynomials of degree $\leq n$ and with complex coefficients, it is an open question if for $\lambda > 0$ there exists finite the limit

$$\lim_{n \to \infty} n^\lambda E_n(|z|^\lambda)(\overline{\mathbb{D}}_1) \quad ?$$

Moreover, since the continuous non-analytic function $f_\lambda(z) = |z|^\lambda$, $z \in \overline{\mathbb{D}}_1$ is obviously convex (transforming any disk of center origin and radius $0 < r < 1$ into a convex set), denoting

$$E_n^{(CONV)}(f_\lambda)(\overline{\mathbb{D}}_1)$$

$$= \inf\{\|P - f_\lambda\|_{\overline{\mathbb{D}}_1} ; P \in \mathcal{P}_n(\mathbb{C}), P(0) = 0, P'(0) \neq 0, P \text{ convex in } \mathbb{D}_1\},$$

it is an open question if there exists finite the limit

$$\lim_{n \to \infty} n^\lambda E_n^{(CONV)}(|z|^\lambda)(\overline{\mathbb{D}}_1) \quad ?$$

Here the convexity of the polynomial P means that we have $Re\left(\frac{zP''(z)}{P'(z)}\right) + 1 > 0$, for all $z \in \mathbb{D}_1$.

Similar problems in the complex case can be posed if we replace $|z|^\lambda$ by $f(z)^\lambda$, $|z| \leq 1$, where $f(z) = z$, if $Re(z) \geq 0$ and $f(z) = -z$ if $Re(z) < 0$.

Bibliography

[1] Abel, U. and Della Vecchia B. (2000) Asymptotic approximation by the operators of K. Balázs and Szabados, *Acta Sci. Math.*(Szeged), **66**, No. 1-2, 137–145.

[2] Abel, U. and Ivan, M. (1999) Some identities for the operator of Bleimann, Butzer and Hahn involving differences, *Calcolo*, **36**, No. 3, 143–160.

[3] Agratini, O. (2000) *Approximation by Linear Operators* (in Romanian), University Press of Cluj, "Babes-Bolyai" University, Cluj-Napoca.

[4] Agratini, O. and Rus, I.A. (2003) Iterates of some bivariate approximation process via weakly Picard operators, *Nonlin. Anal. Forum*, **8**, No. 2, 159–168.

[5] Alexander, J. W. (1915) Functions which map the interior of the unit circle upon simple regions, *Annals of math.*, second series, **17**, 12–22.

[6] Alvarez-Nodarse, R., Atakishieva, M.K. and Atakishiev, N.M. (2002) On q-extension of the Hermite polynomials $H_n(x)$ with the continuous orthogonality property on $\mathbb{R}$, *Bol. Soc. Mat. Mexicana*, **8**, No. 3, 127–139.

[7] Altomare, F. (1989) Limit semigroups of Bernstein-Schnabl operators associated with positive projections, *Annali della Scuola Normale Superiore di Pisa, Classe di Scienze*, Sér. 4, vol. **16**, No. 2, 259–279.

[8] Altomare, F. (1996) On some sequences of positive linear operators on unbounded intervals, *Approximation and Optimization*, Vol. I, Cluj-Napoca, Transilvania, Cluj-Napoca, pp. 1–16.

[9] Altomare, F. and Campiti, M. (1994), *Korovkin-Type Approximation Theory and its Applications*, De Gruyter Studies in Mathematics, **17**, Walter de Gruyter and Co., Berlin.

[10] Altomare, F. and Mangino, E. (1999), On a generalization of Baskakov operator, *Rev. Roumaine Math. Pures Appl.*, **44**, No. 5–6, 683–705.

[11] Altmare, F. and Raşa, I. (1998), Feller semigroups, Bernstein-type operators and generalized convexity associated with positive projections, *New develpments in aproximation theory*, Dortmund, 1998, Birkhauser, Basel, 1999, pp. 9–32.

[12] Anastassiou, G.A. and Gal, S.G. (1999) Some shift-invariant integral operators, univariate case revisited, *J. Comp. Anal. Appl.*, **1**, No. 1, 3–23.

[13] Anastassiou, G.A. and Gal, S.G. (1999) On some differentiated shift-invariant integral operators, univariate case revisited, *Adv. Nonlinear Var. Inequal.*, **2**, No. 2, 71–83.

[14] Anastassiou, G.A. and Gal, S.G. (2006) Approximation of vector-valued functions by polynomials with coefficients in vector spaces and applications, *Demonstr. Math.*, **39**, No. 3, 539–552.

[15] Anastassiou, G.A. and Gal, S.G. (2007) On the best approximation of vector valued

functions by polynomials with coefficients in vector spaces, *Annali di Matematica Pura ed Applicata*, **186**, No. 2, 251–265.

[16] Anastassiou, G.A. and Gal, S.G. (2006) Geometric and approximation properties of generalized singular integrals in the unit disk, *J. Korean Math. Soc.*, **43**, No. 2, 425–443.

[17] Anastassiou, G.A. and Gal, S.G. (2006) Geometric and approximation properties of a complex Post-Widder operator in the unit disk, *Appl. Math. Letters*, **19**, Issue 4, 303–402.

[18] Anastassiou, G.A. and Gal, S.G. (2006) Geometric and approximation properties of some singular integrals in the unit disk, *J. Ineq. Appl.*, Article ID 17231, 19 pages.

[19] Anastassiou, G.A. and Gal, S.G. (2006) Geometric and approximation properties of some complex rotation-invariant integral operators in the unit disk, *J. Comp. Anal. Appl.*, **8**, No. 4, 357–368.

[20] Anastassiou, G.A. and Gal, S.G. (2008) Geometric and approximation properties of some complex Sikkema and spline operators in the unit disk, *J. Concrete and Applicable Mathematics*, **6**, No. 2, 177–188.

[21] Anastassiou, G.A. and Gonska, H. (1995) On some shift-invariant integral operators, univariate case, *Ann. Pol. Math.*, **LXI**, No. 3, 225–243.

[22] Anderson, J.M. and Clunie, J. (1985) Isomorphisms of the disk algebra and inverse Faber sets, *Math. Z.*, **188**, 545–558.

[23] Andreian Cazacu, C. (1971) *Theory of Functions of Several Complex Variables* (Romanian), Edit. Didact. Pedag., Bucharest.

[24] Andrica, D. (1985) Powers by Bernstein's operators and some combinatorial properties, in : *Itinerant Seminar on Functional Equations, Approximation and Convexity*, Cluj-Napoca, Preprint No. **6**, pp. 5–9.

[25] Andrievskii, V.V. (2006) *Constructive Function Theory on Sets of the Complex Plane through Potential Theory and Geometric Function Theory*, Surveys in Approximation Theory, **2**, 1–52.

[26] Andrievskii, V.V., Belyi, V. I. and Dzjadyk, V. K. (1998) *Conformal Invariants in the Constructive Theory of Functions of a Complex Variable* (Russian), Naukova Dumka, Kiev.

[27] Andrievskii, V.V., Belyi, V. I. and Dzjadyk, V. K. (1995) *Conformal Invariants in Constructive Theory of Functions of Complex Variable* (English, translated from Russian by D. N. Kravchuk), Advanced Series in Mathematical Science and Engineering, World Federation Publishers Company, Atlanta, GA.

[28] Angeloni, L. and Vinti, G. (2006) Convergence in variation and rate of approximation for nonlinear integral operators of convolution type, *Result. Math.*, **49**, 1–23.

[29] Angeloni, L. and Vinti, G. (2005) Rate of approximation for nonlinear integral operators with application to signal processing, *Differential Integral Equations*, **18**, No. 8, 855–890.

[30] Anghelutza, Th. (1924) Une remarque sur l'integrale de Poisson, *Darboux Bull.*, **48**, No. 2, 138–140.

[31] Aral, A. (2006) On the generalized Picard and Gauss Weierstrass singular integrals, *J. Comp. Anal. Appl.*, **8**, No. 3, 246–261.

[32] Aral, A. and Gal, S.G. (2009) q-generalizations of the Picard and Gauss-Weierstrass singular integrals, *Taiwanese J. Math.*, **12**, No. 9, 2501–2515.

[33] Balázs, K. (1975) Approximation by Bernstein type rational functions, *Acta Math. Acad. Sci. Hungar.*, **26**, 123–134.

[34] Balázs, K. and Szabados, J. (1982) Approximation by Bernstein type rational functions, II, *Acta Math. Acad. Sci. Hungar.*, **40(3-4)**331–337.

[35] Baskakov, V.A. (1957) An example of a sequence of linear positive operators in the space of continuous functions (Russian) *Dokl. Akad. Nauk SSSR*, **113**, 249–251.

[36] Bărbosu, D. (2004) Kantorovich-Stancu Type Operators, *J. Ineq. Pure Appl. Math.*, vol. **5**, Issue 3, Article 53 (electronic).

[37] Berens H. and Lorentz, G.G. (1972) Inverse theorems for Bernstein polynomials, *Indiana J. Math.*, **21**(8), 693–708.

[38] Bernstein, S.N. (1912-1913) Démonstration du théorème de Weierstrass fondée sur le calcul de probabilités, *Commun. Soc. Math. Kharkow*, **13**. No. 2, 1–2.

[39] Bernstein, S.N. (1935) Sur la convergence de certaines suites des polynômes, *J. Math. Pures Appl.*, **15**, NO. 9, 345–358.

[40] Bernstein, S.N. (1936) Sur le domaine de convergence des polynômes, *C. R. Acad. Sci. Paris*, **202**, 1356–1358.

[41] Bernstein, S.N. (1943) On the domains of convergence of polynomials (in Russian), *Izv. Akad. Nauk. SSSR*, ser. math., **7**, 49–88.

[42] Bernstein, S.N. (1932) Complétement a l'article de E. Voronowskaja, *C.R. Acad. Sci.*, U.R.S.S., 86–92.

[43] Bernstein, S.N. (1926) *Leçns sur les Propriétés Extrémales et la Meilleure Approximations des Fonctions Analytiques d'Une Variable Réelle*, Gauthier-Villars, Paris.

[44] Bernstein, S.N. (1914) Sur la meilleure approximation de $|x|$ par les polynomes des degrés donnés, *Acta Math.*, **27**, 1–57.

[45] Bernstein, S.N. (1938) On the best aproximation of $|x|^p$ by polynomials of very high degree, *Izv. Math. Nauk. SSSR*, 169–180.

[46] Bleimann, G, Butzer, P.L. and Hahn, L. (1980) A Bernstein-type operator approximating continuous functions on the semi-axis, *Indag. Math.*, **42**, 255–262.

[47] Borwein, P. and Erdélyi, T. (1996) Sharp extensions of Bernstein's inequality to rational spaces, *Mathematika*, **43**, No. 2, 412–423.

[48] Bruj, I. and Schmieder, G. (1999) Best approximation and saturation on domains bounded bu curves of bounded rotation, *J. Approx. Theory*, **100**, 157–182.

[49] Brutman, L. and Gopengauz, I. (1999) On divergence of Hermite-Fejér interpolation to $f(z) = z$ in the complex plane, *Constr. Approx.*, **15**, 611–617.

[50] Brutman, L., Gopengauz, I. and Vértesi, P. (2000) On the domain of divergence of Hermite-Fejér interpolating polynomials, *J. Approx. Theory*, **106**, 287–290.

[51] Butzer, P. (1953) Linear combinations of Bernstein polynomials, *Canadian J. Math.*, **5**, 559–567.

[52] Butzer, P. (1953) On two-dimensional Bernstein polynomials, *Canad. J. Math.*, **5**, 107–113.

[53] Butzer, P.L. and Nessel, R.J. (1971) *Fourier Analysis and Approximation, Vol. I, One-Dimensional Theory*, Birkhauser, Basel.

[54] Choi, J.H., Kim, Y.C. and Saigo, M. (2002) Geometric properties of convolution operators defined by Gaussian hypergeometric functions, *Integral Transforms Spec. Funct.*, **13**, No. 2, 117–130.

[55] Cimoca, G. and Lupaş, A. (1967) Two generalizations of the Meyer-König and Zeller operator, *Mathematica(Cluj)*, **9(32)**, No. 2, 233–240.

[56] Derriennic, M.M. (1978) Sur l'approximation des fonctions d'une ou plusieurs variables par des polynomes de Bernstein modifiés et application au probléme de moments, *Thése de 3e cycle*, Univ. de Rennes.

[57] Derriennic, M.M. (1981) Sur l'approximation des fonctions integrable sur $[0, 1]$ par des polynomes de Bernstein modifiés, *J. Approx. Theory*, **31**, 325–343.

[58] Derriennic, M.M. (1985) On multivariate approximation by Bernstein type operators, *J. Approx. Theory*, **45**, 155–166.

[59] Derriennic, M.M. (2005) Modified Bernstein polynomials and Jacobi polynomials in q-calculus, *Rend. Circ. Mat. Palermo*, (2), Suppl. No. **76**, 269–290.

[60] DeVore, R.A. and Lorentz, G.G. (1993) *Constructive Approximation : Polynomials and Splines Approximation*, vol. **303**, Springer-Verlag, Berlin, Heidelberg.

[61] Dinghas, A. (1951) Über einige Identitäten vom Bernsteinschen Typus, *Det Koneglige Norske Videnskabers Selkab*, **24**, No. 21, 96–97.

[62] Dyn'kin, E.M. (1981) The rate of polynomial approximation in the complex domain, in : *Lecture Notes in Mathematics*, vol. **864**, Complex Analysis and Spectral Theory Seminar, Leningrad 1979/80, Complex Analysis and Spectral Theory (Havin, V. P., Nikol'skii, N. K., Eds.), Springer-Verlag, Berlin/Heidelberg, pp. 1–142.

[63] Ditzian, Z. (1985) Rate of approximation of linear processes, *Acta Sci. Math. (Szeged)*, **48** (1-4), 103–128.

[64] Ditzian, Z. and Totik, V. (1987) *Moduli of Smoothness*, Springer-Verlag, New York-Berlin-Heidelberg-London-Paris-Tokyo.

[65] Dressel, F.G., Gergen, J.J. and Purcell, W.H. (1963) Convergence of extended Bernstein polynomials in the complex plane, *Pacific J. Math.*, **13**, No. 4, 1171–1180.

[66] Dubois, D. and Prade, H. (1987) Fuzzy numbers: An overview, Analysis of fuzzy information, vol.1 : Math. Logic, CRC Press, Boca Raton, 3–39.

[67] Durrmeyer, J.L. (1967) Une formule d'inversion de la transformée de Laplace. Applications à la théorie des moments, *Thése de 3e cycle*, Fac. des Sciences de l'Université de Paris.

[68] Dziok, J. and Srivastava, H.M. (2003) Certain subclasses of analytic functions associated with the generalized hypergeometric functions, *Integral Transforms Spec. Funct.*, **14**, 7–18.

[69] Dzjadyk, V.K. (1977) *Introduction to the Theory of Uniform Approximation of Functions by Polynomials* (Russian), Nauka, Moscow.

[70] Faber, G. (1903) Über polynomische Entwicklungen, *Math. Ann.*, **57**, 398–408.

[71] Farcaş, M. (2006) About the coefficients of Bernstein multivariate polynomial, *Creative Math.*, **15**, 17–20.

[72] Favard, J. (1944) Sur les multiplicateurs d'interpolation, *J. Math. Pures Appl.*, **23**, No. 9, 219–247.

[73] Fejér, L. (1904) Untersuchungen über Fourier Reihen, *Math. Ann.*, **58**, 51–69.

[74] Frerick, L. and Müller, J. (1994) Polynomial approximation and maximal convergence on Faber sets, *Constr. Approx.*, **10**, 427–438.

[75] Gaier, D. (1977) Approximation durch Fejér-Mittel in der Klausse A. Mitt. *Math. Sem. Giessen*, **123**, 1–6.

[76] Gaier, D. (1987) *Lectures on Complex Approximation*, Birkhauser, Boston.

[77] Gal, S.G. (2008) *Shape Preserving Approximation by Real and Complex Polynomials*, Birkhauser Publ., Boston, Basel, Berlin.

[78] Gal, S.G. (2008) Voronovskaja's theorem and iterations for complex Bernstein polynomials in compact disks, *Mediterr. J. Math.*, **5**, No. 3, 253–272.

[79] Gal, S.G. (2009) Exact orders in simultaneous approximation by complex Bernstein polynomials, *J. Concr. Applic. Math.*, **7**, No. 3, 215–220.

[80] Gal, S.G. (2009) Approximation by complex Bernstein-Stancu polynomials in compact disks, *Results in Mathematics*, vol. **53**, under press.

[81] Gal, S.G. (2008) Exact orders in simultaneous approximation by complex Bernstein-Stancu polynomials, *Revue d'Anal. Numér. Théor. de L'Approx. (Cluj-Napoca)*, **37**, No. 1, 47–52.

[82] Gal, S.G. (2007) Approximation and geometric properties of some complex Bernstein-

Stancu polynomials in compact disks, *Revue D'Anal. Numér. Théor. de L'Approx. (Cluj-Napoca)*, **36**, No. 1, 67–77.

[83] Gal, S.G. (2008) Approximation and geometric properties of complex Favard-Szász-Mirakian operators in compact diks, *Comput. Math. Appl.*, **56**, 1121–1127.

[84] Gal, S.G. (submitted) Approximation of analytic functions without exponential growth conditions by complex Favard-Szász-Mirakjan operators.

[85] Gal, S.G. (2008) Voronovskaja's theorem and the exact degree of approximation for the derivatives of complex Riesz-Zygmund means, *General Mathematics*, (Sibiu), **16**, No. 4, 61–71.

[86] Gal, S.G. (2001) Convolution-type integral operators in complex approximation, *Comput. Methods and Function Theory*, **1**, No. 2, 417–432.

[87] Gal, S.G. (submitted) Voronovskaja's theorem, shape preserving properties and iterations for complex q-Bernstein polynomials.

[88] Gal, S.G. (2008) Approximation and geometric properties of some nonlinear complex integral convolution operators, *Integral Transforms and Special Functions*, **19**, No. 5, 367–375.

[89] Gal S.G. (2009) Generalized Voronovskaja's theorem and approximation by Butzer's combinations of complex Bernstein polynomials, *Results in Mathematics*, vol. **53**, under press.

[90] Gal, S.G. (submitted) Approximation by complex Baskakov operators in compact disks and semidisks.

[91] Gal, S.G. (2008) Approximation by complex Bernstein-Kantorovich and Stancu-Kantorovich polynomials and their iterates in compact disks, *Revue D'Anal. Numér. Théor. de L'Approx. (Cluj-Napoca)*, **37**, No. 2, 159–168.

[92] Gal, S.G. (2005) Remarks on the approximation of normed spaces valued functions by some linear operators, in :*Mathematical Analysis and Approximation Theory*, Proceedings of the 6th Romanian-German Seminar on Approximation Theory and its Applications, RoGer 2004 (Gavrea I. and Ivan M. eds.), Mediamira Science Publisher, Cluj-Napoca, pp. 99–109.

[93] Gal, S.G. (2006) Geometric and approximate properties of convolution polynomials in the unit disk, *Bull. Inst. Math. Acad. Sin. (N.S.)*, **1**, no. 2, 307–336.

[94] Gal, S.G. (1998) Degree of approximation of continuous functions by some singular integrals, *Revue D'Analyse Numér. Théor. Approx.* (Cluj), **27**, No. 2, 251–261.

[95] Gal, S.G. (2008) Shape preserving approximation by complex polynomials in the unit disk, *Bull. Inst. Math. Acad. Sin. (N.S.)*, **3**, no. 2, 323–337.

[96] Gal, S.G. (2009) Remarks on the strong approximation by Taylor series in the unit disk, *Anal. Univ. Oradea, fasc. math.*, under press.

[97] Gal, C.S., Gal, S.G. and Goldstein, J.A. (2008) Evolution equations with real time variable and complex spatial variables, *Complex Variables and Elliptic Equations*, **53**, No. 8, 753–774.

[98] Ganzburg, M.I. (2008) *Limit Theorems of Polynomial Approximation with Exponential Weights*, Memoirs of the American Mathematical Society, vol. **192**, No. **897**, Providence, Rhode Island.

[99] Gasper, G. and Rahman, M. (1990) *Basic Hypergeometric Series*, Cambridge University Press, Cambridge.

[100] Goldstein, J.A. (1985) *Semigroups of Linear Operators and Applications*, Oxford University Press.

[101] Gonska, H., Kacsó, D. and Pițul, P. (2006) The degree of convergence of over-iterated positive linear operators, *J. Appl. Funct. Anal.*, **1**, 403–423.

[102] Gonska, H., Pițul, P. and Rașa, I. (2006) On peano's form of the Taylor remainder,

Voronovskaja's theorem and the commutator of positive linear operators, in : *Proceed. Intern. Conf. on "Numer. Anal. and Approx. Theory", NAAT, Cluj-Napoca,* Casa Cartii de Stiinta, Cluj-Napoca, pp. 55–80.

[103] Gonska, H., Pițul, P. and Rașa, I. (2007) Over-iterates of Bernstein-Stancu operators, *Calcolo*, **44**, No. 2, 117–125.

[104] Gonska H. and Zhou, X.-L. (1995) The strong converse inequality for Bernstein-Kantorovich operators, *Computers Math. Applic.*, **30**, 103–128.

[105] Graham, I. and Kohr, G. (2003) *Geometric Function Theory in One and Higher Dimensions*, Pure and Applied Mathematics, Marcel Dekker, **255**, New York.

[106] Guo, S., Li, C. and Liu, X., *Pointwise approximation for linear combinations of Bernstein operators*, J. Approx. Theory, **107**(2005), 109–120.

[107] He, F. (1983–1984) The powers and their Bernstein polynomials, *Real Analysis Exchange*, **9**, No. 2, 578–583.

[108] Hildebrandt, T.H. and Schoenberg, I.J. (1933) On linear functional operations and the moment problem for a finite interval in one or several dimensions, *Ann. Math.*, **34**(2), 317–328.

[109] Hille, E. and Phillips, R.S. (1957) *Functional Analysis and Semigroups*, Amer. Math. Soc., Colloq. Publ., Vol. **31**.

[110] Jakimovski, A. and Leviatan, D. (1969) Generalized Szász operators for the infinite interval, *Mathematica(Cluj)*, **34**, 97–103.

[111] Kanas, S. and Srivastava, H.M. (2000) Linear operators associated with k-uniformly convex functions, *Integral Transforms Spec. Funct.*, **9**, No. 2, 121–132.

[112] Kantorovitch, L.V. (1930) Sur certains développements suivant les polynômes de la forme de S. Bernstein, I, II, *C.R. Acad. Sci. URSS*, 563–568, 595–600.

[113] Kantorovitch, L.V. (1931) Sur la convergence de la suite de polynômes de S. Bernstein en dehors de l'interval fundamental, *Bull. Acad. Sci. URSS*, 1103–1115.

[114] Karlin, S. and Ziegler, Z. (1970) Iteration of positive approximation operators, *J. Approx. Theory*, **3**, 310–339.

[115] Kelisky, R.P. and Rivlin, T.J. (1967) Iterates of Bernstein polynomials, *Pacific J. Math.*, **21**, No. 3, 511–520.

[116] Knoop, H.B. and X.-L. Zhou, X.-L. (1994) The lower estimate for linear positive operators, (II), *Results in Math.*, **25**, 315–330.

[117] Kohr, G. (2003) *Basic Topics in Holomorphic Functions of Several Complex Variables*, University Press, Cluj-Napoca.

[118] Kohr, G. and Mocanu, P.T. (2005) *Special Chapters of Complex Analysis* (in Romanian), University Press, Cluj-Napoca.

[119] Kopotun, K.A., Leviatan, D. and Shevchuk, I.A. (2006) Coconvex approximation in the uniform norm-the final frontier, *Acta Math. Hungarica*, **110**, No. 1–2, 117–151.

[120] Kövari, T. and Pommerenke, Ch., (1967) On Faber polynomials and Faber expansions, *Math. Zeitschr.*, **99**, 193–206.

[121] Leonte, A. and Virtopeanu, I. (1986) Study on the rest in an approximation formula of two variables functions by means of a Favard-Szász type operator (Romanian), *Studia Univ. "Babeș-Bolyai", ser. math.*, **31**, No. 3–4, 10–15.

[122] Leindler, L. (1990) Sharpenning of Steckin's theorem to strong approximation, *Anal. Math.*, **16**, 27–38.

[123] Leindler, L. and A. Meir, A. (1992) Sharpening of Steckin's theorem to strong approximation with exponent p, *Anal. Math.*, **18**, 111–125.

[124] Leviatan, D. and Shevchuk, I.A. (1995) Counterexamples in convex and higher order constrained approximation, *East J. Approx.*, **1**, 391–398.

[125] Lorentz, G.G. (1986) *Bernstein Polynomials*, 2nd edition, Chelsea Publ., New York.

[126] Lorentz, G.G. (1987) *Approximation of Functions*, Chelsea Publ., New York.

[127] Lupaş, A. (1967) Some properties of the linear positive operators, I, *Mathematica(Cluj)*, **9(32)**, 77–83.

[128] Lupaş, A. (1967) Some properties of the linear positive operators, II, *Mathematica(Cluj)*, **9(32)**, 295–298.

[129] Lupaş, A. (1966) On Bernstein power series, *Mathematica(Cluj)*, **8(31)**, 287–296.

[130] Lupaş, A. (1974) Some properties of the linear positive operators, III, *Revue d'Analyse Numer. Théor. Approx.*, **3**, 47–61.

[131] Lupaş, L. and Lupaş, A. (1987) Polynomials of binomial type and approximation operators, *Stud. Univ. "Babes-Bolyia", Math.*, **32**, No. 4, 60–69.

[132] Lupaş, A. and Müller, M. (1967) Approximationseigenschaften der Gammaoperatoren, *Math. Zeitschr.*, **98**, 208–226.

[133] Marinescu, G. (1970) *Treatise of Functional Analysis*, (Romanian), vol. **I**, Academic Press, Bucharest.

[134] May, C.P. (1976) Saturation and inverse theorems for combinations of a class of exponential type operators, *Cand. J. Math.*, **28**No. 6, 1124–1150.

[135] Mejlihzon, A.Z. (1948) On the notion of monogenic quaternions (in Russian), *Dokl. Akad. Nauk SSSR*, **59**, 431–434.

[136] Meyer-König, W. and Zeller, K. (1960) Bernsteinsche Potenzreihen, *Studia Math.*, **19**, 89–94.

[137] Mirakjan, G.M. (1941) Approximation des fonctions continues au moyen de polynômes de la forme $e^{-nx}\sum_{k=0}^{n}C_{k,n}x^{k}$ (French), *Dokl. Akad. Nauk SSSR*, **31**, 201–205.

[138] Mocanu, P. T., Bulboacă, T. and Sălăgean, Gr. St. (1999) *Geometric Function Theory of Univalent Functions*, (in Romanian), Science Book' s House, Cluj-Napoca.

[139] Moisil, Gr. C. (1931) Sur les quaternions monogènes, *Bull. Sci. Math. (paris)*, **LV**, 168–174.

[140] Moldovan, G. (1974) Discrete convolutions for functions of several variables and linear positive operators (Romanian), *Stud. Univ. "Babes-Bolyai", ser. math.*, **19**, No. 1, 51–57.

[141] Muntean, I. (1973) *Functional Analysis* (Romanian), vol. **1**, "Babes-Bolyai" University Press, Faculty of Mathematics-Mechanics, Cluj.

[142] Murugusundaramoorthy, G. and Magesh, N. (2007) Starlike and convex functions of complex order involving the Dziok-Srivastava operator, *Integral Transforms Spec. Funct.*, **18**, No. 6, 419–425.

[143] Nagel, J. (1982) Asymptotic properties of powers of Kantorovič operators, *J. Approx. Theory*, **36**, 268–275.

[144] Natanson, I.P. (1955) *Konstruktive Funktionentheorie*, Akademie-Verlag.

[145] Obradović, M. (1997) Simple sufficient conditions for univalence, *Mat. Vesnik*, **49**, 241–244.

[146] Ostrovska, S. (2003) q-Bernstein polynomials and their iterates, *J. Approx. Theory*, **123**, 232–255.

[147] Ostrovska, S. (2008) q-Bernstein polynomials of the Cauchy kernel, *Appl. Math. Comp.*, **198**, No. 1, 261–270.

[148] Ostrovska, S. (2007), The approximation of logarithmic function by q-Bernstein polynomials in the case $q > 1$, *Numer. Algorithms*, **44**, No. 1, 69–82.

[149] Phillips, G.M. (1997) Bernstein polynomials based on q-integers, *Ann. Numer. Math.*, **4**, No. 1–4, 511–518.

[150] Pólya, G. and Schoenberg, I.J. (1958) Remarks on de la Vallée Poussin means and convex conformal maps of the circle, *Pacific J. Math.*, **8**, 295–333.

[151] Pommerenke, Ch. (1959) On the derivative of a polynomial, *Mich. Math. J.*, **6**, 373–375.

[152] Ponnusamy, S. and Ronning, F. (1999) Geometric properties for convolutions of hypergeometric functions and functions with the derivative in a halfplane, *Integral Transforms Spec. Funct.*, **8**, No. 1-2, 121–138.

[153] Ponnusamy, S., Singh, V. and Vasundhra, P. (2004) Starlikeness and convexity of an integral transform, *Integral Transforms Spec. Funct.*, **15**, No. 3, 267–280.

[154] Ponnusamy, S. and Vasundhra, P. (2007) Starlikeness of nonlinear integral transforms, *J. of Analysis*, **15**, 195–210.

[155] Pop, O. (2004) About the generalization of Voronovskaja's theorem, *Annals Univ. Craiova, Math. Comp. Sci. Ser.*, **31**, 79–84.

[156] Pop, O. (2007) The generalization of Voronovskaja's theorem for exponential operators, *Creative Math. and Inf.*, **16**, 54–62.

[157] Prajapat, J.K., Raina, R.K. and Srivastava, H.M. (2007) Some inclusion properties for certain subclasses of strongly starlike and strongly convex functions involving a family of fractional integral operators, *Integral Transforms Spec. Funct.*, **18**, No. 9, 639–651.

[158] Radon, J. (1919) Über die Randwertaufgaben beim logarithmischen Potential, *Sitz.-Ber. Wien Akad. Wiss.*, Abt. IIa, **128**, 1123–1167.

[159] Raşa, I. (1999) On Soardi's Bernstein operators of second kind, in : *Analysis, Functional Equations, Approximation and Convexity*, Proceed. Conf.(Lupsa, L. and Ivan, M. eds.), Carpatica Press, Cluj-Napoca, pp. 264–271.

[160] D. Răducanu, D. and H.M. Srivastava, H.M. (2007) A new class of analytic functions defined by means of a convolution operator involving the Hurwitz-Lerch zeta Function, *Integral Transforms Spec. Funct.*, **18**, No. 12, 933–943.

[161] Ruschewey, S. and Salinas, L.C. (1992) On the preservation of periodic monotonicity, *Constr. Approx.*, **8**, 129–140.

[162] Sewell, W.E. (1942) *Degree of Approximation by Polynomials in the Complex Domain*, Annals of Mathematics Studies, no. **9**, Princeton University Press.

[163] Sikkema, P. C. (1970) On some linear positive operators, *Indagationes Mathem.*, **32**, 327–337.

[164] Sikkema, P. C. (1983) Approximation with convolution operators, in : *Second Edmonton Conference in Approximation Theory*, Edmonton, Alberta, pp. 353–368.

[165] Sikkema, P. C. (1981) Fast approximation by means of convolution operators, *Indagationes Mathem.*, **43**, No. 4, 431–434.

[166] Sikkema, P. C. (1983) Approximation with convolution operators depending on two parameters, in : *Approximation Theory IV* (College Station, Texas), Academic Press, New York, pp. 679–684.

[167] Sikkema, P. C. (1979) On the exact degree of local approximation with convolution operators, *Indagationes Mathem.*, **41**, No. 3, 337–351.

[168] Soardi, P. (1991) Bernstein polynomials and random walks on hypergroups, in : *Probability Measures on Groups, X*, (H. Herbert ed.), Proceed. 10th Oberwolfach Conference, 1990), Plenum Publ. Corp., New York, pp. 387–393.

[169] Srivastava, H.M. and Attiya, A.A. (2007) An integral operator associated with the Hurwitz-Lerch zeta function and differential subordination, *Integral Transforms Spec. Funct.*, **18**, No. 3, 207–216.

[170] Srivastava, H.M., Murugusundaramoorthy, G. and Sivasubramanian, S. (2007) Hypergeometric functions in the parabolic starlike and uniformly convex domains, *Integral Transforms Spec. Funct.*, **18**, No. 7, 511–520.

[171] Stewart, I. and Tall, D. (2002) *Complex Analysis*, Cambridge University Press, Cambridge.

[172] Stancu, D.D. (1977) *Course of Numerical Analysis* (in Romanian), Faculty of Mathematics and Mechanics, "Babes-Bolyai" University, Cluj.

[173] Stancu, D.D. (1969) On a generalization of Bernstein polynomials (in Romanian), *Studia Univ. "Babes-Bolyai", ser. math.*, **14**, No. 2, 31–44.

[174] Stancu, D.D. (1968) Approximation of functions by a new class of linear polynomial operators, *Rev. Roumaine Math. Pures Appl.*, **13**, No. 8, 1173–1194.

[175] Stancu, D.D. (1959) De l'approximation par des polynomes du type Bernstein, des fonctions de deux variables (Romanian), *Commun. Acad. Republ. Popul. Romine*, **9**, 773–777.

[176] Stancu, D.D. (1972) Approximation of functions by means of some new classes of linear polynomial operators, in : *Numerische Methoden der Approximationstheorie*, Proc. Conf. Math. Res. Inst. Oberwolfach, 1971 (L. Colatz, G. Meinardus eds.), Birkhäuser, Basel, 187–203.

[177] Stancu, D.D. (1986) On a new class of multivariate linear positive approximating operators, *Studia Univ. "Babes-Bolyai", ser. Math.*, **31**, No. 4, 55–64.

[178] Stancu, D.D. (1970) Probabilistic methods in the theory of approximation of functions of several variables by linear positive operators, in : *Approximation Theory*, Proc. Sympos. Lancaster, 1969 (A. Talbot ed.), Academic Press, pp. 329–342.

[179] Stancu, D.D. (1970) Approximation of functions of two and several variables by a class of polynomials of Bernstein-type (Romanian), *Stud. Cerc. Math.* (Bucharest), **22**, 335–345.

[180] Stancu, D.D. (1978) Approximation of bivariate functions by means of some Bernstein-type operators, in : *Multivariate Approximation*, Proc. Sympos. Univ. Durham, 1977 (D.C. Handscomb ed.), Academic Press, pp. 189–208.

[181] Stancu, D.D. (1985) Bivariate approximation by some Bernstein-type operators, in : *Proc. Colloq. on Approx. and Optimiz.*, Cluj-Napoca, 1984, pp. 25–34.

[182] Stancu, D.D. (1969) A new class of uniform approximating polynomials in two and several variables, in : *Proc. Conf. on Constructive Theory of Functions* (G. Alexis and S.B. Stechkin eds.), Akad. Kiadó, Budapest, pp. 443–455.

[183] Stancu, D.D. and Vernescu, A.D. (1999) Approximation of bivariate functions by means of classes of operators of Tiberiu Popoviciu type, *Math. Rep.* (Bucharest), **1(51)**, No. 3, 411–419.

[184] Stancu, F. (1970) On te approximation of functions of one and two variables by Baskakov's operators (Romanian), *Stud. Cerc. Mat.*, **22**, 531–542.

[185] Stechkin, S.B. (1978) On the approximation of periodic functions by the de la Vallée Poussin sums, *Anal. Math.*, **4**, 61–74.

[186] Suetin, P.K. (1998) *Series of Faber Polynomials*, Gordon and Breach, Amsterdam.

[187] Suffridge, T.J. (1966) Convolution of convex functions, *J. Math. Mech.*, **15**, 795–804.

[188] Szász, O. (1950) Generalization of S.N. Bernstein's polynomials to the infinite interval, *J. Research, National Bureau of Standards*, **45**, 239–245.

[189] Tomescu, I. (1981) *Problems of Combinatorics and Graph Theory*, Edit. Didact. Pedag., Bucharest.

[190] Tonne, P.C. (1969) On the convergence of Bernstein polynomials for some unbounded analytic functions, *Proc. Am. Math. Soc.*, **22**, 1–6.

[191] Totik, V. (1984) Uniform approximation by positive operators on infinite intervals, *Anal. Math.*, **10**, 163–182.

[192] Totik, V. (1994) Approximation by Bernstein polynomials, *Amer. J. Math.*, **114**, 995–1018.

[193] Totik, V. (1982) Approximation by convolution operators, *Analysis Math.*, **8**, No. 2, 151–163.

[194] Videnskii, V. S. (2004) On q-Bernstein polynomials and related positive linear operators (in Russian), in : *Problems of Modern Mathematics and Mathematical Education*, St.-Petersburg, pp. 118–126.

[195] Videnskii, V. S. (2005) On some classes of q-parametric positive linear operators, in : *Selected Topics in Complex Analysis. The S. Ya. Khavinson Memorial Volume* (Eiderman, V. Ya. et al. eds), Operator Theory : Advances and Applications, vol. **158**, Birkhauser, Basel, pp. 213–222.

[196] Vlaic, G. (1999) On a bivariate multiparameter approximation operator of D.D. Stancu, *Math. Rep.* (Bucharest), **1(51)**, No. 3, 473–479.

[197] Wang, H. (2007) Voronovskaja-type formulas and saturation of convergence for q-Bernstein polynomials for $0 < q < 1$, *J. Aprox. Theory*, **145**, 182–195.

[198] Wang, H. and Wu, X.Z. (2008) Saturation of convergence for q-Bernstein polynomials in the case $q \geq 1$, *J. Math. Anal. Appl.*, **337**, 744–750.

[199] Wright, E.M. (1930) The Bernstein approximation polynomials in the complex plane, *J. London Math. Soc.* **5**, 265–269.

[200] Wood, B. (1968) On a generalized Bernstein polynomial of Jakimovski and Leviatan, *Math. Zeitschr.*, **106**, 170–174.

[201] Wu, Congxin and Zengtai, Gong (2000) *On Henstock integral of fuzzy-number-valued functions*, I, *Fuzzy Sets and Systems*, **115**, no. 3 , 377–391.

[202] Xiang, X., He, Q. and Yang, W. (2007) Convergence rate for iterates of q-Bernstein polynomials, *Analysis in Theory and Appl.*, **23**, No. 3, 243–254.

[203] Xie, L.S. (2005) Pointwise simultaneous approximation by combinations of Bernstein operators, *J. Approx. Theory*, **137**, 1–21.

[204] Zadeh, L.A. (1965) *Fuzzy Sets*, Inform. and Control, **8**, 338–353.

Index